P9-CKX-676

# METHODS IN CELL BIOLOGY

## VOLUME XIX

*Chromatin and Chromosomal Protein Research. IV*

*Methods in Cell Biology*

Series Editor: DAVID PRESCOTT

DEPARTMENT OF MOLECULAR, CELLULAR AND DEVELOPMENTAL BIOLOGY
UNIVERSITY OF COLORADO
BOULDER, COLORADO

# Methods in Cell Biology

## VOLUME XIX

*Chromatin and Chromosomal Protein Research. IV*

*Edited by*

GARY STEIN and JANET STEIN

DEPARTMENT OF BIOCHEMISTRY AND
MOLECULAR BIOLOGY
UNIVERSITY OF FLORIDA
GAINESVILLE, FLORIDA

LEWIS J. KLEINSMITH

DIVISION OF BIOLOGICAL SCIENCES
UNIVERSITY OF MICHIGAN
ANN ARBOR, MICHIGAN

1978

ACADEMIC PRESS • New York San Francisco London
A Subsidiary of Harcourt Brace Jovanovich, Publishers

ACADEMIC PRESS, INC.
111 Fifth Avenue, New York, New York 10003

*United Kingdom Edition published by*
ACADEMIC PRESS, INC. (LONDON) LTD.
24/28 Oval Road, London NW1

LIBRARY OF CONGRESS CATALOG CARD NUMBER: 64-14220

ISBN 0-12-564119-2

PRINTED IN THE UNITED STATES OF AMERICA

# CONTENTS

## Part B. Enzymic Components of Nuclear Proteins—Protein Substrates

## Part D. Chromatin Transcription and Characterization of Transcripts

# LIST OF CONTRIBUTORS

*Numbers in parentheses indicate the pages on which the author's contributions begin.*


J. ALLAN, Department of Biophysics, King's College, London University, London, England (387)

W. FRENCH ANDERSON, Molecular Hematology Branch, National Heart and Lung Institute, National Institutes of Health, Bethesda, Maryland (339)

SUSAN M. ASTRIN, The Institute for Cancer Research, The Fox Chase Cancer Center, Philadelphia, Pennsylvania (359)

CORRADO BAGLIONI, Department of Biological Sciences, State University of New York at Albany, Albany, New York (215)

C. STUART BAXTER, Department of Environmental Health, Kettering Laboratory, University of Cincinnati Medical Center, Cincinnati, Ohio (95)

GRAEME I. BELL, Department of Biochemistry and Biophysics, University of California, San Francisco, San Francisco, California (1)

EBO BOS, Physiologisch-Chemisches Institut I, Philipps-Universität Marburg, Marburg/Lahn, West Germany (197)

BERNDT B. BRUEGGER, Department of Chemistry, University of California, Los Angeles, Los Angeles, California (153)

PAUL BYVOET, Department of Pathology, University of South Florida College of Medicine, and Veterans Administration Hospital, Tampa, Florida (95)

DONALD B. CARTER, Department of Biochemistry, University of North Carolina, Chapel Hill, North Carolina (175)

CHI-BOM CHAIE, Department of Biochemistry, University of North Carolina, Chapel Hill, North Carolina (175)

G. J. COWLEY, Department of Biophysics, King's College, London University, London, England (387)

DIPAK K. DUBE, Institute for Cancer Research, The Fox Chase Cancer Center, Philadelphia, Pennsylvania (27)

ALBERT K. DUNLAP, Department of Chemistry, University of California, Los Angeles, Los Angeles, California (153)

EGON DURBAN, Fels Research Institute and Department of Biochemistry, Temple University School of Medicine, Philadelphia, Pennsylvania (59)

PEGGY H. EFIRD, Department of Biochemistry, University of North Carolina, Chapel Hill, North Carolina (175)

S. J. FEY, Department of Biophysics, King's College, London University, London, England (387)

OSCAR FRICKE, Department of Chemistry, University of California, Los Angeles, Los Angeles, California (153)

DIETER GALLWITZ, Physiologisch-Chemisches Institut I, Philipps-Universität Marburg, Marburg/Lahn, West Germany (197)

R. S. GILMOUR, The Beatson Institute for Cancer Research, Bearsden, Glasgow, Scotland (373)

H. J. GOULD, Department of Biophysics, King's College, London University, London, England (387)

RICHARD M. HALPREN, Department of Chemistry, University of California, Los Angeles, Los Angeles, California (153)

BARBARA A. HAMKALO, Departments of Molecular Biology and Biochemistry, and of Developmental and Cell Biology, University of California, Irvine, Irvine, California (287)

SANGDUK KIM, Fels Research Institute and Department of Biochemistry, Temple University School of Medicine, Philadelphia, Pennsylvania (59, 69, 79, 191)

VALERIE M. KISH, Department of Biology, Hobart and William Smith Colleges, Geneva, New York (101)

LEWIS J. KLEINSMITH, Division of Biological Sciences, University of Michigan, Ann Arbor, Michigan (101, 119, 161)

MURIEL W. LAMBERT, Department of Pathology, College of Medicine and Dentistry of New Jersey, New Jersey Medical School, Newark, New Jersey (43)

THOMAS A. LANGAN, Department of Pharmacology, University of Colorado Medical Center, Denver, Colorado (127, 143)

HYANG WOO LEE, Fels Research Institute and Department of Biochemistry, Temple University School of Medicine, Philadelphia, Pennsylvania (59)

A. LICHTLER, Department of Biochemistry and Molecular Biology, University of Florida, Gainesville, Florida (237)

C. C. LIEW, Department of Clinical Biochemistry, Banting Institute, University of Toronto, Toronto, Ontario, Canada (51, 89)

LAWRENCE A. LOEB, Institute for Cancer Research, The Fox Chase Cancer Center, Philadelphia, Pennsylvania (27)

D. MARYANKA, Department of Biophysics, King's College, London University, London, England (387)

WILLIAM F. MARZLUFF, JR., Department of Chemistry, Florida State University, Tallahassee, Florida (317)

MASAMI MURAMATSU;[1] Department of Biochemistry, Tokushima University School of Medicine, Tokushima, Japan (301)

ARTHUR W. NIENHUIS, Molecular Hematology Branch, National Heart and Lung Institute, National Institutes of Health, Bethesda, Maryland (339)

TIMOTHY NILSEN, Department of Biological Sciences, State University of New York at Albany, Albany, New York (215)

SAMUEL NOCHUMSON, Fels Research Institute and Department of Biochemistry, Temple University School of Medicine, Philadelphia, Pennsylvania (69)

TOSHIO ONISHI, Department of Biochemistry, Tokushima University School of Medicine, Tokushima, Japan (301)

G. CHRISTIAN OVERTON,[2] Department of Biology, Johns Hopkins University, Baltimore, Maryland (273)

WOON KI PAIK, Fels Research Institute and Department of Biochemistry, Temple University School of Medicine, Philadelphia, Pennsylvania (59, 69, 79, 191)

DOROTHY E. PUMO,[3] Division of Biological Sciences, University of Michigan, Ann Arbor, Michigan (119)

RONALD H. REEDER, Department of Embryology, Carnegie Institution of Washington, Baltimore, Maryland (333)

WILLIAM J. RUTTER, Department of Biochemistry and Biophysics, University of California, San Francisco, San Francisco, California (1)

DAVID F. SAYER, Department of Pathology, University of South Florida College of Medicine, and Veterans Administration Hospital, Tampa, Florida (95)

[1] *Present address*: Department of Biochemistry, Cancer Institute, Toshima-ku, Tokyo, Japan.

[2] *Present address*: The Wistar Institute, Philadelphia, Pennsylvania.

[3] *Present address*: Department of Molecular, Cellular, and Developmental Biology, University of Colorado, Boulder, Colorado.

[4] *Present address*: Forensic Biology Laboratory, Institute of Medical and Veterinary Science, Adelaide, South Australia.

A. C. SCOTT,[4] Department of Biochemistry, University of Adelaide, Adelaide, South Australia (257)

ROBERTS A. SMITH, Department of Chemistry, University of California, Los Angeles, Los Angeles, California (153)

HANS STAHL, Physiologisch-Chemisches Institut, I, Philipps-Universität Marburg, Marburg/Lahn, West Germany (197)

ALAN W. STEGGLES, Molecular Hematology Branch, National Heart and Lung Institute, National Institutes of Health, Bethesda, Maryland (339)

G. S. STEIN, Department of Biochemistry and Molecular Biology, University of Florida, Gainesville, Florida (109, 237, 379)

J. L. STEIN, Department of Immunology and Medical Microbiology, University of Florida, Gainesville, Florida (109, 237, 379)

LLOYD A. STOCKEN, Department of Biochemistry, University of Oxford, Oxford, England (167)

GEORGE P. STUDZINSKI, Department of Pathology, College of Medicine and Dentistry of New Jersey, New Jersey Medical School, Newark, New Jersey (43)

D. SURIA, Department of Clinical Chemistry, Banting Institute, University of Toronto, Toronto, Ontario, Canada (89)

J. A. THOMSON, Department of Biochemistry and Molecular Biology, University of Florida, Gainesville, Florida (109)

C. L. THRALL,[5] Department of Biochemistry and Molecular Biology, University of Florida, Gainesville, Florida (237)

ELIZABETH C. TRAVAGLINI, Institute for Cancer Research, The Fox Chase Cancer Center, Philadelphia, Pennsylvania (27)

PABLO VALENZUELA, Department of Biochemistry and Biophysics, University of California, San Francisco, San Francisco, California (1)

LEE A. WEBER, Department of Biological Sciences, State University of New York at Albany, Albany, New York (215)

ERIC S. WEINBERG, Department of Biology, Johns Hopkins University, Baltimore, Maryland (273)

FANYELA WEINBERG, Department of Biochemistry and Biophysics, University of California, San Francisco, San Francisco, California (1)

J. R. E. WELLS, Department of Biochemistry, University of Adelaide, Adelaide, South Australia (257)

GOLDER N. WILSON, Molecular Hematology Branch, National Heart and Lung Institute, National Institutes of Health, Bethesda, Maryland (339)

[5] *Present address*: Department of Molecular Medicine, The Mayo Clinic, Rochester, Minnesota.

# PREFACE

During the past several years considerable attention has been focused on examining the regulation of gene expression in eukaryotic cells with emphasis on the involvement of chromatin and chromosomal proteins. The rapid progress that has been made in this area can be attributed largely to development and implementation of new, high-resolution techniques and technologies. Our increased ability to probe the eukaryotic genome has far-reaching implications, and it is reasonable to anticipate that future progress in this field will be even more dramatic.

We have attempted to present, in four volumes of *Methods in Cell Biology*, a collection of biochemical, biophysical, and histochemical procedures that constitute the principal tools for studying eukaryotic gene expression. Contained in Volume 16 are methods for isolation of nuclei, preparation and fractionation of chromatin, fractionation and characterization of histones and nonhistone chromosomal proteins, and approaches for examining the nuclear-cytoplasmic exchange of macromolecules. Volume 17 deals with further methods for fractionation and characterization of chromosomal proteins, including DNA affinity techniques. Also contained in Volume 17 are methods for isolation and fractionation of chromatin, nucleoli, and chromosomes. Volume 18 focuses on approaches for chromatin fractionation, examination of physical properties of chromatin, and immunological as well as sequence analysis of chromosomal proteins. In this volume (Volume 19) enzymic components of nuclear proteins, chromatin transcription, and chromatin reconstitution are described. This volume also contains a section on methods for studying histone gene expression.

In compiling these four volumes we have attempted to be as inclusive as possible. However, the field is in a state of rapid growth, prohibiting us from being complete in our coverage.

The format generally followed includes a brief survey of the area, a presentation of specific techniques with emphasis on rationales for various steps, and a consideration of potential pitfalls. The articles also contain discussions of applications for the procedures. We hope that the collection of techniques presented in these volumes will be helpful to workers in the area of chromatin and chromosomal protein research, as well as to those who are just entering the field.

We want to express our sincere appreciation to the numerous investigators who have contributed to these volumes. Additionally, we are indebted to Bonnie Cooper, Linda Green, Leslie Banks-Ginn, and the staff at Academic Press for their editorial assistance.

GARY S. STEIN
JANET L. STEIN
LEWIS J. KLEINSMITH

# Part A. Enzymic Components of Nuclear Proteins—Nucleic Acid Substrates

# Chapter 1

# *Isolation and Assay of Eukaryotic DNA-Dependent RNA Polymerases*

PABLO VALENZUELA, GRAEME I. BELL, FANYELA WEINBERG, AND WILLIAM J. RUTTER

*Department of Biochemistry and Biophysics, University of California, San Francisco, San Francisco, California*

## I. Introduction

Eukaryotic cells contain multiple RNA polymerases. Three forms (I, II, III) were originally described in rat liver (*1*), sea urchin (*1*), and yeast (*2*) on the basis of their different chromatographic and enzymic properties. The available evidence indicates that RNA polymerase I is localized in the nucleolus (*3*) and is involved in the synthesis of large ribosomal RNA (*4–6*). Polymerase II is nucleoplasmic and appears to be responsible for the transcription of heterogeneous nuclear RNA, presumably messenger RNA (*4–7*). RNA polymerase III, which is eluted from DEAE-Sephadex at high salt concentrations, is responsible for the synthesis of tRNA and 5 S ribosomal RNA (*5,8*). The most convincing demonstration of the existence of three distinct classes of RNA polymerases has been achieved by direct comparison of the subunit structure of the three enzymes from the same tissue. These studies, which have been carried out in mouse plasmacytoma ($MOPC_{315}$) cells (*9*) and in yeast (*10*), have shown that each enzyme has a unique subunit structure. Therefore, the enzymes are composed largely of products from different genes and are not artifacts of isolation or interconvertible assemblies of subunits representing variations of a common molecule. Similar studies have been carried out with enzymes I and II from various other sources (*11*). Evidence obtained by high-resolution chromatography and native gel electrophoresis indicates that each class of RNA polymerases is heterogeneous (*11,12*). Whether these molecular species have distinct functions; or indeed whether they exist in the living cell, has yet to be resolved.

The existence in eukaryotic cells of multiple forms of RNA polymerase with distinct subunit compositions, subcellular localization, and specific functions suggests that gene expression in eukaryotic cells is regulated, at least in part, via distinct enzymes that specifically transcribe different classes of genes.

The purification of eukaryotic RNA polymerases in sufficient quantities for structural and transcriptive studies has proved to be a difficult task. The polymerase content in higher cells is quite low and, unlike the bacterial polymerase the enzymes are difficult to extract since they occur tightly bound to chromosomal components, such as DNA, RNA, histones, and acidic proteins. Eukaryotic RNA polymerases are complex structures composed of multiple subunits, and losses of activity are usually encountered owing to the apparently delicate nature of these molecules.

This chapter describes the most recent and general procedures used in the purification of eukaryotic nuclear RNA polymerases. Special emphasis will be given to those enzymes which have been extensively purified. Detailed methods will be presented for the purification of yeast RNA polymerases, which have been selected because of the convenience of their source, the authors' interest and experience, and the possible general applicability of this procedure for the purification of polymerases from other sources.

## II. Purification: General Principles

### A. Sources

A major problem in the isolation of eukaryotic RNA polymerases is the limited availability of many eukaryotic tissues and the relative paucity of these enzymes in the cell. This problem can be in part obviated by choosing as starting material sources that contain increased amounts of RNA polymerases per gram of tissue. In this context, important sources for enzyme purification are rapidly proliferating tumor cells, such as HeLa (*13*), $MOPC_{315}$ (*14*), and Krebs ascites (*15*), and embryos, such as those from *Xenopus laevis* (*16*) and *Drosophila melanogaster* (*17*). Calf thymus, a tissue particularly rich in nuclear material, is also a good source of animal RNA polymerases (*18–20*). Wheat germ, which is inexpensive and easy to store, appears to be an excellent source for the isolation of RNA polymerase II (*21,22*). Yeast is also an excellent source. This material, which is readily available at low cost in bulk amounts, is suitable for the purification of milligram quantities of the three nuclear RNA polymerases (*10,23–26*). Small amounts of extensively purified RNA polymerases have been obtained also from other lower eukaryotes, e.g., *Mucor rouxii* (*27*), *Physarum polycephalum* (*28*), and *Dictyostelium discoideum* (*29*). Attempts to increase the cellular levels of polymerases (in yeast, for example) by genetic manipulation may be of interest in the future.

For the purification of polymerases associated with particular organelles, such as nuclei, mitochondria, or chloroplasts, the initial step in the purification usually involves the isolation of the particular organelle under conditions in which the polymerase is retained. Generally, RNA polymerases have been purified from isolated nuclei. Previous isolation of nucleoli has been particularly important in the purification of RNA polymerase I from $MOPC_{315}$ (*30*), rat liver (*15,31*), and Krebs ascites (*15a*). In other instances, however, it has been possible to take advantage of tissues which are particularly rich in nuclei or RNA polymerases, such as calf thymus (*32*) and wheat germ (*22*). In these cases the previous isolation of nuclei is unnecessary. Despite the inconvenience of starting with less pure material, purification of the enzymes from the whole tissue has advantages in that it avoids cellular fractionation and prevents uncontrolled losses of activity due to a low recovery of nuclei or leakage of soluble enzymes that are not bound to the chromatin.

## B. Solubilization

It is of primary importance in any purification procedure that solubilization and recovery of enzyme in the initial extract be optimal. In 1960, Weiss first demonstrated the existence of nuclear RNA polymerase activity *in vitro* in a particulate fraction (*33*). Early attempts to release the enzyme in a soluble form were unsuccessful because of low enzyme yields and lack of stability (*34*). Roeder and Rutter (*1*) developed a simple method for solubilization of the nuclear RNA polymerases. Their procedure involves sonication of isolated nuclei after lysis in a medium of high ionic strength (0.3 *M* ammonium sulfate) that releases the enzyme from chromatin. Using this method, several groups of investigators have succeeded in extracting the enzymes from a variety of vertebrate, plant, insect, and fungal tissues in quantities that are sufficient for preliminary analysis and in some cases for extensive purification (see Table I) (*9,11,15,17–20,22,24,25,30,35–37*). Another method which also has been used to solubilize RNA polymerases is the homogenization of the whole tissue or organelles in buffers of low ionic

TABLE I

ILLUSTRATIVE EXAMPLES OF EXTENSIVELY PURIFIED EUKARYOTIC RNA POLYMERASES[a]

| Enzyme | Purification procedure | Specific activity[b] (units/mg) | Yield[c] (mg) | References |
|---|---|---|---|---|
| Calf thymus and rat liver II | Nuclei, DEAE-Sephadex, P-cellulose, DNA-Sepharose, sucrose gradient | 500 | 0.80 | *18* |
| Calf thymus I | Protamine sulfate, DEAE-cellulose, Sepharose 6B, DEAE-Sephadex, glycerol gradient | 200 | 1.0 | *19* |
| Calf thymus II | Protamine sulfate, DEAE-cellulose P-cellulose hydroxyapatite, P-cellulose, glycerol gradient | 270 | 3.5 | *20* |
| Rat liver I | Nucleoli, DEAE-Sephadex, P-cellulose, sucrose gradient | 98 | 1.8 | *15* |
| Hen liver II | Nuclei, protamine sulfate, DEAE-cellulose, P-cellulose, hydroxyapatite, P-cellulose, glycerol gradient | 200 | 0.4 | *36* |
| Hen oviduct II | Nuclei, protamine sulfate, DEAE-cellulose, P-cellulose, hydroxyapatite, P-cellulose, glycerol gradient | 220 | 0.7 | *36* |
| Krebs ascites I | Nucleoli, DEAE-Sephadex, P-cellulose, sucrose gradient, DNA-cellulose | 94 | 3.7 | *15* |
| Mouse myeloma ($MOPC_{315}$) I | Nucleoli, CM-Sephadex, DEAE-Sephadex, P-cellulose, sucrose gradient, P-cellulose | 190 | 5.5 | *30* |
| Mouse myeloma ($MOPC_{315}$) II | Nuclei, P-cellulose, DEAE-cellulose, sucrose gradient, DEAE-Sephadex | 213 | 1.4 | *30* |
| Mouse myeloma ($MOPC_{315}$) III | DEAE-cellulose, DEAE-Sephadex, CM-Sephadex, P-cellulose, sucrose gradient, P-cellulose | 130 | 2.6 | *9* |
| *Drosophila melanogaster* II | Polyethylene glycol, DEAE-cellulose batch, glycerol gradient, DEAE-Sephadex | 70 | 1.0 | *17* |
| *Bombyx mori* III | DEAE-cellulose, DEAE- | | | |

(*cont.*)

TABLE I (*cont.*)

| Enzyme | Purification procedure | Specific activity[b] (units/mg) | Yield[c] (mg) | References |
|---|---|---|---|---|
| | Sephadex, CM-Sephadex, P-cellulose, sucrose gradient | 350 | 0.8 | *37* |
| Wheat germ II | Polymin P, ammonium sulfate, DEAE-cellulose, P-cellulose | 2.5 | 30 | *22* |
| *Saccharomyces cerevisiae* I | P-cellulose batch, DEAE-cellulose batch, DEAE-cellulose, glycerol gradient | 300 | — | *25* |
| *Saccharomyces cerevisiae* I | P-cellulose batch, DEAE-cellulose batch, DEAE-Sephadex, sucrose gradient | 250 | 12 | *37* |
| *Saccharomyces cerevisiae* II | Protamine sulfate, ammonium sulfate, Sephadex G-50, DEAE-cellulose, glycerol gradient, P-cellulose | 1000 | 6.7 | *24* |
| *Saccharomyces cerevisiae* II | Polymin P, DEAE-cellulose, P-cellulose, sucrose gradient | 440 | 9 | *37* |
| *Saccharomyces cerevisiae* III | Polymin P, DEAE-cellulose, DEAE-Sephadex, DNA-cellulose | 140 | 6 | *37* |

[a]There are other purified RNA polymerases reported in the literature (*11,35*). This table includes a representative sample of enzymes from normal animal cells, tumor cells, insects, plants, and fungi.

[b]Units are roughly corrected from the original literature values to mean nanomoles of UMP incorporated into RNA in 10 minutes.

[c]Yield corresponds to milligrams of purified enzyme per 1 kg of starting material.

strength followed by incubation for various lengths of time at either 4°C or 37°C. In several instances, this treatment has been complemented with sonication. The various procedures for solubilization have been comprehensively reviewed and tabulated by Jacob (*34*).

It is usually difficult to measure and compare the amount of enzyme present in the original tissue with that obtained after solubilization. RNA polymerases seem to exist *in vivo* in at least two functional states: one is active toward the endogenous chromatin template (engaged enzyme), and the other is inactive and can be measured only with exogenous templates (*15*,

*38*). Activity measurements of the extracts are quite unreliable because of the presence of endogenous DNA and high concentrations of RNA and hydrolytic enzymes. The possible presence of various stimulatory or inhibitory proteins in the homogenate, which are subsequently lost during fractionation, make it also difficult to estimate the original amount of enzyme directly from the yield. The amount of RNA polymerase II can also be determined by binding to labeled α-amanitin (*32*). Immunological assays with purified antibodies against each enzyme may be of importance in the future in determining the precise amount of each enzyme in an extract.

## C. Removal of DNA

A necessary early step in the purification of RNA polymerases is the removal of polynucleotides that might bind to the enzymes and hinder subsequent purification. This has been accomplished by a variety of methods. Dilution of the solubilized enzymes in the extract to a lower salt concentration (0.1 *M* ammonium sulfate) followed by high speed centrifugation to remove the aggregated chromatin, as originally recommended by Roeder and Rutter (*1*), has been extensively used (*14, 18*). This procedure may lead to losses of activity due to enzyme reaggregation with DNA (especially serious for polymerase III) at low ionic strength. This can be partially avoided by previous sonication or increasing the salt concentration (*15*). Precipitation of the extract with protamine sulfate and selective elution of the enzyme with increasing salt concentrations has been used in the purification of bacterial polymerases (*39, 40*). Protamine sulfate precipitation of nucleic acids at high ionic strength has been employed in the purification of polymerases from rat liver (*41*), calf thymus (*20*), and hen liver and oviduct (*36*). The use of protamine sulfate is often difficult to reproduce, since precipitation is sensitive to minor alterations in ionic strength and varies with the commercial batch of protamine employed. Zillig and collaborators circumvented these problems for *Escherichia coli* RNA polymerase by introducing a procedure which utilizes an initial precipitation of the extract with Polymin P, a polycationic synthetic polymer, followed by differential elution (*42*). The use of Polymin P is highly reproducible. The RNA polymerases bind to the Polymin P more strongly than to protamine sulfate, and thus the procedure is less subject to variations in ionic strength. However, Polymin P is a potent inhibitor of RNA polymerase activity so that it must be removed, or sufficiently diluted, prior to enzyme assay. Polymin P has been successfully used in the purification of wheat germ RNA polymerase II (*22*) and yeast RNA polymerases II and III (*37*). Phase extraction with polyethylene glycol, originally used to remove nucleic acids during the purification of DNA polymerase from *Bacillus subtilis* (*43*), has been used in the purification of *E. coli* RNA polymerase (*44*) and *Drosophila melanogaster* RNA polymerase II (*17*).

In addition to the above procedures, substantial but not complete separation of nucleic acids can be achieved by the proper use of ion-exchange resins. Partial removal of nucleic acids is obtained by batchwise treatment of extracts with DEAE-cellulose at high salt concentrations (nucleic acids remain bound to the resin) (*45,46*) or by selective batchwise adsorption of the enzyme to phosphocellulose (nucleic acids are largely free) (*25,26*). Ion-filtration or sievorptive chromatography has also been used for the simultaneous removal of nucleic acids and for purification. This procedure, developed by Kirkegaard *et al.* (*47*) and introduced for the purification of RNA polymerases by Goldberg *et al.* (*48*), combines ion-exchange and gel filtration in the same chromatographic step. By properly adjusting the pH and the ionic strength of a DEAE-Sephadex column, it is possible to elute purified enzyme in the first column volume after the excluded peak of noninteracting macromolecules. Thus, the enzyme is eluted in the sieving range. Under the proper conditions, nucleic acids interact more strongly with the ion-exchanger, and more than 90% of the DNA can be separated from the enzyme by this procedure. Ion-filtration chromatography has been applied to the purification of RNA polymerases I from rat liver, Krebs ascites (*15*), and yeast (*26*).

## D. Use of Ion-Exchange Resins

### 1. DEAE-Sephadex and DEAE-Cellulose

The three classes of eukaryotic RNA polymerases can be separated using DEAE-Sephadex chromatography (*1–3*). When the column is developed with an ammonium sulfate gradient, RNA polymerases I, II, and III are eluted in sequence at increasingly higher ionic strengths (0.08–0.12 *M*, 0.15–0.25 *M*, and 0.25–0.35 *M*, respectively). This procedure has been successfully applied to extracts from many organisms and is valid for all except *Physarum* (*28*). Sergeant and Krsmanovic (*49*) found that this elution sequence is altered with DEAE-cellulose; polymerase III elutes together with polymerase I at 0.1–0.12 *M* ammonium sulfate, and polymerase II elutes at 0.25–0.3 *M*, as with DEAE-Sephadex. The reasons for this apparently anomalous behavior are still unknown. Several reports failing to find RNA polymerase III in cell extracts may be traced to the use of DEAE-cellulose chromatography for resolution. Lack of polymerase III in extracts prepared from nuclei may also be due to the tendency of this enzyme to leach out of the nuclei into the cytoplasm (*14,50–52*). As shown in Table I, DEAE-cellulose chromatography has been employed as a major step in the extensive purification of polymerases II from calf thymus (*20*), mouse plasmacytoma cells ($MOPC_{315}$) (*30*), *Drosophila* embryos (*17*), wheat germ (*22*), hen oviduct and liver (*36*), and yeast (*24,37*).

The different elution properties of RNA polymerase III between DEAE-cellulose and DEAE-Sephadex have been exploited in the purification of RNA polymerase III from mouse plasmacytoma cells (*53*), *Bombyx mori* (*37*), and yeast (*37*).

### 2. Phosphocellulose

RNA polymerases are acidic proteins that bind to DEAE-Sephadex and DEAE-cellulose at pH 8 with varying affinity. However, they also absorb to phosphocellulose at the same pH, indicating that the positively charged regions of the molecule that react with DNA are available for the interaction with phosphate groups of the resin. Most of the purification procedures for RNA polymerases have taken advantage of this dual property (see Table I). Chromatography on phosphocellulose often yields the greatest enrichment for polymerase activity. The order of elution from phosphocellulose of the amphibian oocyte (*12*) and yeast (*26*) RNA polymerases is II, III, and I. The batchwise absorption of polymerase I to phosphocellulose from a cell extract is conveniently used for the purification of yeast polymerase I (*25,26*).

### 3. Batchwise versus Analytical Ion-Exchange Chromatography

Enzyme purification can be carried out batchwise or in columns. Whenever it is possible, batchwise procedures are preferred. They have the advantage of being rapid, avoiding excessive dilution of the enzymes, and can be scaled up virtually without limit. Batch procedures have been employed in the purification of RNA polymerases from *E. coli* (*45*), calf thymus (*20*), yeast (*23,25,26*), and *Drosophila* larvae (*17*). For the same reasons, step elution from a column is preferred to gradient elution when the extent of purification can be slightly sacrificed (*19,20,22,37*).

### 4. Heterogeneity

Ion-exchange resins have been a valuable tool for the detection and resolution of various chromatographically distinct forms of RNA polymerases within a given class. DEAE-Sephadex column chromatography separates (a) two forms of polymerase I ($I_A$ and $I_B$, in order of their elution from a cation-exchange resin) from sea urchin (*1*), yeast (*2*), *Xenopus laevis* (*16*) HeLa cells (*13*), and other sources (*12*), (b) two forms of polymerase II ($II_A$ and $II_B$) from calf thymus (*20*), *Xenopus laevis* (*16*), and yeast (*54*), and (c) two or three forms of RNA polymerase III ($III_A$, $III_B$, and $III_C$) from mouse plasmacytoma cells (*53*), rat liver (*51*), and HeLa cells (*13,55*). CM-Sephadex column chromatography separates also two forms of polymerase I from amphibian cells (*16*) and calf thymus (*19*). Phosphocellulose column chromatography separates two forms of polymerase I from rat liver (*31, 56, 57*), and yeast (*79*).

## E. Affinity Chromatography

RNA polymerases bind readily to nucleic acids as a result of the specific affinity of regions of the enzymes with the phosphate groups of the template. DNA, RNA, and synthetic polynucleotides may be prepared for column chromatography and successfully used for the purification of various enzymes that show this property. Nucleic acids can be physically entrapped in a matrix of cellulose (*58, 59*), polyacrylamide (*60*), or agarose (*61*). Covalent attachment to agarose (*62, 63*) or cellulose (*64, 65*) is also possible. Matrix-bound denatured DNA column chromatography has been successfully applied to the purification of several eukaryotic RNA polymerases, including rat liver and calf thymus polymerase II (*18*), Krebs ascites polymerase (*15*), and yeast polymerases I and III (*10, 26*). The order of elution from denatured calf thymus DNA-cellulose may be different from that from phosphocellulose, and in yeast it is II, I, and III vs. II, III, I, respectively. Yeast RNA polymerase III is eluted from denatured DNA-cellulose at 0.7 *M* KCl, and it is efficiently purified by this technique. The higher affinity of polymerase III for DNA is consistent with the higher resistance to salt inhibition (presumably due to lack of enzyme binding to DNA) shown by the enzymes from *Xenopus* (*16*), kidney cells (*52*), calf thymus (*55*), yeast (*10*) and others (*12*). This suggests that DNA-cellulose chromatography, a very efficient step in the purification of yeast polymerase III, may be applied to the purification of this enzyme from other sources.

Ribosomal RNA and poly(U), covalently bound to agarose beads, have been used in the purification of rat liver and mouse ascites RNA polymerase I (*15*) and in the isolation of a peptide affecting transcription that copurifies with polymerase I from rat liver (*15*). Heparin bound to agarose has been introduced as an effective affinity absorbent for the purification of *E. coli* holo RNA polymerase (*67*). This new procedure may simplify the purification of eukaryotic enzymes.

## F. Resolution by Size

Prokaryotic as well as eukaryotic RNA polymerases are multimeric proteins of high molecular weight, in the range of 400,000–600,000. Because of this property, glycerol or sucrose gradient centrifugation has been advantageously used in many procedures for extensive purification of RNA polymerases from eukaryotes (see Table I). This method does not allow processing of large amounts of protein, and it is therefore the choice as a last step (Table I). It also has the advantage that the enzyme is obtained concentrated and in the presence of high concentrations of stabilizing agents, such as glycerol or sucrose.

Gel filtration has also been utilized (*23,68*) and may be preferred over density-gradient centrifugation in large-scale preparations of polymerases.

## G. Enzyme Purity

The purity of RNA polymerases is monitored by (a) increase in the specific activity of the enzyme, (b) decrease in the number of protein bands in native or sodium dodecyl sulfate (SDS) gel electrophoresis, and (c) decrease in contaminating nucleolytic, proteolytic, or other activities. All these methods are necessary to fully assess the purity of a preparation. Specific activities are frequently not comparable from one laboratory to another. The polymerase assay is complex and highly dependent on the nature and state of the template and the particular assay conditions. Furthermore, activity may be affected by low concentrations of protein contaminants. A homogeneous enzyme may nevertheless have a low specific activity due to the template employed or to the loss of part of the enzyme or a protein factor during the purification. Low activities of purified polymerases are frequently observed with double-stranded DNA or chromatin templates as opposed to denatured DNA or synthetic polynucleotide templates. Assay with different templates is therefore often informative. The specific activity of the most highly purified eukaryotic RNA polymerases is of the order of the purified bacterial enzymes and varies between 100 and 1000 nmoles of UMP incorporated into RNA per milligram of protein in 10 minutes at 30°C (see Table I).

A highly purified polymerase preparation should give only one (or more in case of proved heterogeneity) protein band in native polyacrylamide gel electrophoresis. The sensitivity of this measure of purity depends on the precision with which the electrophoresis is run. Large loads and efficient staining and destaining procedures should be used, otherwise numerous impurities of different mobility may escape unnoticed.

SDS gel electrophoresis cannot be initially used as a criterion of purity. Most of the highly purified eukaryotic RNA polymerases are composed of many different subunits (see Table I for particular references); it is not clear whether the peptides identified in the various enzymes are subunits, as opposed to impurities or products of proteolytic degradation. However, once the subunit composition of the functional polymerase has been established, then SDS polyacrylamide gel electrophoresis is undoubtedly one of the most rigorous techniques to monitor its purity during a preparation.

Roeder (*12*) has described the criteria for establishing whether the polypeptides copurifying with RNA polymerase are subunits.

Before they can be used for transcriptional studies, highly purified RNA polymerases should be tested for various enzyme contaminants that may interfere with the particular assay and generate artifactual results. These activities include the exo and endo DNases, RNase, RNase H, protein kinases, polynucleotide phosphorylase, proteases, and phosphatases.

### H. Stability and Storage

RNA polymerases are notoriously unstable, and various precautions are necessary to preserve the activity throughout the purification procedure. First, the enzymes should be kept as concentrated as possible. Long dialysis and excessively large ion-exchange columns should be avoided. Step elutions or steep salt gradients result in better recoveries than shallow salt gradients (*22,30*). The enzymes seem to be fairly stable as ammonium sulfate suspensions, and this is a convenient way of storing them between purification steps. In density gradient centrifugation, the activity recovered is a direct function of the concentration of the enzyme load (*26,30,69*). Proper amounts of the heavy-metal chelating agent, ethylenediaminetetraacetic acid, reducing agents such as dithiothreitol or 2-mercaptoethanol, and protein structure-stabilizing agents such as glycerol or ethylene glycol added to the buffers greatly improve the stability of the enzymes. In several instances, when the protein concentration is extremely low, such proteins as bovine serum albumin (*18,30*) or $\gamma$-globulins (*36*) can be added to stabilize the enzymes.

The purified RNA polymerases appear to be fairly stable at low temperatures in the presence of high concentrations of polyalcohols. Enzymes are conveniently stored in 30–50% glycerol at $-70°C$ or under liquid $N_2$ or unfrozen in 50% glycerol or 50% polyethylene glycol at $-20°C$ (see references of Table I for details on particular enzymes).

## III. Purification of Yeast RNA Polymerases I, II, and III

The following pages describe a convenient procedure for the large-scale simultaneous purification of yeast RNA polymerases I, II, and III. This method incorporates the concepts previously discussed for the isolation of RNA polymerases and may eventually be applied to the purification of the enzymes from other eukaryotic sources.

## A. Assay

The assay of DNA-dependent RNA polymerases measures the incorporation of an appropriately labeled ribonucleoside triphosphate into an RNA product upon direction from a deoxynucleotide template. The incubation mixture contains in 0.060 ml: 50 m*M* Tris-HCl (pH 8.0); 1.6 m*M* $MnCl_2$ or 10 m*M* $MgCl_2$; 0.5 m*M* ATP, CTP, and GTP; 0.01 m*M* UTP; 0.5 μCi of [$^3$H]UTP; 10 m*M* 2-mercaptoethanol, and 10–20 μg of native and/or denatured calf thymus DNA.

### 1. Reagents

Four milliliters of a reaction cocktail suitable for 200 assays is made from the following components:

| | |
|---|---|
| Tris-HCl, 1 *M*, pH 8.0 | 0.600 ml |
| $MnCl_2$, 0.1 *M* | 0.200 ml |
| ATP, CTP, and GTP, 0.03 *M* each | 0.200 ml |
| UTP, 0.005 *M* | 0.025 ml |
| [$^3$H]UTP, 0.5 mCi/ml | 0.200 ml |
| 2-Mercaptoethanol, 10% | 0.025 ml |
| DNA (calf thymus), 2 mg/ml | 2.000 ml |
| $H_2O$ | 0.750 ml |

### 2. Procedure

An incubation mixture is prepared which contains 20 μl of the reaction cocktail plus 20 μl of enzyme properly diluted in 0.05 *M* Tris-HCl (pH 8.0)–25% glycerol–0.1 m*M* EDTA–0.01 *M* 2-mercaptoethanol plus 20 μl of the appropriate KCl or $(NH_4)_2SO_4$ solution. For assay of RNA polymerase in crude extracts, the incubation mixture also included 0.2 μg of crystalline pyruvate kinase and 70 m*M* phosphoenolpyruvic acid. This reaction mixture, containing nonsaturating levels of UTP (0.01 m*M*), is used in routine assays. However, when the specific activity of the different fractions is determined, the UTP concentration is increased to 0.6 m*M*. After incubation for 10 minutes at 30°C, a 50-μl aliquot is withdrawn from each tube and applied directly onto Whatman DE-81 filter disks. The filters are washed seven times with 5% $Na_2HPO_4$, twice with water, twice with ethanol, and dried. Radioactivity is measured by immersing the disks in 3 ml of a solution of 4 gm of Omnifluor (New England Nuclear) per liter in toluene. One unit of activity corresponds to the incorporation of 1 nmole of UMP into RNA in 10 minutes under the above conditions. The specific activity of UTP in the assay mixture is determined on an aliquot of the reaction mixture (without enzyme) which is spotted on a disk, dried, and counted. Owing to the higher counting efficiency of [$^3$H]UMP in RNA than [$^3$H]UTP alone, samples should be subjected to hydrolysis and solubilization by overnight treatment with 250 $\mu$l of a 1:6 mixture of water and NCS (Nuclear Chicago Solubilizer), and the resulting solutions be counted in toluene fluor.

When assayed with saturating amounts of native DNA the optimal salt concentration for polymerase I is approximately 0.05 *M* $(NH_4)_2SO_4$ or 0.1 *M* KCl; for polymerase II it is about 0.1 *M* $(NH_4)_2SO_4$ or 0.25 *M* KCl, and for polymerase III, it is a broad optimum between 0.05 and 0.2 *M* $(NH_4)_2SO_4$ or 0.1–0.4 *M* KCl. The assay described above used $Mn_n^{2+}$ (1.6 m*M*) as the divalent metal ion, but $Mg^{2+}$ (5–10 m*M*) can also be used. With $Mg^{2+}$ the activity of polymerase II is substantially lower and that of polymerase I and III is less markedly affected. Use of denatured instead of native DNA greatly increases the activity of polymerase II (*12, 70*).

The differential inhibition of eukaryotic RNA polymerases by $\alpha$-amanitin can be exploited in the determination of the activity of a given enzyme in a mixture. In animal cells, polymerase I is resistant to 2 mg/ml of $\alpha$-amanitin whereas polymerases II and III are inhibited at 0.2 $\mu$g/ml and 100 $\mu$g/ml of $\alpha$-amanitin, respectively (*12*). In yeast, however, RNA polymerase III is resistant to 2 mg/ml whereas polymerases II and I can be inhibited at 20 $\mu$g/ml and 2000 $\mu$g/ml, respectively (*10*).

## B. Purification of Yeast Polymerases

### 1. Reagents and Materials

Extraction buffer: 0.05 *M* Tris-HCl (pH 8.0)–0.01 *M* $MgCl_2$–0.5 m*M* EDTA–0.01 *M* 2-mercaptoethanol–0.3 *M* $(NH_4)_2SO_4$–10% glycerol–1 m*M* phenylmethylsulfonyl fluoride–1% dimethylsulfoxide

Buffer A: 0.02 *M* Tris-HCl (pH 8.0)–0.5 m*M* EDTA–0.01 *M* 2-mercapto-ethanol–10% glycerol–1 m*M* phenylmethylsulfonyl fluoride–1% dime-thylsulfoxide

Buffer B: 0.02 *M* Tris-HCl (pH 8.4)–0.5 m*M* EDTA–0.01 *M* 2-mercapto-ethanol–1 m*M* phenylmethylsulfonyl fluoride–1% dimethylsulfoxide

Buffer C: 0.02 *M* Tris–HCl (pH 8.0)–0.5 m*M* EDTA–0.01 *M* 2-mercapto-ethanol–25% glycerol–1m*M* phenylmethylsulfonyl fluoride–1% di-methylsulfoxide

Polymin P (5% w/v) solution: Commercial 50% solution diluted 10 times and adjusted to pH 8.0 with HCl

Phosphocellulose (Whatman P-11) precycled and equilibrated with buf-fer A containing 0.15 *M* $(NH_4)_2SO_4$

DEAE-cellulose (Whatman DE-52) precycled and equilibrated with buffer B containing 0.1 *M* $(NH_4)_2SO_4$

DEAE-Sephadex (A-25) precycled and equilibrated with buffer C con-taining 0.1 *M* $(NH_4)_2SO_4$

DNA-cellulose prepared with denatured calf thymus DNA by the method of Alberts and Herrick (*71*).

Yeast cells (*Saccharomyces cerevisiae*) strain Fl in late log phase, obtained commercially. The cells are harvested direct from the fermenter, washed by resuspending first in cold distilled water and then in 4 liters of extraction buffer per 2000 gm, wet weight, of cells. The washed cells are suspended in a small amount of extraction buffer, and the heavy paste is added dropwise to liquid $N_2$ and stored at $-80°C$. Cells prepared in this way can be stored for several months with no noticeable change in RNA polymerase activity. The addition of phenylmethylsulfonyl fluoride to all the buffers is necessary to prevent proteolysis of the enzymes (*54, 72*).

### 2. Purification of Yeast RNA Polymerase I

*a. Preparation of Yeast Extract.* The yeast extract is prepared by a modification of the method of Bhargava and Halvorson (*73*). Unless otherwise stated, all RNA polymerase purification steps are performed at 0–4°C. The cells are disrupted by crushing the frozen pellets through a precooled 200-ml Eaton pressure cell (*74*) at a pressure of 10,000 psi. The broken cells are stirred with a 2:1 ratio (w/v) of extraction buffer, and the pH is adjusted to 8 (pH paper) with solid Tris. The homogenate is centrifuged for 60 minutes at 27,000 *g* (Sorvall rotor GSA), and the supernatant is pooled to give about 5 liters of extract per 2 kg of yeast cells. On the average this fraction contains 30 mg of protein per milliliter. At this stage, owing to the high amounts of RNA and the presence of hydrolytic enzymes, the RNA polymerase assay is not reliable and gives about 22,000 total units/1000 gm of cells (fraction I-1).

*b. Treatment with Phosphocellulose.* Approximately 5 liters of extract are added to 1200 gm (wet weight) of phosphocellulose previously equilibrated in buffer A containing 0.15 *M* ammonium sulfate. After thorough mixing the ammonium sulfate concentration is reduced to 0.15 *M* by addition of 1 volume of buffer A (5 liters). The suspension is slowly stirred for 30 minutes and then filtered through filter paper in a Büchner funnel without allowing the phosphocellulose to dry. The cake is suspended in 8 liters of buffer A containing 0.15 *M* ammonium sulfate stirred for 10 minutes and collected by filtration. This procedure is repeated three more times. After the last filtration the phosphocellulose cake is suspended in 1.5 liters of buffer A containing 0.4 *M* ammonium sulfate, stirred slowly for 30 minutes, and filtered as above. About 1.7 liters of filtrate are obtained, which contain RNA polymerase I with a specific activity of 15 units per milligram of protein (fraction I-2).

*c. Treatment with DEAE-Cellulose.* Fraction I-2 (1.7 liters) is added to 1200 gm (wet weight) of DEAE-cellulose previously equilibrated in buffer B containing 0.1 *M* ammonium sulfate. After thorough mixing, the ammonium sulfate concentration is lowered to 0.1 *M* by addition of 3 volumes of buffer B. The suspension is gently stirred for 30 minutes, filtered as above, and washed four times with 4 liters of buffer B containing 0.1 *M* ammonium sulfate. RNA polymerase is eluted by suspending and stirring the cellulose cake in 1.4 liters of buffer B containing 0.3 *M* ammonium sulfate for 30 minutes. After filtration, 1.5 liters of enzyme solution with a specific activity of 50–80 units/mg of protein are obtained (fraction I-3). At this stage the enzyme can be conveniently stored by precipitating it with ammonium sulfate (see below) and dissolving in buffer C.

*d. Ion-Filtration Chromatography in DEAE-Sephadex.* RNA polymerase (fraction I-3) is precipitated by addition of 35 gm of solid ammonium sulfate/100 ml of solution. After dissolution of the salt, the precipitate is left overnight at 4°C and collected by centrifugation (27,000 *g* for 30 minutes in Sorvall HB4 rotor). The pellet is dissolved in 20 ml of buffer C containing 0.35 *M* ammonium sulfate and loaded onto a column of 4 × 50 cm packed with DEAE-Sephadex A-25. The column is previously equilibrated with buffer C containing 0.1 *M* ammonium sulfate. The column is developed with the same buffer containing 0.35 *M* ammonium sulfate (Fig. 1). The active fractions are pooled (60 ml). The enzyme obtained has a specific activity of 140–180 units per milligram of protein (fraction I-4).

*e. Sucrose Gradient Centrifugation.* Fraction I-4 is precipitated by overnight dialysis against buffer B saturated with ammonium sulfate. The precipitate is collected as above and dissolved in a final volume of 7.2 ml with

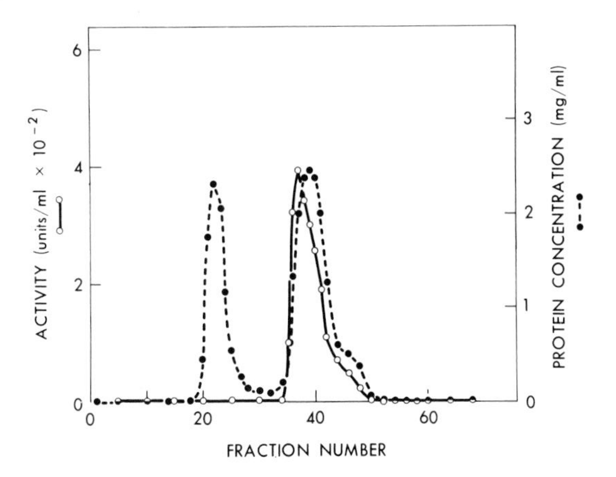


FIG. 1. DEAE-Sephadex ion-filtration column chromatography of yeast RNA polymerase I. Fraction I-3 was concentrated by ammonium sulfate precipitation and loaded onto a column (4 × 50 cm) of DEAE-Sephadex A-25. The column was previously equilibrated and eluted as described in the text. Fractions of 5 ml were collected.

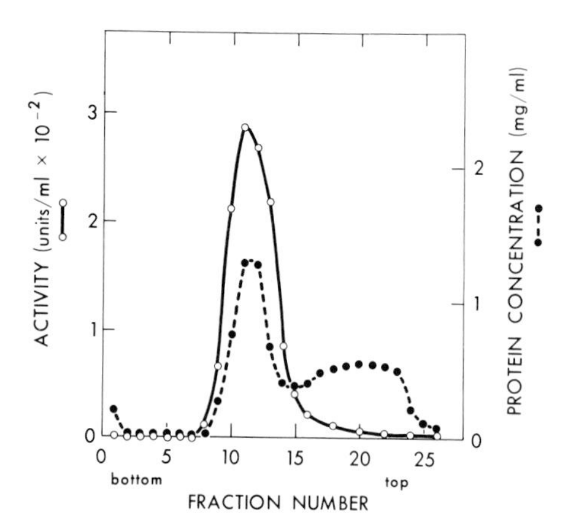


FIG. 2. Sucrose-gradient centrifugation of yeast RNA polymerase I. Enzyme from fraction I-4 was concentrated by ammonium sulfate precipitation and loaded on sucrose gradients which were prepared, loaded, and centrifuged as described in the text. Fractions of 0.5 ml were collected from the bottom of the tubes. Active fractions of all gradients were pooled to give fraction I-5.

buffer A containing 0.2 *M* KCl. Aliquots of 0.6–1.0 ml are layered on 11.8 ml of linear 5 to 20% (w/v) sucrose gradients in 0.05 *M* Tris-HCl (pH 8.0)–15% glycerol–0.5 m*M* EDTA–0.5 *M* KCl–0.02 *M* 2-mercaptoethanol. The gradients are centrifuged at 40,000 rpm for 28 hours at 4°C (Beckman SW-41 rotor). A profile of the sucrose density centrifugation is shown in Fig. 2. The enzyme pooled from the active fractions is essentially pure and has a specific activity of 200–300 units/mg and a protein concentration of 0.7–1.0 mg/ml. It is stored at −80°C (fraction I-5).

### 3. Purification of Yeast RNA Polymerase II

*a. Fractionation with Polymin P.* About 9 liters of extract after adsorption of polymerase I with phosphocellulose (fraction II-III-1) are precipitated by the addition of 0.07 volume (approximately 630 ml) of 5% Polymin P solution with rapid stirring. After 5 minutes, the precipitate is collected by centrifugation (10 minutes, 9000 rpm, Sorvall rotor GS3), and the supernatant is discarded. RNA polymerases II and III are extracted from the pellet with 2.8 liters of buffer A containing 0.3 *M* ammonium sulfate. After centrifugation the protein in the supernatant is precipitated with solid ammonium sulfate (35 gm/100 ml); the pellet is dissolved and diluted with buffer A so that the final ammonium sulfate concentration is 0.15 *M* (fraction II-III-2).

The optimal amount of Polymin P required to precipitate yeast RNA polymerases II and III depends on the concentration of salt, protein, and nucleic

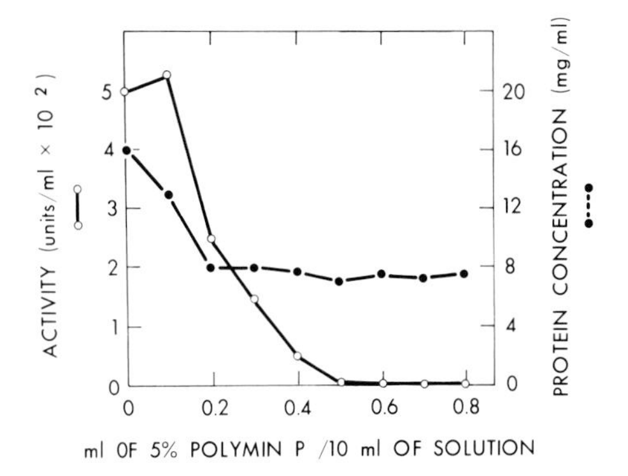


FIG. 3. Precipitation of yeast RNA polymerases II and III with Polymin P. The indicated volume of 5% Polymin P was added to 10 ml of extract (fraction II-III-1) with rapid stirring. After centrifugation (10,000 rpm for 10 minutes, Sorvall rotor HB-4), the supernatants were precipitated with saturated ammonium sulfate. After centrifugation, as above, the pellets were dissolved in buffer A and assayed for protein concentration and RNA polymerase activity.

acids in the extract (*22*). Optimal conditions for yeast enzymes were established by trial experiments. The results are shown in Fig. 3. Addition of 5% Polymin P to a final concentration of about 0.3% precipitates all the RNA polymerases, the nucleic acids and about 50% of the protein of the extract. The optimal conditions for the elution of RNA polymerases II and III were also determined from trial experiments which are summarized in Fig. 4. A

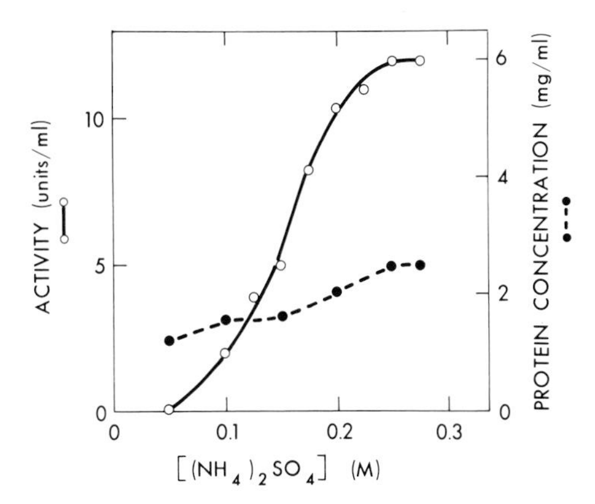


FIG. 4. Extraction of yeast RNA polymerases II and III from the Polymin P precipitate by increasing salt concentration. To 50 ml of extract was added 3.5 ml of 5% Polymin P with rapid stirring. The resulting suspension was divided into 5-ml aliquots and centrifuged (10,000 rpm for 10 minutes, Sorvall rotor HB-4). The supernatants were discarded, and the pellets were resuspended in buffer A containing the indicated concentrations of ammonium sulfate. After centrifugation as above, protein and RNA polymerase activity were assayed in the supernatant.

buffer containing 0.3 *M* ammonium sulfate releases all the RNA polymerase activity into the supernatant, with only 10% of the protein.

The Polymin P step results in about 20-fold purification of the mixture of yeast polymerases II and III with practically quantitative yield. Nucleic acids are largely removed since the 280/260 absorbance ratio increases from 0.6 to about 1.2.

*b. DEAE-Cellulose Chromatography.* Fraction II-III-2 (1100 ml) is applied to a 1000-ml column of DEAE-cellulose previously equilibrated with buffer A containing 0.15 *M* ammonium sulfate. The column is washed with approximately 1 liter of buffer A, 0.15 *M* ammonium sulfate (until the absorbance at 280 nm is less than 0.05) and then step-eluted with 0.3 *M* ammonium sulfate in buffer A. Active fractions are pooled and precipitated by dialysis against buffer B containing 35 gm of ammonium sulfate per 100 ml. The precipitate is collected and dissolved in 300 ml of buffer A (fraction II-3). A typical experiment is shown in Fig. 5. A recently poured column of 8 cm diameter allows very fast flow rates (300 ml per hour) so that the experiment can be carried out in a few hours. If desired, this column step can be easily adapted to a batch procedure. This step results in the complete separation of polymerases II and III and in about 7-fold purification of polymerase II, with a yield of 72% (Table II). The 280/260 absorbance ratio of this fraction is 1.60.

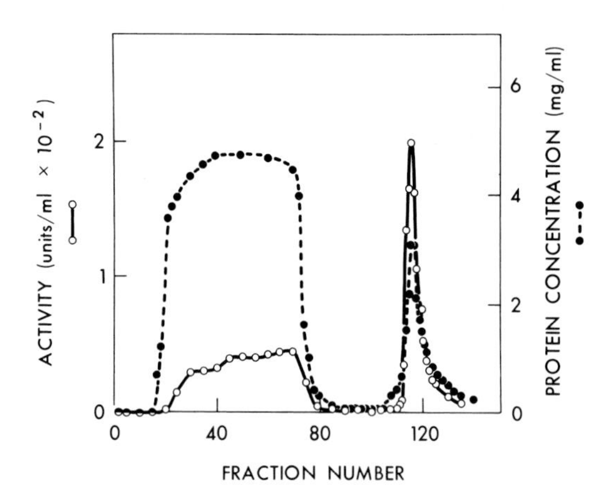


FIG. 5. DEAE-cellulose column chromatography of yeast RNA polymerases II and III. About 1100 ml of fraction II-III-2 containing approximately 5.5 gm of protein were applied to a 8 × 20 cm column of DEAE-cellulose. The column was equilibrated, loaded, washed, and step-eluted as described in the text. Fractions of 20 ml were collected at a flow rate of 300 ml per hour. Tubes 18 to 80 are the flow-through protein fractions (III-3) which contained only RNA polymerase III as determined by its complete resistance to high concentrations of $\alpha$-amanitin (*10*). Tubes 108 to 122 contained RNA polymerase II (fraction II-3).

TABLE II

SUMMARY OF THE PURIFICATION OF YEAST RNA POLYMERASES I, II, AND III

| Purification step | Volume (ml) | Total protein (mg) | Activity (units) | Specific activity (units/mg) | Yield (%) |
|---|---|---|---|---|---|
| RNA Polymerase I | | | | | |
| Extract (I + II + III) | 4500 | 121,500 | 43,000 | 0.35 | — |
| P-cellulose batch | 1730 | 1550 | 20,500 | 13.2 | 100 |
| DEAE-cellulose batch | 1500 | 270 | 15,000 | 55 | 73 |
| Ion filtration (DEAE-Sephadex) | 58 | 69 | 10,000 | 145 | 49 |
| Sucrose-gradient centrifugation | 34 | 24 | 5100 | 212 | 25 |
| RNA Polymerase II | | | | | |
| Extract (after P-cellulose) | 9300 | 111,600 | 32,000 | 0.9 | 100 |
| Polymin P fractionation | 1100 | 5500 | 32,000 | 7.6 | 106 |
| DEAE-cellulose chromatography | 380 | 456 | 22,900 | 50 | 72 |
| P-cellulose chromatography | 64 | 77 | 16,000 | 230 | 50 |
| Sucrose gradient centrifugation | 30 | 28 | 8000 | 440 | 25 |
| RNA Polymerase III | | | | | |
| DEAE-cellulose (flow-through) | 1000 | 4300 | 8000 | 2.3 | 100 |
| DEAE-Sephadex chromatography | 140 | 294 | 3760 | 12.8 | 47 |
| DNA-cellulose chromatography | 6 | 12 | 2020 | 140 | 25 |

[a]The data reported in this table were taken from one representative experiment. Two kilograms of frozen yeast cells (wet weight) were used as starting material. One unit corresponds to 1 nmole of UMP incorporated in 10 minutes at 30°C.

*c. Phosphocellulose Chromatography.* Fraction II-3 is diluted in small portions with buffer C to a concentration of ammonium sulfate of 0.03 *M* and applied to a 100-ml column of phosphocellulose previously equilibrated with buffer C containing 0.04 *M* ammonium sulfate (Fig. 6). The column is washed with approximately 120 ml of buffer C with 0.04 *M* ammonium sulfate until the absorbance at 280 nm is less than 0.01 and step-eluted with 0.13 *M* ammonium sulfate in buffer C as shown in Fig. 6. Active fractions are pooled giving 64 ml of enzyme with 220 units per milligram of protein (fraction II-4). This step, which can be also adapted to a batch procedure, affords a purification of about 4-fold with a yield of 70% (Table II). Polymerase II binds weakly to phosphocellulose, and it is necessary to carefully adjust the salt concentration by conductivity measurements (<6 mmho) of the load and the wash to obtain the maximum possible purification.

*d. Sucrose-Gradient Centrifugation.* Fraction II-4 is precipitated by overnight dialysis against buffer B saturated with ammonium sulfate. The precipitate is collected by centrifugation and dissolved in a final volume of 6 ml

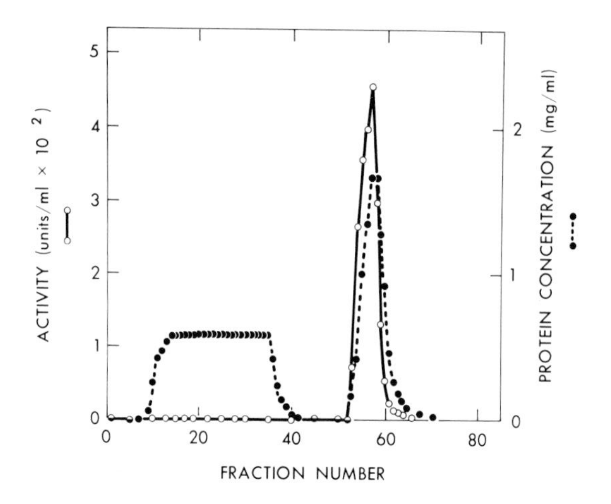


FIG. 6. Phosphocellulose column chromatography of yeast RNA polymerase II. Fraction II-3 containing approximately 450 mg of protein was applied to a 3 × 15 cm column of phosphocellulose. The column was equilibrated, loaded, washed, and step-eluted as described in the text. Fractions of 8 ml were collected. Tubes 53 to 60 contained the bulk of RNA polymerase II activity and were pooled to give fraction II-4.

with buffer A containing 0.2 *M* KCl. Aliquots of 0.8 to 1.0 ml are layered on 11.8 ml linear 5 to 20% (w/v) sucrose gradients in 0.05 *M* Tris-HCl (pH 8.0)–15% glycerol–0.5 m*M* EDTA–0.5 *M* KCl–0.01 *M* 2-mercaptoethanol. The gradients are centrifuged at 40,000 rpm for 32 hours at 4°C (Spinco SW-41 rotor). A profile of the sucrose density centrifugation is shown in Fig. 7. The enzyme pooled from the active fractions is apparently homogeneous and has

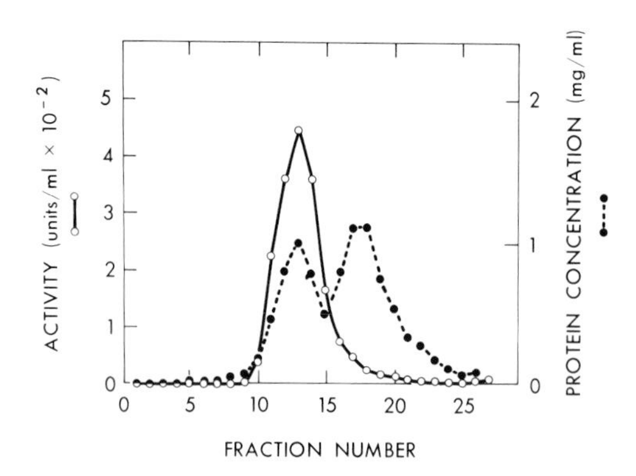


FIG. 7. Sucrose-gradient centrifugation of yeast RNA polymerase II. Fraction II-4 was concentrated by ammonium sulfate precipitation and loaded on sucrose gradients, which were prepared, loaded, and centrifuged as described in the text. Fractions of 0.5 ml were collected from the bottom of the tubes. Active fractions of all gradients are pooled to give fraction II-5.

a specific activity of 400–450 units/mg and a protein concentration of 0.6–0.8 mg/ml. It is stored at −80°C (fraction II-5).

### 4. Purification of Yeast RNA Polymerase III

*a. DEAE-Sephadex Chromatography.* The active flow-through fractions of DEAE-cellulose chromatography (fraction III-3), approximately 1000 ml, are pooled and loaded onto a 800-ml column of DEAE-Sephadex A-25 previously equilibrated with 0.15 *M* ammonium sulfate in buffer A. The column is washed with 800–1000 ml of 0.20 *M* ammonium sulfate in buffer A until the absorbance at 280 nm is less than 0.020 and step-eluted with 0.35 *M* ammonium sulfate in buffer A. Details are shown in Fig. 8. Active fractions are pooled to give a solution of about 2 mg/ml with a specific activity of 12 units/mg (fraction III-4).

*b. DNA-Cellulose Chromatography.* Fraction III-4 is precipitated by overnight dialysis against buffer B saturated with ammonium sulfate; the pellet is dissolved in buffer C until the ammonium sulfate concentration is 0.05 *M* and loaded onto a 120-ml column of DNA-cellulose (0.3 mg of denatured DNA per milliliter of resin). The column is washed with buffer C containing 0.1 *M* KCl and eluted with about 360 ml of a linear gradient of 0.1 *M* to 1.0 *M* KCl in buffer C (Fig. 9). Fractions containing RNA polymerase activity are pooled and precipitated by overnight dialysis against buffer B saturated with ammonium sulfate. The pellet is dissolved in buffer

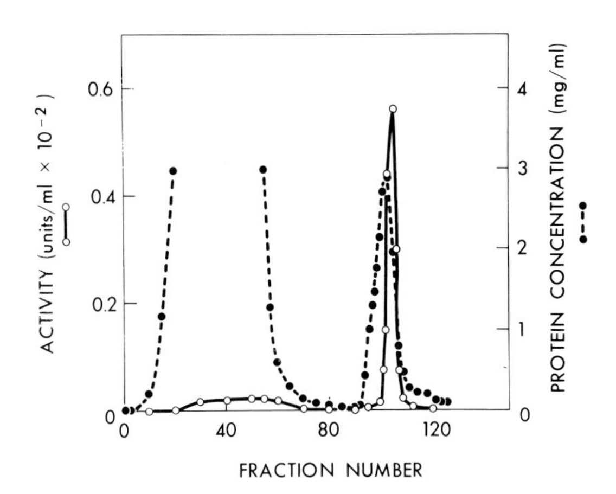


Fig. 8. DEAE-Sephadex column chromatography of yeast RNA polymerase III. The flow-through from DEAE-cellulose chromatography (fraction III-3) was applied to a 8 × 16 cm column of DEAE-Sephadex A-25. The column was equilibrated, loaded, and step-eluted as described in the text. Fractions of 20 ml were collected. Tubes 101 to 107, containing the bulk of RNA polymerase III activity, were pooled to give fraction III-4.

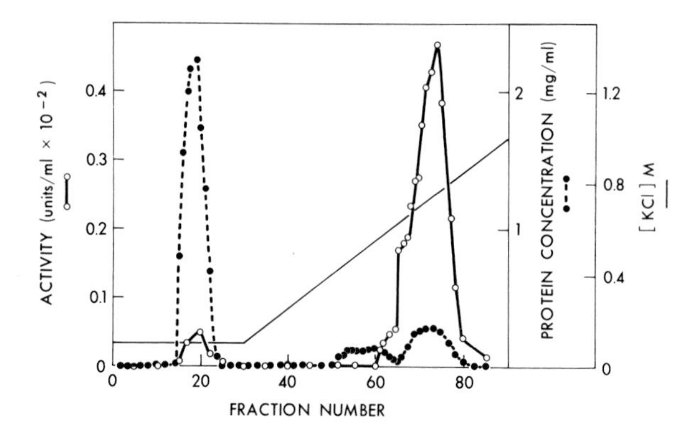


FIG. 9. DNA-cellulose column chromatography of yeast RNA polymerase III. Fraction III-4 was concentrated by ammonium sulfate precipitation and applied to a 3 × 20 cm column of denatured DNA-cellulose. The column was equilibrated, loaded, and eluted as described in the text. Fractions of 8 ml were collected. Tubes 65 to 77 were pooled and concentrated with ammonium sulfate to give fraction III-5.

C to a concentration of 2 mg/ml. The enzyme has a specific activity of 140 units/ml (Table I) and is about 80% pure. Complete purification is achieved by a sucrose.gradient centrifugation step as described for polymerases I and II.

### 5. COMMENTS ON THE PURIFICATION

The purification and yield of each step are summarized in Table II. A final purification of approximately 2000-, 2500-, and 3000-fold is obtained for enzymes I, II, and III, respectively. The yields of enzymes reported here represent minimal amounts recoverable, since the commercial cells used in these experiments are nutritionally limited to less than 10% of their maximal growth rate. Rapidly growing cells would be expected to contain higher amounts of enzymes (*75*). This method has been also applied in this laboratory to the purification of enzymes from other *S. cerevisiae* strains, A364A and $\alpha\alpha$AA 5178 $1C^2 \times 2B^2$, grown in complete or phosphate-depleted media. With minor modifications this procedure can be scaled down 100 times allowing the complete purification of the three yeast polymerases from 20 gm of cells (*76a*).

### 6. STRUCTURE OF THE PURIFIED ENZYMES

The enzymes prepared by this procedure are apparently homogeneous by the criteria previously described. SDS–acrylamide gel electrophoresis indicates that each enzyme is structurally unique and is composed of two large and various smaller molecular weight subunits (*10, 76*), (Fig. 10). Polymerase I contains polypeptides of 185,000, 137,000, 48,000, 44,000, 41,000,

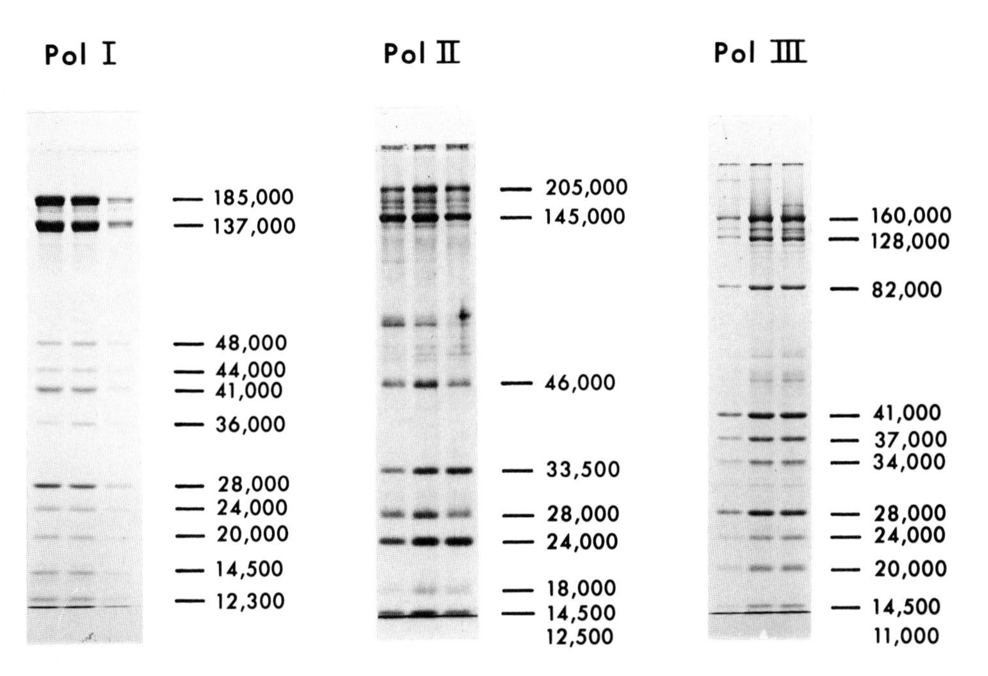


FIG. 10. Putative subunit composition of yeast RNA polymerases I, II, and III. Active fractions from sucrose gradients were subjected to sodium dodecyl sulfate gel electrophoresis as described elsewhere (*26*). Gels are 11%, 10%, and 10% acrylamide, respectively. The 12,500 dalton polypeptide of polymerase II and the 11,000 dalton polypeptide of polymerase III were not resolved under these conditions. Each of the enzymes has a polypeptide of about 9000 daltons, which is visible only in 15% acrylamide gels.

36,000, 28,000, 24,000, 20,000, 14,500, and 12,300 daltons. Polymerase II contains polypeptides of 205,000 (175,000), 145,000, 46,000, 33,500, 28,000, 24,000, 18,000, 14,500, and 12,500 daltons. Polymerase III contains polypeptides of 160,000, 128,000, 82,000, 41,000, 37,000, 34,000, 28,000, 24,000, 20,000, 14,500, and 11,000 daltons. In addition, each of the enzymes has a polypeptide of approximately 9000 daltons, which is visible only in 15% acrylamide gels. The 175,000 dalton polypeptide of polymerase II is a degradation product of the 205,000 dalton subunit (*54*). Some of these putative subunits appear to be less tightly bound to the enzyme complex and dissociate under certain conditions during sucrose-gradient sedimentation (*77*), native gel electrophoresis, or phosphocellulose chromatography (*10, 26, 54*). Subunits of 41,000 and 20,000 seem to be common for polymerases I and III, and subunits of 28,000, 24,000, and 14,500 to be common to polymerases I, II, and III (*76, 78*). Some of these subunits are phosphorylated (*66, 76a*). All yeast polymerases are probably Zn-containing metalloenzymes, as has been shown for polymerases I and II (*80, 81*).

## Acknowledgments

The research carried out in this laboratory referred to in this article was supported by grants from the National Institutes of Health (GM21830 and GM19527).

## References

1. Roeder, R. G., and Rutter, W. J., *Nature* (*London*) **224**, 234 (1969).
2. Roeder, R. G., Doctoral Dissertation, University of Washington, Seattle (1969).
3. Roeder, R. G., Rutter W. J., *Proc. Natl. Acad. Sci. U.S.A.* **65**, 675 (1970).
4. Blatti, S. P., Ingles, C. J., Lindell, T. J., Morris, P. W., Weaver, R. F., Weinberg, F., and Rutter, W. J., *Cold Spring Harbor Symp. Quant. Biol.* **35**, 649 (1970).
5. Weinmann, R., and Roeder, R. G., *Proc. Natl. Acad. Sci. U.S.A.* **71**, 1790 (1974).
6. Zybler, E., and Penman, S., *Proc. Natl. Acad. Sci. U.S.A.* **68**, 2861 (1971).
7. Suzuki, Y., and Giza, P. E., *J. Mol. Biol.*, **107**, 183 (1976).
8. Weinmann, R., Raskas, H., and Roeder, R. G., *Proc. Natl. Acad. Sci. U.S.A.* **71**, 3426 (1974).
9. Sklar, V. E. F., Schwartz, L. B., and Roeder, R. G., *Proc. Natl. Acad. Sci. U.S.A.* **72**, 348. (1975).
10. Valenzuela, P., Hager, G. L., Weinberg, F., and Rutter, W. J., *Proc. Natl. Acad. Sci. U.S.A.* **73**, 1024 (1976).
11. Chambon, P., *Annu. Rev. Biochem.* **43**, 613 (1975).
12. Roeder, R. G., *in* "RNA Polymerase" (M. Chamberlin and R. Losick, eds.) Cold Spring Harbor Lab., Cold Spring Harbor, New York, **285**, (1976).
13. Hossenlopp, P., Wells, D., and Chambon, P., *Eur. J. Biochem.* **58**, 237 (1975).
14. Schwartz, L. B., Sklar, V. E. F., Jaehning, J. A., Weinmann, R., and Roeder, R. G., *J. Biol. Chem.* **249**, 5889 (1974).
15. Goldberg, M., Perriard, J. C., and Rutter, W. J., *Biochemistry* **16**, 1655 (1977).
16. Roeder, R. G., *J. Biol. Chem.* **248**, 241 (1974).
17. Greenleaf, A. L., and Bautz, E. K. F., *Eur. J. Biochem.* **60**, 169 (1975).
18. Weaver, R. F., Blatti, S. P., and Rutter, W. J., *Proc. Natl. Acad. Sci. U.S.A.* **68**, 2994 (1971).
19. Gissinger, F., and Chambon, P., *Eur. J. Biochem.* **28**, 277 (1972).
20. Kedinger, C., and Chambon, P., *Eur. J. Biochem.* **28**, 283 (1972).
21. Jendrisak, J. J., and Becker, W. M. *Biochem. J.* **139**, 771 (1974).
22. Jendrisak, J. J., and Burgess, R., *Biochemistry* **14**, 4639 (1975).
23. Ponta, H., Ponta, U., and Wintersberger, E., *Eur. J. Biochem.* **29**, 110 (1972).
24. Dezélée, S., and Sentenac, A., *Eur. J. Biochem.* **34**, 41 (1973).
25. Buhler, J. M., Sentenac, A., and Fromageot, P., *J. Biol. Chem.* **249**, 5963 (1974).
26. Valenzuela, P., Weinberg, F., Bell, G. I., and Rutter, W. J., *J. Biol. Chem.* **251**, 1464 (1976).
27. Young, H. A., and Whiteley, H. R., *J. Biol. Chem.* **250**, 479 (1975).
28. Gornicki, S. Z., Vuturo, S. B., West, T. V., and Weaver, R. F., *J. Biol. Chem.* **249**, 1792 (1974).
29. Pong, S. S., and Loomis, W. F., *J. Biol. Chem.* **248**, 3933 (1973).
30. Schwartz, L. B., and Roeder, R. G., *J. Biol. Chem.* **249**, 5898 (1974).
31. Coupar, B. E. H., and Chesterton, C. J., *Eur. J. Biochem.* **59**, 25 (1975).
32. Cochet-Meilhac, M., Nuret, P., Courvalin, J. C., and Chambon, P., *Biochim. Biophys, Acta* **353**, 185 (1974).
33. Weiss, S. B., *Proc. Natl. Acad. Sci. U.S.A.* **46**, 1020 (1960).
34. Jacob, S. T., *Prog. Nucleic Acid Res. Mol. Biol.* **13**, 93 (1973).

35. Chambon, P., *in* "The Enzymes" (P. D. Boyer, ed.), 3rd ed., Vol. 10, p. 261. Academic Press, New York (1974).
36. Krebs, G., and Chambon, P., *Eur. J. Biochem.* **61**, 15 (1976).
37. Sklar, V. E. F., Jaehning, J. A., Gage, P. L., and Roeder, R. G., *J. Biol. Chem.* **251**, 3794 (1976).
37. Valenzuela, P. *et al.* (1977). This article.
38. Yu, F. L., *Biochem. Biophys. Res. Commun.* **64**, 1107 (1975).
39. Chamberlin, M., and Berg, P., *Proc. Natl. Acad. Sci. U.S.A.* **48**, 81 (1962).
40. Berg, D., Barret, K., and Chamberlin, M., *in* "Methods in Enzymology" (K. Moldave and L. Grossman, eds.), Vol. 21, Part D, p. 500. Academic Press, New York (1971).
41. Mandel, J. L., and Chambon, P. *FEBS Lett.* **15**, 175. (1971).
42. Zillig, W., Zechel, K., and Halbwachs, H., *Hoppe-Seyler's Z. Physiol. Chem.* **351**, 221 (1976).
43. Okasaki, T., and Kornberg, A., *J. Biol. Chem.* **239**, 259 (1964).
44. Babinet, C., *Biochem. Biophys. Res. Commun.* **36**, 639 (1967).
45. Mangel, W. F., *Arch. Biochem. Biophys.* **163**, 172 (1974).
46. Weinmann, R., Jaehning, J. A., Raskas, H. J., and Roeder, R. G., *J. Virol.* **17**, 114 (1976).
47. Kirkegaard, L., Johnson, T., and Bock, R., *Anal. Biochem.* **50**, 122 (1972).
48. Goldberg, M., Perriard, J. C., Hager, G., Hallick, R. B., and Rutter, W. J., *Basic Life Sci.* **3**, 241 (1974).
49. Sergeant, A. and Krsmanovic, V., *FEBS Lett.* **35**, 331 (1973).
50. Seifart, K. H., Benecke, B. J., and Juhasz, P., *Arch. Biochem. Biophys.* **151**, 519 (1972).
51. Seifart, K. H., and Benecke, B. J., *Eur. J. Biochem.* **53**, 293 (1975).
52. Austoker, J. L., Beebee, T. J. C., Chesterton, C. J., and Butterworth, P. H. W., *Cell* **3**, 227 (1975).
53. Sklar, V. E. F., and Roeder, R. G., *J. Biol. Chem.* **251**, 1064 (1976).
54. Dezélée, S., Wyers, F., Sentenac, A., and Fromageot, P., *Eur. J. Biochem.* **65**, 543 (1976).
55. Weil, P. A., and Blatti, S. P., *Biochemistry* **15**, 1500 (1976).
56. Chesterton, C. J., and Butterworth, P. H. W., *Eur. J. Biochem.* **19**, 232 (1971).
57. Muramatsu, M., Onishi, T., Matsui, T., Kawabata, C., and Tokugawa, S., *Proc. FEBS Meet. 9th, 1974* Vol **33**, p. 325 (1974).
58. Alberts, B. M., Amodio, F. J., Jenkins, M., Gutman, E. D., and Ferris, F. J., *Cold Spring Harbor Symp. Quant, Biol.* **33**, 289 (1968).
59. Litman, R. M., *J. Biol. Chem.* **243**, 6222 (1968).
60. Cavalieri, L. F., and Carroll, E., *Proc. Natl. Acad. Sci. U.S.A.* **67**, 807 (1970).
61. Schaller, H., Nusslein, C., Bonhoeffer, F. J., Kurz, C., and Nietzchmann, I., *Eur. J. Biochem.* **26**, 474 (1972).
62. Poonian, M. S., Schlabach, A. J., and Weissbach, A., *Biochemistry* **10**, 242 (1971).
63. Arndt-Jovin, D. J., Jovin, T., Bahr, W., Frischauf, A., and Marquart, M., *Eur. J. Biochem.* **54**, 411 (1975).
64. Gilham, P. T., *Biochemistry* **7**, 2809 (1968).
65. Noyes, B. E., and Stark, G. E., *Cell* **5**, 301 (1975).
66. Bell, G. I., Valenzuela, P., and Rutter, W. J., *Nature* (*London*) **261**, 429 (1976).
67. Sternbach, H., Engelhardt, R., and Lezius, A. G., *Eur. J. Biochem.* **60**, 51 (1975).
68. Burgess, R. R., *J. Biol. Chem.* **244**, 6160 (1969).
69. Sugden, B., and Keller, W., *J. Biol. Chem.* **248**, 3777 (1973).
70. Gissinger, F., Kedinger, C., and Chambon, P., *Biochimie* **56**, 319 (1974).
71. Alberts, B. M., and Herrick, G. *in* "Methods in Enzymology" (K. Moldave and L. Grossman, eds.), Vol. 21, Part D, p. 198. Academic Press, New York (1971).

*72.* Pringle, J. E., *Methods Cells Biol.* **12**, 149 (1975).
*73.* Bhargava, M. M., and Halvorson, H. O., *J. Cell Biol.* **49**, 423 (1971).
*74.* Eaton, N. R., *J. Bacteriol.* **83**, 1359 (1962).
*75.* Sebastian, J., Mian, F., and Halvorson, H. O., *FEBS Lett.* **34**, 159 (1973).
*76.* Buhler, J. M., Iborra, F., Sentenac, A., and Fromageot, P., *J. Biol. Chem.* **251**, 1712 (1976).
*76a.* Bell, G. I., Valenzuela, P., and Rutter, W. J. *J. Biol. Chem.* **252**, 3082 (1977).
*77.* Valenzuela, P., Bell, G. I., and Rutter W. J., *Biochem. Biophys. Res. Commun.* **71**, 26 (1976).
*78.* Valenzuela, P., Bell, G. I., Weinberg, F., and Rutter, W. J., *Biochem. Biophys. Res. Commun.* **71**, 1319 (1976).
*79.* Huet, J., Buhler, J. M., Sentenac, A. and Fromageot, P. *Proc. Natl. Acad. Sci. U.S.A.* **72**, 3034 (1975).
*80.* Auld, D. S., Atsuya, I., Campino, C. and Valenzuela, P. *Biochem. Biophys. Res. Commun.* **69**, 548 (1976).
*81.* Lattke, L. and Weser, U., *FEBS Lett.* **65**, 288 (1976).

# Chapter 2

# *Isolation and Characterization of DNA Polymerases from Eukaryotic Cells*

DIPAK K. DUBE, ELIZABETH C. TRAVAGLINI, AND LAWRENCE A. LOEB

*Institute for Cancer Research,*
*The Fox Chase Cancer Center,*
*Philadelphia, Pennsylvania*

## I. Introduction

The present chapter is concerned with the isolation and characterization of DNA polymerases from eukaryotic cells. Our goal is to describe briefly these DNA polymerases and to provide a methodology by which procedures for assaying and isolating DNA polymerases from different tissues may be worked out. More detailed information about the specific DNA polymerases which have been identified in different eukaryotic cells can be obtained from several recent reviews (*1–7*).

## II. Nomenclature

Recent studies indicate that there are several classes of DNA polymerases in eukaryotic cells. In the case of mammalian cells, there is widespread agreement that three different DNA polymerases can be distinguished in addition to a mitochondrial DNA polymerase. These enzymes have been recently designated as DNA polymerase-$\alpha$, -$\beta$, -$\gamma$, and mitochondrial (mt) (*2,8*) (Table I). It seems likely that this classification scheme may eventually be applicable to the DNA polymerases in most eukaryotic cells.

Several mammalian DNA polymerases have been omitted from Table I since they do not fit into the classification scheme just described—for example, the enzyme referred to as terminal transferase, terminal deoxy-

TABLE I

NOMENCLATURE AND PROPERTIES OF MAMMALIAN DNA POLYMERASES

| Property | DNA polymerase-$\alpha$ | DNA polymerase-$\beta$ | DNA polymerase-$\gamma$ | Mitochondrial DNA polymerase |
|---|---|---|---|---|
| Molecular weight | $1.2–2.2 \times 10^5$ | $0.4 \times 10^5$ | $1.1 \times 10^5$ | $1.0–1.1 \times 10^5$ |
| Sedimentation coefficient | 6–8 S | 3.3–3.5 S | 6.1–6.3 S | ~6 S |
| *N*-Ethylmaleimide inhibition | Sensitive | Insensitive | Sensitive | Insensitive |
| Template utilization | | | | |
| a. "Activated" DNA | + + + | + + + | + + | + + + |
| b. $(dA)_n \cdot oligo(dT)$ | + | + + | + + + | + + + |
| c. $(rA)_n \cdot oligo(dT)$ | ± | + | + + | – |
| d. $(rC)_n \cdot oligo(dG)$ | – | – | ± | – |
| Subcellular location | Nuclear and cytoplasmic | Nuclear | Cytoplasmic and nuclear | Mitochondrial |
| Nuclease activity | None | None | Not known | No endonuclease |

nucleotidyltransferase (*9*). This polymerase is found only in thymus or thymus-derived tissues and does not copy polynucleotide templates, but polymerizes deoxynucleoside triphosphates onto the 3'-OH end of an oligonucleotide primer. Also, omitted from Table I are the DNA polymerases of viral origin, "reverse transcriptases," which are present in RNA tumor viruses (*10,11*) or in cells infected with these viruses, and the polymerases which are induced by infection of animal cells with DNA viruses (*12,13*). In the latter case, it is not clear whether the virus-induced DNA polymerases represent unique polymerases coded by the DNA virus or are modifications of host cell polymerases.

## III. Description of the Different Eukaryotic DNA Polymerases and Their Probable Function

The various DNA polymerases in eukaryotic cells, particularly in mammalian cells, have been classified according to their size, cellular localization, and template preference (Table I). A particular function for each of these different DNA polymerases has yet to be clearly defined. In bacteria, a knowledge of the function of the different DNA polymerases in cellular metabolism has been obtained from studies of conditional mutants. Conditional mutants in DNA replication are not available for most eukaryotic cells. Thus, information concerning the function of the different eukaryotic DNA polymerases has only been obtained by studying the relationship of the DNA polymerases to changes in the DNA synthetic activity of cells and to changes during the cell cycle. This type of evidence is not as convincing as that obtained for the bacterial polymerases.

### A. DNA Polymerase-$\alpha$

The class of DNA polymerase that appears to be most abundant in dividing mammalian cells is DNA polymerase-$\alpha$. These enzymes are of large molecular weight, having a sedimentation coefficient of approximately 6–8 S. They require sulfhydryl groups for activity as measured by their sensitivity to *N*-ethylmaleimide. DNA polymerase-$\alpha$ preferentially uses activated DNA as a template; its activity is minimal with poly(rA)·oligo(dT) as a template (*3*). It is frequently referred to as the "cytoplasmic" DNA polymerase, since it is most easily detected in, and can be readily purified from, cytoplasmic extracts. The apparent paradox between its intracellular localization and its site of action in the nucleus probably reflects an artifact of isolation. It should be noted that a number of investigators have shown that DNA polymerase-α can be localized in nuclei isolated under particular conditions and recovered in nuclei using various media (*6,14–16*). In rapidly dividing cells, this enzyme can be localized in nuclei by a variety of techniques (*1,14,16–18*).

Recent evidence suggests that DNA polymerase-α participates in DNA replication. From an analysis of the cell cycle in sea urchin embryos, evidence has accumulated (*14*) which indicates that, at the time of each cell division, DNA polymerase-α becomes associated with nuclei. When a variety of nondividing eukaryotic cells are stimulated to divide, the increase in DNA polymerase activity primarily reflects an increase in DNA polymerase-α. For example, when human lymphocytes are stimulated to divide by phytohemagglutinin, total cellular DNA polymerase activity increases 20- to 100-fold (*19*). This increase in activity is proportional to the increased rate of incorporated thymidine into DNA (*20*) and appears to reflect the *de novo* synthesis of DNA polymerase-α (*16–18,21,22*). The ability of DNA polymerase-α to utilize ribonucleotide primers (*23,24*) is also in accord with a replicative role for this enzyme.

## B. DNA Polymerase-β

DNA polymerase-β is a low-molecular-weight enzyme and is usually identified on the basis of its sedimentation coefficient in sucrose gradients and its resistance to 10 m*M* *N*-ethylmaleimide (*25*). DNA polymerase-β is usually isolated from nuclei. The enzyme has been purified to homogeneity and shown to consist of a single polypeptide chain of molecular weight 43,000–45,000. Structural and immunologic studies have clearly demonstrated that α- and β-polymerases share few homologies (*1–5*). Analyses of different eukaryotic cells for polymerase-β have been carried out by Chang (*25*). Polymerase-β has been found to be present in most multicellular organisms but is absent in bacteria, plants, and protozoans, suggesting it may have evolved during the development of metazoans.

Bertazzoni *et al.* (*26*) have presented evidence that polymerase-β activity in lymphocytes correlates with the lymphocytes' ability to carry out DNA repair.

## C. DNA Polymerase-γ

DNA polymerase-γ is the most recently discovered of the mammalian DNA polymerases and has been found in most cells examined (*2,27*). It accounts for only 1% of the total cellular DNA polymerase activity in mammalian cells and has been reported to be present both in the nucleus and cytoplasm of mammalian cells. DNA polymerase-$\gamma$ has a molecular weight of about 110,000 and prefers poly(rA)·oligo(dT) as a template, particularly in the presence of $Mn^{2+}$. The enzyme requires sulfhydryl groups for activity as measured by its sensitivity to *N*-ethylmaleimide.

Two properties of $\gamma$-polymerase are of particular biological interest. First, it has a very low $K_M$ for dNTPs, about 0.5 $\mu M$ (*2*). It is reasonable to assume that an enzyme with such a low $K_M$ may play a very important part in DNA synthetic activity in the cells, since deoxynucleoside triphosphate substrates are present in low concentrations in cells. Second, DNA polymerase-$\gamma$ increases about 2-fold prior to or during the S phase of the cell cycle (*2,28*).

## D. Mitochondrial and Choroplast DNA Polymerases

Mitochondria appear to contain their own genome and exist semiautonomously in eukaryotic cells. A considerable amount of evidence has accumulated for distinct DNA polymerases being components of these subcellular organelles. DNA polymerases have been extensively purified from mitochondria (*29–31*). The amount of mtDNA polymerase activity corresponds to only about 1% of the total DNA polymerase activity in eukaryotic cells. The DNA polymerase present in mitochondria seems to have many properties and template preferences which differ from those of other DNA polymerases. It has been purified to homogeneity from HeLa cells by Radsak *et al.* (*30*). These authors have shown that it is composed of two subunits of molecular weight 45,000 and 60,000. This polymerase has been shown to contain no endonuclease activity. However, it has not been tested for exonuclease activity. It has not been established by immunological criteria whether the mitochondrial polymerase is uniquely different from other cellular DNA polymerases, nor has it been demonstrated that the enzyme is coded for by mtDNA. An interesting clue to its function may be the observation that mtDNA polymerase activity increases 50–200 times upon irradiation of *Tetrahymena* by UV light; this observation suggests that the enzyme is involved in DNA repair (*32*) and in maintaining the integrity of the mitochondrial genome.

Chloroplasts also appear to contain a unique DNA polymerase. DNA polymerase has been purified from chloroplasts of *Euglena gracilis*, and initial data indicate that it is a component of these organelles (*33*).

A number of DNA polymerases that have been isolated from eukaryotic cells do not clearly fall into the classification scheme described in Table I. A list of such polymerases is given in Table II (*34–42*), and detailed information regarding these polymerases can be obtained from the accompanying references.

TABLE II

OTHER EUKARYOTIC DNA POLYMERASES

| No. | Source | Characteristics | References |
|---|---|---|---|
| 1. | *Drosophila melanogaster* (embryos) | High MW; copies poly(rA)·oligo(dT) | *34* |
| 2. | KB cells (nuclei) | MW 70,000 | *35* |
| 3. | *Vinca rosea* (tissue culture) | High M.W. polymerase | *36* |
| 4. | Erythroid hyperplastic bone marrow | Eukaryotic polymerase with 3′→5′ exonuclease | *37* |
| 5. | *Euglena gracilis* | Two high MW polymerases | *38,39* |
| 6. | Smut fungus, *Ustilago maydis*, a temperature sensitive DNA polymerase purified from a *ts* mutant | MW 110,000 with a 3′→5′ exonuclease | *40,41* |
| 7. | Yeast | Two high MW DNA polymerases ~150,000 | *42* |

## IV. Common Characteristics of DNA Polymerases and a General Assay for Polymerase Activity

### A. General Characteristics

*In vitro*, all DNA polymerases have similar requirements for activity, deoxynucleoside triphosphates complementary to the template, a divalent cation such as $Mg^{2+}$ or $Mn^{2+}$, and a template-primer complex (*7*). Thus, a general assay can be employed for measuring polymerase activity in crude extracts.

The activity of these enzymes is usually measured by determining the incorporation of radioactive deoxynucleotides into acid-insoluble material. One unit of enzyme activity is defined as the incorporation of 1 nmole of radioactive deoxynucleotide per hour. The specific activity of the enzyme is defined as units per milligram of protein. In crude extracts it is necessary to carry out assays for short periods of time because of nucleases, proteases, and nucleotidases and to extrapolate the results to that which would occur in 1 hour. As with any enzyme assay, all components of the reaction should be present in saturating amounts and the rate of incorporation must be linear during the period of incubation and proportional to enzyme concentration.

### B. Assay for DNA Polymerase Activity in Crude Extracts

The components of a reaction mixture frequently utilized in our laboratory for crude extracts are tabulated below. All solutions are made at the concentrations designated and stored at −20°C except for the enzyme extract, which should be stored at −70°.

| Components of the reaction | Parts | Example, 50 μl assay (μl) |
|---|---|---|
| "Activated" calf thymus DNA, 1 mg/ml | 2 | 10 |
| Tris-maleate (pH 7.8), 1 *M* | 1 | 5 |
| [$^3$H]- or [α-$^{32}$P]dTTP, 0.5 m*M* (about 10 cpm/pmole) | 1 | 5 |
| dATP, 1 m*M* | 1 | 5. |
| dCTP, 1 m*M* | 1 | 5 |
| dGTP, 1 m*M* | 1 | 5 |
| $MgCl_2$, 0.1 *M* | 1 | 5 |
| Dithiothreitol, 20 m*M* | 0.5 | 2.5 |
| KCl, 2 *M* | 0.5 | 2.5 |
| Enzyme extract | 1 | 5 |

Activated calf thymus DNA is made by hydrolyzing native DNA with pancreatic deoxyribonuclease. A convenient way to obtain maximal DNA template activity is to follow the course of hydrolysis of the DNA by assaying the DNA for its ability to serve as a template for a purified DNA polymerase (*43*).

Since several polymerases are inhibited significantly by as little as 2% ethanol, it is frequently necessary to remove ethanol from these labeled nucleotides, when high concentrations of radioactive deoxynucleotide triphosphates are used, by evaporation with $N_2$ or by lyophilization.

Assays are carried out in a total volume of 25–300 μl. By varying the concentration of the individual components, the enzyme extract may occupy up to 50% of the volume of each reaction mixture. Incubation should be started by adding the enzyme to a mixture of the other components of the reaction. For short periods of incubation, it is desirable to preincubate the reaction mixture without the enzyme for 2–3 minutes at the desired temperature. The total amount of protein in each assay must be less than 25 μg, otherwise the precipitate may clog the filter during the washing procedure.

The reaction is terminated by the addition of 2 ml of 1 *M* $HClO_4$ containing 20 m*M* sodium pyrophosphate and 0.1 ml of calf thymus DNA (1 mg/ml). The precipitate is collected onto glass-fiber filters obtained from Whatman or Schleicher and Schuell, washed with 10 ml of cold $H_2O$, followed by 3 ml of ethanol, dried, and counted by liquid scintillation spectroscopy. With small amounts of enzyme it is frequently necessary to use substrates of a higher specific activity than that routinely used and to wash the precipitate one or more times prior to collecting it on a filter. Washing is carried out by dissolving the precipitate in 0.5 ml of 0.2 *M* NaOH containing 0.05 *M* sodium pyrophosphate and reprecipitating with acid (*44*). When collecting the precipitate onto glass-fiber disks, it is important to prewet the disk to prevent the sequestering by capillary action of nonincorporated nucleotides onto peripheral areas of the filter disk.

## C. Assays for Specific DNA Polymerases

As the polymerases are further purified, more specific assays, based on the characteristics of each polymerase, may be employed. The reaction mixtures for assaying each enzyme are listed below:

| Components | Parts |
|---|---|
| Assay for DNA polymerase-$\alpha$ | |
| "Activated" calf thymus DNA, 1 mg/ml | 1 |
| Tris-HCl, pH 7.4, 1 *M* | 1 |
| [$^3$H]- or [$\alpha$-$^{32}$P]dTTP, 0.5 m*M* (about 20 cpm/pmole) | 1 |
| dATP, 1.0 m*M* | 0.5 |
| dCTP, 1.0 m*M* | 0.5 |
| dGTP, 1.0 m*M* | 0.5 |
| $MgCl_2$ , 0.1 *M* | 0.8 |
| $\beta$-Mercaptoethanol, 20 m*M* | 0.5 |
| KCl, 2 *M* | 0.5 |
| Enzyme fraction | 3.7 |
| Assay for DNA polymerase-$\beta$ | |
| "Activated" calf thymus DNA, 1 mg/ml | 1 |
| Tris-HCl, pH 8.4, 1.0 *M* | 1 |
| [$^3$H]- or [$\alpha$-$^{32}$P]dTTP, 0.5 m*M* (about 100 cpm/pmole) | 1 |
| dATP, 1.0 m*M* | 0.5 |
| dCTP, 1.0 m*M* | 0.5 |
| dGTP, 1.0 m*M* | 0.5 |
| $MgCl_2$, 0.1 *M* | 0.8 |
| Enzyme fraction | 3.7 |
| Assay for DNA polymerase-$\gamma$ | |
| Poly(rA)·$(dT)_{12-18}$, 0.25 mg/ml | 1 |
| Tris-HCl, pH 7.5, 1 *M* | 1 |
| [$^3$H]dTTP, 0.5 m*M* (about 100 cpm/pmole) | 1 |
| *n*-Dithiothreitol, 20 m*M* | 1 |
| $MnCl_2$, 10 m*M* | 1 |
| KCl, 1.0 *M* | 1 |
| Enzyme fraction | 4 |

It is important to note that $\beta$-polymerase differs from $\alpha$-polymerase with respect to its pH optimum (*45*). Also, it does not require a sulfhydryl reagent and KCl, whereas both $\alpha$- and $\gamma$-polymerases do (*45*). Poly(rA)·oligo(dT) in the presence of $Mn^{2+}$ is the most effective template for polymerase-$\gamma$.

## V. Purification of DNA Polymerases

Each of the eukaryotic DNA polymerases can be isolated by separate procedures. Bollum *et al.* (*45*) have described general methods for the purification of several nucleotidyl polymerizing enzymes from calf thymus. Chang (*46*) has described the purification of a homogeneous β-polymerase from calf thymus. Loeb (*47*) isolated a high-molecular-weight DNA polymerase from nuclei of sea urchin embryos. Sedwick *et al.* (*48*) have isolated three nonmitochondrial DNA polymerases from KB cells. Weissbach and his co-workers established a method for the isolation of DNA polymerase-α (*2,49*) and mtDNA polymerase (*30*) from HeLa cells. Lewis *et al.* (*50*) have developed a method that allows one to isolate and separate the three cellular DNA polymerases, viz. α, β, and γ, and also the viral "reverse transcriptase" from cultured cells by a single purification procedure. This method is outlined in Fig. 1. They used cultures of human lymphoblasts infected with a primate C-type RNA tumor virus as their source of material. In this procedure, the fraction containing mtDNA polymerase was separated from the initial cell homogenate by differential centrifugation. The DNA polymerase associated with Epstein–Barr virus, which is presumably present in lymphoblast cultures, was not found in their isolation. We have found this general procedure satisfactory for the purification of DNA polymerase-α and -β from cultures of phytohemagglutinin-stimulated normal human lymphocytes and from human placenta.

For the separation and isolation of different DNA polymerases from a given tissue, one can follow the methods given schematically in Fig. 1. However, when starting purification, one has to be careful about the choice of material, the method of homogenization, the type of subcellular fractionation, the stability of the DNA polymerase, and the removal of nucleic acids. These points are discussed below in detail.

### A. Choice of Tissue Material

Until recently, DNA polymerases have been purified extensively from a limited number of eukaryotic cells. This is due mainly to either the unavailability of many tissues or the limited amounts of total polymerase activity present in them. Rapidly proliferating cells are rich in DNA polymerase-α activity and, hence, are usually the material of choice for the purification of this polymerase. It is also desirable to use a tissue of uniform cell type since it has not been established whether DNA polymerases are cell specific.

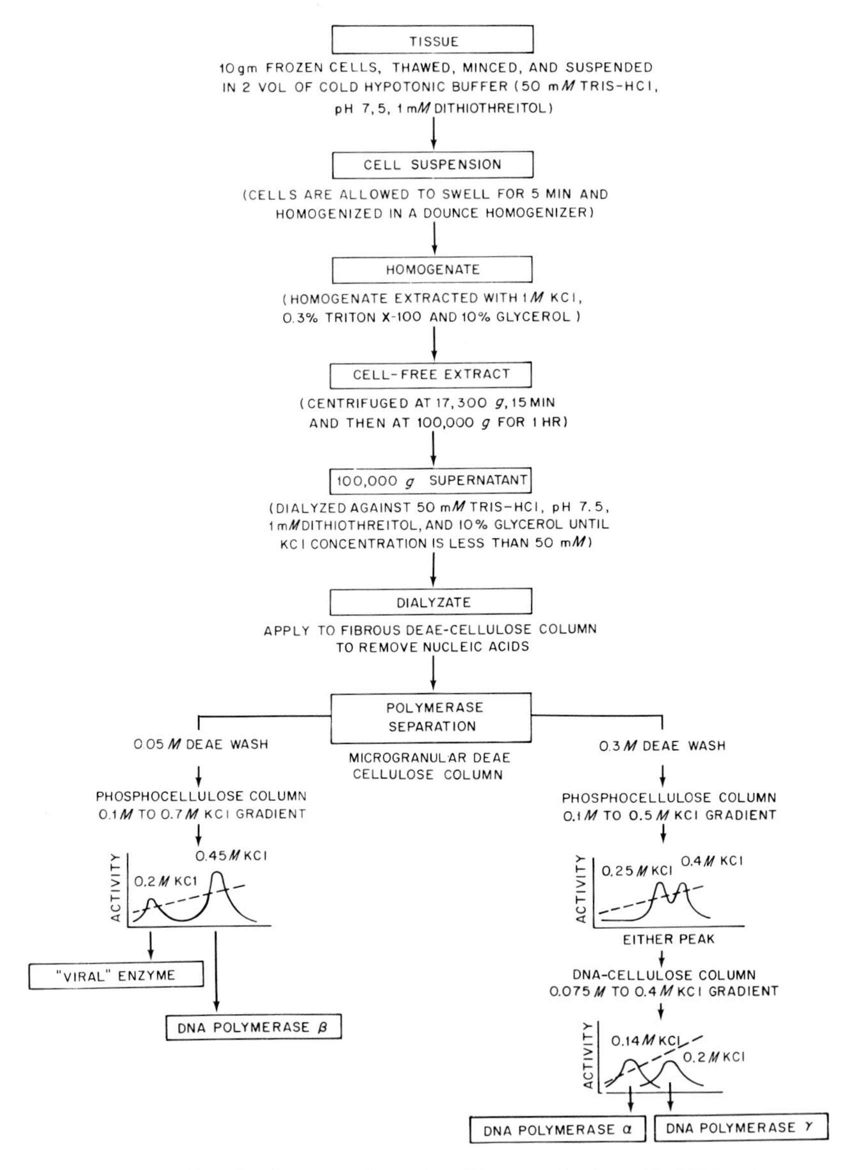


FIG. 1. A procedure for the separation and purification of eukaryotic DNA polymerases. This purification is adapted from that given by Lewis *et al.* (*50*) with modifications with permission of Elsevier Scientific Publishing Company, Amsterdam.

## B. Subcellular Fractionation

Since different DNA polymerases are contained in different cell organelles, the initial step in purification occasionally involves isolation of particular organelles under conditions in which the polymerase is retained in them. For example, a nuclear isolation procedure resulted in a 3-fold enrichment in DNA polymerase when nuclei from rapidly dividing sea urchin embryos were used as a source of DNA polymerase (Table III). Unfortunately, subcellular fractionation usually involves differential centrifugation, which is not always practical for large-scale purifications.

## C. Stability

Some of the DNA polymerases from eukaryotic cells are extremely unstable. The enzyme may lose activity very rapidly, especially in the latter stages of purification, when the protein concentration is extremely low. Addition of bovine serum albumin during the isolation procedure may ameliorate this problem. Usually DNA polymerases can be stabilized by the use of glycerol, sucrose, or ethylene glycol in all buffers throughout the entire purification procedure. In the case of DNA polymerases from sea urchin (*47*) and salmon testes (*51*), polyglycols have been found to be very efficient in preventing the thermal denaturation of the enzyme. The mechanism of such protection is yet to be clarified.

TABLE III

STEPS IN THE PURIFICATION OF DNA POLYMERASE FROM NUCLEI OF SEA URCHIN EMBRYOS

| | Fraction and step | Total activity[a] (units $\times 10^{-3}$) | Specific activity (units/mg protein) | Ratio of DNase to polymerase |
|---|---|---|---|---|
| I | Whole embryo extract[b] | 135 | 5.3 | 141 |
| II | Nuclei | 84.0 | 17.0 | 45 |
| III | Phase separation | 70.8 | 45.6 | |
| IV | Ammonium sulfate | 51 | 54.3 | 33 |
| V | Phosphocellulose | 22.8 | 296.4 | 21 |
| VI | DEAE-cellulose | 15 | 471.6 | 7 |
| VII | Hydroxyapatite | 9.6 | 936.0 | 5 |
| VIII | Gel filtration | 6.6 | 1620.0 | $<0.05$ |

[a] Polymerase activity was determined with "activated" calf thymus DNA as a template; the methods for purification and assay are given in Ref. *47*.

[b] Purification was carried out starting with 1200 ml of hatched sea urchin embryos.

## D. Removal of Nucleic Acids

Nucleic acids have a high binding affinity for certain polymerases and thus cause these polymerases to have spurious chromatographic properties; therefore, prior to extensive purification, it is usually desirable to separate these nucleic acids from the polymerases. As illustrated in Fig. 1, DNA can be removed by adsorption onto a DEAE-cellulose column. This step is frequently used as the first chromatographic step in the purification procedure. Alternative procedures are: (a) digestion of the cell-free extract with pancreatic deoxyribonuclease, (b) separation of nucleohistones from polymerase by selective precipitation, and (c) separation of DNA polymerase by phase extraction with aqueous solvents as described by Albertsson (*52*). Digestion with deoxyribonuclease introduces an additional protein contaminant. Nucleohistone precipitation may also remove some of the enzyme. In the phase procedure, cell-free extracts are vigorously mixed with an aqueous solution of polyethylene glycol (Carbowax 6000) and Dextran 500 in 4 *M* NaCl. The resultant aqueous phases are separated by centrifugation. For example, in the case of sea urchin embryos, more than 95% of the DNA may be found in the lower Dextran phase, while most of the DNA polymerase activity may be recovered in the upper polyethylene glycol phase (*47*). This results in a 3-fold increase in the specific activity of DNA polymerase (Table III). However, it may be that in this procedure polymerase-$\beta$ is sequestered into the Dextran phase, because it is not found during the subsequent purification steps. Polymerase-$\beta$ has been found in sea urchin embryos when other purification methods (*53*) are used; therefore phase separation may not be applicable for the isolation of DNA polymerase-$\beta$.

## E. Selective Protein Precipitation

The large losses in total polymerase activity incurred upon selective protein precipitation early in the purification mitigates against its utility with cells that are in short supply. Thus, it is seldom advantageous to use ammonium sulfate, manganese, ethanol, or acetone precipitation. It is best to start purification via column chromatographic procedures as early as possible.

## F. Chromatography

### 1. DEAE-Cellulose

As an initial chromatographic step, DEAE-cellulose offers some distinct advantages. In the procedure of Lewis *et al.* (*50*), a cell-free extract in a low-salt buffer is rapidly passed through a fibrous DEAE-cellulose column

(DE-23). All the DNA polymerases, free of most of the nucleic acids, can then be eluted from the column with 0.3 *M* KCl. The extract is desalted by dialysis, and the dialyzate is then adsorbed onto a DE-52 column. By batch elution with 0.05 *M* KCl, $\beta$-polymerase and the putative viral enzyme may be recovered; with 0.3 *M* KCl, the $\alpha$- and $\gamma$-polymerases can be subsequently eluted (Fig. 1).

### 2. PHOSPHOCELLULOSE

Chromatography on phosphocellulose is often the most selective step in the purification of DNA polymerases. At neutral pH, phosphocellulose bears a net negative charge and therefore might be imagined to resemble DNA. At a pH of about 6.5 to 7.0, most proteins are either not adsorbed onto phosphocellulose or are easily eluted at low ionic strength. Thus, in the purification of sea urchin polymerase, this step resulted in a 5-fold enrichment (Table III). In the purification of $\alpha$- and $\beta$-polymerase from calf thymus glands, it resulted in a 10–11-fold enrichment (*45*). Using a 0.1 to 0.7 *M* KCl linear gradient on phosphocellulose, Lewis *et al.* (*50*) were able to separate polymerase from the putative viral enzyme (Fig. 1). Also, DNA polymerase can be sequestered from large volumes of extracts when the ionic strength of the solution is low, and at pH 6.5 using phosphocellulose by a batch procedure. After mixing for 1 hour, the phosphocellulose is collected by low speed centrifugation and the polymerase is then eluted from the phosphocellulose at high ionic strength. At times, this is the preferred method for a rapid concentration of DNA polymerase from dilute solutions.

### 3. GEL FILTRATION (SEPHADEX G-100 OR G-200)

Gel filtration is of particular advantage in purifying DNA polymerase-$\alpha$, since the latter is larger than most other proteins. Unfortunately, this enzyme may be irreversibly adsorbed to Sephadex by nonspecific binding sites. One method to alleviate this problem is to prerun a solution of serum albumin through the column prior to purification of the polymerase (*47*). If gel filtration is used, the determination of the void volume of the column with Blue Dextran should be avoided, since insoluble particles of Blue Dextran may stick on top of the column, binding the polymerase very tightly, and thus selectively resulting in large losses of the enzyme (see below).

### 4. AFFINITY CHROMATOGRAPHY

DNA polymerases have been purified using affinity chromatography on columns of DNA-cellulose (*54*) or DNA-agarose (*55*). Lewis *et al.* (*50*) separated DNA polymerase-$\alpha$ and -$\gamma$ on DNA cellulose using a gradient of 0.075 to 0.4 *M* KCl. DNA polymerase-$\alpha$ binds to Blue Dextran (*56*), and tight interactions between sea urchin nuclear DNA polymerase and Blue Dextran have been reported. Recent evidence (*57*) suggests that *Escherichia coli* DNA polymerase I possesses in its secondary structure a dinucleotide fold that causes it to bind to Blue Dextran-Sepharose. The enzyme may be eluted from Blue Dextran-Sepharose columns with 1 m*M* dGTP. Therefore, Blue Dextran-Sepharose columns may be of future value in the purification and separation of eukaryotic DNA polymerases.

When DNA-cellulose is used to purify the polymerases, care should be taken that any DNA which may preferentially bind to the polymerase be removed prior to assaying the polymerase for specific template activity or using it for fidelity experiments.

## G. Homogeneity

Extensively purified eukaryotic DNA polymerases have been available in very limited amounts. So far, no DNA polymerase from either a prokaryotic or eukaryotic source has been isolated in crystalline form.

Although DNA polymerase-$\alpha$ is the predominant species in most eukaryotic cells, it has not yet been purified to homogeneity. DNA polymerase-$\beta$ has been purified to homogeneity by Chang (*58*) and mtDNA polymerase by Radsak *et al.* (*30*). DNA polymerase-$\gamma$ has been only partially purified. In fact, in order to purify an eukaryotic polymerase to homogeneity, it may be necessary to start with several kilograms of tissue; even then, one can anticipate a yield of only a few milligrams at most.

To check whether an enzyme preparation is free from any contaminating protein, the enzyme should be subjected to polyacrylamide gel electrophoresis under denaturing conditions. Gel electrophoresis should be carried out with different amounts of protein. Unless the enzyme is close to homogeneity, it is essential to demonstrate that the predominant protein in nondenaturing gels is indeed the active enzyme. For example, DNA polymerase-$\alpha$ from sea urchin embryos has been extensively purified and exhibits a single symmetrical peak upon chromatography on different adsorbents (*47*). Its activity per milligram of protein is constant upon rechromatography of the most purified fraction on DEAE, and only one migrating band is discernible by staining with amido black after electrophoresis in polyacrylamide gels. However, three protein bands are observed after electrophoresis in polyacrylamide gels under denaturing conditions. It is not clear whether these multiple bands result from storage or proteolytic attack, or represent actual contaminants. Homogeneous *E. coli* DNA polymerase I has been shown to exhibit multiple bands upon prolonged storage at $-70°C$.

With the extensive purification of eukaryotic DNA polymerases, it has been possible to determine which reactions are catalyzed by the polymerases themselves and to begin to define the catalytic sites on these enzymes.

## VI. Conclusion

It is not clear how many DNA polymerases are present in eukaryotic cells or whether different tissues have different polymerases. The role of each of these enzymes in either DNA replication or repair has not been established unequivocally. An understanding of the relationships between the different polymerases, their mechanism for catalysis, and their cellular function awaits future experimentation. Extensive purification will be required as a prerequisite for answering many of these questions.

ACKNOWLEDGMENTS

This study was supported by grants from the National Institutes of Health (CA-11524, Ca-12818) and the National Science Foundation (BMS74-06751), by grants to this Institute from the National Institutes of Health (CA-06927, RR-05539), and by an appropriation from the Commonwealth of Pennsylvania.

REFERENCES

1. Loeb, L. A., *in* "The Enzymes" (P. Boyer, ed.), 3rd ed., Vol. 10, p. 173. Academic Press, New York (1974).
2. Weissbach, A., *Cell* **5**, 101 (1975).
3. Bollum, F. J., *Prog. Nucleic Acid Res. Mol. Biol.* **15**, 109 (1975).
4. Wu, A. M., and Gallo, R. C., *Crit. Rev. Biochem.* **3**, 289 (1975).
5. Holmes, A. M., and Johnston, I. R., *FEBS Lett.* **60**, 233 (1975).
6. Fansler, B. S., *Int. Rev. Cytol., Suppl.* **4**, 363 (1974).
7. Kornberg, A., "DNA Synthesis." Freeman, San Francisco, California (1974).
8. Weissbach, A. D., Baltimore, F., Bollum, F. J., Gallo, R., and Korn, D., *Eur. J. Biochem.* **59**, 1 (1975).
9. Yoneda, M., and Bollum, F. J., *J. Biol. Chem.* **240**, 3385 (1965).
10. Temin, H. M., and Mizutani, S., *Nature (London)* **226**, 1211 (1970).
11. Baltimore, D., *Nature (London)* **226**, 1209 (1970).
12. Keir, H. M., Subak-Sharpe, H., Shedden, W. I. H., Watson, D. H., and Wildy, P., *Virology* **30**, 154 (1966).
13. Weissbach, A., Hong, S. C., Aucker, J., and Muller, R., *J. Biol. Chem.* **248**, 6270 (1973).
14. Loeb, L. A., Fansler, B., Williams, R., and Mazia, D., *Exp. Cell Res.* **57**, 298 (1969).
15. Foster, D. N., and Gurney, T., *J. Cell Biol.* **63**, 103 (1974).
16. Lynch, W. E., Surrey, S., and Lieberman, I., *J. Biol. Chem.* **250**, 8179 (1975).
17. Lynch, W. E., and Lieberman, I., *Biochem. Biophys. Res. Commun.* **52**, 843 (1973).
18. Lynch, W. E., Short, J., and Lieberman, I., *Cancer Res.* **36**, 901 (1976).
19. Loeb, L. A., Agarwal, S. S., and Woodside, A. M., *Proc. Natl. Acad. Sci. U.S.A.* **61**, 827 (1968).
20. Loeb, L. A., and Agarwal, S. S., *Exp. Cell Res.* **66**, 299 (1971).
21. Agarwal, S. S., and Loeb, L. A., *Cancer Res.* **32**, 107 (1972).
22. Mayer, R. J., Smith, R. G., and Gallo, R. C., *Blood* **46**, 509 (1975).
23. Chang, L. M. S., and Bollum, F. J., *Biochem. Biophys. Res. Commun.* **46**, 1354 (1972).
24. Holmes, A. M., Hesslewood, I. P., and Johnston, I. R., *Eur. J. Biochem.* **43**, 487 (1974).
25. Chang, L. M. S., *Science* **191**, 1183 (1976).

26. Bertazzoni, V., Stefanini, M., Pedralinor, G., Giulotto, E., Nuzzo, F., Falaschi, A., and Spadari, S., *Proc. Natl. Acad. Sci. U.S.A.* **73**, 785 (1976).
27. Fridender, F., Fry, M., Bolden, A., and Weissbach, A., *Proc. Natl. Acad. Sci. U.S.A.* **69**, 452 (1972).
28. Yoshida, S., Ando, T., and Kondo, T., *Biochem. Biophys. Res. Commun.* **60**, 1193 (1974).
29. Kalf, G. F., and Ch'ih, J. J., *J. Biol. Chem.* **243**, 4904 (1968).
30. Radsak, K., Knopf, K. W., and Weissbach, A., *Biochem. Biophys. Res. Commun.* **70**, 559 (1976).
31. Tibbetts, C. J. B., and Vinograd, J., *J. Biol. Chem.* **248**, 3367 (1973).
32. Westeraard, O., and Pearlman, R., *Exp. Cell Res.* **54**, 309 (1969).
33. Keller, S. J., Biedenbach, S. A., and Meyer, R. R., *Biochem. Biophys. Res. Commun.* **50**, 620 (1973).
34. Kanakar, J. D., Mangulies, L., and Changatt, E., *J. Biol. Chem.* **250**, 8657 (1975).
35. Wang, T. S. F., Fisher, P., Sedwick, W. D., and Korn, D., *J. Biol. Chem.* **250**, 5270 (1975).
36. Gardner, J. M., and Kado, C. I., *Biochemistry* **15**, 688 (1976).
37. Byrnes, J. J., Downey, K. M., Black, V. L., and So, A. G., *Biochemistry*, **15**, 2817 (1976).
38. McLennan, A. G., and Keir, H. M., *Biochem. J.* **151**, 227 (1975).
39. McLennan, A. G., and Keir, H. M., *Biochem. J.* **151**, 239 (1975).
40. Jeggo, P. A., Unrau, P., Banks, G. R., and Holliday, R., *Nature (London), New Biol.* **242**, 14 (1973).
41. Banks, G. R., and Yarranton, G., *Eur. J. Biochem.* **62**, 143 (1976).
42. Wintersberger, U., and Wintersberger, E., *Eur. J. Biochem.* **13**, 11 (1970).
43. Fansler, B., and Loeb, L. A. *in* "Methods in Enyzmology" (L. Grossman and K. Moldave, eds.), Vol. 29, p. 55. Academic Press, New York (1974).
44. Battula, N., Dube, D. K., and Loeb, L. A., *J. Biol. Chem.* **250**, 8404 (1975).
45. Bollum, F. J., Chang, L. M. S., Tsiapalis, C. M., and Dorson, J. W., *in* "Methods in Enzymology" (L. Grossman and K. Moldave, eds.), Vol. 29, p. 70. Academic Press, New York (1974).
46. Chang, L. M. S., *in* "Methods in Enzymology" (L. Grossman and K. Moldave, eds.), Vol. 29, p. 81. Academic Press, New York (1974).
47. Loeb, L. A., *J. Biol. Chem.* **244**, 1672 (1969).
48. Sedwick, E. D., Wang, T. S. F., and Korn, D., *in* "Methods in Enzymology" (L. Grossman and K. Moldave, eds.), Vol. 29, p. 89. Academic Press, New York (1974).
49. Spadari, S., and Weissbach, A., *J. Biol. Chem.* **249**, 5809 (1974).
50. Lewis, B. J., Abrell, J. W., Smith, R. G., and Gallo, R. C., *Biochim. Biophys Acta* **349**, 148 (1974).
51. Tarr, H. L. A., and Gardner, L., *Can. J. Biochem.* **47**, 19 (1971).
52. Albertsson, P. A., *Arch. Biochem. Biophys.* **98**, Suppl. 1, 264 (1962).
53. dePetrocellis, B. D., Parisi, E., Filosa, S., and Capasso, A., *Biochem. Biophys. Res. Commun.* **68**, 954 (1976).
54. Berger, H., Huang, R. C. C., and Irvin, J. L., *J. Biol. Chem.* **246**, 7275 (1971).
55. Poonian, M. S., Schlabach, A. J., and Weissbach, A., *Biochemistry* **10**, 424 (1971).
56. Brissac, C., Ruvheton, M., Brunel, C., and Jenateur, P. H., *FEBS Lett.* **61**, 38 (1976).
57. Thompson, S. T., Cass, K. H., and Stellwagen, E., *Proc. Natl. Acad. Sci. U.S.A.* **72**, 669 (1975).
58. Chang, L. M. S., *J. Biol. Chem.* **248**, 3789 (1973).

# Chapter 3

## *Methods for Assessment of DNase Activity*

MURIEL W. LAMBERT AND GEORGE P. STUDZINSKI

*Department of Pathology,*
*College of Medicine and Dentistry of New Jersey,*
*New Jersey Medical School,*
*Newark, New Jersey*

## I. Introduction

Deoxyribonucleases (DNases) are of widespread occurrence both in prokaryotes and in the nucleus and cytoplasm of eukaryotes. Eukaryotic nuclear DNases have been found in the nucleoplasm (*1–3*), in the nucleolus (*3*), and associated with chromatin (*4–7*), in particular with the nonhistone chromatin proteins (*5*).

A number of different methods for assaying DNase activity have been described. Those prior to 1962 have been reviewed by Kurnick (*8*). Some of these assays are based on such changes in characteristics of the substrate as viscosity (*9,10*), UV absorption (*11–13*), sedimentation velocity (*14–21*), affinity for methyl green (*22,23*), biological activity (*24–26*), or the ability to be retained on filters (*17,27,28*). Other assays measure liberation of acid-soluble material either from radioactively labeled DNA or poly (dA·[$^3$H]dT) (*29–39*), or spectrophotometrically from unlabeled substrate (*3,40–45*). A fluorometric (*46*) and several polyacrylamide gel assays have also been described (*47–51*). Another method measures, with a pH-stat, the liberation of hydrogen ions as a result of DNase action on DNA (*9, 52–53*).

Many of these assays detect specific types of DNases whereas others have more general applications. Both endonuclease and exonuclease activity can be determined using methods that measure liberation of acid-soluble material from the substrate. Endonuclease activity alone can be assayed using any of the other procedures listed above. A few of the more recently described methods will be discussed briefly.

A sensitive DNase assay uses alkaline sucrose gradient centrifugation of

DNA preincubated with the enzyme (*14–21*). Enzymic activity is determined by a change in the relative sedimentation velocity due to single-strand breaks in the DNA. This assay has the limitation that only a small number of samples can be centrifuged and analyzed at a time. Another sensitive assay takes advantage of the fact that circular DNA, after being nicked by the enzyme, has an increased affinity for ethidium bromide (*46*). Binding of the ethidium bromide to DNA is measured fluorometrically. An important disadvantage of this assay is its lack of specificity. As soon as the closed DNA circle is opened by endonucleolytic hydrolysis, it becomes sensitive to exonucleases that may be present in the preparation and can cause a change in fluorescence. Substances used in enzyme studies as well as ionic changes can also modify the fluorescence.

Several assays have been developed using polyacrylamide gels. In one of these (*47*), the enzyme is separated on a DNA–polyacrylamide gel. The gel is then incubated under conditions appropriate for DNase activity, after which proteins and DNA fragments are removed electrophoreticaly, and the gel is stained for residual DNA. The resulting nuclease pattern is recorded densitometricaly. This assay is useful for a rapid preliminary identification of DNases. Another assay (*48–51*) assesses DNase activity by measuring the release of fragments of isotopically labeled DNA from a DNA–polyacrylamide gel suspension incubated with the enzyme. This method has been used to distinguish between endo- and exonuclease activity. Endonucleases release larger acid-insoluble DNA fragments ($4 \times 10^5$ daltons) whereas exonucleases liberate mainly acid-soluble material.

Most of the studies cited above have assayed DNases isolated from animal tissues or from rapidly growing microbial cultures. An increasing amount of work, however, is being carried out on DNases isolated from mammalian cell cultures. In order to assay these enzymes, without the expense of large-scale cell culture, a more sensitive method is needed. An assay has been developed by Studzinski and Fischman (*54*) which permits detection of DNase activity at levels lower than any previously reported. This method, which is described below, is based on the observation of Sarkar (*55*) and of Aposhian and Kornberg (*56*) that the priming activity of DNA for DNA polymerase is increased by mild treatment of DNA with pancreatic DNase.

## II. Assay Method

The assay is divided into two procedures: DNA primer activation and DNA polymerase reaction.

## A. DNA Primer Activation

### 1. Reagents

Reaction mixture (total volume 65 μl): 10 μg of calf thymus DNA dissolved in 20 m*M* KCl; 5 m*M* $MgCl_2$; 10 m*M* Tris-maleate (pH 7.5); DNase in 50 μl of 1 *M* Tris-HCl (pH 7.5).

### 2. Procedure

A dilution of DNase is added to start the reaction. Kanamycin sulfate (150 μg/ml) may be added to inhibit bacterial growth during incubation. Reaction tubes and blanks (containing all components except DNase) are incubated at 37°C with gentle agitation for 3 hours. The tubes are then heated to 60°C for 10 minutes to inactivate the DNase. Each tube is placed on ice and used directly for the DNA polymerase reaction described below.

## B. DNA Polymerase Reaction

This procedure was adapted from a method of Fansler and Loeb (*57*).

### 1. Reagents

Reaction mixture (total volume 240 μl): 10 nmoles each of dATP, dGTP and dCTP; 0.5 nmole of dTTP; 0.5 μCI of $[^3H]$ dTTP (sp. act. 46 Ci/mmole); 5 μmoles of $MgCl_2$; 0.35 μmoles of β-Mercaptoethanol; 25 μmoles of Trismaleate buffer (pH 8.0).

DNA polymerase, 0.12 U

Denatured calf thymus DNA, 0.1 ml (1 mg/ml in 0.02 *M* KCl)

Perchloric acid, 0.5 *M*, containing 0.01 *M* sodium pyrophosphate (PCAP)

Sodium hydroxide, 0.2 *M*

### 2. Procedure

The reaction mixture (240 μl) is added to the primer activation tubes. The reaction is started by adding 10 μl of a solution of DNA polymerase, usually 0.12 U of the *Micrococcus luteus* preparation. The tubes are placed in a water bath at 37°C for 10 minutes with gentle agitation. The reaction is terminated by the addition of 0.1 ml of 1 mg of denatured calf thymus DNA per milliliter in 0.02 *M* KCl (denatured by placing in boiling water for 20 minutes) followed by 4 ml of cold 0.5 *M* perchloric acid containing 10 m*M* PCAP. After 10 minutes at 0°C the precipitate is pelleted by centrifugation at 5000 *g* for 20 minutes, and the supernatant fraction is drawn off. The pellet is dissolved in 0.5 ml of 0.2 *M* sodium hydroxide and reprecipitated with 2 ml of cold PCAP. After 10 minutes at 0°C, the precipitate is collected on Whatman

GF/C glass-fiber filters saturated with cold PCAP. The filter is washed twice with 5 ml of cold PCAP and then transfered to a scintillation vial and dried; 10 ml of toluene scintillation fluid (Liquifluor-New England Nuclear) are added. Radioactivity is determined by a liquid scintillation counter. One unit of DNase activity is defined as the amount of enzyme which increases priming activity of DNA by 100% in 3 hours. By expressing activity as a percentage increase over that obtained using untreated DNA, the results become independent of such variables as the nature of the DNA polymerase, the amount of DNA used as a primer, or the amount of [$^3$H]dTTP added. Specific activity is expressed as units of DNase activity per milligram of protein.

## C. Characteristics of the Assay

An example of the results obtained using the assay is shown in Fig. 1. This assay is linear between 1 and 10 pg of pancreatic DNase I, with either native or denatured DNA as primer. Denatured DNA is a better primer before nuclease treatment, but native DNA is a better primer after DNase I treatment. Thus the known preference of DNase I for native over denatured DNA is retained under the conditions of primer activation. Less than 1 pg of DNase leads to activation, but it is not proportional to the amount of

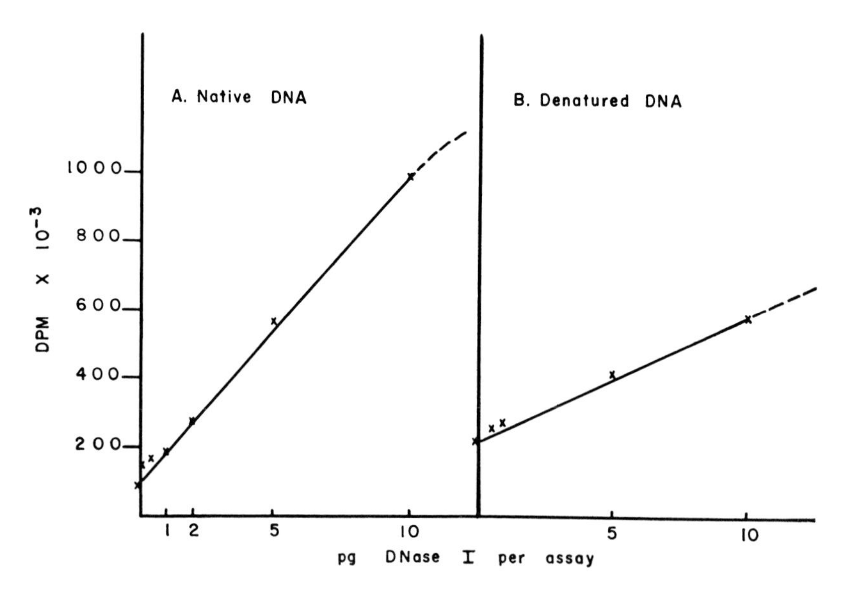


FIG. 1. Activation of calf thymus DNA primer by pancreatic DNase I. DNA polymerase reaction was performed with *Micrococcus luteus* enzyme as described in the text. Reprinted from Studzinski and Fischman (*54*) by permission.

enzyme used. When more than 25 pg of DNase are used in the assay, there is a decrease in the priming activity of the DNA. This is probably due to excessive breakdown of DNA.

## D. Sensitivity

With the DNA primer activation assay, 0.1–20 pg of DNase I can be detected. This is one order of magnitude more sensitive than the assay which measures hydrolysis of tritiated poly(dA·dT) (*32*) or that which measures binding of ethidium bromide to circular DNA (*46*). It is four orders of magnitude more sensitive than the filter retention assay of Geiduschek and Daniels (*27*) and also more sensitive than the modification of this filter method by Zimmerman (*28*).

## E. Specificity

Only those enzymes that are known to break down phosphodiester bonds with production of 3′-hydroxyl groups show activity with this assay (*54*). This method primarily detects endonucleases but also detects exonuclease activity at high enzyme concentrations. It is not clear whether this is due to endonuclease activity in the exonuclease preparation or whether exonucleolytic attack makes 3′-hydroxyl groups more readily available for polymerase action.

## F. Applications

This assay has been used to examine DNase activity associated with HeLa cell nuclei (*2,6*). Two DNA endonucleases from HeLa cell nucleoplasm (*2*) and four associated with HeLa cell chromatin (*2,6*) have been isolated which increase the priming activity of DNA for DNA polymerase. This method has also been used to assay an endonuclease which copurifies with calf thymus DNA polymerase and introduces only a limited number of single-stranded nicks into calf thymus DNA (*58*). An assay using this principle has recently been employed to measure the activity of an enzyme, from *Bacillus subtilis*, which enhances the priming of $\gamma$-irradiated T7 DNA for *Micrococcus* DNA polymerase (*59,60*).

## G. Advantages and Limitations

The sensitivity of this assay has already been discussed. It is a simple method, a large number of samples can be assayed simultaneously, the reagents used are available commercially, and no special equipment is required. This assay is not suitable for detection of DNases in general, since it cannot detect those which hydrolyze phosphodiester bonds with liberation of 3′-phosphoryl groups. The DNases that can increase the priming activity of DNA do this in only a limited concentration range. Therefore, it is necessary to perform the assay at several concentrations until the linear range is determined.

Enzymes other than DNases, such as alkaline phosphatase, do not change the priming activity of DNA, but their presence can interfere with the polymerase reaction. Similarly, high electrolyte concentrations interfere with the polymerase reaction. Assay of DNases in crude enzyme preparations, therefore, require careful controls to exclude such effects. The presence of exonuclease III and DNA phosphatase (*29*) could also lead to liberation of 3′-hydroxyl groups, and this possibility should be excluded.

## Acknowledgments

This work was aided by PHS Grant CA20043-02 from the National Cancer Institute and by Grant IN-92 from the American Cancer Society.

## References


*1.* Lindahl, T., Gally, J. A., and Edelman, G. M., *J. Biol. Chem.* **244**, 5014 (1969).
*2.* Churchill, J. R., Urbanczyk, J., and Studzinski, G. P., *Biochem. Biophys. Res. Commun.* **53**, 1009 (1973).
*3.* Cordis, G. A., Goldblatt, P. J., and Deutscher, M. P., *Biochemistry* **14**, 2596 (1975).
*4.* Swingle, K. F., Cole, L. J., and Bailey, J. S., *Biochim. Biophys. Acta* **149**, 467 (1967).
*5.* O'Connor, P. J., *Biochem. Biophys. Res. Commun.* **35**, 805 (1969).
*6.* Urbanczyk, J., and Studzinski, G. P., *Biochem. Biophys. Res. Commun.* **59**, 616 (1974).
*7.* Ishida, R., Akiyoshi, H., and Takahashi, T., *Biochem. Biophys. Res. Commun.* **56**, 703 (1974).
*8.* Kurnick, N. B., *Methods Biochem. Anal.* **9**, (1962).
*9.* Thomas, C. A., Jr., *J. Am. Chem. Soc.* **78**, 1861 (1956).
*10.* Le Pacq, J. B., and Bourgoin, D., *Biochim. Biophys. Acta* **80**, 173 (1964).
*11.* Kunitz, M., *J. Gen. Physiol.* **33**, 349 (1950).
*12.* Price, P. A., Lui, T.-Y., Stein, W. H., and Moore, S., *J. Biol. Chem.* **244**, 917 (1969).
*13.* Zimmerman, S. B., and Coleman, N. F., *J. Biol. Chem.* **246**, 309 (1971).
*14.* Baril, E., Brown, O., and Laszlo, J., *Biochem. Biophys. Res. Commun.* **43**, 754 (1971).
*15.* Kirtikar, D. M., Slaughter, J., and Goldthwait, D. A., *Biochemistry* **14**, 1235 (1975).
*16.* Kirtikar, D. M., Dipple, A., and Goldthwait, D. A., *Biochemistry* **14**, 5549 (1975).
*17.* Mechali, M., and De Recondo, A.-M., *Eur. J. Biochem.* **58**, 461 (1975).
*18.* Hayase, E., Shibata, T., and Ando, T., *Biochem. Biophys. Res. Commun.* **62**, 849 (1975).
*19.* Bacchetti, S., and Benne, R., *Biochim. Biophys. Acta* **390**, 285 (1975).
*20.* Greenfield, L., Simpson, L., and Kaplan, D., *Biochim. Biophys. Acta* **407**, 365 (1975).
*21.* Kroeker, W. D., and Fairley, J. L., *J. Biol. Chem.* **240**, 3773 (1975).
*22.* Kurnick, N. B., *Arch. Biochem.* **29**, 41 (1950).
*23.* Kurnick, N. B., and Sandeen, G., *Biochim. Biophys. Acta* **39**, 226 (1960).
*24.* Young, E. T., II, and Sinsheimer, R. L., *J. Biol. Chem.* **240**, 1274 (1965).

25. Lacks, S., and Greenberg, B., *J. Biol. Chem.* **250**, 4060 (1975).
26. Singh, S., and Ray, D. S., *Biochem. Biophys. Res. Commun.* **67**, 1429 (1975).
27. Geiduschek, E. P., and Daniels, A., *Anal. Biochem.* **11**, 133 (1965).
28. Zimmerman, B. K., *J. Biol. Chem.* **241**, 2035 (1966).
29. Richardson, C. C., Lehman, I. R., and Kornberg, A., *J. Biol. Chem.* **239**, 251 (1964).
30. Shortman, K., and Lehman, I. R., *J. Biol. Chem.* **239**, 2964 (1964).
31. Naber, J. E., Schepman, A. M. J., and Rörsch, A., *Biochim. Biophys. Acta* **99**, 307 (1965).
32. Lindahl, T., Gally, J. A., and Edelman, G. M., *Proc. Natl. Acad. Sci. U.S.A.* **62**, 597 (1969).
33. Klenow, H., and Henningsen, I., *Proc. Natl. Acad. Sci. U.S.A.* **65**, 168 (1970).
34. Koh, J. K., Waddell, A., and Aposhian, H. V., *J. Biol. Chem.* **245**, 4698 (1970).
35. Vovis, G. F., and Buttin, G., *Biochim. Biophys. Acta* **224**, 29 (1970).
36. Lindahl, T., *in* "Methods in Enzymology" (K. Moldave and L. Grossman, eds.), Vol. 21, Part D, p. 148. Academic Press, New York (1971).
37. Ball, W. D., and Rutter, W. J., *J. Exp. Zool.* **176**, 1 (1971).
38. Yamada, M., Nagao, M., Miwa, M., and Sugimura, T., *Biochem. Biophys. Res. Commun.* **56**, 1093 (1974).
39. Greth, M.-L., and Chevallier, M.-R., *Biochim. Biophys Acta* **390**, 168 (1975).
40. McDonald, M. R., *in* "Methods in Enzymology" (S. P. Colowick and N. O. Kaplan, eds.), Vol. 2, p. 437. Academic Press, New York (1955).
41. Eaves, G. N., and Jeffries, C. D., *J. Bacteriol.* **85**, 273 (1963).
42. Lazarus, H. M., and Sporn, M. B., *Biochemistry* **6**, 1386 (1967).
43. Slor, H., and Hodes, M. E., *Arch. Biochem. Biophys.* **139**, 172 (1970).
44. Dulaney, J. T., and Touster, O., *J. Biol. Chem.* **247**, 1424 (1972).
45. Bartholeyns, J., Peeters-Joris, C., Reychler, H., and Baudhuin, P., *Eur. J. Biochem.* **57**, 205 (1975).
46. Paoletti, C., and Le Pecq, J. B. *in* "Methods in Enzymology" (K. Moldave and L. Grossman, eds.), Vol. 21, Part D, p. 255, Academic Press, New York (1971).
47. Boyd, J. B., and Mitchell, H. K., *Anal. Biochem.* **13**, 28 (1965).
48. Melgar, E., and Goldthwait, D. A., *J. Biol. Chem.* **243**, 4401 (1968).
49. Friedberg, E. C., and Goldthwait, D. A., *Proc. Natl. Acad. Sci. U.S.A.* **62**, 934 (1969).
50. Hewish, D. R., and Burgoyne, L. A., *Biochem. Biophys. Res. Commun.* **52**, 475 (1973).
51. Lavin, M. F., Kikuchi, T., Counsilman, C., Jenkins, A., Winzor, D. J., and Kidson, C., *Biochemistry* **15**, 2410 (1976).
52. Douvas, A., and Price, P. A., *Biochim. Biophys. Acta* **395**, 201 (1975).
53. Price, P. A., *J. Biol. Chem.* **250**, 1981 (1975).
54. Studzinski, G. P., and Fischman, G. J. *Anal. Biochem.* **58**, 449 (1974).
55. Sarkar, N. K., *Arch. Biochem. Biophys.* **93**, 328 (1961).
56. Aposhian, H. V., and Kornberg, A., *J. Biol. Chem.* **237**, 519 (1962).
57. Fansler, B. S., and Loeb, L. A., *in* "Methods in Enzymology" (L. Grossman and K. Moldave, eds.), Vol. 29, Part E, p. 53. Academic Press, New York (1974).
58. Wang, E.-C., Henner, D., and Furth, J. J., *Biochem. Biophys. Res. Commun.* **65**, 1177 (1975).
59. Noguti, T., and Kada, T., *Biochim. Biophys. Acta* **395**, 284 (1975).
60. Noguti, T., and Kada, T., *Biochim. Biophys. Acta* **395**, 294 (1975).

# Part B. Enzymic Components of Nuclear Proteins—Protein Substrates

# Chapter 4

## *Assessment of Acetylation in Nonhistone Chromosomal Proteins*

C. C. LIEW

*Department of Clinical Biochemistry,*
*Banting Institute, University of Toronto,*
*Toronto, Ontario, Canada*

## I. Introduction

Postsynthetic modification of proteins such as acetylation, hydroxylation, methylation, and phosphorylation is believed to be involved in the organization of chromatin and to be associated with changes in template activity (*1–24*). Acetylation of proteins was found at times of gene activation and protein synthesis (*1,6,25–27*). There is increasing evidence that α-amino terminal (i.e., external) acetylation of proteins occurs at the early phase of protein synthesis (*26–28*) whereas ε-amino-group acetylation of protein (i.e., internal) occurs both in the cytoplasm and the nucleus (*28*). An enzyme which catalyzes histone acetylation has been partially purified and found to be located in the nucleus (*29–33*).

Allfrey and his co-workers (*2,19,20*) provided the first evidence that there is a temporal correlation between an increase in RNA synthesis and histone acetylation. Other investigators have reported that the acetylation of histone is closely associated with chromatin assembly (*7,8*), the cell cycle (*22,34*), and hormonal as well as drug action (*6,11,15,35*). Recently, acetylation of nonhistone chromosomal proteins (NHCP) was reported (*15,36*). It was found that an increase of acetylation in NHCP occurred in liver regeneration

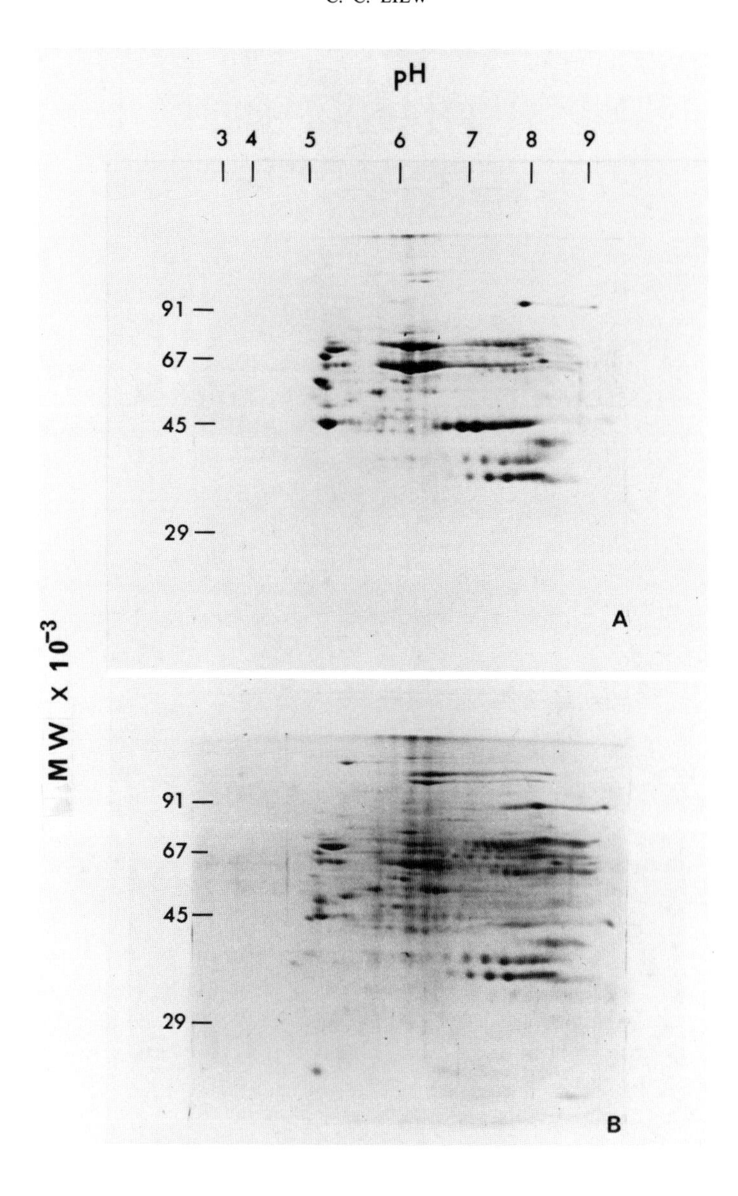


FIG. 1. Two grams of heart tissue (panel A) were homogenized in a polytron homogenizer (Brinkman, 10 seconds at setting 4) in 10 volumes of medium A[0.25 *M* sucrose, 10 m*M* Tris-HCl pH 8.0–3 m*M* $MgCl_2$–0.1 m*M* phenylm ethylsulfonyl fluoride (PMSF)] and centrifuged at 1000 *g* for 10 minutes. The crude pellet was suspended in medium B (0.1% Triton X-100 in medium A), and this suspension was filtered through one layer of nylon and two sieves with meshes 120 and 300, respectively. After centrifugation, the nuclear pellet was suspended once with medium B, centrifuged, and finally resuspended in Medium C (2.2 *M* sucrose in medium

and hormonal stimulation (*15,35,36*). However, acetate within cells is actively metabolized into amino acid residues, e.g., glycine. An assessment of acetyl groups which are covalently linked with proteins should be established as follows: (a) Acetyl groups in the proteins can be removed by acid hydrolysis. (b) Acetylation of protein is independent of protein synthesis. (c) Acetylation of protein should occur at physiological pH (e.g., 7.4) and be enzymic, whereas nonspecific acetylation (e.g., nonenzymic) should be eliminated. (d) Amino acid residues which are covalently linked with acetyl groups (e.g., acetyllysine) should be delineated.

---

A) and underlayered with 5 ml of medium C. Nuclei were isolated by centrifugation at 113,000 *g* for 1 hour on a Beckman SW-27 rotor.

Similarly, 5 gm of liver tissue (panel B) were homogenized in medium A by) a Teflon pestle homogenizer. After centrifugation the crude pellet was suspended in medium A and filtered through four layers of nylon. After centrifugation the nuclear pellet was suspended with medium B, centrifuged again, and finally suspended in medium C.

Chromatin was prepared from the isolated nuclei by lysing in 25 ml of 12.5% glycerol–1 m*M* Tris-HCl (pH 8.0)–0.1 m*M* EDTA–0.5 m*M* DTT–0.1 m*M* PMSF. The swollen chromatin was pelleted by centrifugation at 12,000 *g* for 10 minutes and this was repeated once more.

Nonhistone chromosomal proteins (NHCP) were then isolated by the following steps: (a) extraction of histones with 0.25 *N* HCl, (b) exposure to chloroform-methanol, and (c) suspension in 100 m*M* Tris-HCl (pH 8.4)–10 m*M* EDTA–0.14 *M* MSH (TEM) and an equal volume of TEM-saturated phenol for extraction of NHCP. NHCP were then concentrated by dialysis and finally dissolved in 8 *M* urea–0.02 *M* Tris-HCl (pH 8.4)–0.02 *M* glycine–3 m*M* MSH as described previously (36).

Fractionation of NHCP was carried out by two-dimensional polyacrylamide-gel electrophoresis as follows: The protein sample was mixed with acrylamide solution (38.4% acrylamide–1.6% *N,N'*-methylene bisacrylamide, w/v), ampholine [40%, w/v, (pH 3.5–10)], and ammonium persulfate (0.125% in 10 *M* urea, freshly prepared) in the proportions 1.25:0.5:0.25:3.0 (by volume) and the mixture was immediately pipetted into an acid-washed glass tube (2.5 × 90 mm) to a height of 68 mm. After overlayering with water, the gel was allowed to polymerize for 30 minutes. The gels were subsequently electrofocused under similar conditions as described previously (*24*).

The SDS-polyacrylamide slab gel electrophoresis was modified from the method of Laemmli (*40*). The following stock solutions were used to prepare the second-dimensional slab gel:

(a) 38.94% (w/v) acrylamide–1.06% (w/v) *N,N'*-methylene bisacrylamide; (b) 1.5 *M* Tris-HCl (pH 8.8); (c) 0.5 *M* Tris-HCl (pH 6.8); (d) 1.0% (w/v) SDS; (e) 0.08% (w/v) ammonium persulfate freshly prepared; and (f) 2.0% (w/v) TEMED.

The lower, small-pore gel was prepared according to the following proportions by volume: 1(a):1(b):0.4(d):1.5(e):0.1(f); and the large-pore gel (i.e., cap gel) needed 0.45(a):1.0 (c):0.4(d):1.5(e):0.2(f) and 0.45 part of distilled water as described previously (*39*).

After electrophoresis the slab gels were fixed with 10% trichloroacetic acid and stained with 0.1% Coomassie Blue–50% ethanol–10% acetic acid for 4 hours. Destaining of the gels was carried out in 25% ethanol–10% acetic acid. The gel was finally stored in 7% acetic acid.

## II. Definitions of Nonhistone Chromosomal Proteins

Chromatin is a DNA–protein complex with trace amounts of RNA. A precise definition of chromatin is obscure. Chromatin in our laboratory was prepared essentially according to the method of Reeder (*37*). In general, physical features, such as UV spectrum and composition of chromatin, should be defined. However, the criteria for a "native" chromatin could be defined as follows: (a) these should be a dispersed state of the interphase nuclei; (b) the RNA populations that are transcribed from chromatin should be identical to the RNA in isolated nuclei or present under *in vivo* conditions; and (c) the activity of endogenous DNA-dependent RNA polymerases should be essentially the same as in intact nuclei. De Pomerai *et al.* (*38*), examined six methods that were widely used and concluded that chromatin prepared by the method of Reeder (*37*) provided the most optimal condition for RNA transcription *in vitro.* Chromatin was prepared from heart and liver nuclei, and their protein components were analyzed by two-dimensional gel electrophoresis (*24,36,39*). The chromatin proteins were first fractionated according to their isoelectric points and subsequently separated on the basis of their molecular weights by sodium dodecyl sulfate (SDS) gel electrophoresis. This system provides an additional criterion for examining the heterogeneous components of chromosomal proteins. As shown in Fig. 1 (A and B) (*24,36,39*) more than 100 NHCP could be identified.

## III. Preferred Methods for the Isolation of Nonhistone Chromosomal Proteins

Several methods have been developed for the isolation of NHCP. The methods of urea-guanidine extraction (*41*), or phenol (*42*), or SDS solubilization (*43*) of NHCP are the most widely used. The method chosen depends on the individual investigator. We have studied these methods extensively for yield and reproducibility (*24*). The phenol method is the best choice for reproducibility from experiment to experiment. However, recovery of NHCP is relatively low as compared to the other methods. We chose the phenol solubilization procedure for an assessment of acetylated NHCP. The rationale is set forth as follows: (a) Histone can be rapidly and completely removed by acid extraction. It is well known that histone is highly acetylated *in vivo* and *in vitro.* Removal of histone from NHCP is essential. (b) Lipid and phospholipid contamination are eliminated by methanol–chloroform extraction. It has been reported that acetate is rapidly metabolized and incorporated into lipids, phospholipids, and glycoprotein. Exposure of NHCP to these organic solvents is necessary. (c) Nucleic acids are immiscible in the phenol phase. (d) NHCP solubilized in the phenol phase are gradually completely dissolved in either SDS or urea buffer without any precipitation. In most cases the urea–guanidine extraction method gave rise to precipitation of NHCP during dialysis or concentration.

## IV. Identification of Acetylated Nonhistone Chromosomal Proteins

In most reports, acetylation of chromosomal protein was carried out in the histone fraction. The acetylated NHCP has recently been identified (*35,36*) with radioactive acetate as the commonly used isotope for the study.

In general, 10–20 minutes of pulse-labeling of [$^3$H] or [$^{14}$C]acetate in an animal is sufficient. Prolonged labeling with isotope will provide a greater turnover of acetate into amino acids, and identification of acetylated NHCP becomes more complicated. Puromycin may be used prior to radioactive acetate pulse-labeling of animals. The procedure for identification of acetylated NHCP can be carried out in the following way:

### A. Acid Hydrolysis

Acetyl groups of a protein could either be determined as *N*-2, 4-dinitrophenyl acethydrazide by a method involving hydrazinolysis of the protein and subsequent coupling with fluorodinitrobenzene at pH 3, as described by Phillips (*44*), or detected by the chromatographic method of Brown and Hall (*45*) for measuring aliphatic acid anions. [$^3$H]- and [$^{14}$C]acetate covalently linked to protein can be rapidly released by acid hydrolysis. Proteins (100–500 $\mu$g) were dissolved in 2 *M* $H_3PO_4$ and followed by steam distillation. Isotopically labeled acetate was trapped by 0.3 *M* NaOH as described by Gershey *et al.* (*46*). Alternatively, the proteins were dialyzed extensively against water to remove salt and were subsequently hydrolyzed in 6 *N* HCl for 18 hours in a sealed evacuated tube. The hydrolyzed sample was transferred quantitatively to counting vials. Flash evaporation was carried out at 75°C. The drying of the precipitate was repeated twice by dissolving the precipitate in 1 ml of water to remove traces of HCl.

## B. Protease Digestion

Proteins (1–2 mg) were dissolved in 1.0 ml of 0.1 *M* ammonium bicarbonate (pH 8.4) and treated with trypsin in a proportion of 1:1 at 37°C for 12 hours. The suspension was then heated to 60°C for 2 minutes to inactivate trypsin. Pronase was then added, and the mixture was incubated at 37°C for 18 hours. The digested peptides were acidified by 0.1 *N* HCl. Partial separation of acetylated amino acid residues was achieved by chromatography of the suspension on a Bio-Gel P-2 column, using 0.01 *N* HCl as an eluting solvent (*26,47,48*). The fractions which coincided with standards (e.g., *N*-acetyllysine and acetylated serine) were pooled and lyophilized.

## C. Identification of Acetylated Amino Acid Residues by the Amino Acid Analyzer

The acetylated form of amino acid residues could be identified by the amino acid analyzer. The following two methods could be used:

### 1. Direct Identification

The major acetylated amino acid residues in polypeptides which were extensively studied were α-*N*-acetylserine and ε-*N*-acetyllysine (*49*). However, α-amino-acetylated methionine, alanine, threonine, glycine, and aspartic acid were found in proteins that were sequenced (*14*). An identified sample of acetylated amino acids could be compared with the authentic standard, which was eluted at certain times. For example, ε-*N*-acetyllysine was eluted in a Beckman amino acid analyzer (Model 120C) at a time (e.g., 83.7 minutes) when no other amino acid residues were known to elute out.

### 2. Indirect Identification

The acetylated amino acid residues that were eluted from the amino acid analyzer without prior reaction with ninhydrin reagent could be collected and lyophilized. After acid hydrolysis in 2 *M* $H_3PO_4$ and steam distillation, the hydrolysate was analyzed by means of the amino acid analyzer (*48*).

Alternatively, the acetylated amino acid residues obtained from the enzyme digest were subjected to high-voltage electrophoresis as described by Narita (*50*). Authentic standards were co-chromatographed with the acetylated amino acids. The identified acetylated amino acid residues could be eluted out and subjected to acid hydrolysis. The amino acid residues which were covalently modified could be further confirmed by means of the amino acid analyzer.

## V. Remarks

Acetate is rapidly converted into acetyl-CoA as a precursor for carbohydrate, lipid, and protein synthesis. Its radioactive incorporation into protein does not constitute a true acetylation. An assessment of acetylation of proteins should be rigorously proved by the criteria discussed in this article. Currently, it has been established that α-*N*-acetyl-serine occurs at early times of protein synthesis. The α-amino-terminal acetylation of amino acid residues occurs at the ribosomal level and it remains stable while ε-*N*-acetyllysine occurs after the completion of peptide synthesis and occurs in the cytoplasm and the nucleus. Also, it has a rapid turnover and is believed to be associated with the structure and function of the chromatin complex.

## References


*1.* Allfrey, V. G., *in* "Histone and Nucleohistones" (D. M. P. Phillips, ed.), p. 241, Plenum Press, New York (1971).
*2.* Allfrey, V. G., Faulkner, R. M., and Mirsky, A. E., *Proc. Natl. Acad. Sci. U.S.A.* **51**, 786 (1964).
*3.* Allfrey, V. G., Johnson, E. M., Karn, J., and Vidali, G., *in* "Protein Phosphorylation in Control Mechanisms" (F. Huijing and E. Y. C. Lee, eds.), p. 217. Academic Press, New York (1973).
*4.* Allfrey, V. G., Inoue, A., Karn, J., Johnson, E. M., and Vidali, G., *Cold Spring Harbor Symp. Quant. Biol.* **38**, 785 (1973).
*5.* Berlowitz, L., and Pallotta, D. P., *Exp. Cell Res.* **71**, 45 (1972).
*6.* Brown, I. R., and Liew, C. C., *Science* **188**, 1122 (1975).
*7.* Candido, E. P. M., and Dixon, G. H., *Proc. Natl. Acad. Sci. U.S;A.* **69**, 2015 (1972).
*8.* Candido, E. P. M., and Dixon, G. H., *J. Biol. Chem.* **247**, 3868 (1972).
*9.* Gornall, A. G., and Liew, C. C., *Adv. Enzyme Regul.* **12**, 267 (1974).
*10.* Gorovsky, M. A., Pleger, G. L., Keevert, J. B., and Johmann, C. A., *J. Cell Biol.* **57**, 773 (1973).
*11.* Hnilica, L. S., *in* "The Structure and Biological Functions of Histones", p. 79. *Chem. Rubber Publ. Co.*, Cleveland, Ohio (1974).
*12.* Johnson, E. M., and Hadden, J. W., *Science* **187**, 1198 (1975).
*13.* Johnson, E. M., and Allfrey, V. G., *Arch. Biochem. Biophys.* **152**, 786 (1972).
*14.* Jornvall, H., *J. Theor. Biol.* **55**, 1 (1975).
*15.* Jungmann, R. A., and Schweppe, J. S., *J. Biol. Chem.* **247**, 5535 (1972).
*16.* Kleinsmith, L., *in* "Chromosomal Proteins and Their Role in the Regulation of Gene Expression" (G. S. Stein and L. J. Kleinsmith, eds.), p. 45. Academic Press, New York (1975).
*17.* Louie, A. J., Candido, E. P. M., and Dixon, G. H., *Cold Spring Harbor Symp. Quant. Biol.* **38**, 803 (1973).
*18.* Park, W. K., and Kim, S., *Adv. Enzymol.* **42**, 227 (1975).
*19.* Pogo, B. G. T., Allfrey, V. G., and Mirsky, A. E., *Proc. Natl. Acad. Sci U.S.A.* **55**, 805 (1966).
*20.* Pogo, B. G. T., Pogo, A. O., Allfrey, V. G., and Mirsky, A. E., *Proc. Natl. Acad. Sci. U.S.A.*, **59**, 1337 (1968).

*21.* Ruiz-Carrillo, A., Wangh, L. J., Littau, V. C., and Allfrey, V. G., *J. Biol. Chem.* **249**, 7358 (1974).
*22.* Shepherd, G. R., *Biochim. Biophys. Acta* **299**, 485 (1973).
*23.* Stein, G. S., Spelsberg, T. C., and Kleinsmith, L. J., *Science* **183**, 817 (1974).
*24.* Suria, D., and Liew, C. C., *Can. J. Biochem.* **52**, 1143 (1974).
*25.* Libby, P. R., "Estrogen Target Tissues and Neoplasia" (T. L. Dao, ed.), p. 85. Univ. of Chicago Press, Chicago, Illinois (1972).
*26.* Liew, C. C., Haslett G. W., and Allfrey, V. G. *Nature* (*London*) **226**, 414 (1970).
*27.* Pestana, A., and Petot, H. C., *Nature* (*London*) **247**, 200 (1974).
*28.* Ruiz-Carrillo, A., Wangh, L. J. and Allfrey, V. G., *Science* **190**, 117 (1975).
*29.* Candido, E. P. M., *Can. J. Biochem.* **53**, 796 (1975).
*30.* Gallwitz, D., and Sures, I., *Biochim. Biophys. Acta* **263**, 315 (1972).
*31.* Lue, P. F., Gornall, A. G., and Liew, C. C., *Can. J. Biochem.* **51**, 1177 (1973).
*32.* Pestana, A., Sudilovsky, O., and Pitot, H. C., *FEBS. Lett.* **19**, 83 (1971).
*33.* Racey, L. A., and Byvoet, P., *Exp. Cell Res.* **64**, 366 (1971).
*34.* Sanders, L. A., Schechter, N. M., and McCarty, K. S., *Biochemistry* **12**, 783 (1973).
*35.* Liew, C. C., Suria, D., and Gornall, A. G., *Endocrinology* **93**, 1025 (1973).
*36.* Suria, D., and Liew, C. C., *Biochem. J.* **137**, 355 (1974).
*37.* Reeder, R. H., *J. Mol. Biol.* **80**, 229 (1973).
*38.* De Pomerai, D. I., Chesterton, C. J., and Butterworth, P. H. W., *Eur. J. Biochem.* **46**, 461 (1974).
*39.* Jackowski, G., Suria, D., and Liew, C. C., *Can. J. Biochem.* **54**, 9 (1976).
*40.* Laemmli, U.K., *Nature*, (*London*) **227**, 680 (1970).
*41.* Levy, S., Simpson, R. T., and Sober, H. A., *Biochemistry* **11**, 1547 (1972).
*42.* Teng, C. S., Teng, C. T. and Allfrey, V. G., *J. Biol. Chem.* **246**, 3597 (1971).
*43.* Elgin, C. C. R., and Bonner, J., *Biochemistry* **22**, 4440 (1970).
*44.* Phillips, D. M. P. *Biochem. J.* **87**, 258 (1963).
*45.* Brown, F., and Hall, L. P., *Nature* (*London*) **166**, 66 (1950).
*46.* Gershey, E. L., Vidali, G,, and Allfrey, V. G., *J. Biol. Chem.* **243**, 5018 (1968).
*47.* Liew, C. C. and Gornall, A. G., *J. Biol. Chem.* **248**, 977 (1973).
*48.* Vidali, G., Gershey, E. L., and Allfrey, V. G., *J. Biol. Chem.* **243**, 6361 (1968).
*49.* Delange, R. J., and Smith, E. L., *Annu. Rev. Biochem.* **40**, 279 (1971).
*50.* Narita, K., *Biochim, Biophys. Acta* **28**, 184 (1958).

# Chapter 5

# *Purification and Characterization of Protein Methylase I (S-Adenosylmethionine: Protein-Arginine Methyltransferase; EC 2.1.1.23) From Calf Brain*

EGON DURBAN, HYANG WOO LEE, SANGDUK KIM, AND WOON KI PAIK

*Fels Research Institute and Department of Biochemistry, Temple University School of Medicine, Philadelphia, Pennsylvania*

## I. Introduction

This enzyme was first described by Paik and Kim (*1*). It was originally isolated from calf thymus and found to be located primarily in the cytosol. Analysis of endogenous methylated proteins showed that mainly histones were methylated (*1*). The enzyme was found in various organs of the rat and was especially elevated in brain (*2*), thymus, testis, and spleen (*3*). The products of histone methylation by protein methylase I were identified as $N^{G}$-mono, $N^{G}$, $N^{G}$-di-, and $N^{G}$, $N'^{G}$-dimethylarginine (*4,5*).

## II. Assay Method

### A. Principle

Incorporation of *S-adenosyl*-L-[methyl-$^{14}$C]methionine into histone (hot trichloroacetic acid and ethanol-insoluble fraction) is measured under conditions (pH 7.2) favorable for methylation of the guanidino group of arginine residues. Methylation of lysine residues by protein methylase III (*6*) is negligible at pH 7.2 (optimal at pH 9.0). Methylation of carboxyl groups of glutamyl and aspartyl residues by protein methylase II (*7*) may take place when using crude enzyme extracts but can be destroyed by slight alkaline treatment at elevated temperature.

## B. Reagents

Sodium phosphate buffer, 0.5 *M* (pH 7.2)
Sodium phosphate buffer, 0.5 *M* (pH 8.0)
Histone suspension, 30 mg per milliliter of water (histone type II-A of Sigma Chemical Co.)
*S*-Adenosyl-L-[methyl-$^{14}$C] methionine (abbreviated SAM; specific activity 50 mC/mmole) diluted to give 100–150 dpm per picomole and a concentration of $10^{-5}$ *M* or 10 $\mu$*M*.
Enzyme; 20–30 $\mu$g of protein of purified enzyme and 0.5–1.0 mg protein of crude extract to be used per assay
Scintillation cocktail capable of dissolving the ethanol precipitate (commercially available scintillation cocktails adequate for counting aqueous samples were found satisfactory)

## C. Procedure

Place 0.1 ml of 0.5 *M* phosphate buffer (pH 7.2), 0.1 ml of [*methyl*-$^{14}$C] SAM (10 nmoles), 0.1 ml of histone suspension (3 mg), enzyme preparation, and water (to give 0.5 ml total assay volume) in a conical glass centrifuge tube. Incubate at 37°C for 10 minutes. Enzyme preparation heated at 100°C for 5 minutes serves as a blank. The reaction is terminated by addition of 0.5 ml of 30% trichloroacetic acid. To remove acid-soluble free [*methyl*-$^{14}$C] SAM as well as nucleic acids and phospholipids, the tubes containing trichloroacetic acid at 15% final concentration are heated at 90°C for 15 minutes, cooled, centrifuged for 10 minutes in a table-top centrifuge (IEC), and the trichloroacetic acid-insoluble pellets are resuspended in 15% trichloroacetic acid. This trichloroacetic acid wash is repeated twice, followed by treatment with ethanol at 70°C for 10 minutes. When one is dealing with purified enzyme, the precipitate collected by centrifugation is dissolved in a small amount of scintillation cocktail and transferred quantitatively into a scintillation vial with a final volume of 10 ml of scintillation cocktail. However, when crude enzyme extract is used, at this stage the products of protein methylase II [methylated carboxyl groups of glutamyl and aspartyl residues (*8*)] have to be removed. The precipitate from the ethanol treatment is dissolved in 1 ml of 0.5 *M* phosphate buffer (pH 8.0), and incubated at 60°C for 5 minutes. Then 3 ml of 15% trichloroacetic acid are added, and the suspension is kept at room temperature for 15 minutes. The precipitate is collected by centrifugation, washed once with ethanol, and dissolved in 10 ml of scintillation fluid for counting.

### D. Definition of Specific Activity

Specific activity is expressed as picomoles of [*methyl*-$^{14}$C]SAM incorporated into protein per minute and milligram of enzyme protein. Enzyme protein is determined by the method of Lowry *et al.* (*9*).

## III. Purification Procedure

The initial steps (1–3) of the procedure are those described by Paik and Kim (*1*) with minor modification. The final enzyme preparation is free of other protein methylases (II and III).

### A. Step 1

Fresh calf brain obtained from a slaughterhouse is kept frozen until use. Enzyme activity remains unchanged for up to 3 months. Frozen calf brain, 100 gm, is thawed and homogenized in 4 volumes of precooled 0.25 *M* sucrose–1 m*M* EDTA in a Waring Blendor for 20–30 seconds at 30 V. This step and all the following procedures are carried out at 0°–3°C. The crude homogenate is further homogenized in a Teflon–glass homogenizer at 900 rpm and passed through a double layer of cheesecloth followed by centrifugation at 105,000 *g* for 1 hour in a Beckman L2 ultracentrifuge. A lipid layer on top of the supernatant is removed by passage through glass wool.

### B. Step 2. First Ammonium Sulfate Precipitation

Per 100 ml of supernatant, 31.3 gm of crystalline ammonium sulfate (analytical grade) are added slowly under continuous stirring within 15–20 minutes. The solution is allowed to stand in the cold for at least 30 minutes after all ammonium sulfate is dissolved. The precipitate is collected by centrifugation at 18,000 rpm (39,000 *g*) in a Sorvall Superspeed centrifuge and dissolved in approximately 100 ml of cold distilled water.

## C. Step 3. Calcium Phosphate Gel Treatment

Five volumes of calcium phosphate gel (suspension of 17 mg per milliliter distilled water) are added to the enzyme solution under gentle stirring and allowed to stand for 20 minutes for complete adsorption. After centrifugation at low speed, the sedimented gel is washed twice by resuspension in a large volume of cold distilled water and centrifugation. For elution of the enzyme, the final gel sediment is resuspended in 60 ml of 0.25 *M* phosphate buffer (pH 7.2) and allowed to stand for 10–15 minutes. The eluate is recovered by centrifugation at high speed (supernatant).

## D. Step 4. Second Ammonium Sulfate Precipitation

The eluate is precipitated with crystalline ammonium sulfate (31.3 gm/100 ml) as described in step 2. The precipitate collected by centrifugation is dissolved in about 6 ml of 5 m*M* phosphate buffer (pH 6.0) containing 3 mg of dithiothreitol per 100 ml and 10% glycerol. The solution is dialyzed against water for 4–5 hours and subsequently against 5 m*M* phosphate buffer (pH 6.0)–dithiothreitol 3 mg/100 ml–10% glycerol for several hours. Initial dialysis against water is necessary to speed up the dialysis process, which otherwise takes at least overnight and is accompanied by a considerable loss of enzyme activity.

## E. Step 5. Cellex-D Chromatography

Cellex-D resin (Bio-Rad) is prepared by the following washes: (a) 0.25 *N* NaCl–0.25 *N* NaOH, (b) distilled water, (c) 0.25 *N* HCl, and (d) water until free of chloride. The resin is then loaded into a 1.2 × 25 cm column and equilibrated with 5 m*M* phosphate buffer (pH 6.0)–dithiothreitol 3 mg/100 ml–10% glycerol. After equilibration, the dialyzed solution obtained in step 4 is applied to the column and eluted with a linear gradient formed by 100 ml of 5 m*M* phosphate buffer (pH 6.0)–dithiothreitol, 3 mg/100 ml–10% glycerol and 100 ml of 0.8 *M* ammonium sulfate in the same buffer. Flow rate is maintained at 50–60 ml/hour. Protein methylase I activity is recovered between 110 and 130 ml of effluent. Fractions from this range are pooled and concentrated by precipitation with crystalline ammonium sulfate (60% saturation). The solution is stirred for 30 minutes after all ammonium sulfate is dissolved and centrifuged at 39,000 *g* for 10 minutes. The supernatant is discarded. The pellet is dissolved in a small volume of 5 m*M* borate buffer (pH 7.2)–10% glycerol and dialyzed against the same buffer for several hours.

TABLE I

PURIFICATION OF PROTEIN METHYLASE I FROM CALF BRAIN

| Purification steps | Protein (mg) | Specific activity (pmoles/mg/min) | Total activity (pmoles/min) | Purification | |
|---|---|---|---|---|---|
| | | | | -Fold | % Recovery |
| Whole homogenate | 7950 | 0.32 | 2567 | 1 | 100 |
| Supernatant | 1150 | 1.49 | 1713 | 4.6 | 66.7 |
| First $(NH_4)_2SO_4$ precipitation | 517 | 2.93 | 1513 | 9.1 | 58.9 |
| Eluate from calcium phosphate gel | 265 | 5.08 | 1348 | 15.7 | 52.7 |
| Second $(NH_4)_2SO_4$ precipitation | 161 | 7.80 | 1252 | 24.2 | 48.8 |
| Cellex-D | 64.4 | 12.7 | 818 | 39.3 | 31.8 |
| Bio-Gel A-15M | 2.77 | 38.3 | 348 | 118.6 | 13.6 |

## F. Step 6. Chromatography on Bio-Gel A-15M

A Bio-Gel A-15M (Bio-Rad) column (1.3 × 160 cm) is equilibrated with 5 m*M* borate buffer (pH 7.2)–10% glycerol for 3–4 days. A portion of the enzyme preparation obtained in step 5 is applied to the column, usually 1–1.5 ml containing 30–50 mg protein, and eluted with 5 m*M* borate buffer (pH 7.2)–10% glycerol at a flow rate of 10 ml per hour. Of the protein methylase I activity, 95% is eluted in a sharp peak just preceding the main protein peak (50–60 ml effluent). Enzyme from this peak is completely free of other protein methylases, especially protein methylase II which is eluted much later from the column (around 120 ml effluent). The overall purification at this stage is 110-fold with a yield of 13% and a specific activity of 38. Pertinent data on the various steps in a typical procedure are summarized in Table 1.

# IV. Properties

## A. Specificity

The enzyme is quite specific for histone (type II-A, Sigma Chemical Co.), which represents the most effective methyl acceptor for [*methyl*-$^{14}$C] SAM as shown in Table II. The next best methyl acceptor protein (Table II) is myelin basic protein with about 30% substrate activity as compared to that of histone. Miyake and Kakimoto (*10, 11*) suggested that arginine in myelin basic protein is methylated by a different protein methylase than arginine in

TABLE II

RELATIVE EFFICIENCY OF VARIOUS PROTEINS AS SUBSTRATE FOR PROTEIN METHYLASE I

| Protein used as methyl acceptor[a] | % Efficiency |
|---|---|
| Histone type II-A | 100.0 |
| Al protein (myelin basic protein) | 29.9 |
| Polyarginine | 24.9 |
| Fibrinogen | 15.4 |
| Ribonuclease (oxidized; bovine pancreas) | 13.3 |
| Ribonuclease (native; bovine pancreas) | 0 |
| Polylysine | 0 |
| Albumin (bovine serum) | 0 |
| Lysozyme | 0 |

[a] 3 mg were used.

histone. However, this was disputed by Jones and Carnegie (*12*), who reported that the same enzyme catalyzes both methylation of myelin basic protein and histone. To date, this controversy has not been resolved.

## B. Products of the Reaction

Analysis of methylated products was carried out as described by Paik and Kim (*3*). Briefly, methylated protein substrate was hydrolyzed for 48 hours at 110°C in sealed glass tubes (Under vacuum) in 6 *N* HCl. After removing HCl under reduced pressure, the hydrolyzate was analyzed on a Perkin-Elmer KLA-3B amino acid analyzer by chromatography through Bio-Rad A-5 resin. Sodium citrate buffer, 0.38 *N*, (pH 5.84 at 24°C) was used for elution up to the tyrosine and phenylalanine positions, then eulution was continued with 0.35 *N* citrate buffer (pH 4.70). As illustrated in Fig. 1, all methylated arginines were well separated by this method. $N^G$-Monomethylarginine is the main methylated product (> 50%), $N^G$, $N^G$-dimethylarginine comprises 35–45% and $N^G$, $N'^G$-dimethylarginine comprises less than 10% of the methylated end products.

## C. Inhibitors

$Zn^{2+}$ and guanidine-HCl have noticeable inhibitory effect (activity reduced to 68% and 56%, respectively). $Ca^{2+}$, $Mg^{2+}$, $Fe^{3+}$, $K^+$, and EDTA at 1 m*M* had no effect, while dithiothreitol (3 mg/100 ml) slightly stimulated enzyme activity. *N*-Hydroxy-2-aminofluorene is a very powerful inhibitor (72% inhibition at 1.0 m*M*) (*13*). *S*-Adenosyl-L-homocysteine, a product of the methylation reaction, inhibits enzyme activity by 46% ($K_i$ of 2.6 $\mu M$) while other free amino acids such as L-ornithine, L-lysine, and L-arginine had no effect.

## D. Effect of pH

The optimum pH of protein methylase I from calf brain is 7.2 using histone type II-A as substrate which pH was also found to be optimal for protein methylase I from calf thymus (*1*). Activity is very rapidly lost at pH < 7 (at pH 6 practically no activity present) while activity is lost to lesser degree toward alkaline pH (at pH 9 approximately 50% activity present compared to activity at optimum pH).

## E. Isoelectric Point

The isoelectric point is at pH 5.1.

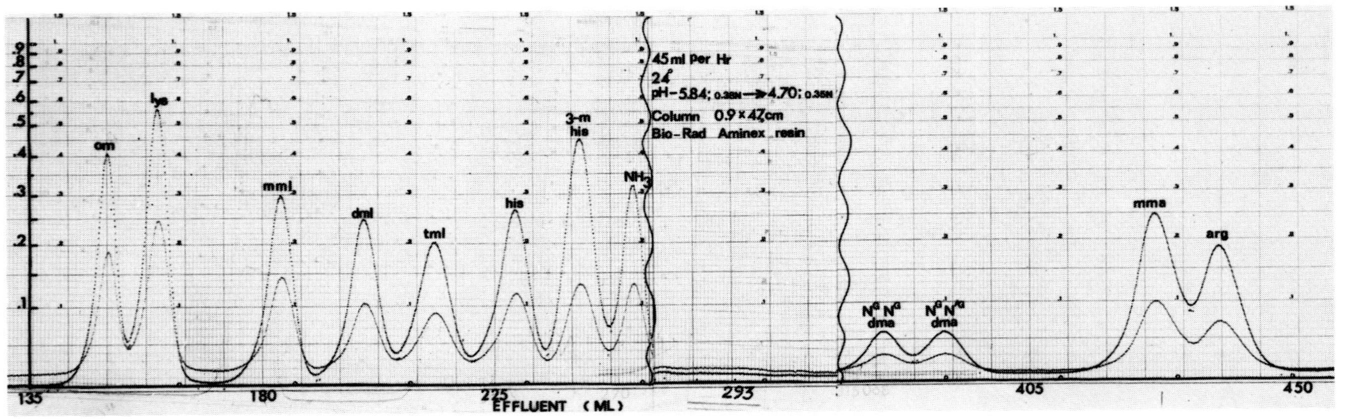


FIG. 1. Chromatographic separation of basic amino acids and their methylated derivatives with an automatic amino acid analyzer, orn, lys, mml, dml, tml, his, 3-m-his, $N^G$ $N^G$ dma, $N^G$ $N'^G$ dma, mma, and arg represent, respectively, ornithine, lysine, ε-*N*-monomethyllsine, ε-*N*-dimethyllsine, ε-*N*-trimethyllsine, histidine, 3-*N*-methylhistidine, $N^G$, $N^G$-dimethylarginine, $N^G$, $N'^G$-dimethylarginine, $N^G$-monomethylarginine, and arginine.

## F. Stability

The purified enzyme can be stored in 10% glycerol in the frozen state. About 25% activity is lost over 3–4 weeks.

## G. Substrate Concentration Effect

The Michaelis constant for *S*-adenosyl-L-methionine is 1.1 $\mu M$. Maximum rate of methyl transfer occurs at a concentration of 3 mg of histone per 0.5 ml of assay mixture.

### References


1. Paik, W. K., and Kim, S., *J. Biol. Chem.* **243**, 2108 (1968).
2. Paik, W. K., and Kim, S., *J. Neurochem.* **16**, 1257 (1969).
3. Paik, W. K., and Kim, S., *Adv. Enzymol.* **42**, 227 (1975).
4. Paik, W. K., and Kim, S., *Arch. Biochem. Biophys.* **134**, 632 (1969).
5. Kakimoto, Y., and Akazawa, S., *J. Biol. Chem.* **245**, 5751 (1970).
6. Paik, W. K., and Kim, S., *J. Biol. Chem.* **245**, 6010 (1970).
7. Kim, S., *Arch. Biochem. Biophys.* **157**, 476 (1973).
8. Kim, S. and Paik, W. K., this volume, Chapter 7.
9. Lowry, O. H., Rosebrough, N. J., Farr, A. L., and Randall, R. J., *J. Biol. Chem.* **193**, 265 (1951).
10. Miyake, M., and Kakimoto, Y., *J. Neurochem.* **20**, 859 (1973).
11. Miyake, M., *J. Neurochem.* **24**, 909 (1975).
12. Jones, G. M., and Carnegie, P. R., *J. Neurochem.* **23**, 1231 (1974).
13. Baxter, C. S., and Byvoet, P., *Cancer Res.* **34**, 1418 (1974).

# Chapter 6

# *The Methylation and Demethylation of Protein Lysine Residues*

SAMUEL NOCHUMSON, SANGDUK KIM, AND WOON KI PAIK

*Fels Research Institute and Department of Biochemistry, Temple University School of Medicine, Philadelphia, Pennsylvania*

## I. Protein Methylase III (*S*-Adenosylmethionine: Protein-Lysine Methyltransferase; EC 2.1.1.43)

The enzyme(s) responsible for methylating the lysine residues of proteins have been termed protein methylase III or, more properly, *S*-adenosylmethionine:protein-lysine methyltransferase. This enzyme has been found to be elevated in rapidly proliferating tissues (*1*). In mammalian cells the enzyme has been shown to be localized within the nucleus associated with the chromatin (*2*). Histones have been implicated as the major protein substrate for the enzyme in studies performed on HeLa S-3 cells (*3*). The products of the reaction are $\epsilon$-*N*-mono-, $\epsilon$-*N*-di, and $\epsilon$-*N*-trimethyllysine. In bacteria, a ribosomal protein methylase III which is responsible for methylating predominantly the L-11 protein of the 50 S ribosomal subunit has been purified 380-fold (*4*). Currently, we are purifying a protein methylase III from *Neurospora crassa*, which is responsible for methylating cytochrome *c*.

### A. Assay Method

#### 1. Principle

The amount of radioactive methyl groups enzymically transferred from *S*-adenosyl-L-[*methyl*-$^{14}$C]methionine into trichloroacetic acid-precipitable material (histones) is measured under conditions favorable for the methylation of the $\epsilon$-amino group of lysine residues (pH 9.0). Methylation of the guanidino group of arginine and carboxyl group of glutamyl and aspartyl residues by protein methylases I and II is negligible at this pH, and both can be destroyed by alkaline treatment at elevated temperatures.

### 2. Reagents

Tris-HCl buffer, 0.5 *M* (pH 9.0)
*S*-Adenosyl-L-[*methyl*-$^{14}$C]methionine, 4.78 nmoles (100 dpm/pmole)
Histone suspension, 20 mg per milliter of water (histone type II-A of Sigma Chemical Co.).
Trichloroacetic acid, 15% and 30%
NaOH, 0.2 *N*
HCl, 0.5 *N*
Enzyme preparation adjusted to 10 mg/ml

### 3. Procedure

One-tenth milliter of 0.5 *M* Tris-HCl buffer, 0.1 ml of histones, 0.1 ml of the enzyme fraction, and 0.1 ml of water were preincubated at 37°C for 3 min. in 12-ml conical centrifuge tubes. Then 0.1 ml of *S*-adenosyl-L-[*methyl*-$^{14}$C]methionine was added to initiate the reaction. After 5 min, the reaction was terminated by the addition of 0.5 ml of 30% trichloroacetic acid. The enzyme fraction heated at 100°C for 5 minutes served as a blank and was treated the same way as the reaction mixture. Nucleic acids were removed by heating the reaction tubes at 90°C for 15 minutes; the tubes were then cooled and centrifuged in a clinical centrifuge (IEC) for 10 minutes. The supernatant was discarded, and the precipitate was washed twice using 10 ml of 15% trichloroacetic acid and a pointed glass rod as an aid in breaking up the precipitate. After the second trichloroacetic acid wash, the precipitate was heated at 70°C for 10 minutes in 10 ml of 95% ethanol and then cooled and centrifuged. In order to remove guanido-*N*-[*methyl*-$^{14}$C]arginines resulting from the action of protein methylase I, the washed histone precipitate was suspended in 1 ml of 0.2 *N* NaOH and heated in a boiling water bath uncovered for 2 hours. The tubes were then cooled and neutralized with 0.4 ml of 0.5 *N* HCl and quantitatively transferred into scintillation vials which contained 10 ml of scintillation solution.

## B. Solubilization and Partial Purification of Protein Methylase III

Frozen or fresh calf thymus may be used. A crude preparation of nuclei was obtained by homogenizing the thymus in 10 volumes of 0.25 *M* sucrose containing 6 m*M* $CaCl_2$ in an electrically driven Teflon–glass homogenizer. The homogenate was filtered through a double layer of cheesecloth, and was centrifuged at 700 *g* for 10 minutes. The pellet was them made into an acetone powder by rehomogenizing it in 10 volumes of acetone cooled to −70 °C. After a uniform suspension was achieved, the acetone homogenate was collected on a Büchner funnel and air-dried at room temperature. Fresh calf thymus, 100 gm, yielded 2 gm of nuclear acetone powder, which can be stored at −5°C for at least 1 year without loss of enzyme activity.

The acetone powder (100 mg) was extracted with 15 ml of ice-cold 0.4% deoxycholic acid for 20 minutes in an electrically driven Teflon–glass homogenizer and the homogenate was centrifuged at 39,000 *g* for 30 minutes at 4°C. The clear supernatant (about 11 ml) was adjusted to a final concentration of 0.01 *M* Tris-HCl (pH 9.0), and 400 mg of dried phosphocellulose was then added and allowed to stir for an additional 5 minutes at 4°C. The phosphocellulose and its bound protein were removed by centrifugation at 39,000 *g* for 10 minutes at 4°C. All the enzyme activity was recovered in the supernatant. All attempts at further purification of the enzyme by classical techniques have been unsuccessful, thus the protein methylase III has only been purified 1.3 times with a 25% yield.

## C. Properties of Calf Thymus Protein Methylase III

### 1. Specificity

The enzyme has a relative specificity toward the various histone fractions with little activity toward nonhistone proteins. This is in marked contrast to a protein methylase III found in *Neurospora crassa* (Table I) which is readily soluble and is extremely specific for horse heart cytochrome *c* and has negligible activity toward various histone fractions.

### 2. Products of the Reaction

The products of the enzyme reaction can be analyzed after hydrolysis in 6 *N* HCl for 48 hours, using the amino acid analyzer as previously described (see Chapter 5, this volume). All three methylated lysines have been found in the same distribution as in *in vivo* experiments (*5*).

### 3. pH Optimum

The enzyme was found to exhibit optimal activity at pH 9.0 in 0.1 *M* Tris-HCl buffer.

### 4. Effect of S-Adenosyl-L-Methionine Concentration

The $K_m$ for *S*-adenosyl-L-methionine was found to be 3.0 $\mu M$, which is similar to that for both protein methylases I and II. Protein methylase III from rat spleen nuclei showed 55% inhibition with the addition of 3.25 $\mu M$ *S*-adenosyl-L-homocysteine. This inhibition by the product *S*-adenosyl-L-

TABLE I

COMPARISON OF THE SPECIFICITY OF PROTEIN METHYLASE III FROM CALF THYMUS NUCLEI WITH PROTEIN METHYLASE III FROM NEUROSPORA CRASSA

| Protein used as a methyl acceptor | Enzyme activity (%) | |
|---|---|---|
| | Calf thymus | *N. crassa* |
| Histone type II-A (mixture) | 100[a] | 2.9[b] |
| Lysine-rich histone | 94 | 6.5 |
| Slightly lysine-rich histone | 47 | 0 |
| Arginine-rich histone | 162 | 0 |
| Polylysine | 44 | 0.2 |
| Ribonuclease (bovine pancreas) | 6.7 | 4.7 |
| Cytochrome *c* (horse heart) | 2.6 | 100 |
| Albumin | 0 | 0 |
| γ-Globulin | 0 | 0 |

[a] Relative activity based on the substrate histone type II-A as 100% and represents 2.48 pmoles of [*methyl*-$^{14}$C] label transferred per minute per milligram of enzyme protein.

[b] Relative activity based on the substrate cytochrome *c* as 100% and represent 15.3 pmoles of *methyl*-$^{14}$C label transferred per minute per milligram of enzyme protein.

homocysteine is a common feature of *S*-adenosylmethionine methyltransferases.

### 5. EFFECT OF VARIOUS METALS AND SOME REDUCING AGENTS

Various divalent cations such as $Ca^{2+}$, $Co^{2+}$, $Mn^{2+}$, $Mg^{2+}$, $Fe^{2+}$, as well as reduced glutathione, cysteine, and EDTA, did not inhibit the enzyme activity at 2 m*M* concentrations. However, $Hg^{2+}$, $Cu^{2+}$, and $Zn^{2+}$ showed 100% inhibition at the same concentration.

### 6. STABILITY OF THE ENZYME

Once the enzyme is extracted from the acetone powder it becomes extremely unstable, so that the enzyme activity disappears on overnight storage at −10°C or 3°C. This problem of instability has not yet been solved and has hindered the further purification of the enzyme.

## II. Histone Demethylating Enzyme (ε-Alkyllysinase or ε-Alkyl-L-Lysine: Oxygen Oxidoreductase; EC 1.5.3.4)

An enzyme that catalyzes the removal of methyl groups from the ε-*N*-amino group of lysine in histones as well as at the free amino acid level has been found in mammalian tissues (*6*). The enzyme reaction utilizes molecular oxygen to oxidize the methyl group to formaldehyde with the production of hydrogen peroxide. This enzyme is known as ε-alkyllysinase (EC 1.5.3.4) and is localized in the mitochondria and is most abundant in rat kidney mitochondria. In rapidly growing hepatomas the enzyme activity is decreased whereas the protein methylase III activity is increased (*1*).

## A. Assay Method

### 1. PRINCIPLE

Initially this enzyme was assayed by measuring the amount of L-lysine formed manometrically using lysine decarboxylase (*7*). However, we have recently developed a radiometric assay in which the radioactive formaldehyde released from ε-*N*-[*methyl*-$^{14}$C]L-lysine is trapped as a formaldemethone derivative and counted in a scintillation counter (*6*). Since the radioactive substrate is not available commercially, it must be synthesized. Thus, ε-*N*-[*methyl*-$^{14}$C]L-lysine can be synthesized chemically by the method of Benoiton (*8*) or by using the chemical reduction of protein with [$^{14}$C]formaldehyde and sodium borohydride. This reaction has been shown to preferentially methylate the ε-amino groups of protein-lysine residues (*9*).

The ε-alkyllysinase has been shown to demethylate both ε-*N*-methyllysine in histones and at the free amino acid level (*6*). However, in the assay method using $^{14}$C-methylated histones as the substrate, there is a marked inhibition above a certain enzyme protein concentration. The reason for this inhibition is not understood, but the inhibition is not present when ε-*N*-[*methyl*-$^{14}$C]lysine is used as the substrate. Therefore, the chemically reduced histones are hydrolyzed in 6 *N* HCl, and the free ε-*N*-[*methyl*-$^{14}$C]lysine is used in measuring the enzyme.

### 2. PREPARATION OF SUBSTRATE

One gram of histone type II-A was dissolved in 12 ml of water, and 6 ml of 0.2 *M* borate buffer (pH 9.0) and 10 mg of $NaBH_4$ were added. Formaldehyde in 0.01-ml aliquots was added at 5-minute intervals at 0°C until a total of 0.05 ml was reached. The formaldehyde was prepared by diluting [$^{14}$C]formaldehyde (specific activity 59 $C_i$/mole) with an equal volume of nonradioactive 37% formaldehyde. After the last addition of formaldehyde, the reaction was allowed to proceed for another 5 minutes at 0°C, at which time the complete reaction mixture was dialyzed against 6 liters of water. Dialysis continued for 24 hours with three changes of water. The dialyzed sample was then lyophilized. The $^{14}C$-methylated lysine histones can be used as a substrate for the histone demethylating enzyme or it can be hydrolyzed in 10 ml of 6 *N* HCl *in vacuo* at 110°C for use in the ε-alkyllysinase assay. A 0.05 *M* solution of nonradioactive ε-*N*-monomethyl-L-lysine (Calbiochemical Co.) was prepared containing a portion of the radioactive methylated histone hydrolyzate to yield approximately 1800 cpm/μmole.

## 3. Reagents

Phosphate buffer, 0.5 *M* (pH 7.2).
Semicarbazide, 0.1 *M* (pH around 7).
Flavin adenine dinucleotide (FAD), 1 m*M*.
Phenazine methosulfate, 1%.
ε-*N*-[*methyl*-$^{14}C$]L-lysine, 0.05 *M* (approximately 1800 cpm/μmole).
Formaldehyde, 0.74% (commercially available solution is diluted 50 times).
5,5-Dimethyl-1,3-cyclohexanedione (dimedon), 0.4%.
NaOH, 2 *N*.
Acetic acid, 2 *N*.
Trichloroacetic acid, 50%.

## 4. Procedure

The reaction mixture contained 0.2 ml of 0.05 *M* ε-*N*-[*methyl*-$^{14}C$]lysine, 0.1 ml of phosphate buffer, 0.1 ml of semicarbazide, 0.1 ml of FAD, 0.1 ml of phenazine methosulfate, and an appropriate amount of enzyme (4–8 mg of protein) in a total volume of 1.5 ml. The blank contained everything as in the reaction mixture, except that the enzyme was inactivated by boiling for 5 minutes. The incubation time was 2 hours at 37°C, at which time 1.5 ml of a 50% solution of trichloroacetic acid was added to stop the reaction. The mixture was then centrifuged in a clinical centrifuge (table top) for 10 minutes and the supernatant was added to 60 ml of dimedon in 250-ml flasks. One milliliter of 0.74% formaldehyde was added as a carrier, giving 78 mg of formaldemethone as the theoretical yield. About 3.5 ml of 2 *N* NaOH and 1.7 ml of 2 *N* acetic acid were then added to form a salt. The precipitate was then collected on a Büchner funnel (i.d. 3 cm) after standing for 30 minutes at room temperature, washed several times with water, and then dried under an infrared lamp. Drying was completed within 30 minutes with the lamp 25 cm above the sample. A portion of the sample was weighed and transferred to a scintillation vial containing 10 ml of scintillation solution. From the radioactivity obtained, the total radioactivity based on 78 mg of formaldemethone was calculated. The enzyme activity was expressed as micromoles of formaldehyde formed per hour per milligram of enzyme protein.

## B. Purification

All procedures were carried out at 0°–4°C.

### 1. Step 1. Triton X-100 Extraction

Five grams of rat kidney from adult animals were homogenized in 9 volumes of 0.25 *M* sucrose–6 m*M* $CaCl_2$ using an electrically driven Teflon–glass homogenizer. The homogenate was filtered through a double layer of cheesecloth and centrifuged at 39,000 *g* for 20 minutes. The precipitate was extracted with 20 ml of 0.5% Triton X-100 for 10 minutes in an ice-cold Teflon–glass homogenizer. The suspension was centrifuged at 39,000 *g* for 30 minutes and the supernatant was saved.

### 2. Step 2. Calcium Phosphate Gel Treatment

Forty milliliters of calcium phosphate gel (17 mg of solid per milliliter, product of Sigma Chemical Co.) was added to the Triton X-100 extract from step 1, and the mixture was centrifuged at 700 *g* for 5 minutes. The precipitate was washed twice with a large volume of cold distilled water and the absorbed enzyme removed by the addition of 10 ml of 0.3 *M* phosphate buffer (pH 7.2). The enzyme was found in the supernatant after removing the calcium phosphate gel by centrifugation at 39,000 *g* for 10 minutes.

### 3. Step 3. Ammonium Sulfate Precipitation

Saturated ammonium sulfate at room temperature was added to the calcium phosphate gel eluate from step 2 in a ratio of 6:10 [$(NH_4)_2SO_4$: eluate]. After about 10 minutes, the precipitate was collected by centrifugation at 39,000 *g* for 10 minutes. If frozen at this stage the enzyme activity remained unchanged for at least 2 weeks.

### 4. Step 4. Heat Treatment

The precipitate was dissolved in 1.8 ml of 5 m*M* $\epsilon$-*N*-monomethyl-L-lysine, and the suspension was heated at 55°C for 6 minutes in a small test tube. After cooling in ice, the sample was centrifuged at 105,000 *g* for 20 minutes and the supernatant saved.

### 5. Step 5. Sephadex G-200 Chromatography

The supernatant from step 4 was charged on a Sephadex G-200 column (1 × 93 cm) equilibrated with 10% glycerol containing 5 m*M* phosphate buffer (pH 7.2), and 1 ml fractions were collected. Fractions 29–32 contained the enzyme activity and were therefore pooled (*6*). The enzyme was purified 15-fold with 5% yield.

TABLE II

DEMETHYLATION ACTIVITY DURING ENZYME PURIFICATION, USING [*METHYL*-$^{14}C$] HISTONE AND ε-*N*-[*METHYL*-$^{14}C$]-L-LYSINE AS SUBSTRATE (*6*)

| Purification step | With [*methyl*-$^{14}C$] histone | | With ε-*N*-[*methyl*-$^{14}C$]L-lysine | | |
|---|---|---|---|---|---|
| | Enzyme activity (pmoles/hr/mg protein) | Purification | Enzyme activity (μmoles/hr/mg protein) | Purification | Ratio of column 2 to column 4 |
| Whole homogenate | 6.5 | 1.0 | 0.155 | 1.0 | 42 |
| Triton X-100 extract | 16.3 | 2.5 | 0.356 | 2.3 | 46 |
| $(NH_4)_2SO_4$ precipitate | 24.7 | 3.8 | 0.711 | 4.6 | 35 |
| Sephadex G-200 eluate | 63.7 | 9.8 | 1.696 | 11.0 | 38 |

## C. Properties

### 1. Copurification of ϵ-*N*-Methyllysine and Histone Demethylating Enzymes

Table II shows that the ratio between [*methyl*-$^{14}$C]histone and ϵ-*N*-mono [*methyl*-$^{14}$C]L-lysine demethylating activities remain constant throughout the purification, suggesting that the two activities are due to a single enzyme. In addition, both demethylating activities decrease at the same rate upon heating at 55°C without the presence of stabilizing substrate (*6*).

### 2. pH Optimum

The enzyme shows optimal activity for demethylating ϵ-*N*-[*methyl*-$^{14}$C] lysine at pH 7.2 using phosphate buffer.

### 3. Effect of Substrate Concentration

The enzyme exhibits classical Michaelis–Menten kinetics and has a $K_m$ of 1.05 m*M* for ϵ-*N*-monomethyl-L-lysine.

### 4. Specificity

The enzyme will also demethylate ϵ-*N*-dimethyl-L-lysine and α-keto-ϵ-methylaminocaproic acid. The ornithine analog, δ-*N*-methyl-L-ornithine was not demethylated.

### 5. Effect of Various Compounds on the Enzyme Activity

The divalent cations $Ni^{2+}$, $Zn^{2+}$, and $Co^{2+}$ were found to be inhibitory at 1.2 m*M* concentration. At 1.3 m*M*, KCN reduced the enzyme activity to 40% and 2,6-dichlorophenol was also found to be inhibitory toward the partially purified enzyme.

## References

*1*. Paik, W. K., Kim, S., Ezirike, J., and Morris, H. P., *Cancer Res.* **35**, 1159 (1975).
*2*. Paik, W. K., and Kim, S., *J. Biol. Chem.* **245**, 6010 (1970).
*3*. Lee, H. W., Paik, W. K., and Borun, T. W., *J. Biol. Chem.* **248**, 4194 (1973).
*4*. Chang, F. N., Cohen, L. B., Navickas, I. J., and Chang, C. N., *Biochemistry* **14**, 4994 (1975).
*5*. Borun, T. W., Pearson, D. B., and Paik, W. K., *J. Biol. Chem.* **247**, 4288 (1972).
*6*. Paik, W. K., and Kim, S., *Arch. Biochem. Biophys.* **165**, 367 (1974).
*7*. Kim, S., Benoiton, L., and Paik, W. K., *J. Biol. Chem.* **239**, 3790 (1964).
*8*. Benoiton, L., *Can. J. Chem.* **42**, 2043 (1964).
*9*. Means, G. E., and Feeney, R. E., *Biochemistry* **7**, 2192 (1968).

# Chapter 7

# *Purification and Assay of Protein Methylase II (S-Adenosylmethionine : Protein-Carboxyl Methyltransferase; EC 2.1.1.24)*

SANGDUK KIM AND WOON KI PAIK

*Fels Research Institute and Department of Biochemistry, Temple University School of Medicine, Philadelphia, Pennsylvania*

## I. Introduction

Protein methylase II (*S*-adenosylmethionine:protein-carboxyl methyltransferase; EC 2.1.1.24) methylates (esterifies) the free carboxyl groups of protein or polypeptide substrates with *S*-adenosyl-L-methionine as methyl donor (*1,2*). A striking feature of the reaction is the formation of an unstable protein-methyl ester which yields methanol nonenzymically in weak aqueous alkaline solution. Owing to this methanol formation from the immediate enzymic product, the enzyme has also been referred to as 'methanol-forming enzyme' (*3,4*). The enzyme is widely distributed in mammalian organs with higher activities in testis, brain, pituitary gland, and blood (*2,3,5*). Subcellular localization within several tissues indicate that it is localized mainly in the cytosol (*4,6*) with the exception of the pituitary gland in which an appreciable amount of the enzyme is particulate bound (*3*).

The amino acid residues which have been esterified *in vivo* have not yet been identified. This is mainly due to the fact that the reaction yields an unstable product that is readily destroyed during acid or base hydrolysis. However, there is strong evidence that the endogenous natural substrates for protein methylase II are various polypeptide hormones in the pituitary gland (*7,8*) and histones (*9*).

## II. Assay Methods

### A. Principle

There are presently three assay methods available for protein methylase II activity. The first method is to determine the amount of radioactive methyl groups from *S*-adenosyl-L-[*methyl*-$^{14}$C]methionine incorporated into trichloroacetic acid-precipitable substrate protein (*1, 10*). The second method takes advantage of the alkali-lability of the enzymically formed ester bond: The sample to be analyzed is distilled at a reduced pressure, and the collected [$^{14}$C]methanol is counted for radioactivity (*11*). The third method involves extraction of [$^{14}$C]methanol with isoamyl alcohol (*3, 12*). The second and third methods are best suited for assaying the enzyme activity with small peptides that are not precipitated by trichloroacetic acid.

### B. Reagents

(a) Citric acid, 0.25 *M* (24 gm/500 ml)
(b) $Na_2HPO_4$, 0.5 *M* (35.5 gm/500 ml)
(c) Ethylenediaminetetraacetic acid disodium salt (EDTA), 0.02 *M* (3.72 gm/500 ml); adjust pH to 6.0 with 1 *N* NaOH
(d) 2-Mercaptoethanol

The citrate-phosphate buffer cocktail is prepared by mixing 6 ml of (a), 10 ml of (b), 8 ml of (c), and 0.08 ml of (d). The pH of the mixture is readjusted to 6.0.

Substrate protein; histone type II-A (Sigma Chemical Co.) at 100 mg/ml. Adjust pH to 6.0.

*S*-Adenosyl-L-[*methyl*-$^{14}$C]methionine (abbreviated SAM); specific activity 50 $mC_i$/mmole diluted to give 100–150 dpm/pmole and a 0.1 m*M* concentration.

Scintillation cocktail; see this volume, Chapter 5.

### C. Procedure

In a 12-ml, thick-wall Pyrex conical centrifuge tube immersed in an ice bucket, 0.15 ml of citrate-phosphate buffer, 0.1 ml of histone suspension, and an appropriate amount of the enzyme preparation are mixed in a final volume of 0.4 ml with water. After equilibration of tubes at 37°C for 3 minutes, the reaction is started by adding 0.1 ml of [*methyl*-$^{14}$C]SAM. The reaction is allowed to proceed for 10 minutes. Two sets of duplicate tubes are prepared, one of which serves as a control. After the incubation, one of the three following methods can be used to quantitate the amount of methyl groups incorporated into the substrate protein.

### 1. Trichloroacetic Acid Precipitation Method

*a. Reagents*

Trichloroacetic acid solution (TCA), 15%.
Chloroform:ether:ethanol mixture (1:2:2, v/v).
Ethanol, 98%
Sodium phosphate buffer, 0.2 *M* (pH 7.2).

*b. Procedure.* The reaction is terminated by the addition of 5 ml of 15% TCA. The mixture is carefully overlayered with ethanol and centrifuged for 15 minutes in a table-top clinical centrifuge. The supernatant is decanted, and the precipitate is washed three times with 8 ml of TCA solution, once with chloroform:ether:ethanol, and once with ethanol. The precipitates in one set of tubes are transferred quantitatively into scintillation vials containing 10 ml of scintillation solution and counted in a scintillation counter. One milliliter of 0.2 *M* phosphate buffer is added to the precipitates in a second set of tubes (serving as control), which are then placed in a boiling water bath for 5 minutes. This treatment decomposes the protein-methyl ester. After heating, the proteins are reprecipitated by addition of TCA solution, then washed once with TCA and once with ethanol. Finally, the radioactivity in the precipitate is counted. The difference in the radioactivity between the heated and unheated tubes is taken as the radioactivity due to protein-methyl ester formed by protein methylase.

### 2. Methanol Distillation Method

*a. Apparatus and Reagents*

Distillation apparatus; a combination of a Claison head with a fractionating column and Perkin triangle (Bantam-ware, K-28500, Kontes Glass Co.)
Water bath, 70–73°C
Distillation flask, 25-ml capacity
Receiving flask, 10 ml capacity
Thermometer
HCl, 1 *N*
NaOH, 1 *N*
Methanol
Octanol

*b. Procedure.* The enzyme reaction is stopped by the addition of 0.1 ml of 1 *N* HCl and the tubes are kept in ice for 5 minutes. The pH of the reaction mixture is then adjusted to 7.2–7.4 with 1 *N* NaOH (about 0.1 ml is required). The mixture is transferred quantitatively to a 25-ml distillation flask with the aid of 1 ml of cold water in two portions. Methanol, 0.4 ml, is added as a carrier. The distillation is carried out under a pressure of 253 mm Hg with the distillation flask in a water bath at 70°–73°C. An addition of one drop of octanol to the mixture prevents foaming. The receiving flask is immersed in ice-cold water. Distillation is performed in three 10-minute periods. After the first 10 minutes of distillation, 0.3 ml of methanol is added and the distillation is continued for another 10 minutes. This process is repeated once more. At the end of third distillation, the condenser is rinsed with a small amount of water and drained into the receiving flask. The combined distillates are then diluted to exactly 5.0 ml with water, and 0.5 ml of this solution is counted for its radioactivity.

*c. Comments.* This method is more suitable for purified enzyme in which radioactive methanol is only derived from protein-methyl ester. It has the advantage that the radioactive methanol can be converted into 3,5-dinitrobenzoate. It should also be mentioned that the critical conditions for experimental success of this method lie in maintaining constant pressure and temperature.

### 3. Methanol Extraction Method

*a. Reagents*

Sodium borate buffer, 0.125 *M* (pH 10.0)

Isoamyl alcohol (3-methyl-1-butanol)

*b. Procedure.* The enzyme reaction is stopped by the addition of 1 ml of borate buffer. The tube is reincubated at 37°C for 1 minute and then cooled in an ice bucket. Five milliliters of isoamyl alcohol are added, and the tube is left in ice for at least 5 minutes. The mixture is then vigorously mixed for 15 seconds in a Vortex mixer. The isoamyl alcohol and water layer are then separated by centrifugation in a table-top clinical centrifuge for 2 minutes. After centrifugation, a 2.0 ml portion of the isoamyl alcohol extract is transferred into two vials each, one of which is used to measure radioactive methanol. The contents of the other vial are evaporated to dryness on a steam bath for 30 minutes. After evaporation, the remaining radioactivity is counted. The difference in radioactivity in these two vials is taken as the enzyme activity. When using a partially purified enzyme preparation, the control tube can be prepared with heated enzyme (100°C for 5 minutes), in which case the steam-bath drying step is not necessary.

*c. Definition of Specific Activity.* The specific activity of the enzyme is defined as pmoles of methyl-$^{14}$C incorporated per minute per milligram of enzyme protein.

## Purification Procedure

### A. Reagents

Sucrose solution: 0.25 *M* sucrose–5 m*M* Tris (hydroxymethyl) aminoethane—3 m*M* $CaCl_2$–1 m*M* EDTA–2.4 m*M* 2-mercaptoethanol; pH of the solution is adjusted to 7.0 with NaOH.

Borate buffer: 5 m*M* sodium borate–5 m*M* EDTA–2.4 m*M* 2-mercaptoethanol. (pH 9.3). This buffer is prepared by dissolving 1.9 gm of sodium borate ($Na_2B_4O_7 \cdot 10\ H_2O$), 1.86 gm of EDTA·$Na_2$ and 2.4 m*M* 2-mercaptoethanol in 800 ml of water, adjusting the pH to 9.3 with NaOH, and diluting to 1000 ml.

Monoethanolamine buffer: 10m*M* monoethanolamine–2.5 m*M* EDTA–2.4 m*M* 2-mercaptoethanol (pH 9.7). This is prepared by dissolving 0.61 gm of monoethanolamine, 0.93 gm of EDTA. $Na_2$ and 2.4 m*M* 2-mercaptoethanol in 800 ml of water, adjusting the pH to 9.7, and diluting to 1000 ml.

Saturated ammonium sulfate solution at 20°C

### B. Procedure

All the procedures are carried out at 4°C (*4*).

#### 1. Step 1. Preparation of Supernatant

Freshly frozen, defatted calf thymus, 500 gm, is thawed and homogenized in 5 volumes of sucrose solution. The homogenate is centrifuged at 39,000 *g* (18,000 rpm in a Sorvall RC2-B) for 60 minutes, and the supernatant is passed through a funnel plugged with glass wool.

#### 2. Step 2. pH 5.1 Treatment

Approximately 2000 ml of the supernatant obtained in step 1 is adjusted to pH 5.1 by dropwise addition of 0.2 *N* acetic acid. Care should be taken to avoid overacidification. The suspension is stirred for 10 minutes and centrifuged at 12,000 *g* for 20 minutes.

#### 3. Step 3. Ammonium Sulfate Precipitation

The pH 5.1 supernatant is brought to 45% saturation of ammonium sulfate by the addition of saturated solution over a period of 30 minutes. After stirring for another 15 minutes, the precipitate is recovered by 30-minute centrifugation at 39,000 *g*. The precipitate is suspended in about 100 ml of the borate buffer, and the pH of the suspension is immediately adjusted to 9.0. The solution is then dialyzed overnight against the same buffer.

TABLE I

PURIFICATION OF PROTEIN METHYLASE II FROM[a] CALF THYMUS

| Purification step | Procedure | Volume (ml) | Protein | | Enzyme activity | | Yield (%) | Purification (fold) |
|---|---|---|---|---|---|---|---|---|
| | | | Mg/ml | Total mg | Specific act.[b] | Total act.[c] | | |
| | Whole homogenate | 2500 | 16.8 | 42,000 | 2.0 | 84,000 | 100 | 1 |
| 1 | Supernatant at 39,000 *g* | 2400 | 5.0 | 12,000 | 7.3 | 87,600 | 104 | 3.7 |
| 2 | pH 5.1 treatment | 2350 | 2.3 | 5400 | 12.3 | 66,400 | 79 | 6.2 |
| 3 | $(NH_4)_2SO_4$ supernatant | 70 | 18.5 | 1300 | 38.2 | 49,700 | 59 | 19 |
| 4 | DEAE-Sephadex A-50 | 34 | 6.1 | 207 | 211 | 43,700 | 52 | 106 |
| 5 | QAE-Sephadex A-50 | 40 | 0.83 | 33 | 682 | 22,500 | 26.8 | 341 |
| 6 | Sephadex G-75 | 5 | 0.75 | 3.75 | 2020 | 7560 | 9 | 1010 |
| | Rechromatography by Sephadex G-75 | 6 | 0.13 | 0.78 | 4300 | 3350 | 4 | 2150 |

[a] Starting from 500 g of calf thymus

[b] Specific activity; picomoles of *S*-adenosyl-[*methyl*-$^{14}C$]methionine used per minute per milligram of enzyme protein.

[c] Total activity; picomoles of *S*-adenosyl-L-[*methyl*-$^{14}C$]methionine used per minute.

### 4. Step 4. DEAE-Sephadex Treatment

The DEAE-Sephadex A-50 treatment is carried out by a batchwise elution. The gel is swollen and is washed several times with the borate buffer. To the enzyme preparation obtained in step 3, an equal volume of the washed gel is added and the mixture is stirred slowly for 30 minutes (about 15 mg of protein per milliliter of preswollen gel). Unadsorbed protein is collected, either by centrifugation or by using a sintered-glass funnel, and saved. The gel is washed four times with 50-ml aliquots of the borate buffer containing 0.02 *M* NaCl. All the washings and the unadsorbed protein are combined, lyophilized, and dialyzed against the borate buffer.

### 5. Step 5. QAE-Sephadex Chromatography

A column of QAE-Sephadex A-50 (2.5 × 20 cm) is equilibrated with the monoethanolamine buffer. The dialyzed sample obtained in step 4 is adjusted to pH 9.8 and applied to the column. The column is washed with 50 ml of the monoethanolamine buffer before the elution of the enzyme is carried out with a NaCl gradient using five chambers of a Buchler Varigrad Instrument; The first and second chambers contain 100 ml each of the monoethanolamine buffer, the third chamber 100 ml of 0.1 *M* NaCl in the buffer, the fourth 100 ml of the buffer, and the fifth 100 ml of 0.25 *M* NaCl in the buffer, pH 9.7. Four mililiters of each fraction are collected at a flow rate of 20 ml per hour. If the flow of the column slows down, a peristaltic pump can be placed between the gradient and the column. Of each fraction, 0.02 is assayed for protein methylase II activity. The active fractions are pooled and lyophilized. Figure 1 shows a typical elution profile.

### 6. Step 6. Chromatography on Sephadex G-75

The sample obtained by step 5 is dissolved in 1 ml of borate buffer and applied to a column of Sephadex G-75 (1.0 × 160 cm) previously equilib-

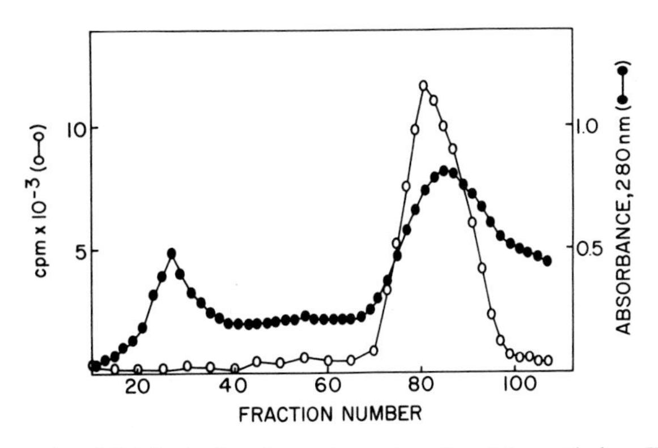


Fig. 1. QAE-Sephadex chromatography of protein methylase II.

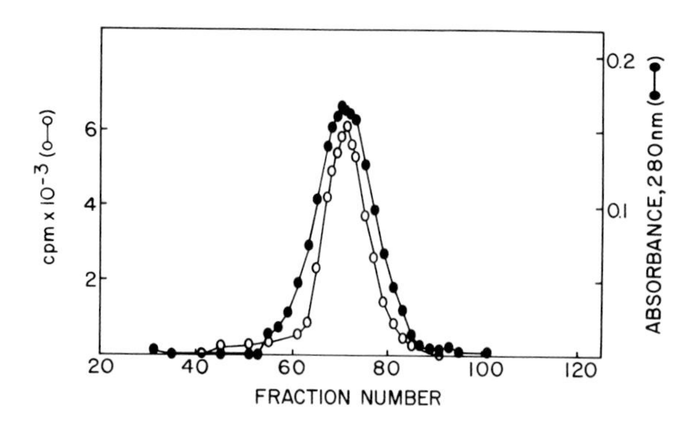


FIG. 2. Sephadex G-75 rechromatography.

rated with the borate buffer. Elution is carried out with the same buffer at a flow rate of 10 ml per hour. One milliliter fractions are collected, and 0.01-ml aliquots are assayed for enzyme activity. The protein peak at this stage of purification does not coincide with the enzyme activity peak. Repetition of step 5 and 6 will considerably improve the purity of the enzyme. Pertinent data on the various steps in a typical procedure are summarized in Table 1. The purified enzyme is homogeneous, showing a single peak on molecular sieve column chromatography (Fig. 2) and a single band on polyacrylamide gel electrophoresis.

## IV. Properties

### A. Stability of the Enzyme

The enzyme is stable in the presence of 50% glycerol at pH 8.0 or in the form of lyophilized powder for at least 2 years. Although acidic pH denatures the enzyme irreversibly, it is stable at pH 8–10 (*4*).

### B. Specificity

The enzyme is highly specific for *S*-adenosyl-L-methionine as the methyl donor. *S*-Adenosyl-L-[ethyl-$^{14}C$]ethionine shows only 2.4% donor efficiency when compared to the methionine derivative. The specificity for polypeptide substrates (methyl acceptor) is very broad. The relative substrate activities of various proteins and polypeptides are shown in Table II. It is noted that denature proteins are far better methyl acceptors; for example, gelatin, F-P-

TABLE II
RELATIVE SUBSTRATE SPECIFICITY FOR PROTEIN METHYLASE II

| Substrate | Relative activity (%)[a] |
|---|---|
| Gelatin | 180 |
| Ribonuclease, (oxidized; bovine pancreas) | 175 |
| γ-Globulin | 110 |
| Histone, type II-A of Sigma | 100 |
| Ribonuclease (native, bovine pancreas) | 88 |
| Cytochrome *c* (horse heart) | 63 |
| Serum albumin (bovine) | 24 |
| Pepsin | 20 |
| Polyglutamic acid | 4 |
| Glutathione (reduced) | 0 |
| Aspartic acid | 0 |
| Glutamic acid | 0 |

[a]Relative activity is based on the activity with histone type II-A of Sigma Chemical Co. as 100%. Ten milligrams of each substrate were used for enzyme assay.

100, and oxidized pancreatic ribonuclease [refer to Kim and Paik (*11, 13*) for further details].

## C. Physical Properties

The purified enzyme from calf thymus has a p*I* value of 4.95. Multiple peaks of the enzyme are found by electrofocusing; p*I* of 5.5, 5.8, 6.0, 6.2, and 6.5 from various sources were obtained (*4, 5*). The enzyme is composed of subunits which reversibly dissociate and associate with guanidinium chloride or urea or with freezing. The most stable and frequently observed size is 25,000, and the smallest size is 16,000. Aggregates of 48,000 have also been noted. All the size classes of the enzyme catalyze the formation of protein-methyl ester.

## D. Catalytic Properties

The optimum pH of the reaction varies depending on the methyl acceptor substrate, although pH 6.0 is observed with most protein substrates (*5*). The Kinetic mechanism of the calf thymus enzyme follows sequential random mechanism (*14*). The $K_m$ value of the enzyme for *S*-adenosyl-L-methionine is 0.9 $\mu M$. The $K_m$ values for methyl acceptor substrates vary greatly depending on the nature of the polypeptide; 0.1 m$M$ for histone, 0.4 m$M$ for pancreatic ribonuclease, 7.7 $\mu M$ for follicle-stimulating hormone, and 3.2 m$M$ for pentapeptide (Phe-Asp-Ala-Ser-Val) (*12*).

## E. Activators and Inhibitors

The enzyme does not require any cofactor. Various metals such as $Ca^{2+}$, $Mn^{2+}$, $Cu^{2+}$, $Zn^{2+}$, $Co^{2+}$, $Fe^{2+}$, $Fe^{3+}$, and $Mg^{2+}$ at 2 m*M* have no effect on the enzyme activity. *p*-Chloromercuribenzoate at 0.4 m*M* inhibits the enzyme activity by 40%; the inhibition can be reversed by 12 m*M* 2-mercaptoethanol (*1*). The enzyme activity is completely lost in 2 *M* urea and 0.5 M guanidinium chloride; however, the enzyme can be reactivated upon dialysis of the denaturant (*4*).

### REFERENCES

*1*. Kim, S., and Paik, W. K., *J. Biol. Chem.* **245**, 1806 (1970).
*2*. Paik, W. K., and Kim, S., *Science* **174**, 114 (1971).
*3*. Diliberto, E. J., Jr., and Axelrod, J., *Proc. Natl. Acad. Sci. U.S.A.* **71**, 1701 (1974).
*4*. Kim, S., *Arch. Biochem. Biophys.* **157**, 476 (1973).
*5*. Kim, S., *Arch. Biochem. Biophys.* **161**, 652 (1974).
*6*. Kim, S., and Paik, W. K., *Biochim. Biophys. Acta* **252**, 526 (1971).
*7*. Axelrod, J., and Daly, J., *Science* **150**, 892 (1965).
*8*. Kim, S., Pearson, D., and Paik, W. K., *Biochem. Biophys. Res. Commun.* **67**, 448 (1975).
*9*. Byvoet, P., *Arch. Biochem. Biophys.* **148**, 558 (1972).
*10*. Kim, S., Wasserman, L., Lew, B., and Paik, W. K., *J. Neurochem.* **24**, 625 (1975).
*11*. Kim, S., and Paik, W. K., *Anal. Biochem.* **42**, 255 (1971).
*12*. Jamaluddin, M., Kim, S., and Paik, W. K. *Biochemistry* **15**, 3077 (1976).
*13*. Kim, S., and Paik, W. K., *Biochemistry* **10**, 3141 (1971).
*14*. Jamaluddin, M., Kims, S., and Paik, W. K., *Biochemistry* **14**, 694 (1975).

# Chapter 8

# *Assessment of Methylation in Nonhistone Chromosomal Proteins*

C. C. LIEW AND D. SURIA

*Department of Clinical Biochemistry,*
*Banting Institute, University of Toronto,*
*Toronto, Ontario, Canada*

## I. Introduction

Methylated amino acids and nucleic acids are known in nature (*1–3*). The first evidence of methylated lysine derivatives in flagella protein of *Salmonella typhimurium* was reported by Ambler and Rees in 1959 (*4*). Subsequently, the identification of ξ-*N*-methyllysine from histones of various species and tissues was demonstrated by Murray (*5*). The extensive studies by Paik and Kim (*1,6–12*) and others (*13–21*) have profoundly established the significance of methylation of protein in most living organisms. So far, the most extensive studies on methylation have dealt with lysine and arginine residues in histone, encephalitogenic basic protein of myelin and contractile proteins (e.g., myosin and actin). The first report in regard to methylation of acid-insoluble nuclear proteins was by Friedman *et al.* in 1969 (*19*).

## II. Properties of Methylated Amino Acids in Proteins

It has been established that methylation of protein is a posttranslational event. The methylation of protein occurs at cytoplasmic and nuclear levels by utilization of methyl donors such as *S*-adenosyl-L-methylmethionine or L-methylmethionine. Incorporation of labeled methylmethionine into protein was inhibited by puromycin. However, when using *S*-adenosyl-L-methylmethionine as a methyl donor, the methylation of protein is insensitive to puromycin. This indicates that *S*-adenosyl-L-methylmethionine acted only as the methylating agent of the protein, but did not participate in the protein synthesis (*1*).

Three types of methylase have been specifically determined (*1*). Protein methylase I (*S*-adenosylmethionine: protein-arginine methyltransferase; EC 2.1.1.23) methylates the guanidine group of arginine residues. Protein methylase II (*S*-adenosylmethionine:protein-carboxyl methyltransferase; EC 2.1.1.24) methylates the free carboxyl group of aspartyl and glutamyl residues. Protein methylase III (*S*-adenosylmethionine:protein-lysine methyltransferase; EC 2.1.1.25) methylates the $\xi$-amino group of lysine residues. A system that allows a simultaneous assay in the same homogenate of these three protein-specific methyltransferases has been reported (*1*). Protein methylases I and II are present in tissue homogenate whereas protein methylase III is exclusively in the nucleus (*8*). Since the methylation involves specific enzymes, and the occurrence of methylated amino acids in nature is extremely diverse, such postsynthetic enzymic modification, might have far reaching effects on the control of chromosomal protein function.

## III. Choice of Methods for the Study of the Methylation of Nonhistone Chromosomal Proteins (NHCP)

### A. Isolation of NHCP

Nuclei were isolated as described in Chapter 4. Preparation of chromatin was according to the method of Reeder (*22*). Solubilization methods of NHCP could be sodium dodecyl sulfate extraction (*23*), high salt dissociation (*24*) or phenol extraction (*25*). The phenol method which was chosen for the solubilization of NHCP has been discussed in the previous chapter. Besides, it is well known that methylation occurs in DNA and RNA of all species (*1,3,26*). Since some NHCP are tightly bound to nuclei acids, phenol extraction provides an excellent solvent for extracting and dissociating NHCP from nucleic acids.

### B. Identification of Methylated Amino Acids

Methods for the analysis of the amino acid residues modified by methylation are now available (*1,5,27,28*). Since these modified amino acids are stable after acid hydrolysis, the acid hydrolysate of the modified protein could be analyzed by an automated amino acid analyzer as described by Paik and Kim (*1*). The method was simplified as follows: A column (0.9 $\times$

35 cm) was packed with Bio-Rad A-5 resin (p 13 $\pm 2$ $\mu$m diameter) using a Perkins-Elmer KLA-3B amino acid analyzer. The protein hydrolysate was first eluted with 0.38 *N* sodium citrate buffer (pH 5.84) at 24°C at a flow rate of 45 ml per hour. After 2 hours (or at the position of tyrosine or phenylalanine) the second buffer of 0.35 *N* sodium citrate (pH 4.70) was introduced. As indicated by the investigators, methylated lysines ($\xi$-*N*-mono-, $\xi$-*N*-di-, and $\xi$-*N*-trimethyllysine), 3-*N*-methylhistidine, and methylated arginines ($N^G$, $N^G$-di, $N^G$, $N'^G$-di, and $N^G$-monomethylarginine) were well separated, except for 1-*N*-methylhistidine overlapping with ammonia. Huszar (*27*) also developed a method similar to that of Paik and Kim. He digested the methylated myosin with trypsin, and then the radioactive methyl groups containing peptides were fractionated on a Dowex-50 column. The methylated amino acid residues were subsequently identified with an automated amino acid analyzer. The analysis of the methylated amino acid residues by the amino acid analyzer was supposed to be superior to high-voltage electrophoresis (*29*) or paper chromatography (*28*), which does not resolve some methylated amino acid residues.

## IV. *In Vivo* Methylation of NHCP

So far, there is no information in regard to methylation of NHCP *in vivo*. We intend to present some evidence on *in vivo* methylation of NHCP in different tissues which are known to be metabolically active, using the proliferating cell type, such as liver and kidney, versus the nonproliferating cell type, such as heart. Either L-methylmethionine or *S*-adenosyl-L-methylmethionine can be used as the methyl donor. However, owing to the high turnover rate of NHCP (*30*), and the finding that only *S*-adenosyl-L-methylmethionine acts as the methylating agent of protein and does not participate in peptide formation (*24*), *S*-adenosyl-L-(*methyl*-$^3$H)methionine was selected as the labeled donor in the study of the *in vivo* methylation of NHCP in different tissues.

Female white Wistar rats, weighing 200 gm, were each injected intraperitoneally with 250 $\mu$Ci of *S*-adenosyl-L-[methyl-$^3$H]methionine (specific activity 8.82 Ci/nmole) for 1 hour of pulse-labeling prior to killing. Nuclei were isolated from liver, kidney, and heart tissues. Chromatin was then prepared by Reeder's method (*22*). Histone and NHCP were fractionated as described in Chapter 4. The specific activities of histone and NHCP which were obtained from heart, liver, and kidney are shown in Table I. It was found that incorporation of $^3$H-methyl-labeled groups into NHCP was signi-

TABLE I

SPECIFIC ACTIVITY OF RAT HISTONE AND NONHISTONE CHROMOSOMAL PROTEINS (NHCP) FOLLOWING ADMINISTRATION OF *S*-ADENOSYL-L-[methyl-$^3$H]-METHIONINE *in Vivo*[a]

| Tissue | Histones (cpm) | NHCP (cpm) |
|---|---|---|
| Heart | 392 | 2040 |
| Kidney | 29,600 | 35,900 |
| Liver | 1250 | 3210 |

[a]The values, expressed as counts per minute per milligram of protein, are the means of two separate experiments.

ficantly higher than into histone. NHCP from kidney were the most highly methylated as compared to the proteins of heart and liver cells. However, the significance of methylation in kidney NHCP remains unknown. The kidney NHCP were subsequently fractionated according to their isoelectric points as reported previously (*31*). This system permits further separation of nonhistones from histones, which have p*I* values greater than 9.5, if histones were not completely removed during the acid extraction. As shown in Fig. 1, most of the NHCP could be methylated. However, two fractions of acidic proteins, which were focused at pH 6.2 and 6.5, were highly methylated.

## V. Concluding Remarks

Covalent modification of proteins (e.g., acetylation, phosphorylation, and methylation) has been implicated in some regulatory role(s) in terms of cellular activity. Characterization and identification of amino acid residues in NHCP which are modified by these reactions have yet to be clearly delineated. In general, the covalent modification of proteins should be independent of polypeptide synthesis. Studies of phosphorylation of NHCP have been extensively investigated in an attempt to explore the cause-and-effect relationship in gene regulation (*32–34*). However, acetylation and methylation of NHCP in relation to gene expression have remained relatively unexplored. One report indicated that methylation of NHCP might correlate with DNA replication in salivary gland by use of autoradiographic studies (*21*). Others demonstrated the incorporation of [*methyl*-$^{14}$C]methio-

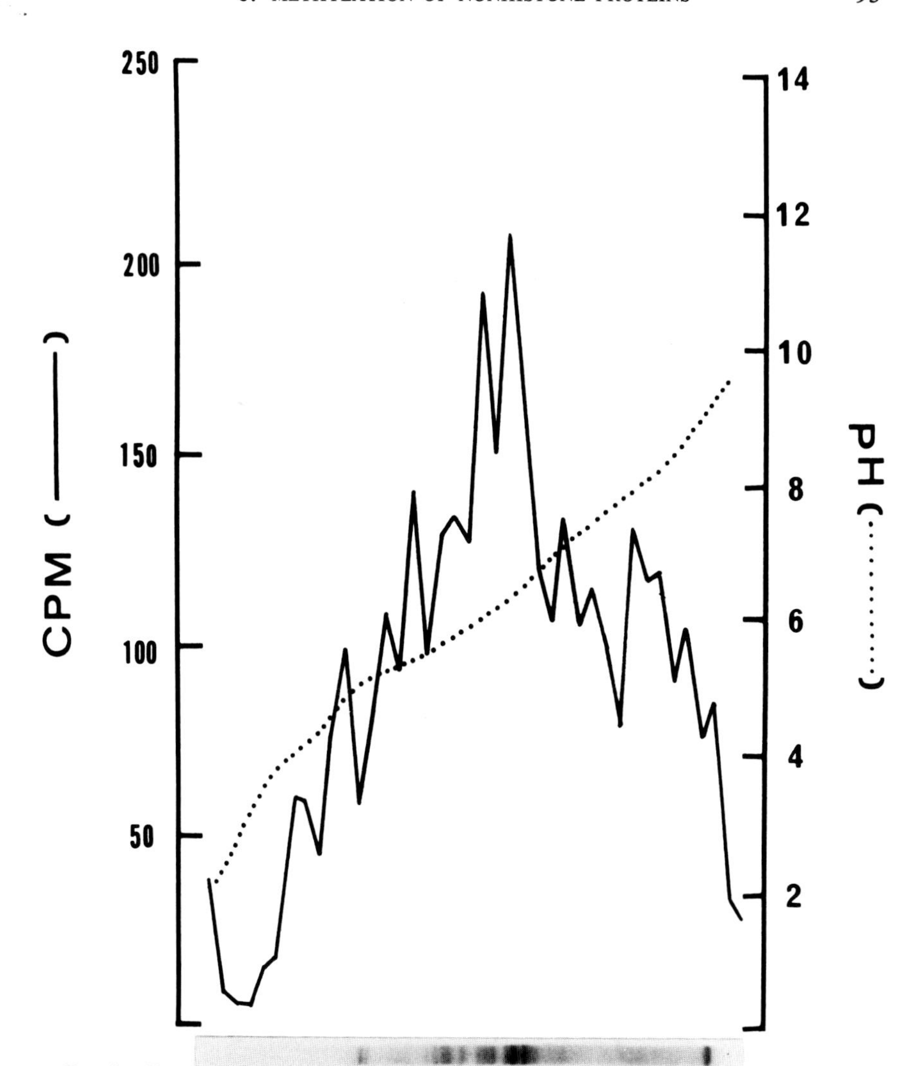


FIG. 1. Fractionation of $^3$H-methylated kidney nonhistone chromosomal proteins (NHCP) by isoelectrofocusing polyacrylamide gel electrophoresis. Kidney NHCP (100 $\mu$g), which were labeled with *S*-adenosyl-L-[*methyl*-$^3$H]methionine *in vivo*, were fractionated by isoelectrofocusing gel electrophoresis. The gel was sliced into 2-mm sections, and radioactivity incorporated into the fractionated proteins was determined. The solid line indicates the $^3$H radioactivity, and the broken line the pH gradients of the gel.

nine into acid-insoluble nuclear proteins after 4 days of consecutive injection in rats (*19*). Although extensive studies of methylation in histones are well demonstrated, the role(s) of this postsynthetic modification of NHCP has yet to be elucidated.

We presented the *in vivo* incorporation of $^{3}H$ groups from *S*-adenosyl-L-[methyl-$^{3}H$]methionine into phenol-soluble NHCP and subsequently fractionated the $^{3}H$-methyl-labeled proteins from kidney by isoelectrofocusing. These results together with the established procedures developed for the investigation of chromosomal proteins may provide some interest for further exploration of the structure and function of NHCP.

## References


1. Paik, W. K., and Kim, S., *Adv. Enzymol.* **42**, 227 (1975).
2. Starr, J. L., and Sells, B. H., *Physiol. Rev.* **49**, 623 (1969).
3. Srinivasan, P. R., and Borek, E., *Prog. Nucleic Acid Res. Mol. Biol.* **5**, 157 (1966).
4. Ambler, R. P., and Rees, W. M., *Nature (London)* **184**, 56 (1959).
5. Murray, K., *Biochemistry* **3**, 10 (1964).
6. Kim, S., and Paik, W. K., *J. Biol. Chem.* **240**, 4629 (1965).
7. Paik, W. K., and Kim, S., *Biochem. Biophys. Res. Commun.* **27**, 479 (1967).
8. Paik, W. K., and Kim, S., *J. Neurochem.* **16**, 1257 (1969).
9. Paik, W. K., and Kim, S., *J. Biol. Chem.* **245**, 88 (1970).
10. Paik, W. K., and Kim, S., *Science* **174**, 114 (1971).
11. Paik, W. K., Lee, H. W., and Lawson, D., *Exp. Gerontol.* **6**, 271 (1971).
12. Paik, W. K., Lee, H. W., and Morris, H. P., *Cancer Res.* **32**, 37 (1972).
13. Allfrey, V. G., Faulkner, R., and Mirsky, A. E., *Proc. Natl. Acad. Sci. U.S.A.* **51**, 786 (1964).
14. Asatoor, A. M., and Armstrong, M. D., *Biochem. Biophys. Res. Commun.* **26**, 168 (1967).
15. Baldwin, G. S., and Carnegie, P. R., *Science* **171**, 579 (1971).
16. Byvoet, P., *Biochim. Biophys. Acta* **238**, 375a (1971).
17. Duerre, J. A., and Chakrabarty, S., *J. Biol. Chem.* **250**, 8457 (1975).
18. Elzinga, M., *Biochemistry* **10**, 224 (1971).
19. Friedman, M., Shull, K. H., and Farber, E., *Biochem. Biophys. Res. Commun.* **34**, 857 (1969).
20. Gershey, E. L., Haslett, G. W., Vidali, G., and Allfrey, V. G., *J. Biol. Chem.* **244**, 4871 (1969).
21. Goodman, R. M., and Benjamin, W. B., *Exp. Cell Res.* **77**, 63 (1973).
22. Reeder, R. H., *J. Mol. Biol.* **80**, 229 (1973).
23. Elgin, S. C. R., and Bonner, J., *Biochemistry* **22**, 4440 (1970).
24. Levy, S., Simpson, R. T., and Sober, H. A., *Biochemistry* **11**, 1547 (1972).
25. Teng, C. S., Teng, C. T., and Allfrey, V. G., *J. Biol. Chem.* **246**, 3597 (1971).
26. Perry, R. P., and Kelley, D. E., *Cell* **1**, 37 (1974).
27. Huszar, G., *J. Mol. Biol.* **94**, 311 (1975).
28. Kakimoto, Y., and Adazawa, S., *J. Biol. Chem.* **245**, 5751 (1970).
29. Reporter, M., and Corbin, J. L., *Biochem. Biophys. Res. Commun.* **43**, 644 (1971).
30. Stein, G. S., Spelsberg, T. C., and Kleinsmith, L. J., *Science* **183**, 817 (1974).
31. Suria, D., and Liew, C. C., *Can. J. Biochem.* **52**, 1143 (1974).
32. Allfrey, V. G., Inoue, A., Karn, J., Johnson, E. M., and Vidali, G., *Cold Spring Harbor Symp. Quant. Biol.* **38**, 785 (1973).
33. Gornall, A. G., and Liew, C. C., *Adv. Enzyme Regul.* **12**, 267 (1974).
34. Kleinsmith, L. J., *in* "Acidic Proteins of the Nucleus" (I. L. Cameron and J. R. Jeter, Jr., eds.), p. 103. Academic Press, New York (1974).

# Chapter 9

# *Methods for the Assessment of Histone Methylation*

PAUL BYVOET AND DAVID F. SAYRE

*Department of Pathology, University of South Florida College of Medicine and Veterans Administration Hospital, Tampa, Florida*

C. STUART BAXTER

*Department of Environmental Health, Kettering Laboratory, University of Cincinnati Medical Center, Cincinnati, Ohio*

## I. Introduction

*N*-methylation of internal histone lysine residues in nuclear chromatin occurs during the last part of S phase and $G_2$ (*1–3*). It involves specific lysine residues in histones H3 and H4, which have been identified as residues 9 and 27 in calf thymus histone H3, and as residue 20 in H4. The methylation reaction is catalyzed by histone lysine methyltransferase(s) and utilizes *S*-adenosylmethionine as methyl donor. The enzyme has been characterized and purified from cell nuclei by Paik and Kim (*4*), who called it methylase III.

## II. Principle

Uptake of radiomethyl into histones can be measured *in vivo* and *in vitro*. *In vitro* measurements can be carried out with isolated nuclei or with the isolated histone methyltransferase, which will methylate free histones in the presence of *S*-adenosylmethionine. This latter system is discussed by Paik and Kim (*5*); see also this volume, Chapter 7. For *in vitro* protocols using isolated nuclei, *S*-adenosylmethionine (Me-$^{3}$H-labeled) should also be used as methyl donor.

For *in vivo* experiments, however, methyl-labeled methionine should be employed. This label has the disadvantage of being available to the cellular synthetic machinery in the form of labeled methionine, as well as labeled *S*-adenosylmethionine. Histones (and other proteins) will therefore become labeled owing to incorporation of labeled methionine and/or radiomethyl.

## III. *In Vivo* Assessment of Histone Methylation

After administration of the appropriate amount of radioactivity in the form of Me-$^3$H or $^{14}$C-labeled l-methionine, tissues are isolated or cells are harvested after the desired time interval.

Nuclei are isolated according to the usual methods (*6*) and washed several times with 20–30 volumes of saline–EDTA [0.08 *M* NaCl–0.02 *M* EDTA (pH 7.4)] by suspension with a Potter–Elvehjem homogenizer and spinning at 2000 *g* for 10 minutes. The pellet can then be extracted according to the method of Johns (*7,8*). Whole histones can be prepared by extracting twice with 4–5 volumes of 0.25 *N* HCl and clarifying the combined extracts at 12,000 *g* for 15 minutes. Histones are precipitated by addition of 6 volumes of acetone and stored overnight at −20°C. The precipitate is collected by centrifugation at 2000 *g* for 15 minutes and subsequently resuspended in acetone–1% HCl and acetone, with centrifugations at 2000 *g* for 15 minutes, and finally air dried.

This histone preparation can then be fractionated by P-60 gel chromatography according to von Holt and co-workers (*9*). This method proved to be very useful in our hands, although alkylation, according to Dixon and Candido (*10*), is recommended to prevent aggregation of those H3 (e.g., calf thymus) molecules that contain two −SH groups. Histones (up to 20 mg) are dissolved in 0.4 ml of 5 *M* urea–20 m*M* dithiothreitol–0.1 *M* sodium borate buffer (pH 8.9) and incubated at room temperature for 90 minutes; then solid iodacetamide is added to a concentration of 0.08 *M*, and incubation is continued for another 90 minutes.

## IV. Uptake of Radiomethyl into Methylated Lysine Derivatives

To distinguish radioactivity due to actual incorporation of radiomethyl into methyllysine derivatives from that due to incorporation of the amino acid methionine, some form of amino acid chromatography is required. In the simplest form, this can be performed on Dowex W-X8 in the ammonia form. After hydrolysis of the histone sample as described below, it is charged to the column (0.9 × 3 cm) in dilute acid (HCl, pH 2.0). Neutral and acidic amino acids are then eluted with water (10 ml) and basic amino acids (containing the radiomethyl) with 3 *N* $NH_4OH$ (5 ml). The radioactivity of the water fraction, which stems from methionine, can be useful as a measure of histone synthetic activity (*11*).

If, in addition, it is desirable to estimate incorporation of radiomethyl into *N*-mono-, di-, and trimethyllysine, more extensive chromatography is indicated:

Histone fractions are hydrolyzed in 5.9 *N* HCl at 110°C for 16 hours in $N_2$-flushed, sealed glass ampoules. The HCl is removed by cryogenic transfer from the hydrolyzates which are then dissolved in dilute HCl of pH 2. A concentrated sucrose solution is added to bring the concentration to 10–20%. The sample is then layered on a 0.9 × 50 cm AA-15 resin (Beckman) column, equilibrated with 0.35 *N* citrate buffer (pH 6.95), which is also the eluting buffer (*12*). Pressure of elution is 17 atm; flow rate is approximately 30 ml/per hour. The column jacket is kept at 28° ± 0.02°. Fractions can be analyzed by ninhydrin, e.g., by means of the Technicon Automated Amino Acid Analyzer detection system. The effluent can be mixed with liquid scintillation cocktail (TX-100, toluene; 1:2, 5.5 gm of PPO, 0.1 gm of POPOP per liter) without noticeable quenching, except in very highly con-

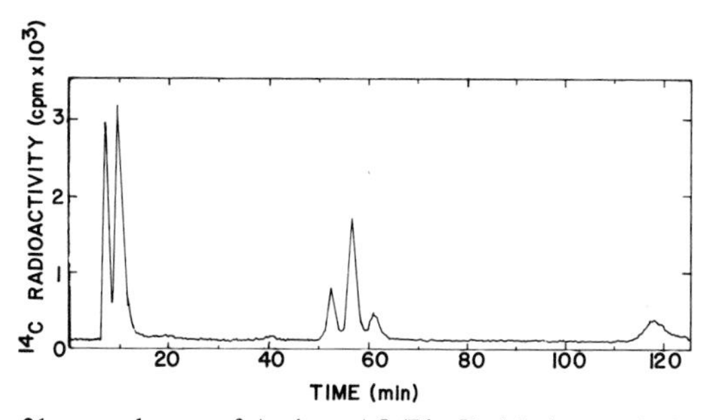


FIG. 1. A 1 × 21 cm column of Aminex A5 (Bio-Rad Laboratories) was eluted at 22°C with 0.35 *N* citrate buffer (pH 6.48) at a flow rate of 1 ml/min (*13, 14*). The column effluent was directed through an anthracene crystal scintillation detector cell installed in a two-channel scintillation spectrometer adapted for flow operations (Packard Tricarb Flow System). The rate meter output for both channels, each set at different full-scale ranges, was recorded using a two-channel linear recorder. The first two major peaks to emerge were tentatively identified as methionine and methionine sulfoxide, respectively. The sulfoxide-derivative peak was considerably diminished if hydrolyses were performed in an inert atmosphere. The next triad of peaks represents monomethyl-, dimethyl-, and trimethyl lysine, respectively. The last peak, emerging near the position of arginine and at the reported position of $N^G$-monomethyl arginine (*13*), was identified as this monomethyl derivative of arginine.

centrated fractions. Since the methylated derivatives are present in low concentration, this generally does not present a problem.

Alternatively, a stream splitting device can be utilized to divert part of the effluent through an analytical system based on the ninhydrin reaction or one of the more sensitive fluorometric detection procedures now available which utilize Fluorescamine® (Roche), or Fluoropa® (Durrum). Radioactivity can of course also be measured continuously in a flow cell, as demonstrated in Fig. 1 (*13, 14*). The insensitivity of this type of detection, however, requires a high level of radioactivity.

In addition to the procedures detailed above, other methods, specifically designed for the separation of methylated lysines, have been described (*15–17*). Most of these methods can also be used to detect methylarginine and methylhistidine, the presence of which has been reported in histones and nonhistone chromosomal proteins (*14, 18–20*).

## V. *In Vitro* Assessment of Histone Methylation

The *in vitro* methylation of histones in isolated nuclei seems for all practical purposes to be very similar to the *in vivo* process. Incorporation takes place only into H3 and, to a lesser extent, into H4, in the form of mono- di-, and trimethyllysine. The main advantage is that, under these conditions, *S*-adenosylmethionine can be used as the radiomethyl donor, and thus incorporation of radioactivity into histones is solely due to *N*-methylation. It is necessary, however, to purify the labeled histones following incubation in the presence of *S*-adenosylmethionine, since under these conditions, active methylation of DNA and nonhistone proteins is taking place as well.

Finally, it should be emphasized that, in a tissue such as rat liver, only a small percentage of the total cell population will incorporate radiomethyl into its histones at any given time, and that the phenomena observed may well be restricted to only those (premitotic) cells in which histone methylation is taking place.

The following is a procedure used in the author's laboratory for measurement of radiomethyl uptake into histones from rat liver nuclei incubated in the presence of methyl-labeled *S*-adenosylmethionine.

Nuclei from 0.5 gm of rat liver, fresh or stored at −60°C with an overlay of 2.1 *M* sucrose–3 m*M* $CaCl_2$, are incubated at 37°C with 2.5 μCi *S*-adenosyl-[*methyl*-$^3$H] L-methionine (8 Ci/mmole) for 15 minutes in a mix containing 0.25 *M* sucrose–0.05 *M* Tris-HCl–0.02 *M* KCl–0.01 *M* $MgCl_2$–0.01 *M* mercaptoethanol (pH 8.5) (at 37°C) (total volume 0.5 ml).

Nuclei from 10 gm liver are gently suspended in a mixture of 2 ml of quintuply concentrated buffer (i.e., 1.25 *M* sucrose–0.25 *M* Tris/HCl–0.1 *M* KCl–0.05 *M* $MgCl_2$–0.05 *M* mercaptoethanol (pH 8.5) (37°C) and 4 ml of water. For each individual incubation mixture, 300-$\mu$l aliquots of this suspension are used, and duplicates are run in each case. Water is added to make the volume up to 400 $\mu$l; 50 $\mu$l of *S*-adenosylmethionine solution (100 $\mu$Ci/2 ml) is added, the mixture is shaken, and the label is rinsed down with 50 $\mu$l of water.

Mixtures are then incubated with shaking at 37°C for 15 minutes. The reaction is stopped by chilling and addition of 2.5 ml NaCl/EDTA (0.08:0.02 *M*) (pH 7.2). The mixtures are centrifuged at 4°C at top speed in a Sorvall GLC-1 centrifuge (3200 rpm = 2000 *g*) for 10 minutes. The pellet is washed twice more with NaCl/EDTA, centrifuging at 2000 *g* after washing, and then extracted with 0.25 *N* HCl (0.6 ml), followed by centrifugation, then extracted further with 0.3 ml of 0.25 *N* HCl. The extracts are combined, centrifuged, and made 25% trichloroacetic acid by addition of 0.3 ml of 100% (w/v) trichloroacetic acid. The tubes are left overnight at 4°C, centrifuged at 2000 *g* for 20 minutes; the residue is washed consecutively with 1% HCl in acetone (1 ml), acetone (1 ml), and ether (1 ml), each step being followed by centrifugation at 2000 *g* for 10 minutes. The residue is dried under vacuum for 30–60 minutes, dissolved in 1 ml of water and stirred vigorously. The tubes are allowed to stand at 4°C for 30 minutes, stirred again (on Vibromixer), and centrifuged at 2000 *g* for 10 minutes; aliquots are tested for radioactivity by addition to 10 ml of toluene-based scintillation cocktail described above. Protein concentration can be measured by either the Hartree method (*21*) or fluorometry (*22*). It is advisable to use the specific radioactivity of histones with respect to radiomethyl if substances other than those listed above are added to the incubation mixture, since they may influence the yield of histone. Some loss of chromatin generally occurs during incubation, probably owing to nuclease activity, and this activity may be inhibited or stimulated by certain conditions. Finally, alternative methods of incubation have been described (*23–25*) with slight variations from the one detailed above.

## REFERENCES


1. Tiddwell, T., Allfrey, V. A., and Mirsky, A. E., *J. Biol. Chem.* **243**, 707 (1968).
2. Shepherd, G. R., Hardin, J. M., and Noland, B. J., *Arch. Biochem. Biophys.* **143**, 1 (1971).
3. Lee, H. W., Paik, W. K., and Borun, T. W., *J. Biol. Chem.* **248**, 4194 (1973).
4. Paik, W. D., and Kim, S., *Science* **174**, 114 (1971).
5. Paik, W. D., and Kim, S., *J. Biol. Chem.* **245**, 6010 (1970).
6. L. Grossman and K. Moldave, eds., "Methods in Enzymology," Vol. 12, Part A, Sect. III. Academic Press, New York, 1967.
7. Johns, E. W., *Biochem. J.* **92**, 55 (1964).

8. Oliver, D., Sommer, K. R., Panyim, S., Spiker, S., and Chalkley, R., *Biochem. J.* **129**, 349 (1971).
9. Bohm, E. L., Strickland, W. N., Strickland, M. J., Thwaits, B. H., van der Westhuyzen, D. R., and von Holt, C., *FEBS Lett.* **34**, 217 (1973).
10. Candido, E. P. M., and Dixon, G. H., *J. Biol. Chem.* **247**, 3868 (1972).
11. Byvoet, P., *Arch. Biochem. Biophys.* **152**, 887 (1972).
12. Lange, H. W., Lower, R., and Hempel, K., *J. Chromatogr.* **76**, 252 (1973).
13. Seely, J. H., Edattel, S. R. and Benoiton, N. L., *J. Chromatogr.* **44**, 618 (1969).
14. Byvoet, P., Shepherd, G. R., Hardin, J. M., and Noland, B. J., *Arch. Biochem. Biophys.* **148**, 558 (1972).
15. Duerre, J. A., and Chakrabarty, S., *J. Biol. Chem.* **250**, 8457 (1975).
16. Honda, B. M., Dixon, G. H., and Candido, E. P. M., *J. Biol. Chem.* **250**, 8681 (1975).
17. Deibler, G. E., and Martenson, R. E., *J. Biol. Chem.* **248**, 2387 (1973).
18. Gershey, E. L., Haslett, G. W., Vidali, G., and Allfrey, V. G., *J. Biol. Chem.* **244**, 4871 (1969).
19. Paik, W. K., and Kim, S., *Biochem. Biophys. Res. Commun.* **40**, 224 (1970).
20. Byvoet, P., *Biochim. Biophys. Acta* **238**, 375a (1971).
21. Hartree, E. F., *Anal. Biochem.* **48**, 422 (1972).
22. Bohlen, P., Stein, S., Dairman, W., and Udenfriend, S., *Arch. Biochem. Biophys.* **155**, 213 (1973).
23. Lee, C. T., and Duerre, J. A., *Nature* (*London*) **251**, 240 (1974).
24. Sekeris, C. E., Sekeris, K. E., and Gallwitz, D., *Hoppe-Seyler's Z. Physiol. Chem.* **348**, 1660 (1967).
25. Thomas, G., Lange, H. W., and Hempel, K., *Hoppe-Seyler's Z. Physiol. Chem.* **353**, 1423 (1972).

# Chapter 10

# *Purification and Assay of Nuclear Protein Kinases*

VALERIE M. KISH

*Department of Biology, Hobart and William Smith Colleges, Geneva, New York*

LEWIS J. KLEINSMITH

*Division of Biological Sciences, University of Michigan, Ann Arbor, Michigan*

## I. Introduction

The phosphorylation of nonhistone chromatin proteins is thought to play a key role in the regulation of gene activity in eukaryotic cells (for reviews, see *1–4*). This chapter is concerned with methods for fractionation and assay of the chromatin-associated protein kinases involved in catalyzing the phosphorylation of these proteins. The general approach employed involves purification of phosphorylated nonhistone chromatin proteins via salt extraction, chromatography on phosphocellulose columns involving a salt/pH step gradient, and assay for protein kinase activity in the absence of exogenous substrate.

## II. Fractionation Procedure

### A. Preparation of Phosphocellulose Columns

Ten grams of phosphocellulose (Whatman P-11) are suspended in 300 ml of 0.5 *N* NaOH and stirred for 30 minutes at 22°C. The fines are removed by suction, and the slurry is transferred to a small Büchner funnel fitted with a circular disk of Whatman No. 1 filter paper. The phosphocellulose is washed on the funnel with distilled water until the pH of the filtrate is 8 as tested by pH paper. The cake is then transferred to another beaker and stirred with 300 ml of 0.5 *N* HCl for 30 minutes at 22°C. The resulting slurry is again washed on the Büchner funnel with water until the filtrate reaches a pH of approximately 4. The phosphocellulose is then suspended in 0.5 *N* NaOH and stirred, and washed again to pH 8 as described above. The resulting cake is suspended in 50 m*M* Tris-HCl (pH 7.5)–0.3 *M* NaCl, and the pH of the solution is carefully adjusted to 7.5 using concentrated HCl. The mixture is stirred at 22°C for several hours, and the pH is checked at intervals. Phosphocellulose washed in this manner is immediately used to pack the column. The excess cellulose is stored at 4°C in a tightly closed screw-capped bottle and can be used for packing subsequent columns within the next week. The day before use, the phosphocellulose is allowed to come to room temperature and the pH is checked prior to packing.

The best separations are obtained with freshly washed preparations of phosphocellulose. Storage of washed phosphocellulose at 4°C for periods of more than 7–10 days results in protein profiles which are not as sharply resolved as those resulting from chromatography on freshly washed cellulose. Washed phosphocellulose is slurried in 50 m*M* Tris-HCl (pH 7.5)–0.3 *M* NaCl in a ratio of approximately 1:2. A 0.9 × 15 cm column is filled approximately halfway with 50 m*M* Tris-HCl (pH 7.5–0.3 *M* NaCl. After removing bubbles from the bottom section, the bottom of the column, which is fitted with a 2-inch section of polyethylene tubing, is clamped off and phosphocellulose is pipetted into the column and allowed to settle for 5 minutes. The phosphocellulose should be finely dispersed, with no visible chunks of material present. The bottom is then slowly opened over a period of 10 minutes, and additional slurry is added as the meniscus recedes from the top of the column. When the level of packed phosphocellulose has reached a height approximately 1 cm from the top of the column, the bottom is closed off and the column is replumbed in the cold room. An elution flask, fitted with a capillary tube and containing starting buffer at room temperature, is then attached to the column at a differential of 65–70 cm, and the column bottom is opened. After approximately 2 hours the differential is lowered to approximately 4 cm, and the column is allowed to temperature equilibrate at 4°C overnight. The column is washed with 75–100 ml of the starting buffer, and the optical density and pH of the eluate are checked prior to protein application. Changes in the permeability of the upper surface of the phosphocellulose usually occur during the overnight wash, as evidenced by the obvious clumping of the cellulose in the top 2–5 mm. To avoid irregular bands caused by uneven permeability of the surface, the top few centimeters of adsorbent are stirred up and removed, leaving a smooth, flat surface. The height of the bed is routinely adjusted to 12–

12.2 cm. It should be pointed out that column flow should not be interrupted at any time from the beginning of the buffer wash to the end of fraction collecting.

## B. Extraction and Fractionation of Nuclear Protein Kinases

The nonhistone chromatin phosphoprotein fraction, with its associated protein kinase activities, is prepared starting from purified nuclei. The soluble proteins of the nuclear sap are removed by suspending the nuclei in 10 m*M* Tris-HCl (pH 7.5) for 20 minutes. The nuclei are collected by centrifugation at 10,000 *g* for 10 minutes, and are then suspended in 1.0 *M* NaCl–20 m*M* Tris-HCl (pH 7.5) to a final protein concentration of 2 mg/ml. The suspension is dispersed using a Polytron homogenizer (Brinkmann) for 20 seconds at setting 3. The resulting viscous solution is mixed with 1.5 volumes of 20 m*M* Tris-HCl (pH 7.5) and the precipitated nucleohistone is removed by centrifugation at 200,000 *g* for 2 hours. Bio-Rex 70 ($Na^+$), which has previously been equilibrated with 0.4 *M* NaCl–20 m*M* Tris-HCl (pH 7.5), is then added to the supernatant at a ratio of 20 mg of Bio-Rex per milligram of protein. After stirring slowly for 10 minutes, the suspension is centrifuged for 10 minutes at 6000 *g* and the supernatant is withdrawn. The resin is washed by suspending it in 10–15 ml of 0.4 *M* NaCl–20 m*M* Tris-HCl (pH 7.5) and centrifuging again for 10 minutes at 6000 *g*. The two supernatants are then combined, and calcium phosphate gel is added at a ratio of 0.46 mg of gel per milligram of protein. After slowly stirring for 20 minutes, the suspension is centrifuged for 5 minutes at 6000 *g*, and the supernatant is discarded. The gel is washed by resuspension in 10–15 ml of 1.0 *M* $(NH_4)_2SO_4$–0.05 *M* Tris-HCl (pH 7.5) using a motor-driven stirrer at 1000 rpm, followed by centrifugation at 6000 *g* for 5 minutes. The supernatant is again discarded, and the gel is then dissolved by gentle homogenization with a Teflon Potter–Elvehjem tissue grinder in 0.3 *M* EDTA (pH 7.5)–0.33 *M* $(NH_4)_2SO_4$ in a ratio of 0.2 ml of solution per milligram of gel. The suspension is allowed to stand for 1 hour in the cold with occasional rehomogenization, and the insoluble residue is then removed by centrifugation for 15 minutes at 33,000 *g*. The supernatant, enriched in nonhistone phosphoproteins and protein kinase activity, is dialyzed overnight against 50 m*M* Tris-HCl (pH 7.5). After concentration at 50 psi of nitrogen in an Amicon ultrafilter equipped with a UM-10 membrane, the proteins are either frozen at −25° C, or are immediately dialyzed for 12–15 hours at 4° C against 100 volumes of 50 m*M* Tris-HCl (pH 7.5)–0.3 *M* NaCl in preparation for fractionation via phosphocellulose chromatography.

The phosphocellulose fractionation procedure is a modification of the technique reported by Takeda *et al.* (5). All procedures are carried out at 4 °C unless otherwise noted. Approximately 1.5–3 mg of the phosphoprotein fraction (670–980 μg/ml) is applied by pipette directly to the top of the phosphocellulose column after the last of the starting buffer enters the adsorbent. The entire surface is covered with sample quickly to permit uniform penetration of the sample into the bed. When the last of the sample enters the cellulose, the surface of the bed and the column wall above it are washed with three 0.5-ml portions of starting buffer, each wash being permitted to sink into the cellulose before the next is applied. After the last buffer wash enters the cellulose, the column is filled and the elution flask containing starting buffer is attached. Fractions of 1.0–1.2 ml are collected at 15–20 ml/hour at a constant differential of 24–30 cm until 25–28 ml of starting buffer have passed through the column. The buffer above the bed is then changed to 50 m*M* Tris (pH 8.1)–0.6 *M* NaCl, and elution is continued for another 34–36 ml. At a total volume of 64–66 ml the last step in the gradient is initiated using 50 m*M* Tris-HCl (pH 8.1)–1.0 *M* NaCl, and an additional 30–35 ml are collected. The protein content of each fraction is measured by ultraviolet adsorbance at 280 nm prior to enzyme assay. Since the protein kinases are extremely labile after fractionation (storage at −70 °C or overnight at 2 °C results in a significant loss of activity), the enzyme assay is carried out immediately.

## C. Assay of Protein Kinase Activity

Protein kinase activity in each fraction is measured in the presence of endogenous substrate immediately after the protein profile is read. The reaction mixture of 0.3 ml contains 13 μmoles of Tris-HCl (pH 7.5); 0.1–2 nmoles of [$\gamma$-$^{32}$P] ATP (770–3650 mCi/mmole); 7.5 μmoles of magnesium acetate, and 0.2 ml of each column fraction. When 3′,5′-cyclic adenosine monophosphate (cAMP) is added to the reaction (0.3 nmole), it is first preincubated for 2 minutes at 30 °C in the presence of protein kinase and buffer before the remaining components of the reaction mixture are added. The reaction is initiated by the addition of magnesium; after 10 minutes in a 30 °C shaking water bath, 3.0 ml of cold 1 m*M* ATP are added followed by 3.3 ml of cold 10% trichloroacetic acid (TCA) containing 3% sodium pyrophosphate. The dilute reaction mixture is then filtered under vacuum through nitrocellulose membrane filters (Schleicher and Schuell, B-6, 0.45 μm, 25 mm diameter) which are presoaked in 1 m*M* ATP for at least 30 minutes at room temperature. Each filter is washed twice with 5 ml of 5% TCA–1.5% sodium pyrophosphate, oven dried, and counted in 5 ml of toluene scintillation fluid.

The final assay conditions were derived on the basis of results of several preliminary experiments. A comparison of nitrocellulose membrane filters with glass fiber filters (Reeve Angle 934-H and 934-AH) in the presence and in the absence of bovine serum albumin as carrier protein demonstrates lower background counts using the nitrocellulose filter; therefore we have routinely used this type of filter in our assay. In order to minimize still further the background radioactivity on nitrocellulose filters, a variety of conditions were employed, with the results summarized as follows: (a) unlabeled ATP is the most effective agent in which to soak the filters; neither potassium phosphate buffer (20 m*M*) nor sodium pyrophosphate (3%) results in any significant diminution of background counts, thus suggesting that the ring structure of the ATP may be important in this regard; (b) terminating the reaction with 5% TCA in the presence or in the absence of 3% sodium pyrophosphate, followed by unlabeled ATP, leads to a high background, whereas adding the ATP first, followed by TCA (our standard procedure) results in low background radioactivity. The fact that the order of addition is critical, coupled with the observation that TCA if used alone results in high background levels, suggests that there may be some interaction between the acid and the nitrocellulose filter causing the increased nonspecific retention of labeled ATP; (c) use of 2 *M* NaCl to stop the reaction (rather than TCA) does not decrease the high background.

It should be pointed out that other methods which monitor the degree of incorporation of $^{32}P$ into proteins have been reported. Comparison of the nitrocellulose filter procedure with a modification of that described by Reimann *et al.* (*6*) involving filter paper washed in cold 10% TCA + 1% sodium pyrophosphate followed by propanol:ether substantiates the superiority of the former procedure. Another method which has been reported (*7*) involves the precipitation of acid-insoluble $^{32}P$-labeled material with 10% TCA followed by dissolution of the centrifuged pellet in 0.1 ml of 1 *N* NaOH and reprecipitation by 5% TCA. This precipitate is then collected on glass fiber filters. However, since the phosphodiester bond is alkali labile it is unclear how this method can be used to monitor $^{32}P$ incorporation, since much of the $^{32}P$ will remain in the supernatant after the second acid precipitation. In conjunction with these studies we discovered that the presence of bovine serum albumin (200 μg added after the TCA precipitation step), though not altering the background level of radioactivity, resulted in a significant decrease in the amount of radioactivity retained after filtration of the samples containing protein kinase. Since attempts to replace the albumin with a phosvitin carrier resulted in elevated backgrounds, we examined the retention of $^{32}P$ in the absence of carrier protein. Under these conditions the observed retention of counts in experimental samples is significantly increased, hence carrier protein is routinely left out of the procedure.

## III. Typical Results

When the purified nonhistone phosphoproteins are chromatographed on phosphocellulose columns and the fractions are monitored for protein kinase activity as described above, a complex profile of heterogeneity is seen which routinely consists of up to a dozen distinct regions of enzyme activity (Fig. 1). This fractionation procedure thus emphasizes the multiplicity of the protein kinases associated with the nonhistone chromatin phosphoprotein fraction, and also demonstrates the presence of endogenous substrate within each column fraction. The properties of these various kinase fractions, including their substrate specificities and responsiveness to regulation by cAMP, have been published in detail elsewhere (*8*).

In order to observe this heterogeneity, however, it is necessary to closely follow the procedures outline here, since other investigators using somewhat different techniques have not seen this level of complexity (*5,9*). Aside from the obvious differences in tissue source and methods of enzyme assay, factors which in themselves may be very important considerations, the most fundamental difference between our work and that of others involves the purity of the starting material used for phosphocellulose chromatography. The highly purified nonhistone phosphoprotein fraction with which

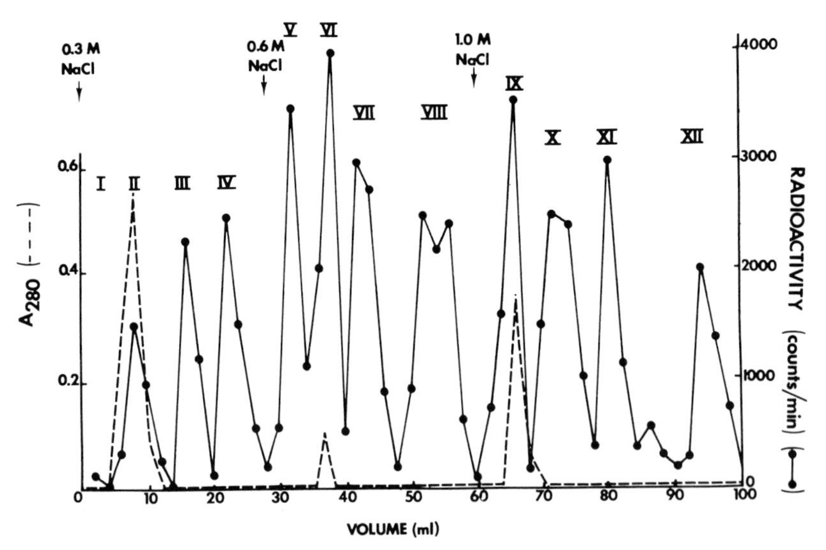


FIG. 1. Phosphocellulose column chromatography of protein kinase activity associated with the nonhistone chromatin phosphoprotein fraction of beef liver. Each column fraction is assayed for protein content by ultraviolet absorption at 280 nm (- - -) and for protein kinase activity using endogenous substrate (●——●).

we start has a very high protein kinase specific activity, thus allowing us to resolve enzyme activities which are not seen when starting with cruder material.

In addition, we have included an additional step in our column elution that increases the complexity of the protein kinase profile by a substantial amount. These enzyme activities are presumably left on the column under conditions employed by others (*5,9*). Another factor that may also be involved is the extreme lability of these kinases, which might allow them to become inactivated by other components in the relatively crude starting material usually used in such studies. Other investigators also routinely employ exogenous substrates, such as casein and histone, when assaying for protein kinase, and we have found that some of our enzyme activities are actually inhibited by these substrates. Thus, the present methodology has several distinct advantages over the techniques previously used, and, although the exact extent of the multiplicity of nuclear protein kinases is not yet known, the present results indicate the existence of a large number of distinctly different enzymes involved in the phosphorylation of chromatin-associated proteins.

## ACKNOWLEDGMENTS

Studies on this subject in our laboratory have been supported by grants from the National Science Foundation (GB-8123, GB-23921, and BMS74-23418).

## REFERENCES


*1.* Kleinsmith, L. J., *in* "Acidic Proteins of the Nucleus" (I. L. Cameron, and J. R. Jeter, Jr., eds.), p. 103. Academic Press, New York, 1974.
*2.* Stein, G. S., Spelsberg, T. C., and Kleinsmith, L. J., *Science* **183**, 817 (1974).
*3.* Kleinsmith, L. J., *J. Cell. Physiol.* **85**, 459 (1975).
*4.* Kleinsmith, L. J., *in* "Chromosomal Proteins and Their Role in the Regulation of Gene Expression" (G. S. Stein, and L. J. Kleinsmith, eds.), p. 45, Academic Press, New York, 1975.
*5.* Takeda, M., Yamamura, H., and Ohga, Y., *Biochem. Biophys. Res. Commun.* **42**, 103 (1971).
*6.* Reimann, E. M., Walsch, D. A., and Krebs, E. G., *J. Biol. Chem.* **246**, 1986 (1971).
*7.* Erlichman, J., Hirsch, A. H., and Rosen, O. M., *Proc. Natl. Acad. Sci. U.S.A.* **68**, 731 (1971).
*8.* Kish, V. M., and Kleinsmith, L. J., *J. Biol. Chem.* **249**, 750 (1974).
*9.* Ruddon, R. W., and Anderson, S. L., *Biochem. Biophys. Res. Commun.* **46**, 1499 (1972).

# Chapter 11

# *Phosphorylation of Nonhistone Chromosomal Proteins by Nuclear Phosphokinases Covalently Linked to Agarose*

J. A. THOMSON,[1] G. S. STEIN,[1] AND J. L. STEIN[2]

## I. Introduction

Among the nuclear proteins is a fraction that is posttranslationally modified by phosphoprotein kinases. This phosphorylation of nuclear proteins has been implicated in the control of gene expression. Supporting research for this has been reviewed (*1,2*), and recent results (*3,4*) have suggested the possibility that activation of histone gene transcription from chromatin of HeLa cells is mediated by a phosphoprotein which loses its ability to activate transcription upon dephosphorylation. Because of the correlation between phosphorylation and gene control, strong interest has arisen concerning the nuclear phosphoproteins and the enzymes responsible for their phosphorylation. However, several problems hinder further investigation of the role of phosphorylation in the control of gene readout. One is the lack of a purified nuclear protein kinase. All preparations of nuclear kinases contain a multitude of protein species and are almost always associated with endogenous substrate. Another problem is that all preparations of phosphorus-rich proteins contain high levels of endogenous kinase activity, although this activity can be removed by heating. Hence, interpretations of the results of studies directed toward determining the effects of the *in vitro* phosphorylation of a specific nuclear substrate fraction by kinases from another fraction are often ambiguous.

A convenient tool to circumvent these problems would be a matrix-attached nuclear protein kinase fraction. This would allow the *in vitro* phos-

[1] Department of Biochemistry and Molecular Biology, University of Florida, Gainesville, Florida.

[2] Department of Immunology and Medical Microbiology, University of Florida, Gainesville, Florida.

phorylation of a protein or proteins with the subsequent separation of substrate and kinases. This communication describes the methodology for attachment of nuclear protein kinases to an agarose matrix, the characteristics of the immobilized enzymes, and utilization of the matrix-bound kinases for phosphorylation of chromosomal proteins.

## II. Methods

### A. Preparation of Phosphoproteins

The source of protein kinase to be used is the nuclear phosphoprotein fraction isolated by the method originally described by Langan (*5*) and employed by many investigators in the study of nuclear phosphoproteins (*4, 6–8*). This protein fraction is fairly easy to obtain, has not been exposed to irreversible denaturing conditions, is rich in phosphoproteins and protein kinases (*5, 6, 8*), and has very little endogenous protease or phosphatase activity (J. A. Thomson, unpublished results). The source of phosphoproteins for all the following experiments was log-phase HeLa $S_3$ cells. The isolation procedure is briefly described in Fig. 1 (*6, 10*).

### B. Attachment of Protein Kinase to Agarose

Approximately 3 mg of phosphoproteins prepared as described above are mixed with 0.1–0.3 gm of cyanogen bromide-activated agarose (Affi Gel 10, Bio-Rad Laboratories) and gently agitated on a circular rotator in 4°C environment for 6–8 hours. The phosphoproteins are in 0.1 *M* sodium phosphate buffer (pH 7.0) at a protein concentration of 0.4–0.7 mg/ml. An equal volume of 2 *M* ethanolamine–0.1 *M* sodium phosphate buffer (pH 7.0) is added, and the mixture is rotated at 4°C for an additional 1–2 hours to bind to any remaining active sites. The solution is poured into a small column (0.9 × 14 cm) and continuously eluted with 0.1 *M* sodium phosphate buffer (pH 7.0)–1 *M* NaCl until the absorbance at 260 is zero (200–300 ml). To avoid any problem of aggregation of nonhistone chromosomal proteins, which serve as substrate for the matrix-bound kinase, *in vitro* phosphorylation is performed in the presence of 2 *M* urea–10 m*M* Tris (pH 8.0). Hence, the agarose-kinase column is equilibrated with 2 *M* urea–10 m*M* Tris (pH 8.0) before use. The agarose-kinase is removed from the column, mixed with the substrate and the incubation medium in a large test tube, and rotated for 10 minutes at 37°C. The reaction is terminated by the addition of solid urea to a final concentration of 6 *M*. The agarose-kinase is then removed from the reaction mixture by eluting the solution through a small column (0.9 × 15 cm). Recovery of substrate can be enhanced by eluting the agarose-kinase column with 5 ml of 5 *M* urea–3 *M* NaCl–10 m*M* Tris (pH 7.5)–0.1 m*M* dithiothreitol, and the eluate is combined with the previously eluted reaction mixture.

## C. Amount of Phosphoproteins Retained during the Coupling Reaction

To determine the percentage of phosphoprotein permanently coupled to the agarose, [$^3$H]leucine labeled phosphoproteins were isolated from HeLa cells. Two liters of HeLa cells (at $5 \times 10^5$ cells/ml) were pelleted (500 *g* for 5 minutes) and resuspended in 100 ml of spinner salts (GIBCO). Two

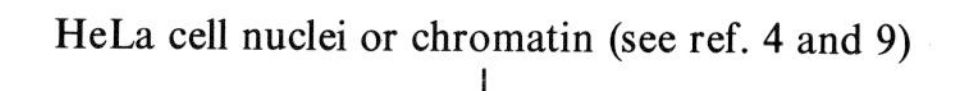

HeLa cell nuclei or chromatin (see ref. 4 and 9)

↓

Pellet homogenized in 1.0 *M* NaCl–20 m*M* Tris (pH 7.5) at a DNA concentration 2 mg/ml

↓

One and one-half volumes of 20 m*M* Tris (pH 7.5) added and mixture homogenized again; solution centrifuged at 80,000 *g* for 1 hour

↓

Supernatant mixed for 5 minutes with 0.1 ml of Bio-Rex 70 (200 mg/ml) per 1 mg of protein mixture centrifuged at 6000 *g* for 5 minutes.

↓

Supernatant mixed for 5 minutes with 0.46 mg of $CaPO_4$ gel per 1 mg of protein, mixture centrifuged at 600 *g* for 5 minutes.

↓

$CaPO_4$ pellet homogenized in 40 ml of 1.0 *M* $(NH_4)_2$ $SO_4$–50 m*M* Tris (pH 7.5) and centrifuged at 6000 *g* for 5 minutes.

↓

$CaPO_4$ gel homogenized in 0.3 *M* EDTA (pH 7.5)–0.33 *M* $(NH_4)_2SO_4$ in a ratio of 0.2 ml per milligram of gel. After sitting the ice for 1 hour with occasional homogenization, the mixture was centrifuged at 33,000 *g* for 15 minutes.

↓

Supernatant dialyzed against 0.1 *M* sodium phosphate buffer (pH 7.0)

↓

Phosphoprotein fraction

FIG. 1. Brief flow diagram of the isolation procedure of the nuclear phosphoproteins. The procedure is identical to that described by Kish and Kleinsmith (*6*), except that a Dounce homogenizer was utilized instead of a Polytron homogenizer. $CaPO_4$ gel was prepared by the method described by Keilen and Hartree (*10*). Bio-Rex 70 (minus 400 mesh purchased from Bio-Rad) was previously equilibrated with 0.4 *M* NaCl–20 m*M* Tris (pH 7.5).

milliliters of fetal calf serum were added along with 500 $\mu$Ci of [$^3$H]leucine. The cells were stirred in a 37°C environment for 1 hour. Two hundred milliliters of spinner salts were then added, and the mixture was centrifuged at 500 *g* for 5 minutes. The cells were washed two additional times with spinner salts, and phosphoproteins were extracted as described (Fig. 1). The resulting specific activity of the phosphoproteins was $3.8 \times 10^5$ cpm/mg. Three milligrams of the $^3$H-labeled phosphoproteins were mixed with 0.1 gm of the activated agarose as described in Section II,B. The column was eluted with 1.0 NaCl–0.1 *M* sodium phosphate until both the absorbance at 260 was zero and no radioactivity was being washed from the column. Examination of all the eluted solutions revealed a total of $1.05 \times 10^5$ cpm, indicating that, of the 3 mg of phosphoproteins, approximately 2.7 mg of proteins (90%) were coupled with the agarose matrix and 0.3 mg was eluted from the column.

## D. Amount of Protein Kinase Activity Retained after Coupling to Agarose

To determine the percentage of protein kinase activity remaining after the phosphoproteins were coupled to agarose, the agarose-kinase (with approximately 3 mg of attached phosphoproteins) was incubated with $\gamma$-$^{32}$P-ATP (specific activity 1 Ci/mmole), 20 m*M* $MgCl_2$, and heated phosphoproteins (1.5 mg) which contained no endogenous kinase activity as substrate. The mixture was incubated and eluted through a column in the aforementioned manner (Section II,B). Of the eluted material, one-half was utilized for determining the amount of $^{32}P\text{-}PO_4$ incorporated [as trichloroacetic acid (TCA)-insoluble radioactivity] and one-half was examined by polyacrylamide gel electrophoresis (see Section II,E). For comparison, an equal amount (~3 mg) of active phosphoproteins was incubated in an identical manner without the addition of the agarose-kinase. TCA-insoluble radioactivity from both samples was determined in the following manner. Three 50-$\mu$l aliquots of each sample were transferred to 12-ml conical glass centrifuge tubes. Ten milliliters of 10% TCA–2% sodium pyrophosphate were added, followed by the addition of 4 mg of bovine serum albumin as carrier. The mixtures were centrifuged for 5 minutes at 700 *g* in an IEC centrifuge. The supernatant was decanted, and the pellet was thoroughly resuspended in 5% TCA–1% sodium pyrophosphate with the use of a glass rod and vigorous vortexing. The solutions were centrifuged and washed in an identical manner two additional times. The resulting pellet was solubilized in 0.4 ml of NCS (Amersham/Searle) and after the addition of 3 ml of a dioxane-toluene scintillation fluid, the samples were counted in a Beckman Scintillation Counter. The TCA-insoluble radioactivity of the sample phosphorylated by the agarose-attached protein kinase was approximately 12.5% of the radioactivity incorporated by an identical amount of active phosphoproteins. Therefore, approximately 12.5% of the endogenous protein kinase activity is retained when the phosphoproteins are coupled to agarose in this manner.

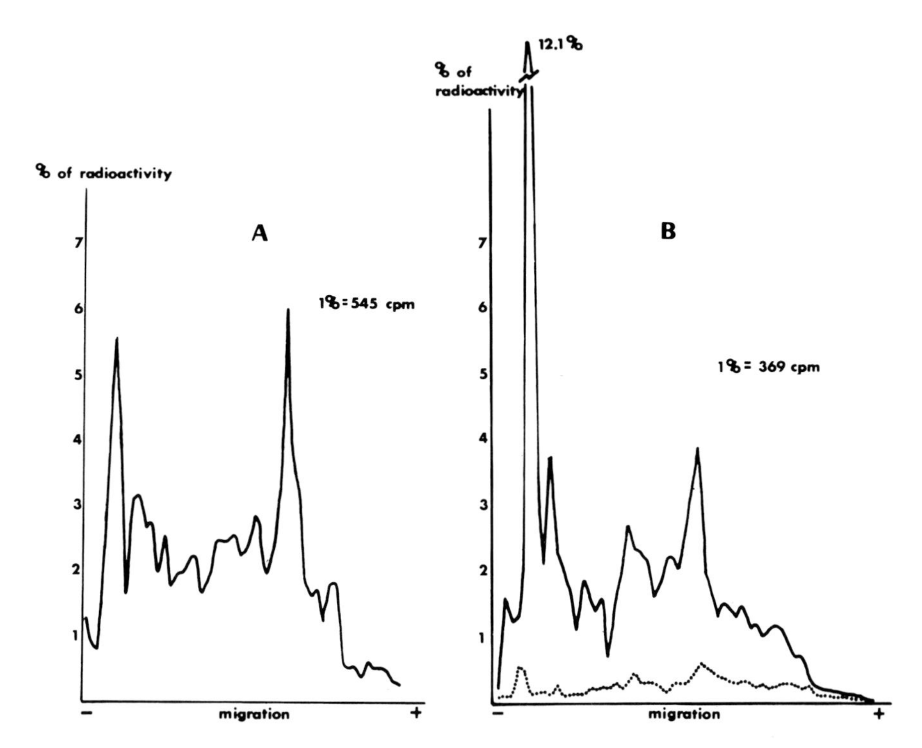


FIG. 2. Comparison of the incorporation of $^{32}P$ into various molecular weight classes of chromosomal phosphoproteins induced by protein kinases and that induced by these protein kinases coupled to agarose. Proteins were incubated with [$^{32}P$]ATP (specific activity 1.0 Ci/mmole ATP) in 2 *M* urea–20 m*M* $MgCl_2$–10 m*M* Tris (pH 7.5) in the presence or in the absence of agarose-bound protein kinases. The reaction was terminated by the addition of solid urea to a final concentration of 6 *M*. The agarose-kinase was removed and protein solutions were dialyzed against 1% sodium dodecyl sulfate–0.1% mercaptoethanol–10 m*M* sodium phosphate buffer (pH 7.0). The proteins were then examined by means of polyacrylamide gel electrophoresis as described elsewhere (11). After staining, the gels were sliced into 1-mm sections and examined for radioactivity. Panel A shows the resulting radioactivity profile induced by endogenous protein kinase activity in the nuclear phosphoproteins. Panel B shows the resulting radioactivity profile after incubation of heated nuclear phosphoproteins (no endogenous kinase activity) in the presence of agarose-bound protein kinases (—) and in the absence of agarose-bound protein kinases (. . .).

## E. Comparison of the Phosphorylation Pattern Induced by Protein Kinases and Protein Kinases Coupled to Agarose

It is possible that different kinases in the phosphoprotein fraction have unequal affinities for agarose attachment or unequal abilities for retaining activity after agarose attachment. Therefore, we felt that it was necessary to compare the phosphorylation pattern induced by active protein kinases endogenous to the phosphoproteins and protein kinases coupled to the agarose matrix. Samples of the eluted protein from the incubations described in Section II,D were examined by polyacrylamide gel electrophoresis [see Krause *et al.* (*11*) for details]. After staining, the gels were sliced into 1-mm fractions and examined for radioactivity. The resulting radioactivity profiles are shown in Fig. 2. The profiles are similar but not identical, probably indicating some specificity in retained activity of the phosphoprotein kinases after coupling to agarose.

## F. Examining the Possibility of Protein Leakage from Agarose-Kinase

To determine the amount, if any, of protein leakage from the agarose-kinase during incubation, the following experiment was performed. [$^3$H]Leucine-labeled phosphoproteins (2.7 mg, specific activity $3.82 \times 10^5$ cpm/mg) were coupled with 0.1 gm of cyanogen bromide-activated agarose in the previously described manner. The agarose-kinase was then incubated for 15 minutes at 37°C with 20 m*M* NaCl, phosphoprotein as substrate, and ATP (3 nmoles). The mixture was eluted through a small column, and the eluate was examined for the presence of radioactivity. The radioactivity found in the eluted mixture was equivalent to 6.9 $\mu$g of protein. This indicated that only 0.25% of the attached phosphoprotein had been released during the incubation. Although any protein leakage is slightly disturbing, this small amount of leakage should not interfere with the interpretation of most experiments.

## G. Determination of Protease Activity Associated with Agarose-Bound Phosphoproteins

We approached the problem of determining the protease activity associated with agarose-bound phosphoproteins in two ways. The first was to examine and compare by the techniques of polyacrylamide gel electrophoresis, heated phosphoproteins (containing no endogenous protease activity) which had been incubated in the presence and in the absence of agarose-bound kinases. As is shown in Fig. 3, the electrophoretic profile of

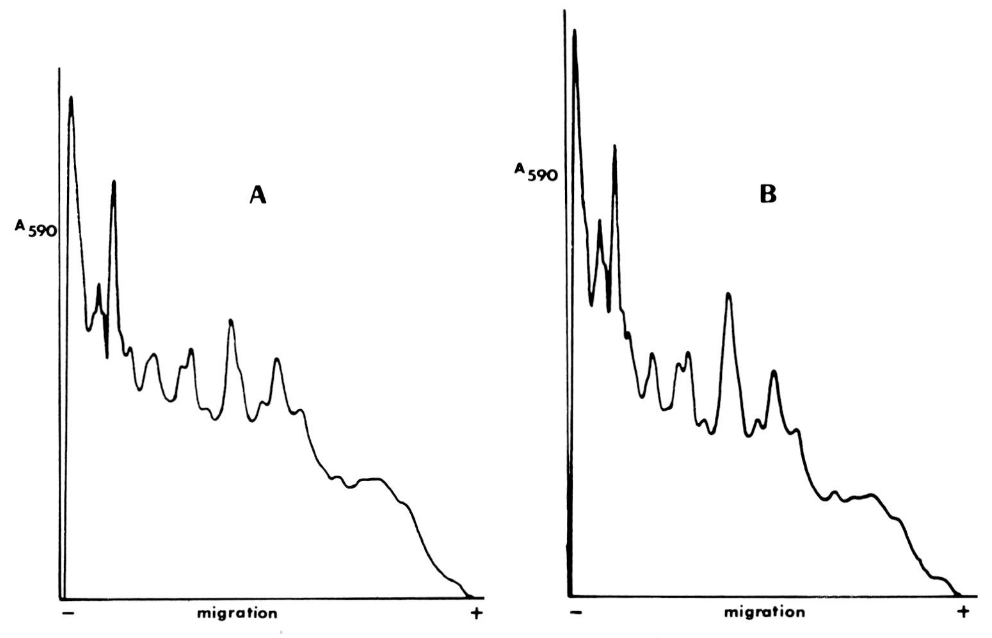


FIG. 3. Electrophoretic profile of nuclear phosphoproteins before and after incubation with agarose-bound protein kinases. Nuclear phosphoproteins were incubated with ATP (30 nmoles) in 2 *M* urea–10 m*M* Tris (pH 7.5)–20 m*M* $MgCl_2$, in the presence of agarose-bound protein kinases. After 10 minutes at 37°C, solid urea was added to a final concentration of 6 *M*. The agarose-kinase was removed by the elution of the mixture through a small column. Proteins were dialyzed against 1% sodium dodecyl sulfate–0.1% mercaptoethanol–10 m*M* sodium phosphate buffer (pH 7.0). Proteins were then examined by means of polyacrylamide-gel electrophoresis as described elsewhere (*11*). Profile A is the electrophoretic profile of the phosphoproteins before incubation; profile B is the electrophoretic profile after incubation with agarose-bound protein kinases.

heated phosphoproteins incubated in the presence of agarose-kinases was virtually identical to that of heated phosphoproteins incubated alone. There was no increase in the low-molecular-weight protein species in the agarose-kinase exposed proteins, the presence of which would be indicative of protease activity.

The second approach was to incubate [$^3$H]leucine-labeled phosphoproteins in the presence and in the absence of agarose-kinase under normal incubation situations. The phosphoproteins recovered from both incubations were then TCA-extracted as described in Section II,B. The TCA supernatants from all washes were examined for TCA-soluble radioactivity. The TCA-soluble radioactivity from the extraction of phosphoproteins exposed to the agarose-kinase was identical to the radioactivity from the extraction of phosphoproteins not exposed to agarose-kinase. Although there are limitations to these methods of determining protease activity, the above results firmly rule out the possibility of gross protease activity contaminating the protein kinase attached to agarose by our procedure.

### H. Determination of Phosphatase Activity Associated with Agarose-Bound Phosphoproteins

Determination of phosphatase activity of the agarose kinase was performed in the following manner. Six milligrams of phosphoprotein were incubated at 37°C for 10 minutes in the presence of [$\gamma$-$^{32}$P] ATP and 20 m*M* $MgCl_2$. The reaction was terminated by the addition of solid urea to a final concentration of 6 *M*. The resulting specific activity of the phosphoprotein with respect to $^{32}$P was $9 \times 10^6$ cmp/mg. The proteins were then heated at 65°C for 10 minutes to remove any endogenous enzymic activity. One half of the [$^{32}$P]phosphoproteins was incubated with 20 m*M* $Mg^{2+}$, ATP (100 nmoles), and agarose-kinase (containing 3 mg of phosphoproteins). The other half was incubated with 20 m*M* Mg, ATP (100 nmoles), and 0.3 gm of activated agarose that had been treated with 1 *M* ethanolamine only. The agarose matrix or agarose-kinase was removed from the samples by elution through small columns, and aliquots of eluted solution from both samples were TCA extracted as described in Section II,B. Both the TCA-soluble radioactivity and the TCA-precipitable radioactivity were identical between [$^{32}$P]phosphoproteins that were exposed to agarose-kinase and those exposed to agarose only. These results indicate that the phosphatase activity of the agarose preparation is negligible.

## III. Conclusions

With increasing interest in nuclear protein phosphorylation, the availability of agarose-coupled protein kinase activity is very desirable. The method of attaching phosphoprotein kinase activity to the commercially available activated agarose, Affi Gel 10, as described in this communication is an extremely effective procedure. The lack of harsh chemicals or extreme pH and the quick coupling time are extremely important for retaining activity of the rather sensitive nuclear protein kinases. Since these kinases lose activity very quickly even when attached to agarose (activity is completely lost after 2 weeks), the easy coupling techniques of cyanogen bromide-activated agarose are beneficial. By using partially purified nuclear phosphoproteins instead of total nonhistone chromosomal proteins for the source of nuclear protein kinase activity, the resulting agarose-protein kinase column is relatively free of protease or phosphatase activity. It should be noted that a small amount of protein leakage does occur; however, this should not alter interpretations of most experimental results. Hence, nuclear protein kinases attached to agarose as described here offer a valuable tool for the study of nuclear protein phosphorylation with numerous applications.

## Acknowledgments

These studies were supported by Grant BMS 75-18583 from the National Science Foundation, and Grants GM 20535 and CA 18875 from the National Institutes of Health.

## References

1. Stein, G. S., Spelsberg, T. C., and Kleinsmith, L. J., *Science* **183**, 817 (1974).
2. Kleinsmith, L. J., *J. Cell. Physiol.* **85**, 459 (1975).
3. Kleinsmith, L. J., Stein, J. L., and Stein, G. S., *Proc. Natl. Acad. Sci. U.S.A.* **73**, 1174 (1976).
4. Thomson, J. A., Stein, J. L., Kleinsmith, L. J., and Stein, G. S., *Science* **194**, 428 (1976).
5. Langan, T. A., *in* "Regulation of Nucleic Acid and Protein Biosynthesis" (V. V. Koningsburger and L. Bosch, eds.), p. 233. Am. Elsevier, New York, 1967.
6. Kish, V., and Kleinsmith, L. J., *J. Biol. Chem.* **249**, 750 (1974).
7. Platz, R. D., Grimes, S. R., Hord, G., Meistrich, M. L., and Hnilica, L. S., *in* "Chromosomal Proteins and Their Role in the Regulation of Gene Expression" (G. S. Stein and L. J. Kleinsmith, eds.), p. 67. Academic Press, New York, 1975.
8. Thomson, J. A., Chiu, J. F., and Hnilica, L. S., *Biochim. Biophys. Acta* **407**, 114 (1975).
9. Stein, G. S., and Farber, J. L., *Proc. Natl. Acad. Sci. U.S.A.*, **69**, 2918 (1972).
10. Keilin, D., and Hartree, E. F., *Proc. R. Soc. London, Ses. B* **124**, 397 (1938).
11. Krause, M. O., Kleinsmith, L. J., and Stein, G. S., *Exp. Cell Res.* **92**, 164 (1975).

# Chapter 12

# *Methods for the Assessment of Nonhistone Phosphorylation (Acid-Stable, Alkali-Labile Linkages)*

DOROTHY E. PUMO[1] AND LEWIS J. KLEINSMITH

*Division of Biological Sciences,*
*University of Michigan,*
*Ann Arbor, Michigan*

## I. Introduction

A large proportion of the nonhistone chromatin proteins are phosphorylated via the esterification of phosphate groups to the hydroxyl group of serine and threonine residues. Work from several laboratories indicates that the phosphorylation of these proteins may play an integral role in the regulation of gene transcription (for reviews, see 1–4). In this chapter we will describe the major approaches available for assessing the occurrence of nonhistone protein phosphorylation.

## II. Methods

### A. Determination of Alkali-Labile Phosphate

It has been known for many years that phosphate groups attached to serine and threonine residues in proteins can be readily released by treatment with mild alkali, but not acid (*5*). Alkali-induced release of phosphate does not occur via hydrolysis, as evidenced by the fact that the isolated amino acids phosphoserine and phosphothreonine are not susceptible

[1]*Present address*: Department of Molecular, Cellular, and Developmental Biology, University of Colorado, Boulder, Colorado.

to having their phosphate groups removed under these conditions. In intact proteins, however, the addition of alkali promotes a $\beta$-elimination reaction (Fig. 1). Since the release of inorganic phosphate under these conditions is relatively specific for phosphoserine and phosphothreonine, the dual criteria of acid stability and alkali lability comprise a generally useful and convenient approach for the routine measurement of protein phosphorylation.

Alkali-labile phosphate can be measured either in whole nuclei or in purified protein fractions. When whole nuclei are used, start with 15–20 mg and follow the entire procedure as written. If, however, the material to be analyzed is a purified protein fraction, eliminate the washing procedures described in this paragraph and skip to the next paragraph, beginning the procedure with the addition of alkali. Wash the nuclei three times by suspending in cold 16% trichloroacetic acid (TCA), collecting them each time by centrifugation at 1000 *g*. Heat the material in 16% TCA for 15 minutes at 90°C, collect by centrifuging at 1000 *g*, and wash once again in 16% TCA. Wash three times more, again centrifuging between washes, first using 1:1 chloroform–methanol, then 2:1 chloroform–methanol containing 1:300 concentrated HCl, and finally ether. Dry the sample under the vacuum.

The resulting protein residue (or 1–5 mg purified protein sample) is dissolved in 2.0 ml of 1.0 *N* NaOH. At this point, a 0.1-ml aliquot may be removed for determining protein concentration. Heat the alkali-dissolved residue in a boiling water bath for 15 minutes, then cool and add 0.5 ml of a mixture containing 4 *N* HCl and 1 *N* $H_2SO_4$. The protein is precipitated by adding 0.1 ml of 0.1 *M* silicotungstic acid reagent (16.5 gm of Fisher Reagent silicotungstic acid made up to 50 ml with 0.1 *N* $H_2SO_4$). Centrifuge at 1000 *g*, and transfer a 2-ml aliquot of the supernatant to a glass-stoppered tube. Add 0.5 ml of 5% ammonium molybdate in 4 *N* $H_2SO_4$, followed by 2.5 ml of a 1:1 isobutanol–benzene solution, shake 15 seconds to extract the phosphomolybdate complex, and centrifuge until there is a clean separation

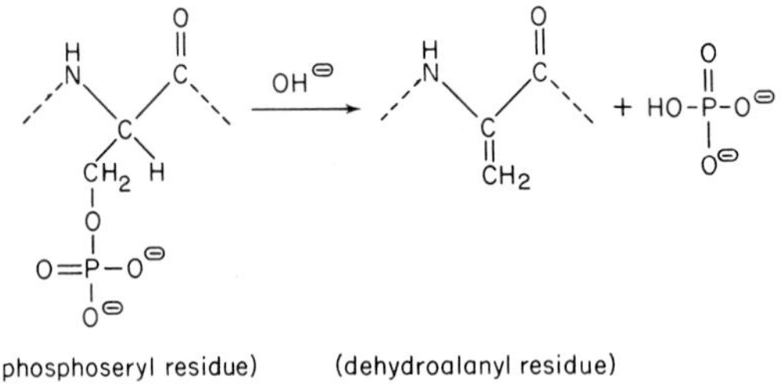

(phosphoseryl residue) (dehydroalanyl residue)

FIG. 1. Diagram of the $\beta$-elimination reaction undergone by protein-bound phosphoserine in the presence of alkali.

of the layers (about 5 minutes). If measuring $^{32}P$, remove a 1-ml portion of the upper phase and count via scintillation methods. To determine total phosphate add 0.45 ml of ethanol–$H_2SO_4$ (49:1 v/v), followed by 0.05 ml of a freshly diluted $SnCl_2$ solution (1 part stock $SnCl_2$ to 200 parts of 1 $N$ $H_2SO_4$), and read the resulting blue color at 660 nm. The stock $SnCl_2$ solution, which is made by dissolving 10 gm of $SnCl_2 \cdot 2H_2O$ in 25 ml of concentrated HCl, is stable for up to 6 months when stored cold and in the dark. Because the final blue color obtained in this reaction is in an organic solvent medium that evaporates quickly, it is best to use stoppered cuvettes to avoid changes in absorbance readings.

As phosphate standards for the assay, use 0.025 to 0.25 $\mu$moles of inorganic phosphate in the original volume of 1.0 $N$ NaOH, and carry them through the entire procedure. It is important that tubes used in this assay be acid washed, and that etched tubes be avoided.

## B. Ion-Exchange Chromatography of Phosphorylated Amino Acids

Ion-exchange chromatography permits the direct analysis of phosphoserine and phosphothreonine residues in phosphorylated proteins. Its major advantage over analysis of alkali-labile phosphate is that it permits unequivocal identification of the phosphorylated amino acids involved, as well as a quantitative measure of their relative abundance. Unfortunately, the acid hydrolysis required to release phosphoserine and phosphothreonine from proteins causes a considerable degradation of these amino acids at the same time. For this reason it is necessary to run known amounts of phosphoserine and phosphothreonine through the entire procedure and correct for their recoveries. This approach may not give an exactly accurate value for degradation, however, since phosphoserine and phosphothreonine within proteins may not be destroyed at exactly the same rate as the corresponding free amino acids employed as controls.

The unknown phosphoprotein sample is suspended in 1 ml of 2 $N$ HCl in an 18 × 150 Pyrex tube. Heat the mid-section of the tube with an oxygen-acetylene torch and form a neck with an internal diameter less than 1.5 mm. Evacuate the tube and shake to avoid bubbling over. A trap should be affixed to a vacuum pump for protection from the HCl. After a 5-minute evacuation of the tube, melt the neck closed and place the prepared sample in an oven at 100°C for 10 hours. After hydrolysis, remove the HCl from the sample. This is most easily done by lyophilization. Again, steps should be taken to avoid HCl uptake by the vacuum pump. It is possible to evacuate the HCl without freezing, but a control must be run to ensure that there is no further binding of free phosphate to nonphosphorylated serine or threonine residues. The dried hydrolyzate is dissolved in 1 ml of 0.05 *N* HCl and is ready to load onto the column.

The column fractionation used is a slight modification of the procedure of Schaffer *et al.* (*6*). The cation-exchange resin AG-50W-X4 $H^+$, 200–400 mesh (form of Dowex 50 obtained from Bio-Rad) is used for the separation of inorganic phosphate, phosphoserine, and phosphothreonine. Prepare the resin by washing for 15 minutes with 6 volumes of 6 *N* HCl. Decant the acid with the fines, and wash the resin 5 times more with 20 volumes of distilled water. These washes are followed by two washes in 20 volumes of 0.05 *N* HCl. The resin is poured into a 0.8 cm column to a height of 37 cm, and the column is eluted with 0.05 *N* HCl. With a pressure head of 35 cm, the flow rate is 14 ml/hour. Calibrate the column with $^{32}P_i$ or $H_3PO_4$ (1 to 3000 dilution of concentrated), and 1–3 μmoles of phosphoserine and phosphothreonine dissolved in 0.05 *N* HCl. Use 1 ml total sample to calibrate column, collecting 110 fractions of 1 ml each.

The locations of the phosphate-containing compounds are determined using a phosphate assay based on the procedure of Ames and Dubin (*7*). All reactions are carried out in acid-washed 13 × 100 test tubes. To 0.1-ml aliquots of each fraction, add 0.025 ml of 12% $Mg(NO_3)_2$–$6H_2O$ in 95% ethanol. Continuously shake the tube over a flame until the brown fumes disappear. At this point, the white precipitate at the bottom of the tube should be very dry. Since there are many tubes when doing the assay of the column fractions, a small amount of moisture may accumulate in the bottom of the tube before the next step. This does not appear to affect the assay. When the tubes are cooled, add 0.3 ml of 0.5 *N* HCl. Cover the tubes with marbles or glass bubbles, and place in a boiling water bath for 15 minutes. This part of the procedure is necessary to ensure complete hydrolysis of any pyrophosphate which may have formed during the heating step. To the cooled tubes add 0.7 ml of the following mixture, which should be made up fresh each day: 1 part 10% ascorbic acid (stable up to 1 month at 4°C) and 6 parts 0.42% ammonium molybdate-$4H_2O$ in 1 *N* $H_2SO_4$. Cover the tubes, incubate 1 hour at 37°C, and read the blue color at 820 nm. As is shown in Fig. 2, this detection method reveals an excellent resolution between phosphoserine, phosphothreonine, and inorganic phosphate.

## C. Sodium Dodecyl Sulfate–Polyacrylamide Gel Electrophoresis

The methods described thus far are useful for assessing the overall level of phosphorylation of a protein sample, but not for determining the degree of phosphorylation of individual polypeptides within a protein sample. Analysis of $^{32}P$-labeled proteins via SDS–polyacrylamide gel electrophoresis has been the most widely used approach for this purpose. We describe here the method most commonly employed in our laboratory (*8*), but it should be emphasized that other SDS–polyacrylamide systems may be used as well.

The procedure described will make 12 cylindrical gels of 10% polyacrylamide that are 6 cm in length. Dialyze samples of $^{32}$P-labeled nonhistone protein overnight in three changes of SDS dialysis buffer. This buffer is made with 10 ml of 10% SDS, 1 ml of β-mercaptoethanol, 11 ml of 0.5 *M* $Na_2HPO_4$, and 9 ml of 0.5 *M* $NaH_2PO_4$, brought to a final volume of 1 liter. The running buffer must be made fresh and diluted 1:1 just prior to use, as it becomes cloudy. The undiluted running buffer contains 15.6 gm of $NaH_2PO_4$, 77.2 gm of $Na_2HPO \cdot 7H_2O$, and 40 ml of 10% SDS stock solution in a final volume of 2 liters. To make the gels, mix 9 ml of acrylamide stock (22.2 gm of acrylamide and 0.6 gm of bisacrylamide in 100 ml of water, filtered with Whatman No. 1 filter paper, stable up to 6 months at 4°C in a dark bottle) and 10 ml of the undiluted running buffer. Deaerate for 10 minutes, then add 1 ml of a 1% solution of ammonium persulfate (made fresh daily) and 25 μl of TEMED (*N*,*N*,*N*′,*N*′-tetramethylethylenediamine), swirling gently. Immediately and carefully pour the mixture into the gel tubes to a height of 6.1 cm. Distilled water is layered quickly on top of the acrylamide. Gels are fully polymerized in 30 minutes and should then be run in diluted gel buffer (1:1 with water) for 20 minutes at 6–8 mA/gel.

Mix 10–100 μl of dialyzed protein sample, containing 10,000–100,000 cpm, with 100 μl of tracking dye-glycerol solution (50 ml glycerol, 3 ml 0.05% bromophenol blue, 5 ml β-mercaptoethanol, 50 ml dialysis buffer). Apply the samples and run the gels at 2–3 mA/gel for 20–30 minutes to

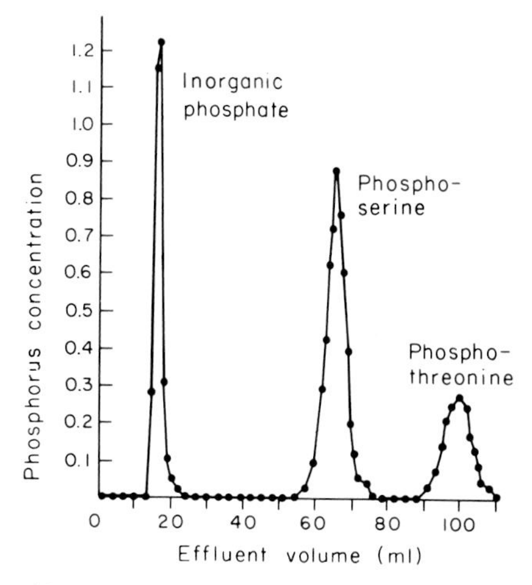


FIG. 2. Separation of inorganic phosphate, phosphoserine, and phosphothreonine obtained by the column chromatography method described in the text.

permit entry into gels, then run for 4 hours at 6–8 mA/gel, or until the tracking dye nears the bottom of the tubes. After electrophoresis, remove the gels from the tubes and put into 7.5% acetic acid. Gels to be stained are treated with 0.025% Coomassie Blue in 5:1:5 methanol–acetic acid–

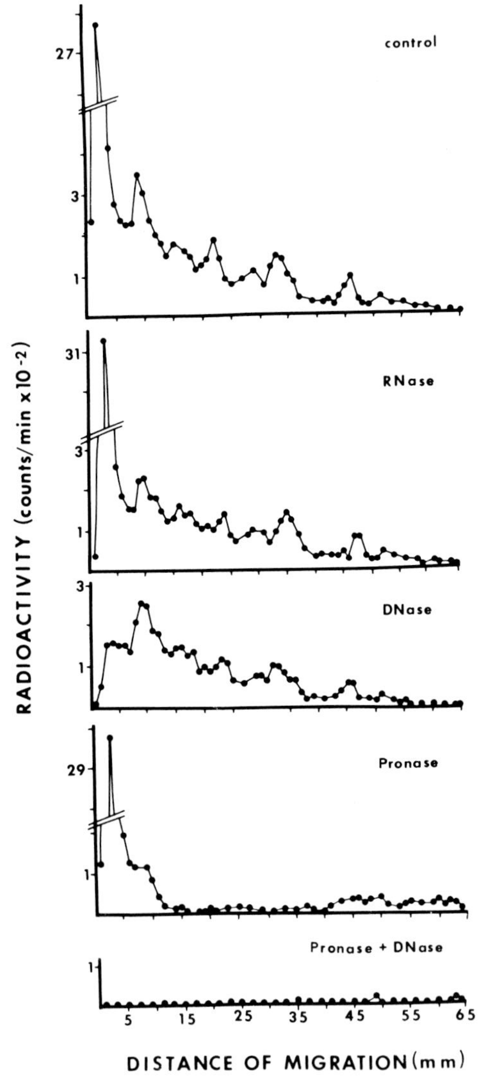


FIG. 3. Results of digesting $^{32}P$-labeled HeLa nonhistone proteins with various enzymes prior to electrophoresis on sodium dodecyl sulfate–polyacrylamide gels. The protein nature of most of the peaks is evidenced by their susceptibility to Pronase but not to DNase or RNase. Peak 1, however, is an exception, being digestible primarily by DNase.

water for 4–6 hours, and destained in 5:1:5 methanol–acetic acid–water. Stained or unstained gels are sliced at 0.5–2.0 mm intervals, dried, and counted in a toluene-based scintillation fluid.

Although it is often assumed that the appearance of $^{32}P$-labeled peaks in SDS–polyacrylamide gels indicates the presence of phosphorylated chromatin proteins, the possibility of contamination by other phosphorus-containing compounds (DNA, RNA, phospholipid, etc.) should be checked for by treatment with appropriate degradative enzymes. Before exposure to SDS, the $^{32}P$-labeled chromatin protein fraction in 0.4 *M* NaCl–0.02 *M* Tris-HCl (pH 7.5) can be incubated with DNase (0.1 $\mu$g per microgram of nonhistone protein), RNase (13 units per microgram of nonhistone protein), or Pronase (200 $\mu$g per microgram of nonhistone protein). After 30 minutes of incubation at 37°C, the material is dialyzed against SDS and electrophoresed as described above. As is shown in Fig. 3, although most of the $^{32}P$-labeled peaks in the nonhistone phosphoprotein fraction isolated from HeLa cells are removed by Pronase treatment as expected, one is found to be sensitive to DNase instead. Hence such enzymic studies may be useful in identifying the presence of $^{32}P$-labeled material that is not of protein origin.

## III. Concluding Remarks

Our major focus in this paper has been the detection of phosphate groups attached to serine and threonine residues in nonhistone chromosomal proteins. Methods for the detection of histone phosphorylation, which includes acid-labile linkages to lysine and histidine as well as the acid-stable linkages to serine and threonine, are described elsewhere in this volume.

Future studies on phosphorylated nonhistone proteins will require the development of methods for fractionation of these proteins under conditions where they retain their biological functions. Although the use of SDS–polyacrylamide gel electrophoresis is useful for overcoming the tendency of nonhistone phosphoproteins to aggregate, the resulting fractionated proteins are denatured and of little further use. Several attempts have been made in this laboratory to remove the SDS from nuclear phosphoproteins, but none of the methods tried were found to be suitably effective. Obviously work needs to be continued not only on the study of nonhistone phosphorylation per se, but also on the development of techniques with which to examine this phosphorylation, especially in suitably purified phosphoproteins.

## Acknowledgments

Studies on this subject in our laboratory have been supported by grants from the National Science Foundation (GB-8123, GB-23921, and BMS-23418).

## References

1. Kleinsmith, L. J., *in* "Acidic Proteins of the Nucleus" (I. L. Cameron, and J. R. Jeter, Jr., eds.), p. 103. Academic Press, New York, 1974.
2. Stein, G. S., Spelsberg, T. C., and Kleinsmith, L. J., *Science* **183**, 817 (1974).
3. Kleinsmith, L. J., *J. Cell. Physiol.* **85**, 459 (1975).
4. Kleinsmith, L. J., *in* "Chromosomal Proteins and Their Role in the Regulation of the Gene Expression" (G. S. Stein, and L. J. Kleinsmith, eds.), p. 45. Academic Press, New York, 1975.
5. Plimmer, R. H. A., and Bayliss, W. M., *J. Physiol. (London)* **33**, 439 (1906).
6. Schaffer, N. K., May, S. C., and Summerson, W. H., *J. Biol. Chem.* **235**, 769 (1953).
7. Ames, B. N., and Dubin, D. T., *J. Biol. Chem.* **235**, 769 (1960).
8. Platz, R. D., Kish, V. M., and Kleinsmith, L. J., *FEBS Lett.* **12**, 38 (1970).

# Chapter 13

# *Methods for the Assessment of Site-Specific Histone Phosphorylation*

THOMAS A. LANGAN

*Department of Pharmacology,*
*University of Colorado Medical Center,*
*Denver, Colorado*

## I. Introduction

It has become clear that cells contain a substantial number of different enzymes that catalyze the transfer of phosphate from ATP or other nucleoside triphosphates to serine or threonine residues of proteins. The diversity of protein kinases and the widely different protein substrates which are phosphorylated by them make it evident that protein kinases are involved in a large variety of functions within cells. However, this very diversity often makes the investigation of protein kinases and protein phosphorylation difficult, since, depending on the methodology used, one type of phosphorylation reaction often interferes with assessment of the presence and function of another.

Among the properties that can be used to distinguish protein kinases, the nature of the substrates phosphorylated is one of the most useful; it is also a property that is likely to be closely related to the function of the protein kinase. The substrate specificity exhibited by various protein kinases involves not only preferential phosphorylation of particular proteins by the enzyme, but also the preferential or specific phosphorylation of particular serine- or theonine-containing sites within the protein molecule. In the case of lysine-rich (H1) histone the same protein is phosphorylated by more than one protein kinase, each of which catalyzes the phosphorylation of a different site or set of sites in the molecule (*1–4*). There is evidence in this case (*5,6*) and in the case of phosphorylase kinase (*7,8*) and glycogen synthetase (*9*) that the phosphorylation of different sites in the same protein molecule may serve different functions. The occurrence of multiple types of phosphorylation reactions on the same protein molecule suggests a good deal of complexity in the mechanisms by which protein phosphorylation influences cell function. It also introduces additional difficulties in distinguishing the types of phosphorylation that occur in cells and in evaluating their functions.

In this laboratory we have developed methodology for the assessment of different types of phosphorylation reactions that occur on H1 histone. The methods involve the isolation of specific phosphopeptides derived from the histone that contain individual phosphorylation sites. The procedures, although somewhat time-consuming, have the following advantages: (a) They allow classification of H1 histone kinases on the basis of the specific phosphorylation reaction catalyzed, rather than on the basis of more arbitrary properties, such as size or position of elution from chromatographic columns. Thus, classification is based on a property that is likely to be related to function. (b) The presence of a particular histone kinase activity in a crude preparation can be readily demonstrated and reasonably quantitated, even in the presence of large amounts of enzymes catalyzing other types of H1 histone phosphorylation. In addition, the isolation of specific phosphopeptides from H1 histone phosphorylated *in vivo* provides evidence for the action of the corresponding enzyme on H1 histone in the cell, and allows investigation of the relationship between individual phosphorylation reactions and cell function and growth. (c) In studies carried out in crude systems or *in vivo*, interference due to contamination of the isolated H1 histone with other phosphorylated proteins can be greatly reduced by subjecting the phosphopeptides derived from H1 to extensive purification steps.

In this paper, methodology for the assessment of site-specific H1 histone phosphorylation by phosphopeptide analysis is described, and its application in several *in vivo* and *in vitro* systems is outlined. In Chapter 14, procedures for isolating three distinct protein kinases catalyzing the phosphorylation of different sites in H1 histone are given.

## II. Analysis of Phosphorylation of Serine Residues 37 and 106 in H1 Histone

Cyclic AMP-dependent protein kinase (HK1) catalyzes the phosphorylation of H1 histone specifically or $Ser_{37}$ (*1, 2, 10*). A second histone kinase (HK2), also found widely distributed in eukaryotic cells, is cyclic nucleotide independent and catalyzes phosphorylation preferentially on $Ser_{106}$ (*2,3*). A third enzyme, originally described by Lake and Salzman (*11*), is found only in growing cells and so is termed here *growth-associated* histone kinase. This enzyme is bound to chromatin and catalyzes cyclic nucleotide independent phosphorylation of a number of serine and threonine residues in the N- and C-terminal regions of the histone (*4, 12*). The phosphorylation sites identified so far in H1 histone are summarized in Fig. 1.

The procedure described below is adequate to measure phosphorylation occurring specifically on $Ser_{37}$ and $Ser_{106}$ in reaction mixtures containing crude and partially purified enzyme preparations, and to distinguish phosphorylation catalyzed by the growth-associated kinase. However, for a more complete analysis of phosphorylation occurring at growth associated sites, the more extensive procedure given in Section III of this Chapter must be employed.

The procedure below is also applicable to the study of H1 histone phosphorylation *in vivo*. In nongrowing tissues, such as adult rat liver, the phosphopeptide analysis serves primarily to identify $Ser_{37}$ phosphorylation (*1*) and to distinguish it from radioactive contaminants normally present in H1 isolated by trichloroacetic acid extraction, since no *in vivo* phosphorylation of $Ser_{106}$ has so far been detected. In the experiments we have carried out to date in growing cells, H1 histone has been extracted directly from whole cells with dilute $H_2SO_4$ (in order to preclude the action of phosphatases) and isolated by differential trichloroacetic acid precipitation. In this case the amount of contamination is much greater, and a preliminary purification of the histone by ion-exchange chromatography is required before the phosphopeptide analysis is carried out. $Ser_{37}$ phosphorylation is distin-

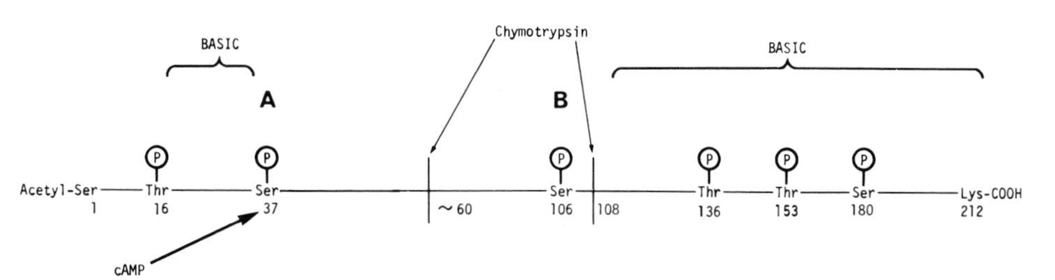


FIG. 1. Phosphorylation sites in lysine-rich (H1) histone. The positions of phosphorylation sites have been assigned by comparing the sequence of isolated tryptic phosphopeptides (*1,3,12*) with the current overall sequence of H1 histone worked out by Cole and co-workers (R. D. Cole, personal communication). Positions 16, 136, 153, and 180 are sites of growth-associated phosphorylation. Since amino acid deletions and insertions occur at various locations in different H1 histones, numerical assignments are not exact for all sequences. However, phosphorylation at sites corresponding to those shown has been observed in H1 histones from all mammalian species examined (calf, rabbit, rat, mouse). The cleavages which give rise to the N-terminal and C-terminal chymotryptic peptides and the location of basic clusters in the histone (*15, 16*) are also indicated.

guished from growth-associated phosphorylation, but again the second procedure is required for a more extensive analysis of growth-associated sites.

## A. Phosphorylation of H1 Histone

Phosphorylation is carried out enzymically with [$\gamma$-$^{32}$P]ATP or *in vivo* with [$^{32}$P]phosphate as the source of label. Examples of phosphorylation conditions and the specific activities and amounts of label employed are given in Chapter 14 and in the specific applications below.

## B. Isolation of Phosphorylated H1 Histone

### 1. Isolation from Enzymic Reaction Mixtures

The reaction is stopped by the addition of trichloroacetic acid to a final concentration of 5%. Enzyme proteins are removed by centrifugation, and H1 histone is recovered in the supernatant. If the amount of enzyme protein is large compared to the amount of histone, recovery will be reduced owing to coprecipitation. Carrier H1 (1–2 mg) may be added to the supernatant if the amount of H1 present is much less than 1 mg. The trichloroacetic acid concentration is then raised to 25% by the addition of 100% (1 gm/ml) trichloroacetic acid, precipitation is allowed to occur on ice for 10–15 minutes, and H1 histone is collected by centrifugation at 3300 *g* for 15 minutes. The histone is dissolved as completely as possible in 6 ml of $H_2O$ with warming to 37° C, reprecipitated by the addition of 2 ml of 100% trichloroacetic acid, and chilled and centrifuged as above. The precipitate is redissolved and reprecipitated two additional times, washed once with acidified acetone (0.5 ml conc. HCl per 100 ml), twice with acetone, and dried in a vacuum desiccator.

### 2. Isolation from Rat Liver

Livers are frozen by immersion in liquid nitrogen at the time of removal and stored at −20°C. Each liver (5–10 gm) is homogenized at 4°C in a Waring Blendor for 1 minute at 85 V and 4 minutes at 45 V in 175 ml of 0.14 *M* NaCl–0.01 *M* sodium citrate (pH 6.0). A 1-ml aliquot of the homogenate is frozen immediately for subsequent measurement of the specific activity of ATP or liver inorganic phosphate pools. The homogenate is filtered through a single layer of Miracloth (Chicopee Mills, Inc.) and centrifuged 15 minutes at 3300 *g*. The sediment (crude chromatin) is suspended in 140 ml of homogenizing medium and recentrifuged. To extract H1 histone, the crude chromatin is blended in 47.5 ml of $H_2O$ for 1 minute at 120 V in a semimicro blender container, and 2.5 ml of 100% trichloroacetic acid is added during the next minute of blending. Blending is continued for an additional 30 seconds and the suspension centrifuged 5 minutes at 3300 *g*. The trichloroacetic acid concentration of the supernatant is adjusted to 18% by the addition of 0.16 volume of 100% trichloroacetic acid, and precipitation is allowed to take place on ice for 10–15 minutes before centrifuging for 20 minutes at 7000 *g*. The precipitate is dissolved in 6 ml of $H_2O$ with warming to 37°C, reprecipitated twice with 25% trichloroacetic acid, washed once with acidified acetone, twice with acetone and dried, as described above. The yield of H1 from 10 gm of rat liver is approximately 1 mg.

### 3. Isolation from Cultured Cells

The medium is aspirated and cells are frozen by immediately placing culture dishes on blocks of Dry Ice. The frozen cells may be stored at −20°C. A total of 19 ml of cold 0.4 *N* $H_2SO_4$ containing 3 mg of carrier H1 histone is added to ten 100-mm dishes, and the cells are removed by scraping and homogenized in a tight-fitting glass homogenizer. After centrifugation for 15 minutes at 3300 *g*, 1 ml of 100% trichloroacetic acid is added to the supernatant and the precipitate removed by centrifuging for 15 minutes at 37,000 *g*. (It is essential to obtain a water-clear supernatant at this step in order to avoid contamination of the final H1 histone product.) The trichloroacetic acid concentration of the supernatant is adjusted to 18% by the addition of 0.16 volume of 100% trichloroacetic acid, and the remainder of the isolation is carried out as described above.

The isolated H1 histone is purified by ion-exchange chromatography, using an elution procedure based on the gradient method of Kinkade and Cole (*13*). The procedure is sufficient to separate contaminants although it does not fully resolve Hl into its subcomponents. Hl histone, 1–3 mg, is dissolved in 0.3 ml of $H_2O$ and 2.7 ml of 7% guanidinium chloride (Mann, Ultrapure) in 0.1 *M* Na–$PO_4$ (pH 6.8) is added. Chromatography is carried out at room temperature. The sample is applied to a 0.9 × 15 cm column of Bio–Rex 70 resin, 200–400-mesh (Bio-Rad Laboratories), equilibrated with 7% guanidinium chloride–0.1 *M* Na–$PO_4$ (pH 6.8) and the column eluted at a flow rate of 2.9 ml per hour with a linear gradient of 7 to 14% guanidinium chloride–0.1 *M* Na–$PO_4$ (pH 6.8) (total gradient volume 140 ml). H1 histone components are eluted between 60 and 90 ml. The column is regenerated for re-use by washing with 35 ml of 40% guanidinium chloride–0.1 *M* Na–$PO_4$ (pH 6.8) and reequilibrating with 7% guanidinium chloride–0.1 *M* Na–$PO_4$ (pH 6.8). Fractions containing the eluted histone are pooled and dialyzed vs $H_2O$ overnight at 4°C. The histone is precipitated by adding trichloroacetic acid to a final concentration of 25%, and is washed once with acidified acetone, twice with acetone and dried, as above.

## C. Tryptic Digestion of H1 Histone

The following conditions are suitable for digestion of up to 2 mg of H1 histone. The isolated histone is dissolved in 0.15 ml of 0.067 $M$ $NH_4HCO_3$ and 0.005 ml of freshly prepared trypsin solution (Worthington, 5 mg/ml in $H_2O$) is added. Digestion is carried out for 1 hour at 37°C and any insoluble material is removed by centrifugation. (The insoluble material typically contains negligible amounts of radioactivity.) The clear supernatant may be kept on ice several hours before electrophoresis, or stored frozen.

## D. Phosphopeptide Analysis

The nature of the sites phosphorylated in preparations of H1 histone is determined by comparison of tryptic phosphopeptides derived from the sample under investigation with marker phosphopeptides derived from H1 histone phosphorylated in specific sites by purified histone kinases.

### 1. Preparation of Radioactive Marker Phosphopeptides

H1 histone is phosphorylated by purified enzymes with $[\gamma\text{-}^{32}P]$ATP under the conditions given in Chapter 14. Tryptic or chymotryptic digestion and isolation of the peptides is carried out by the methods described in this paper. For isolation of tryptic phosphopeptides from Hl histone phosphorylated by purified HK1 and HK2 (containing $Ser_{37}$ and $Ser_{106}$, respectively) electrophoresis at pH 7.9 as described below is generally sufficient. Both peptides move toward the anode. The peptide containing $Ser_{106}$ migrates approximately 1.6 times more rapidly than the peptide containing $Ser_{37}$ (see Fig. 2). Marker phosphopeptides derived from Hl phosphorylated by purified growth-associated kinase may be used as mixtures, or peptides containing individual growth-associated phosphorylated sites may be isolated by the methods given later in this paper, as appropriate.

### 2. High-Voltage Paper Electrophoresis at pH 7.9

As noted above, this procedure separates the phosphopeptides containing $Ser_{37}$ and $Ser_{106}$. However, if any growth-associated phosphorylation is present, or if the Hl histone contained significant amounts of nonhistone impurities, the chromatographic step described below must also be employed for accurate assessment of phosphorylation at $Ser_{37}$ and $Ser_{106}$, and for detection of growth-associated phosphorylation.

Commonly used electrophoresis papers (e.g., Whatman 3MM) typically contain large amounts of a material that interferes with subsequent cellulose thin-layer chromatography of eluates from the paper. The use of the paper and washing procedure described here is necessary to obtain reliable

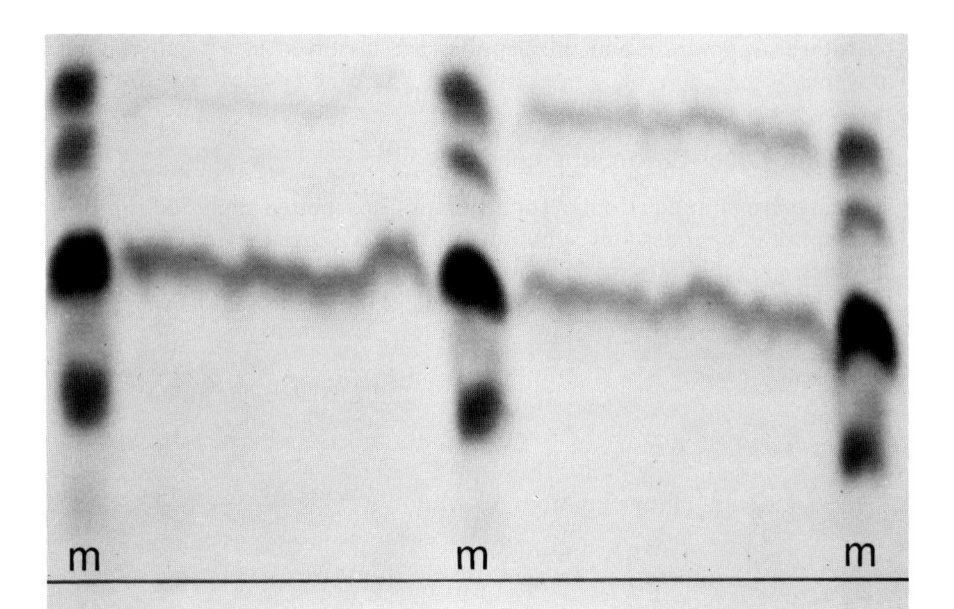


FIG. 2. Analysis of histone kinase activities present in crude and purified fractions of calf liver. Tryptic phosphopeptides from H1 histone phosphorylated by purified calf liver HK1 (left) and by a crude 105,000 *g* supernatant from calf liver (right) were separated by electrophoresis at pH 7.9. Mixtures of marker peptides are designated by *m*. The phosphorylation of H1 almost exclusively at $Ser_{37}$ by the purified enzyme preparation is shown by the presence of a single phosphopeptide corresponding to the marker peptide containing this site. The crude extract contains approximately equal amounts of HK1 and HK2 activity, as shown by the presence of the additional faster moving phosphopeptide, containing $Ser_{106}$. Phosphorylation was carried out under standard assay conditions (see Chapter 14) with 50 $\mu$g of 105,000 *g* supernatant and 0.002 unit of purified HK1, using [$^{32}$P]ATP of specific activity 55,000 cpm/nmole.

chromatography of eluted peptides. Schleicher and Schuell (S & S) No. 589 paper is autoclaved for 90 minutes in 0.05 *M* $NH_4OH$, washed by transferring three times to fresh deionized water, soaked overnight in 95% ethanol, and air-dried.

Tryptic digests (0.15 ml) are applied in 6-cm bands to an origin near the cathode of paper previously wetted with 0.06 *M* $NH_4HCO_3$. Appropriate marker peptides are applied adjacent to samples. Electrophoresis is carried out on a water-cooled flat plate (Savant Instruments) for 30 minutes at 100 V/cm. Wicks leading to buffer vessels are isolated from electrophoresis papers by four layers of Cellophane dialysis membrane. Phosphopeptides are located by autoradiography on Kodak blue brand X-ray film, and bands corresponding in mobility to marker phosphopeptides containing $Ser_{37}$ and $Ser_{106}$ are eluted from the paper with 0.01 *N* HCl·HCl is removed.

by rotary evaporation and the peptides are dissolved in a small volume of water.

## E. Thin-Layer Chromatography

Eluates from the electrophoretic separation are spotted alongside appropriate marker peptides on cellulose thin layers (Eastman No. 6064) and chromatographed in pyridine/*n*-butanol/glacial acetic acid/$H_2O$, 10/15/3/12. Radioactive peptides are again visualized by autoradiography. The

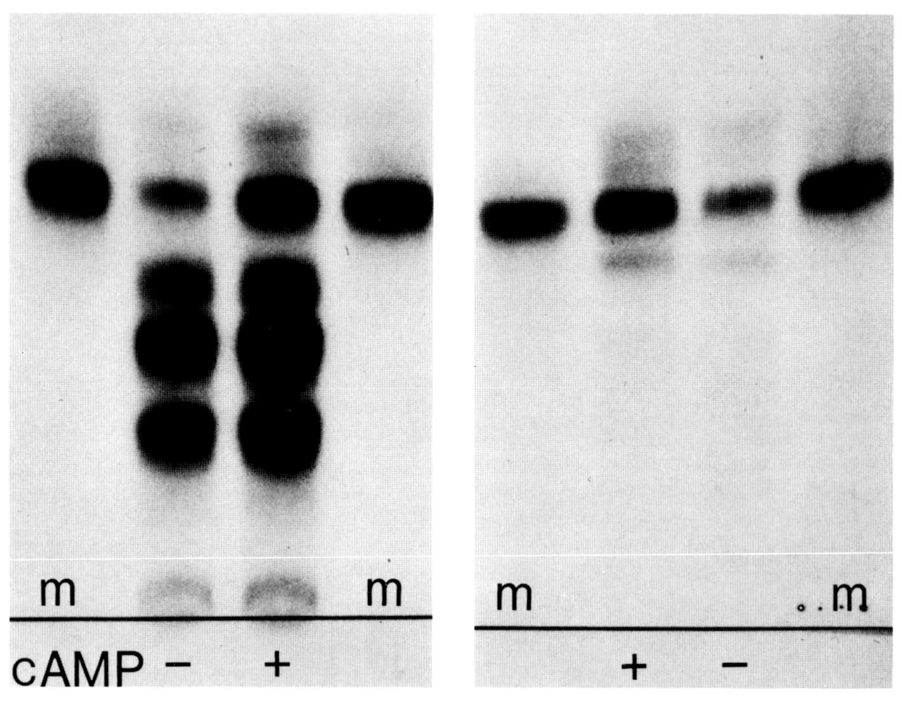


FIG. 3. Analysis of chromatin-bound histone kinase activity in growing and nongrowing cells. Tryptic phosphopeptides from H1 histone phosphorylated by washed chromatin preparations were separated by electrophoresis at pH 7.9 followed by thin-layer chromatography as described in the text. Left: Chromatin from randomly growing mouse L5178Y lymphoma cells; right: chromatin from calf thymus. *m* designates marker peptide containing $Ser_{37}$. HK1 activity, stimulated by cAMP, is present in both chromatin preparations as shown by the phosphopeptides migrating with the marker peptide. The presence of large amounts of growth-associated histone kinase in the lymphoma chromatin, not significantly stimulated by cAMP, is indicated by the phosphopeptides of lower chromatographic mobility. Phosphorylation was carried out under standard assay conditions (see Chapter 14) with chromatin containing 100 $\mu$g of protein as the source of enzyme, using [$^{32}$P]ATP of specific activity 80,000 cpm/nmole. cAMP, 10 $\mu M$, was present where indicated.

peptide containing $Ser_{37}$ has a chromatographic mobility approximately 1.4 times greater than the peptide containing $Ser_{106}$. Phosphorylation of growth-associated sites is best revealed by the presence of peptides that migrate together with the peptide containing $Ser_{37}$ on electrophoresis at pH 7.9, but have a lower chromatographic mobility (see Figs. 3 and 4). Phosphopeptides from individual growth-associated sites are not resolved by these procedures, however. A quantitative indication of the phosphorylation of $Ser_{37}$ and $Ser_{106}$ in H1 histone can be obtained at this point by counting the appropriate peptide spots from the chromatogram.

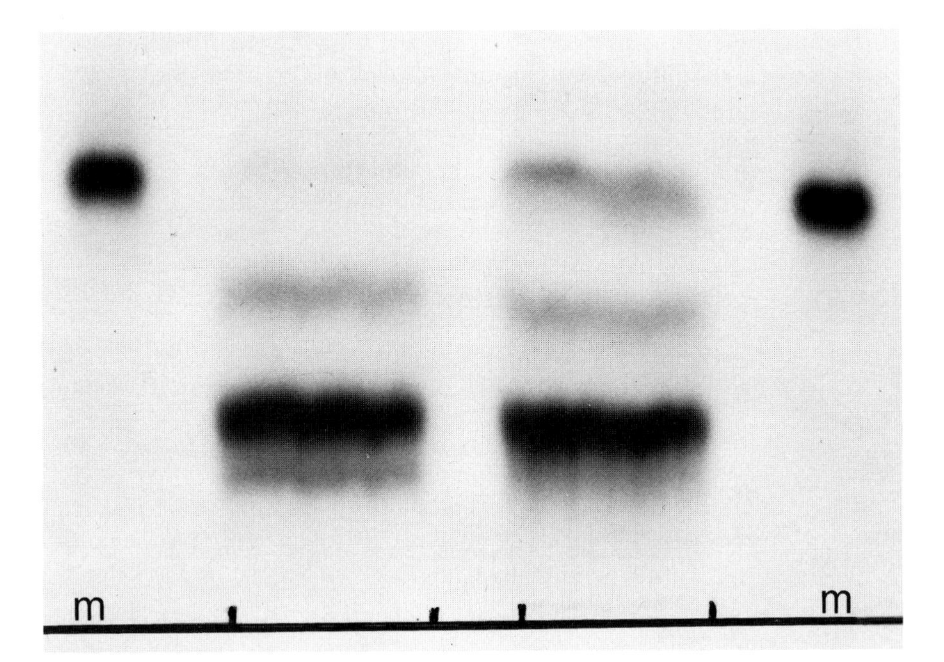


FIG. 4. Analysis of site-specific H1 histone phosphorylation occurring *in vivo* in Reuber hepatoma cells. Tryptic phosphopeptides from H1 histone phosphorylated in control cells (left) and in cells with elevated cAMP levels produced by cholera toxin treatment (*14*) (also B. Leichtling, J. Wimalasena, and W. D. Wicks, personal communication) (right) were separated by electrophoresis at pH 7.9 followed by thin-layer chromatography as described in the text. *m* designates marker peptide containing $Ser_{37}$. cAMP stimulation of phosphorylation at $Ser_{37}$ in the cholera toxin-treated cells is evident from the increase in labeled phosphopeptide corresponding to marker peptide. The phosphorylation of growth-associated sites, contained in the phosphopeptides of lower chromatographic mobility, is unaffected by elevated cAMP levels. Reuber H35 hepatoma cells (twenty 100-mm dishes) were labeled with [$^{32}$P]phosphate (160 $\mu$Ci/ml) for 2.5 hours in serum-free low phosphate (10 $\mu M$) medium. Cholera toxin (0.1 $\mu$g/ml) was added to ten of the dishes 30 minutes after [$^{32}$P]phosphate. H1 histone was isolated and purified by ion-exchange chromatography as described in the text.

In H1 histone samples in which the ratio of growth-associated phosphorylation to $Ser_{37}$ phosphorylation is extremely high (e.g., in rapidly growing cells), it is necessary to check the peptide containing $Ser_{37}$ for contamination by material derived from growth-associated sites by a further electrophoretic separation, if an accurate assessment of $Ser_{37}$ phosphorylation is important. Electrophoresis at pH 1.9 in the system described below is employed for this purpose.

## F. Examples of Applications

Details of the use of the above procedures to analyze protein kinase activities in crude and purified enzyme preparations and to distinguish cAMP-regulated H1 phosphorylation from growth-associated phosphorylation *in vivo* are given in Figs. 2–4.

# III. Analysis of Growth-Associated Phosphorylation in H1 Histone

In order to facilitate a more complete analysis of the growth-associated phosphorylation sites in H1 histone, we have employed chymotryptic digestion, as described by Bustin and Cole and co-workers (*15, 16*) to cleave the histone into several large peptides derived from different regions of the histone, which are then analyzed separately for phosphorylation sites. All the growth-associated phosphorylation sites occur in chymotryptic peptides derived from the N-terminus and C-terminus of the histone, and are absent from the central hydrophobic region (Fig. 1).

## A. Chymotryptic Digestion of H1 Histone

After phosphorylation and isolation as described in the preceding section, the histone is dissolved in 0.025 *M* Tris (pH 8.0) at a concentration of 10 mg/ml or less, and digested for 1 hour at 37°C with 1.5 $\mu$g of chymotrypsin (Worthington) per milligram of H1. Insoluble material is removed by centrifugation, and gel filtration as described below is begun immediately.

## B. Isolation of Chymotryptic Peptides

The chymotryptic digest is applied to a 0.9 × 180 cm column of Sephadex G-100, 40–120 $\mu$m beads, equilibrated with 0.02 *N* HCl. The column is eluted with 0.02 *N* HCl at a flow rate of 2.8 ml/hour, and radioactivity and peptide absorption at 218 or 230 nm are monitored in the effluent. Alternatively, a 2.5 × 250 cm column operated at a flow rate of 20 ml/hour may be used for samples up to 60 mg. Three major peptide peaks are obtained (Fig. 5); these, in order of elution, are derived from the C-terminal 105 residues, the N-terminal 60 or so residues, and the central hydrophobic region of H1 histone (*15,16*) (see also Fig. 1). Only the first two peaks contain significant radioactivity, unless the histone is also phosphorylated at $Ser_{106}$. Peak tubes are pooled, and HCl is removed by repeated evaporation to dryness in a rotary evaporator.

## C. Phosphopeptide Analysis of the C-Terminal Region of H1 Histone

### 1. High-Voltage Electrophoresis at pH 7.9

The chymotryptic peptide eluting first from the Sephadex G-100 column is digested with trypsin, and the digest is subjected to electrophoresis at pH 7.9, as described above. Marker peptide containing $Ser_{37}$ of H1 histone is

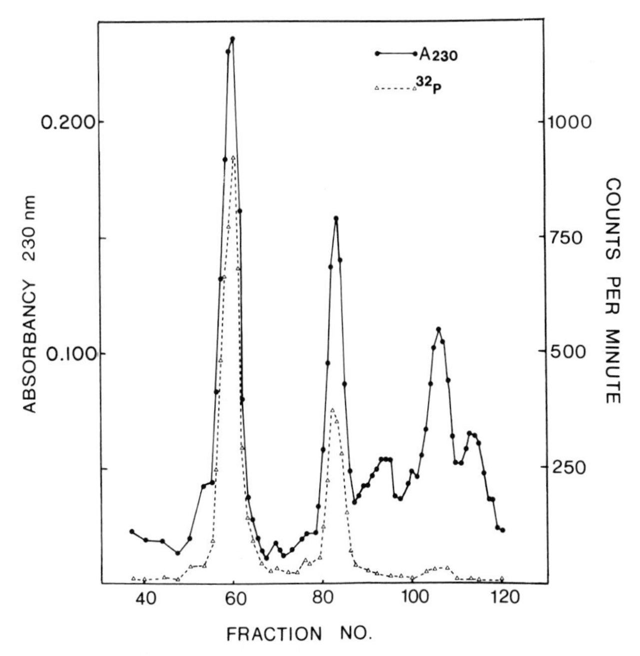


FIG. 5. Sephadex G-100 chromatography of a chymotryptic digest of calf thymus H1 histone phosphorylated with growth-associated histone kinase. Histone (18 mg) containing 2046 nmoles of [$^{32}$P]phosphate (461 cpm/nmole) was digested with chymotrypsin and chromatographed on a 2.5 × 250 cm column of Sephadex G-100 as described in the text. Ten-milliliter fractions were collected and radioactivity in 0.1-ml aliquots was determined.

employed, and the broad band of radioactive peptides with a mobility similar to this marker is eluted. Other radioactive peptides in the digest have not yet been examined further.

#### 2. High-Voltage Electrophoresis at pH 1.9

Eluates from the electrophoretic separation at pH 7.9 are applied to an origin near the anode of S & S No. 589 paper previously wetted with formic acid-acetic acid buffer (pH 1.9) (25 ml of 88% formic acid plus 87 ml of glacial acetic acid per liter). Electrophoresis is carried out as described above at 100 V/cm for 30 minutes. Three radioactive peptides, containing the major growth-associated phosphorylation sites in the C-terminal region of H1 histone, migrate toward the cathode. The sequences determined for these phosphopeptides from calf thymus H1 (*12*) are, in order of increasing electrophoretic mobility: (a) Lys-Ala-Thr-Gly-Ala-Ala-ThrPO$_4$-Pro-Lys; (b) Val-Ala-Lys-SerPO$_4$-Pro-Lys; (c) Lys-ThrPO$_4$-Pro-Lys. Comparison of these sequences with the current overall sequence of Hl derived by Cole and co-workers (R. D. Cole, personal communication)[1] indicates that the phosphorylation sites occur at Thr$_{136}$, Ser$_{180}$, and Thr$_{153}$, respectively. The sequences surrounding the phosphorylation sites in the C-terminus of mammalian H1 histones are similar to, but somewhat more varied than the Lys-SerPO$_4$-Pro-Lys sequences shown in trout H1 by Dixon (*17*).

### D. Phosphopeptide Analysis of the N-Terminal Region of H1 Histone

#### 1. Gel Filtration on Sephadex G-50

The second chymotryptic peptide to elute from the Sephadex G-100 column is digested with trypsin as described above and applied to a 1.5 × 50 cm column of Sephadex G-50 equilibrated with 0.02 *N* HCl. The column is eluted with 0.02 *N* HCl at a flow rate of 5 ml/hour and radioactivity and peptide absorption at 218 nm are monitored in the effluent. A typical elution pattern is shown in Fig. 6. In this pattern, the peaks of absorption at 218 nm represent peptides derived primarily from the unphosphorylated carrier H1 in the preparation. The major phosphopeptide peak is eluted more rapidly than the carrier peptides. This has been found to be due to blockade of tryptic cleavage at a lysine residue adjacent to the major growth-associated phosphorylation site in the N-terminal region of H1 histone. Amino acid analyses and partial sequencing (*12*) of the components of this peak

[1] Our study and evaluation of specific phosphorylation sites in H1 histone has been greatly aided by unpublished information on the overall sequence of H1 histone provided at frequent intervals by Dr. Cole.

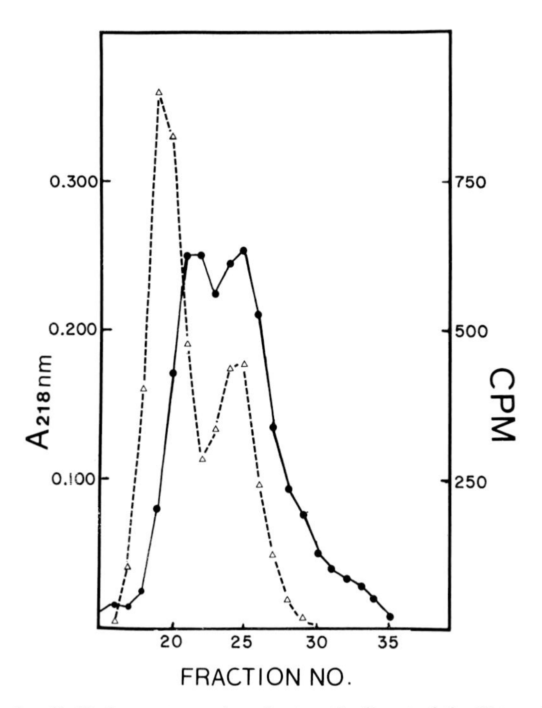


FIG. 6. Sephadex G-50 chromatography of a tryptic digest of the N-terminal region from H1 histone phosphorylated with growth-associated histone kinase. Phosphorylated peptides ($2.1 \times 10^5$ cpm total) derived from 0.5 mg of H1 were chromatographed in the presence of carrier peptides derived from 5 mg of histone as described in the text. Fractions of 2.5 ml were collected, and radioactivity in 0.05-ml aliquots was determined. ●——●, $A_{218}$; △---△, $^{32}P$.

(resolved by the precedures given below) show that it is composed of a family of closely similar phosphopeptides derived from the heterogeneous amino acid sequences identified by Rall and Cole (*18*) at the N-terminus of H1 histones.

The second phosphopeptide peak eluting from the G-50 column contains peptides derived from at least one additional growth-associated phosphorylation site in the N-terminal region of H1 histone. This site has not been detected in all subfractions of H1 histone. It has not been studied further. The tryptic peptide containing $Ser_{37}$ of H1 histone is also eluted in this peak.

## 2. High Voltage Electrophoresis at pH 7.9

The fractions containing the major phosphopeptide peak eluted from Sephadex G-50 are pooled and HCl is removed by repeated rotary evaporation. Electrophoresis at pH 7.9 on washed S & S No. 589 paper is carried out as described above. Three phosphopeptide bands moving toward the anode are resolved.

### 3. Cellulose Thin-Layer Chromatography

The phosphopeptide bands resolved by electrophoresis at pH 7.9 are eluted and subjected to cellulose thin-layer chromatography in the system described above. A total of 5 major phosphopeptides are resolved by the combined electrophoretic and chromatographic separation (Fig. 7). Each of these peptides contains approximately 19 amino acids, derived as noted above from the heterogeneous sequences at the N-terminus of H1 histone.

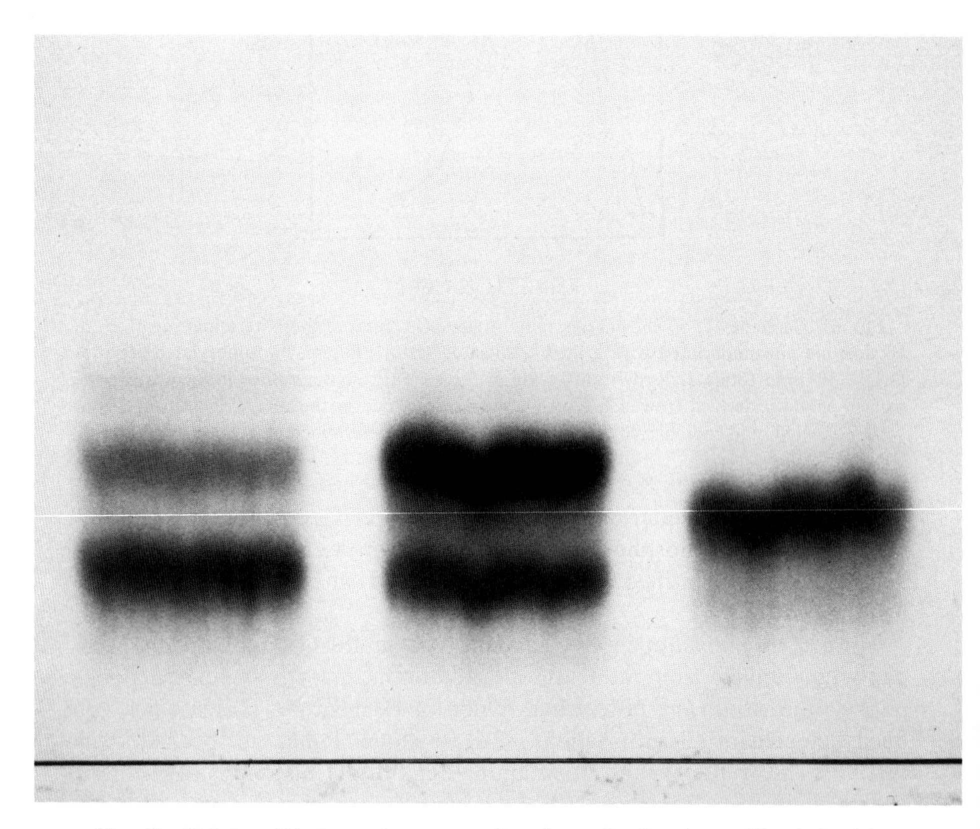

Fig. 7. Cellulose thin-layer chromatography of tryptic phosphopeptides derived from the N-terminal region of H1 histone phosphorylated with growth-associated histone kinase. Eluates of the three phosphopeptide bands resolved by electrophoresis at pH 7.9 were chromatographed as described in the text. Peptides of lowest electrophoretic mobility at left; intermediate at center; fastest at right. The phosphorylated histone was a sample of subcomponent 3 of calf thymus H1.

Depending on the particular sample of phosphorylated H1, the relative amounts of the peptides may vary. The partial sequence (Thr, Ser, $Glx_2$, $Pro_4$, $Ala_6$) Lys-Thr$PO_4$-Pro-Val-Lys has been determined for one of these peptides from calf thymus H1 (*12*). The phosphorylated threonine residue occurs at position 16. Indirect evidence indicates that the phosphorylated residue in other peptides of this group is located in a similar position; in some cases this residue is serine. In contrast to these observations with mammalian H1, phosphorylation in the N-terminal region of trout H1 has not been observed (*17*).

## E. Applications

The extended procedure given here for analysis of growth-associated phosphorylation sites has been applied to enzymically phosphorylated preparations of calf, rabbit, and rat thymus and mouse ascites cell H1 histone, and to the four subcomponents of calf and rabbit H1. In all cases the four phosphorylation sites described here have been identified. However, in subcomponents 3 and 4 of calf and rabbit H1, phosphorylation of the position at $Ser_{180}$ is relatively low. The procedure has also been used to show that positions at $Thr_{16}$, $Thr_{136}$, $Thr_{153}$, and $Ser_{180}$ are major sites of growth-associated phosphorylation *in vivo* in Reuber hepatoma and mouse ascites tumor cells. It is currently being applied to the study of H1 histone phosphorylation in several cultured cell lines under various conditions of growth and cyclic nucleotide stimulation.

ACKNOWLEDGMENT

This work was supported by Grant CA 12877 of the National Cancer Institute, U.S. Department of Health, Education, and Welfare.

REFERENCES

1. Langan, T. A., *Proc. Natl. Acad. Sci. U.S.A.* **64**, 1276 (1969).
2. Langan, T. A., *N. Y. Acad. Sci.* **185**, 166 (1971).
3. Langan, T. A., *Fed. Proc. Fed. Am. Soc. Exp. Biol.* **30**, 1089 (1971).
4. Langan, T. A., and Hohmann, P. *Fed. Proc., Fed. Am. Soc. Exp. Biol.* **33**, 1597 (1974).
5. Langan, T. A., and Hohmann, P., *in* "Chromosomal Proteins and Their Role in the Regulation of Gene Expression" (G. S. Stein and L. J. Kleinsmith, eds.), p. 113. Academic Press, New York, 1975.
6. Hohmann, P., Tobey, R. A., and Gurley, L. R., *J. Biol. Chem.* **251**, 3685 (1976).
7. Cohen, P., Watson, D. S., and Dixon, G. H., *Eur. J. Biochem.* **51**, 79 (1975).
8. Yeaman, S. J., and Cohen, P., *Eur. J. Biochem.* **51**, 93 (1975).
9. Nimmo, H. G., and Cohen, P., *FEBS Lett.* **47**, 162 (1974).
10. Langan, T. A., Rall, S. C., and Cole, R. D., *J. Biol. Chem.* **246**, 1942 (1971).
11. Lake, R. S., and Salzman, N. P., *Biochemistry* **11**, 4817 (1972).
12. Langan, T. A., *Fed. Proc., Fed. Am. Soc. Exp. Biol.* **35**, 1623 (1976).

13. Kinkade, J. M., Jr., and Cole, R. D., *J. Biol. Chem.* **241**, 5790 (1966).
14. Cuatrecasas, P., *Adv. Cyclic Nucleotide Res.* **5**, 79 (1975).
15. Bustin, M., Rall, S. C., Stellwagen, R. H., and Cole, R. D., *Science* **163**, 391 (1969).
16. Bustin, M., and Cole, R. D., *J. Biol. Chem.* **245**, 1458 (1970).
17. Dixon, G. H., Candido, E. P. M., Honda, B. M., Louie, A. J., MacLeod, A. R., and Sung, M. T., *Ciba Found. Symp.* [N.S.] **28**, 229 (1975).
18. Rall, S. C., and Cole, R. D., *J. Biol. Chem.* **246**, 7175 (1971).

# Chapter 14

## *Isolation of Histone Kinases*

THOMAS A. LANGAN

*Department of Pharmacology,*
*University of Colorado Medical Center,*
*Denver, Colorado*

## I. Introduction

The procedures described below allow the purification of three distinct histone kinases, each of which catalyzes the phosphorylation of H1 histone at a specific site or group of sites, as described in Chapter 13. The resulting enzymes are not purified to homogeneity. However, each preparation is substantially free of the other histone kinases and is low in protease and phosphatase activity. The purified enzymes are suitable for the preparation of substantial quantities (batches of 50 mg are quite practical) of H1 histone phosphorylated in specific sites for sequencing studies (*1–3*), for physical studies of H1-DNA interactions (*4,5*), and for use as substrates for the detection and characterization of histone phosphatases (*6*).

## II. Assay of Histone Kinase Activity

All the enzyme activities described here may be assayed under the same conditions. Reaction mixtures (0.25 ml) contain Tris-chloride buffer, pH 7.5, 12.5 $\mu$moles; $MgCl_2$, 1.25 $\mu$moles; dithiothreitol (DTT), 0.25 $\mu$mole; Hl histone (or other protein substrate), 250 $\mu$g; and $\gamma$-[$^{32}$P]ATP, 0.125 $\mu$mole. For routine assay of enzyme activity, enzyme sufficient to catalyze transfer of 0.2–2.0 nmoles of phosphate is employed, using [$^{32}$P]ATP of specific activity 1000–10,000 cpm/nmole. The reactions are initiated by addition of [$^{32}$P]ATP, incubated for 20 minutes at 37°C, and terminated by the addition of 2 ml of 28% trichloroacetic acid. The tubes are chilled on ice, and acid-insoluble material is collected on HA Millipore filters, washed with a total of 30–35 ml of 25% trichloroacetic acid, dried, and counted. Incorporation of radioactivity into enzyme proteins is determined in separate reaction mixtures containing no substrate (substrate added after trichloroacetic acid), and subtracted. A unit of enzyme activity is defined as the amount of enzyme catalyzing the transfer of 1 μmole of phosphate per hour to H1 histone under the above conditions.

## III. Isolation of Histone Kinase 1 (Cyclic AMP-Dependent Protein Kinase)

This enzyme has now been purified to homogeneity in a number of laboratories (e.g., *7–9*). The procedure described here is modified from an earlier method (*10*) and provides a useful partially purified preparation with characteristics as noted above.

### A. Initial Extract

Frozen calf liver (Pel-Freeze Biologicals), 60 gm, is homogenized in 250 ml of 50 m*M* potassium phosphate for 90 seconds at 120 V in a Waring Blendor. The pH of this buffer and all others used in the preparation is 7.5, unless indicated otherwise. The homogenate is centrifuged for 20 minutes at 35,000 *g*, and the resulting supernatant is centrifuged again for 1 hour at 100,000 *g*.

### B. Calcium Phosphate Gel Fractionation

The initial extract (237 ml) is diluted with an equal volume of water, and 237 ml of calcium phosphate gel suspension[1], 12 mg/ml, is added slowly with stirring. Stirring is continued for an additional 5 minutes, and the suspension is centrifuged for 5 minutes at 1500 *g*. The gel is washed twice by resuspension in 200 ml of 80 m*M* potassium phosphate and centrifugation as above. Enzyme is extracted with 200 ml of 0.5 *M* potassium phosphate, and the gel is removed by centrifuging for 15 minutes at 35,000 *g*. This and all subsequent solutions used in the preparation contain 1 m*M* DTT. Solid ammonium sulfate is added to the gel extract to 60% saturation (390 gm/liter), and the precipitate is collected by centrifuging for 15 minutes at 35,000 *g* and dissolved in 150 ml of 50 m*M* potassium phosphate. This

[1]Prepared according to Keilin and Hartree (*11*).

solution may be frozen and stored without significant loss of activity for up to 6 months.

### C. Alumina Gel Fractionation

A suspension of alumina C$\gamma$ gel (Sigma), 0.75 gm dry weight in 600 ml of 1 m*M* DTT, is added slowly with stirring to the solution obtained in the preceding step. The mixture is stirred for an additional 5 minutes and centrifuged at 3300 *g* for 10 minutes. The gel is extracted with 240 ml of 0.125 *M* potassium phosphate and removed by centrifuging as above.

### D. Ammonium Sulfate Fractionation

The alumina gel extract is fractionated by the addition of solid ammonium sulfate to 30% saturation (176 gm/liter), and the precipitate is removed by centrifuging for 15 minutes at 35,000 *g*. The ammonium sulfate concentration of the supernatant is raised to 47% saturation by the addition of 107 gm/liter, and the precipitate is collected by centrifugation as above and dissolved in 30 ml of 50 m*M* potassium phosphate.

### E. DEAE-Cellulose Chromatography

The solution obtained above is diluted to 160 ml with 1 m*M* DTT and applied to a 2.5 × 6 cm column of DEAE-cellulose (Serva) equilibrated with 25 m*M* potassium phosphate. The column is eluted with successive 50-ml portions of 0.10, 0.15, and 0.25 *M* potassium phosphate, and the column regenerated with 1 *M* phosphate buffer. The active fractions usually elute with the 0.15 *M* buffer. They are concentrated to 2–4 ml by ultrafiltration on an Amicon XM-50 filter and dialyzed 18 hours against 500 ml of 5 m*M* Tris buffer (pH 7.8). The yield is 7–12 mg of protein of specific activity 0.8–1.2 units/mg when assayed in the absence of cyclic AMP (cAMP). Stimulation by cAMP is approximately 3-fold. The enzyme can be quick-frozen in small aliquots in ethanol–Dry Ice and stored for up to 6 months without substantial loss of activity.

## IV. Isolation of Histone Kinase 2

This enzyme, like cAMP-dependent protein kinase, is widely distributed in animal tissues (*12*). It is distinct from the catalytic subunit of cAMP-dependent protein kinase as shown by the fact that it catalyzes phosphorylation of a separate site in the histone (*2, 13*), and is unaffected by the protein inhibitor of cAMP-dependent kinase (*13*). Calf thymus is used as a starting material because it contains a relatively high proportion of HK2.

### A. Initial Extract

Frozen calf thymus from newborn calves (Pel-Freeze Biological), 17 gm, is homogenized in 10 volumes of 50 m*M* potassium phosphate for 90 seconds at 120 V in a Waring Blendor. The pH of this buffer and all others used in the preparation is 7.5. The homogenate is centrifuged 20 minutes at 35,000 *g*, and any floating lipid material is carefully removed by aspiration.

### B. Calcium Phosphate Gel Fractionation

Of the initial extract, 150 ml is diluted with 225 ml of 1 m*M* DTT; 75 ml of calcium phosphate gel suspension[1], 12 mg/ml, is added slowly with stirring. Stirring is continued for an additional 5 minutes, and the gel is collected by centrifugation at 1500 *g* for 5 minutes. Enzyme is extracted with 120 ml of 0.25 *M* potassium phosphate, and the gel is removed by centrifugation for 10 minutes at 35,000 *g*. This and all subsequent solutions used in the preparation contain 1 m*M* DTT.

### C. Ammonium Sulfate Fractionation

The calcium phosphate gel extract is fractionated by the addition of solid ammonium sulfate to 45% saturation (277 gm/liter), and the precipitate is removed by centrifuging for 15 minutes at 35,000 *g*. The ammonium sulfate concentration of the supernatant is raised to 65% saturation by the addition of 134 gm/liter, and the precipitate is collected by centrifuging as above and dissolved in 20 ml of 20 m*M* potassium phosphate.

### D. DEAE-Cellulose Chromatography

The solution obtained above is diluted to 180 ml with 1 m*M* DTT and chromatographed on DEAE-cellulose as described in the procedure for preparation of HK1. The active fractions elute with the 0.25 *M* buffer. They are concentrated to 2–4 ml by ultrafiltration on an Amicon UM-1 filter. The yield is 4–5 mg of protein of specific activity 2–4 units/mg. After mixing with an equal volume of glycerol, the concentrated enzyme can be stored for periods in excess of 1 year at −20°C without substantial loss of activity.

## V. Isolation of Growth-Associated Histone Kinase

This enzyme is found bound to chromatin (*14*) and is present only in growing cells. Ehrlich ascites tumor cells are used as a source of relatively inexpensive, rapidly growing tissue.

### A. Growth of Ehrlich Ascites Cells

Swiss–Webster mice are inoculated intraperitoneally with 0.25 ml of cell suspension taken from the peritoneal cavity of a tumor-bearing mouse. Cells are harvested 1 week later by collecting peritoneal fluid and centrifuging cells 5 minutes at 500 *g*. The unwashed cells are frozen and stored at −70°C.

### B. Preparation and Extraction of Crude Chromatin

Frozen ascites cells, 40 gm, are homogenized in 360 ml of 75 m*M* NaCl–24 m*M* EDTA (pH 8.0) for 1 minute at 85 V and 4 minutes at 45 V, and the homogenate is centrifuged for 15 minutes at 1500 *g*. Care is taken to collect foamed material, which contains a large proportion of the enzyme. The sediment is washed once in 450 ml of homogenizing medium and twice in 450 ml of 50 m*M* Tris buffer (pH 8.0); 15 ml of 4 *M* NaCl are added, and the mixture is blended with a Polytron homogenizer (Brinkman, Inc.) for 20 seconds at full speed.

The salt concentration is then lowered to 0.4 *M* by the slow addition, with stirring, of 90 ml of 20 m*M* Tris buffer (pH 8.0), and the aggregated suspension is centrifuged for 45 minutes at 250,000 *g*. The supernatant is collected and 1 *M* potassium phosphate is added to give a final phosphate concentration of 0.2 *M*. The pH of the potassium phosphate used in this and all subsequent steps of the preparation is 7.5.

### C. Ammonium Sulfate Fractionation

In addition to growth-associated histone kinase, the chromatin extract obtained in the preceding step contains substantial amounts of HK1 (cAMP-dependent protein kinase), which is not easily separated by a variety of fractionation procedures. In order to achieve separation, cAMP-dependent protein kinase is dissociated into its regulatory and catalytic subunits by the addition of cAMP (final concentration 10 $\mu M$) to the extract. The dissociated catalytic subunit of HK1 is then readily separated from the growth-associated histone kinase by the following procedure. Solid ammonium sulfate is added to the extract to 17.5% saturation (99.5 gm/liter), and the precipitate is removed by centrifuging for 15 minutes at 35,000 *g*. The ammonium sulfate concentration of the supernatant is raised to 35% saturation by the addition of 106 gm/liter and the precipitate is collected by centrifuging as above. It is dissolved as completely as possible in 15 ml of 25 m*M* potassium phosphate by stirring for several hours, and allowed to stand overnight. Inactive, insoluble material is removed by centrifugation for 10 minutes at 3300 *g*. The supernatant, which contains the bulk of the growth-associated histone kinase activity, has been shown by the phosphopeptide analysis procedures described in Chapter 13 to be practically free of activity due to HK1 and its catalytic subunit (phosphorylation of $Ser_{37}$ in H1 histone) and HK2 (phosphorylation of $Ser_{106}$).

## D. Calcium Phosphate Gel Fractionation

Nine volumes of 25 m*M* potassium phosphate are added to the fraction obtained in the previous step, and 3.75 ml of calcium phosphate gel suspension, 12 mg/ml, is added slowly with stirring. Stirring is continued for an additional 5 minutes, and the gel is collected by centrifugation at 1500 *g* for 5 minutes. The gel is washed once by resuspension in 15 ml of 0.1 *M* potassium phosphate and collected by centrifugation as above. Enzyme is extracted with 12 ml of 0.2 *M* potassium phosphate and the gel removed by centrifuging 10 minutes at 35,000 *g*. The yield is 8–12 mg of protein of specific activity 0.6–0.8 unit/mg. The enzyme can be quick-frozen in small aliquots in ethanol–Dry Ice and stored for up to 6 months without substantial loss of activity. Activity is lost with repeated freezing and thawing or upon dialysis or exposure of the enzyme preparation to solutions of low ionic strength.

TABLE I

RELATIVE ACTIVITY OF HISTONE KINASES TOWARD VARIOUS HISTONE SUBSTRATES[a]

| Histone substrates | Histone kinase 1 | Histone kinase 2 | Growth-associated histone kinase |
|---|---|---|---|
| H1 | 100 | 100 | 100 |
| H2B | 141 | 17 | 4.6 |
| H3 | 3.6 | 0.8 | 2.0 |
| H2A + H4 | 2.0 | 1.6 | 0.2 |

[a] Activity was determined under the standard assay conditions given in the text.

## VI. Substrate Specificity of Histone Kinases

The relative activity of the three histone kinase preparations towards various histone substrates is shown in Table I. Activity towards casein, phosvitin, and bovine serum albumin is less than 2% of the activity with H1 histone in each case. The specific sites in H1 histone phosphorylated by the enzymes are described in Chapter 13.

## VII. Utilization of Histone Kinases for the Preparation of Enzymically Phosphorylated H1 Histones

### A. Preparation of H1 Histone Phosphorylated at $Ser_{37}$ and $Ser_{106}$

Each reaction mixtures (5 ml) contains Tris-chloride buffer, pH 7.5, 250 $\mu$moles; $MgCl_2$, 25 $\mu$moles; DTT, 5 $\mu$moles; H1 histone, 25 mg; and ATP as indicated below. Phosphorylation of $Ser_{37}$ is carried out with 0.8 unit of purified HK1 in the presence of 5 $\mu M$ cAMP. For phosphorylation of $Ser_{106}$, 1.25 units of HK2 are employed. The reactions are started by the addition of 2.5 $\mu$moles of ATP and the mixtures are incubated at 37°C for 4 hours. H1 histone phosphorylated at both $Ser_{37}$ and $Ser_{106}$ is prepared by including both enzyme preparations plus cAMP in the reaction mixtures, and adding ATP at zero time (3 $\mu$moles) and 90 minutes (1 $\mu$mole). For each enzymically phosphorylated histone preparation, a control preparation is carried through an identical procedure of incubation and isolation, except that ATP is omitted from the reaction mixtures.

The reactions are stopped by the addition of trichloroacetic acid to a final concentration of 5%, and enzyme protein is removed by centrifuging 5 minutes at 3300 *g*. An equal volume of 50% trichloroacetic acid is added to the supernatant, the tubes are chilled on ice for 15 minutes and H1 histone is recovered by centrifuging 10 minutes at 3300 *g*. The histone is redissolved in 6 ml of $H_2O$ and reprecipitated by the addition of 2 ml of 100% trichloroacetic acid, chilling, and centrifuging as above. This step is repeated once. The histone is washed twice with acidified acetone (0.5 ml conc. HCl per 100 ml) and once with acetone, and dried in a vacuum desiccator.

## B. Preparation of Hl Histone Phosphorylated at Growth-Associated Phosphorylation Sites

Each reaction mixture (10 ml) contains Tris-chloride buffer, pH 7.5, 500 μmoles; $MgCl_2$, 50 μmoles; DTT, 5 μmoles; Hl histone, 5 mg; purified growth-associated histone kinase, 0.6 unit; and ATP as indicated below. Under the above reaction conditions, the volume of enzyme solution added [in 0.2 *M* potassium phosphate (pH 7.5)] must not exceed 1 ml, owing to the inhibitory effect of elevated ionic strength on the activity of the growth-associated histone kinase. If larger volumes of enzyme solution are required, the Tris buffer may be omitted and up to 1.8 ml of enzyme solution added. The reactions are started by the addition of 5 μmoles of ATP, and the mixtures are incubated for 4 hours at 37°C. The reactions are terminated by the addition of trichloroacetic acid to a final concentration of 5%, and the histone is isolated as described in the section above. Control histone preparations, carried through an identical procedure of incubation and isolation except for the omission of ATP from the reaction mixture, are made in parallel with each phosphorylated preparation.

## C. Purification of Phosphorylated and Control Hl Histone Preparations

After isolation from the enzymic reaction mixtures, phosphorylated and control histone preparations are purified by the gradient method of Kinkade and Cole (*15*). The samples (up to 75 mg) are dissolved in 6 ml of $H_2O$, and 9 volumes of 7% guanidinium chloride are added. This and all other guanidinium chloride solutions employed contain 0.1 *M* sodium phosphate (pH 6.8). The samples are applied to a 2.5 × 10 cm column of Bio-Rex 70 resin (200–400-mesh) (Bio-Rad Laboratories) equilibrated with 7% guanidinium chloride, and histone is eluted with a linear gradient of 7% to 14% guanidinium chloride (400 ml total volume) at a flow rate of 16.6 ml/hour. Thirty-minute fractions are collected and absorbancy at 230 nm is monitored. This gradient separates impurities and degraded histone, but does not resolve Hl into its subcomponents. The fractions containing histone (major peak emerging after fraction 20) are pooled and dialyzed vs $H_2O$ and the Hl histone is recovered by precipitation with trichloroacetic acid at a final concentration of 25%. The histone is washed once with acidified acetone and twice with acetone, and dried in a vacuum desiccator. The column is regenerated for re-use by washing with 40% guanidinium chloride.

## D. Phosphate Content of Phosphorylated Histones

The phosphate content of the histone preparations is determined after dephosphorylation with alkali by the procedure described previously (*6*), or in a semimicro modification of this procedure employing alkaline phosphatase to release phosphate. For assay of histone phosphorylated in growth-associated sites the use of alkaline phosphatase is essential, since phosphate esterified to threonine in those preparations is partially resistant to alkali.

Histone containing 1.5–15 nmoles of phosphate is incubated for 1 hour at 37°C in 100 $\mu$l of 15 m*M* $NH_4HCO_3$ with 50 $\mu$g of *Escherichia coli* alkaline phosphatase (Worthington). Control experiments with $^{32}$P-labeled histone have shown that dephosphorylation is complete under these conditions. The reaction is stopped by the addition of 100 $\mu$l of 0.4 *N* $H_2SO_4$ and 10 $\mu$l of 0.01 *M* silicotungstic acid in 0.1 *N* $H_2SO_4$. Fifty microliters of 5% ammonium molybdate in 4 *N* $H_2SO_4$ and 300 $\mu$l of isobutanol–benzene 1:1 is added, the mixture is emulsified with a vortex mixer, and the phases are separated by a brief centrifugation at room temperature. Then 200 $\mu$l of the solvent phase is transferred to a tube containing 90 $\mu$l of ethanol-concentrated sulfuric acid 49:1, and 10 $\mu$l of stannous chloride reagent[2] is added. The resulting color is read at 660 nm in a Zeiss spectrophotometer in covered semimicro cuvettes. Standards of inorganic phosphate are carried through the entire procedure of incubation, extraction, and color development.

The phosphate content of phosphorylated histone preparations are typically 45–50 nmoles/mg for Hl phosphorylated at $Ser_{37}$ or $Ser_{106}$, and 90–100 nmoles/mg for histone phosphorylated at both these sites. These levels correspond to close to 1 and 2 moles of phosphate per mole of Hl histone, respectively. Hl histone phosphorylated to saturation with growth-associated histone kinase typically contains 100–110 nmoles of phosphate per milligram (2.0–2.2 moles per mole of histone). Control H1 preparations, as well as H1 histone when isolated from calf thymus or other slow-growing or nongrowing tissues, contain less than 0.2 mole of phosphate per mole of histone.

ACKNOWLEDGMENT

This work has supported by Grant CA 12877 of the National Cancer Institute, U.S. Department of Health, Education, and Welfare.

We thank Dr. Walden Roberts for gifts of mice bearing Ehrlich ascites tumors.

[2]Stannous chloride stock solution (10 gm of $SnCl_2 \cdot 2H_2O$ plus 25 ml of concentrated HCl, stored in the dark at 4°C) is diluted 1:200 with 1 *N* $H_2SO_4$ shortly before use.

## References


1. Langan, T. A., *Proc. Natl. Acad. Sci. U.S.A.* **64**, 1276 (1969).
2. Langan, T. A., *Fed. Proc., Fed. Am. Soc. Exp. Biol.* **30**, 1089 (1971).
3. Langan, T. A., *Fed. Proc., Fed. Am. Soc. Exp. Biol.* **35**, 1623 (1976).
4. Adler, A. J., Schaffhausen, B., Langan, T. A., and Fasman, G. D., *Biochemistry* **10**, 909 (1971).
5. Adler, A. J., Langan, T. A., and Fasman, G. D., *Arch. Biochem. Biophys.* **153**, 769 (1972).
6. Meisler, M. H., and Langan, T. A., *J. Biol. Chem.* **244**, 4961 (1961).
7. Beavo, J. A., Bechtel, P. J., and Krebs, E. G., *in.* "Methods Enzymology" (J. G. Hardman and B. W. O'Malley, eds.), Vol. 38, p. 299. Academic Press, New York, (1974).
8. Erlichman, J., Rubin, C. S., and Rosen, O. M., *J. Biol. Chem.* **248**, 7607 (1973).
9. Tao, M., *in* "Methods in Enzymology" (J. G. Hardman and B. W. O'Malley, eds.), Vol. 38, p. 315. Academic Press, New York, (1974).
10. Langan, T. A., *Science* **162**, 579 (1968).
11. Keilin, D., and Hartree, E. F., *in* "Methods in Enzymology" (S. P. Colowick and N. O. Kaplan, eds.), Vol. 1, p. 98. Academic Press, New York, (1955).
12. Langan, T. A., *Ann. N. Y. Acad. Sci.* **185**, 166 (1971).
13. Vandepeute, J., and Langan, T. A., In preparation. (1977).
14. Lake, R. S., and Salzman, N. P., *Biochemistry* **11**, 4817 (1972).
15. Kinkade, J. M., Jr., and Cole, R. D., *J. Biol. Chem.* **241**, 5790 (1966).

# Chapter 15

## *Chromosomal Protein Phosphorylation on Basic Amino Acids*

ROBERTS A. SMITH, RICHARD M. HALPERN, BERNDT B. BRUEGGER, ALBERT K. DUNLAP, AND OSCAR FRICKE

*Department of Chemistry,*
*University of California, Los Angeles,*
*Los Angeles, California*

### I. General Considerations

Phosphorylation of chromosomal proteins has generally been measured either after isolation of phosphorylated chromosomal proteins from the nucleus or after the action of a chromosomal protein kinase in an *in vitro* system. Almost totally, these procedures involve one or the other step carried out at an acidic pH value such that the only surviving phosphoryl linkage is found on the hydroxyl group of serine or threonine. Usually the protein is precipitated, generally by acid, followed by suitable washing or extraction to remove nonprotein phosphorus components. Sometimes the protein is dissolved or suspended in a suitable medium followed by removal of nonprotein phosphorus components by ion exchange or extraction procedures. Almost invariably these procedures involve the use of dilute mineral acid solutions in order to avoid chromosomal protein aggregation and/or proteolytic degradation. Such procedures obviously preclude the detection and isolation of phosphorylated components labile to the acidic pH values employed.

N-phosphorylated compounds have been recognized as phosphoryl donors since Fiske and SubbaRow (*1*) identified *N*-phosphorylcreatine. The occurrence of *N*-phosphoryl linkages in proteins was first discovered by Boyer and his colleagues (*2*) with their isolation of N-phosphorylated succinyl-CoA synthase from mitochondria. The very rapid identification of the N-phosphorylated species as 3-phosphohistidine was quickly followed by the demonstration of other N-phosphorylated enzymes, such as the phosphoenolpyruvate transferase of Roseman (*3*) and the phosphoramidatehexose transferase system described by Stevens-Clark *et al.* (*4*).

The single most outstanding property of N-phosphorylated compounds is their extreme sensitivity to acidic pH and their relative stability under basic conditions (*5–7*). Most N-phosphorylated species have half-lives of less than a few minutes in $10^{-2}$ *M* mineral acid at room temperature, while they are virtually stable at alkaline pH values. This property is in contrast to that of phosphorylated peptide-bound serine, which is quite stable at normally employed acidic pH values, but undergoes rapid $\beta$-elimination at alkaline pH values (*8*). Free phosphoserine, on the other hand, is relatively more base stable (*9*). It is this difference in properties between the *O*-phosphoryl and *N*-phosphoryl linkages in proteins that forms the basis of procedures we have used thus far for isolation and demonstration of such *N*-posphoryl linkages in histones (*10, 11*).

The procedures described herein permit the detection of *N*-phosphoryl amino acids in chromosomal proteins based upon isolation of $^{32}P$-labeled *N*-phosphoryl substances. The $^{32}P$-labeled chromosomal proteins are isolated either after enzymic reactions have taken place utilizing $\gamma$-$^{32}P$-labeled ATP as phosphoryl donor or directly from whole animal tissues excised and fractionated after rapid sacrifice of animals following administration of $^{32}P_i$ intraperitoneally. We have used the procedures directly most with rat liver preparations and/or rat mammary tumor Walker-256 carcinosarcoma or directly from enzyme reaction mixtures.

## II. Phenol Extraction Procedure

### A. Principles

Chromosomal proteins and lipids are extracted into phenol, while $^{32}P_i$ and/or [$\gamma^{32}P$]ATP are removed by careful washing of the phenol layer. $^{32}P$ left in the phenol extract gives a measure of the total covalently bound protein and lipid $^{32}P$. For further elucidation of the nature of the phosphorylated material, the radioactivity can be precipitated from phenol by acetone and the water-insoluble lipid phosphate extracted from the precipitate by organic solvents. The phosphate subsequently released by acid hydrolysis is a measure of the N-phosphorylated amino acid, and phosphoserine and phosphothreonine are measured as the acid-stable fraction.

The following procedure is slightly modified from that of Boyer and Bieber (*12*).

## B. Procedure

Incubation mixtures containing from 0.5 to 20 mg of chromosomal proteins in 1–2 ml were stopped with 2 ml of 88% phenol (adjusted to pH 8 with 0.01 *M* sodium phosphate buffer). Each mixture was added to a 40-ml centrifuge tube containing an additional 8 ml of the same solution. The solution was mixed thoroughly with 25 ml of a washing buffer [0.01 *M* EDTA–0.01 *M* sodium pyrophosphate (pH 8.3)] and centrifuged at about 2500 rpm in a swinging-bucket clinical centrifuge. The upper aqueous layer was removed by careful aspiration. Wash buffer was again added, and the washing procedure was repeated six times. The $^{32}P$-containing proteins were then precipitated by the rapid addition of about 5 volumes of cold acetone. The mixture was usually stored at −20° C for several hours or overnight, and the precipitate was ultimately collected by centrifugation. Any lipids were removed by successive extraction with about 5 ml of chloroform–methanol (1:1, v:v), 5 ml chloroform–methanol–water (20:10:1, v:v:v), and 5 ml of methanol, in each case centrifuging in a clinical centrifuge to separate the residue from the organic solvents. The protein residue can be treated with 3 ml of 0.3 *N* trichloroacetic acid–1 m*M* sodium phosphate for 3 minutes in a boiling water bath. The mixture is cooled and the proteins removed by centrifugation in a clinical centrifuge. An aliquot of the supernatant fluid is counted, giving a measure of the acid-labile *N*-phosphoryl protein phosphate. The protein residue is washed twice with about 10 ml of phosphate buffer, and the $^{32}P$ remaining with the precipitate is used as a measure of the acid-stable phosphoprotein.

In our experience, very little $^{32}P$-phospholipid is formed in the short enzymic incubation mixtures, and often we dispense with the solvent extraction procedure for such mixtures. In such cases, phosphoserine and phosphothreonine are eliminated, since we usually stop the reaction by making the mixture 0.15 *M* in sodium hydroxide and heating it to 60°C for 15 minutes prior to the phenol extraction. Thus, $^{32}P$ remaining in the phenol layer is a direct measure of the amount of N-phosphorylated protein formed.

# III. Isolation of [$^{32}P$]Phosphohistones from Whole Animal Tissues

Rats were sacrificed by decapitation between 1 and 3 hours after the intraperitoneal administration of $^{32}P_i$ in 0.9% NaCl solution. The tissues of choice (usually liver, regenerating liver, or Walker-256 carcinosarcoma) were rapidly excised and placed in ice cold 0.25 *M* sucrose–0.05 *M* Tris·HCl (pH 7.5)–0.05 *M* NaCl. After the adhering connective tissues were removed, the nuclei were isolated by a modification of the method of Chauveau *et al.* (*13*). All procedures were performed at ice bath temperatures. The tissues were minced with scissors in an isotonic solution containing 0.25 *M* sucrose and 1 m*M* $MgCl_2$. The minced tissue was homogenized with a Potter–Elvehjem homogenizer using a motor-driven pestle (0.005–0.007 inch clearance) for 40 strokes. The homogenate was filtered through one layer of coarse cheesecloth, followed successively by filtration through one layer of fine cheesecloth and eight layers of fine cheesecloth. The nuclei were pelleted by centrifugation at 1000 *g* for 10 minutes, and then resuspended in a solution containing 1 m*M* $MgCl_2$ and 2.2 *M* sucrose. The suspension was centrifuged at 40,000 *g* for 90 minutes. The pellet was washed twice with a solution of 1 *M* sucrose–1 m*M* $MgCl_2$ by homogenation with a loose-fitting Teflon pestle, followed by recentrifugation at 3000 *g* for 10 minutes. Chromatin was prepared from the nuclei by washing them with a 0.15 *M* NaCl solution. The chromatin was subsequently washed 3 times with 0.05 *M* Tris·HCl buffer (pH 7.5) followed by three washings with 0.15 *M* NaCl.

The total chromosomal proteins were isolated by dissociation of the purified chromatin for 4 hours with constant stirring in an ice bath in 2.5 *M* guanidinium hydrochloride–0.1 *M* sodium phosphate buffer (pH 7). The DNA was removed by centrifugation at 90,000 *g* for 10 hours.

In separate aliquots, the phosphoserine linkages were almost totally removed by treatment at 60°C for 10 minutes with 0.1 *N* NaOH (the acid-

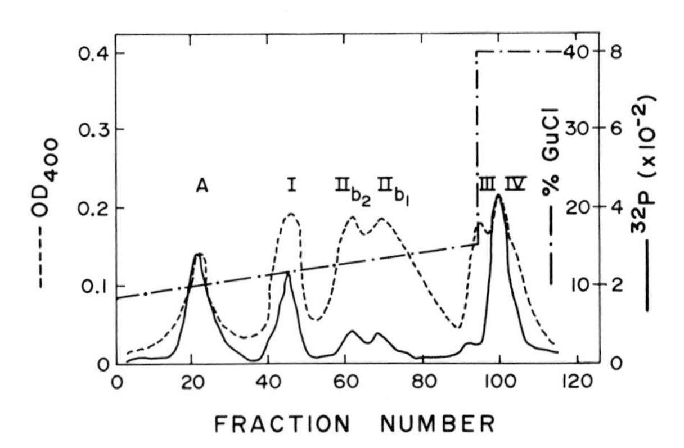


FIG. 1. Chromatography of acid-labile histone phosphates and acidic proteins on Bio-Rex 70 resin. Incorporation of radioactive inorganic phosphate, isolation of histones and acidic proteins, removal of acid-stable protein phosphates, and chromatographic separation were performed as described (*11*). Calf thymus histones (3 mg) were added as carriers during column chromatography. – – –, $OD_{400}$ ($Cl_3CCOOH$ turbidity); —, radioactivity ($^{32}P$ incorporated *in vivo*).

labile phosphoryl linkages were eliminated by treatment of an aliquot at 60°C for 10 minutes with 1 *N* HCl). This treatment was followed by dialysis overnight against two changes of 2 liters each of 0.01 *M* sodium phosphate buffer (pH 7).

Separation of the chromosomal proteins was carried out on a Bio-Rex 70 column (0.6 × 60 cm). The elution was performed with a total of 50 ml of a linear gradient of 8–13% guanidinium HCl in 0.1 *M* phosphate buffer (pH 6.8) prior to elution with 40% guanidinium hydrochloride in the same buffer essentially as described by Bonner *et al.* (*14*). For most of our experiments, we also added 3 mg of whole calf thymus histone to the column so that the appropriate histone peak could easily be identified by the TCA turbidity method (*14*). Radioactivity was determined by liquid scintillation spectrometry. A typical result showing the distribution of radioactivity associated with the chromosomal proteins eluted from a Bio-Rex 70 resin as described above is shown in Fig. 1.

## IV. Degradation of Phosphoproteins and Demonstration of *N*-Phosphoryl Amino Acids

The separate peaks from the Bio-Rex 70 column after a whole tissue run or from an *in vitro* kinase reaction were pooled and dialyzed against two changes of 1 liter each of 0.01 *M* Tris·HCl buffer (pH 8) and concentrated by lyophilization. The residue was usually suspended in 0.1% $NaHCO_3$ (pH 8.5) and incubated with trypsin (on a weight-to-weight ratio of 100:1, protein sample to proteolytic enzyme) for 2 hours at 37°C. After the trypsin incubation, Pronase was added to the reaction mixture (45 proteolytic units per milligram of Pronase) on a 25:1 weight-to-weight protein sample to Pronase ratio, and the incubation was continued at 37°C for 24 hours. This treatment was usually sufficient to accomplish complete hydrolysis to the component amino acids without apparent degradation of P-N bonds.

Alternatively, samples were occasionally hydrolyzed with 3 *M* KOH as follows. To the lyophilized protein sample (usually less than 1 mg), 1 ml of 3 *M* KOH was added and the reaction mixture was sealed in a glass test tube, placed in an autoclave at 15 psi pressure for 3 hours. After cooling and removal of the sample from the test tube, the potassium ions were precipitated by the dropwise addition of 10% perchloric acid to a pH of 7.5. The pH was never allowed to go below that value. The mixture was centrifuged to remove the potassium perchlorate precipitate, and the precipitate was washed once with a small volume of cold water. Both phosphohistidine and $\epsilon$-aminophospholysine survive this procedure, whereas phosphoarginine is destroyed by the basic hydrolysis. All three *N*-phosphoryl amino acids survive the enzymic proteolysis described above.

## V. Chromatographic Identification of *N*-Phosphoryl Amino Acids

Paper partition chromatography was usually performed with Whatman No. 1 or Whatman 3 MM paper, either ascending or descending in a solvent system consisting of isopropanol, ethanol, water, triethylamine (30:30:39:1, $v:v$). The $R_f$ values for the phosphoryl amino acids are reported as a function of phospholysine, which has been assigned a value of 1. They are *O*-phosphoserine, 1.10–1.25; phosphoarginine, 0.90–0.95; 3-phosphohistidine, 0.60–0.62; and 1-phosphohistidine, 0.45.

In addition, since phosphoarginine and phospholysine do not separate

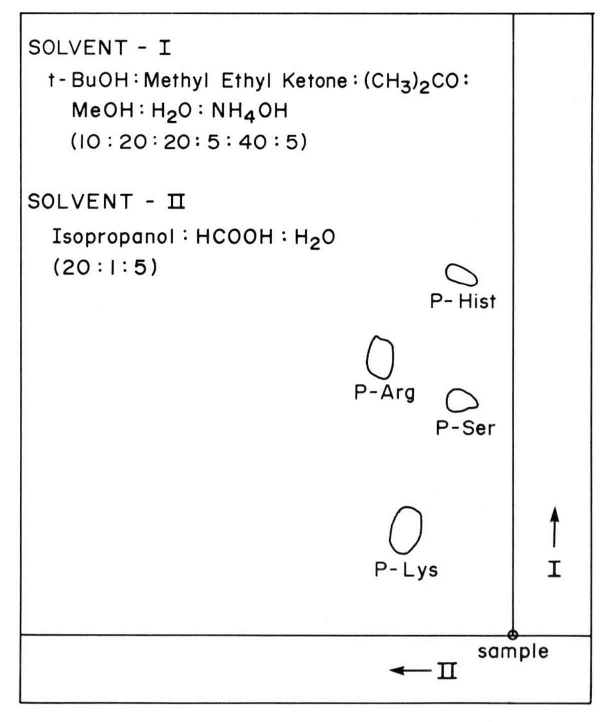


FIG. 2. Two-dimensional thin-layer chromatography of phosphoamino acids on ICN-4-4613 gel glass plates.

well by any of the many paper chromatographic methods we have attempted, a two-dimensional thin-layer chromatographic system has been developed; it is as follows (Fig. 2). The sample is applied to a silica gel glass plate (ICN-404613, 20 × 20 cm) and developed in the first direction with *t*-butanol, methyl ethyl ketone, acetone, methanol, water, conc. $NH_4OH$ (10:20:20:5:40:5, v:v), allowed to dry and developed in the second dimension with isopropyl alcohol, formic acid, $H_2O$ (20:1:5, v:v). A grid is constructed on the plate with pencil, and the dried matrix is removed for counting *N*-phosphoryl amino acids distribution as shown in Fig. 2.

## REFERENCES


1. Fiske, C. H., and SubbaRow, Y., *J. Biol. Chem.* **81**, 629 (1929).
2. Boyer, P. D., Hultquist, D. E., Peter, J. B., Kreil, G., Mitchell, R. A., DeLuca, M., Hinkson, J. W., Butler, L. G., and Moyer, R. W., *Fed. Proc., Fed. Am. Soc. Exp. Biol.* **22**, 1080 (1963).
3. Kundig, W., Ghosh, S., and Roseman, S., *Proc. Natl. Acad. Sci. U.S.A.* **52**, 1067 (1964).
4. Stevens-Clark, J. R., Conklin, K. A., Fujimoto, A., and Smith, R. A., *J. Biol. Chem.* **243**, 4474 (1968).
5. Halmann, M., Lapidot, A., and Samuel, D., *J. Chem. Soc.* p. 1299 (1963).
6. Ratlev, T., and Rosenberg, T., *Arch. Biochem. Biophys.* **65**, 319 (1956).
7. Chanley, J. D., and Feageson, E., *J. Am. Chem. Soc.* **85**, 1181 (1963).
8. Plimmer, R. H. A., and Bayliss, W. M., *J. Physiol.* (*London*) **33**, 439 (1906).
9. Plimmer, R. H. A., *Biochem. J.* **35**, 461 (1941).
10. Smith, D. L. Chen, C.-C., Bruegger, B. B., Holz, S. L., Halpern, R. M., and Smith, R. A., *Biochemistry* **13**, 3780 (1974).
11. Chen, C.-C., Smith, D. L., Bruegger, B. B., Halpern, R. M., and Smith, R. A., *Biochemistry* **13**, 3785 (1974).
12. Boyer, P. D., and Bieber, L. L., *in* "Methods in Enzymology" (R. W. Estabrook and M. E. Pullman, eds.), Vol. 10, p. 768. Academic Press, New York, (1967).
13. Chauveau, J., Moule, Y., and Rouiller, C., *Exp. Cell Res.* **11**, 317 (1956).
14. Bonner, J., Chalkley, G. R., Dahmus, M., Fambrough, D., Fujimura, F., Huang, R. C., Huberman, J., Jensen, R., Marushige, K., Ohlenbusch, H., Olivera, B. M., and Wilholm, J., *in* "Methods in Enzymology" (L. Grossman and K. Moldave, eds.), Vol. 12, Part B, p. 3. Academic Press, New York, (1968).

# Chapter 16

# *Assay of Nonhistone Phosphatase Activity*

LEWIS J. KLEINSMITH

*Division of Biological Sciences,*
*University of Michigan,*
*Ann Arbor, Michigan*

## I. Introduction

Nonhistone proteins are subject to covalent modification via both phosphorylation and dephosphorylation reactions (*1–4*). The protein kinases involved in catalyzing the phosphorylation step have been the subject of extensive investigation (*5–10*), but the enzyme(s) catalyzing the dephosphorylation step have been barely studied. Such enzymes not only are potentially important to the physiological regulation of the state of nonhistone protein phosphorylation, but also are useful as experimental tools for artificially modulating the phosphorylation state of these proteins during chromatin reconstitution studies (*11*). The present chapter is therefore concerned with methods for the assay and isolation of nonhistone phosphatases.

## II. Methods

### A. Preparation of $^{32}P$-Labeled Nonhistone Proteins

$^{32}P$-labeled nonhistone proteins were employed as substrates to assay for nonhistone phosphatase activity. Starting with nuclei prepared according to any of the standard procedures employing dense sucrose, the soluble proteins of the nuclear sap are removed by suspending the nuclei in 10 m*M* Tris·HCl (pH 7.5) for 20 minutes. The nuclei are collected by centrifugation at 10,000 *g* for 10 minutes, and are then suspended in 1.0 *M* NaCl–20 m*M* Tris·HCl (pH 7.5) to a final protein concentration of 2 mg/ml. The suspension is dispersed using a Polytron homogenizer (Brinkmann) for 20 seconds at setting 3. The resulting viscous solution is mixed with 1.5 volumes of 20 m*M* Tris·HCl (pH 7.5), and the precipitated nucleohistone is removed by centrifugation at 20,000 *g* for 2 hours. Bio-Rex 70 ($Na^+$), which has previously been equilibrated with 0.4 *M* NaCl–20 m*M* Tris·HCl (pH 7.5) is then added to the supernatant at a ratio of 20 mg of Bio-Rex per milligram of protein. After stirring slowly for 10 minutes, the suspension is centrifuged for 10 minutes at 6000 *g* and the supernatant is withdrawn. The resin is washed by suspending it in 10–15 ml of 0.4 *M* NaCl–20 m*M* Tris· HCl (pH 7.5) and centrifuging again for 10 minutes at 6000 *g*. The two supernatants are then combined and calcium phosphate gel added at a ratio of 0.46 mg of gel per milligram of protein. After slowly stirring for 20 minutes, the suspension is centrifuged for 5 minutes at 6000 *g*; the supernatant is discarded. The gel is washed by resuspension in 10–15 ml of 1.0 *M* $(NH_4)_2SO_4$–0.05 *M* Tris·HCl (pH 7.5) using a motor-driven stirrer at 1000 rpm, followed by centrifugation at 6000 *g* for 5 minutes. The supernatant is again discarded, and the gel is then dissolved by gentle homogenization with a Teflon Potter–Elvehjem tissue grinder in 0.3 *M* EDTA (pH 7.5)–0.33 *M* $(NH_4)_2SO_4$ in a ratio of 0.2 ml of solution per milligram of gel. The suspension is allowed to stand for 1 hour in the cold with occasional rehomogenization, and the insoluble residue is then removed by centrifugation for 15 minutes at 33,000 *g*. The supernatant, enriched in nonhistone phosphoproteins and protein kinase activity, is dialyzed overnight against 50 m*M* Tris·HCl (pH 7.5).

This protein phosphoprotein fraction is labeled by incubating the following in a reaction mixture of 1.0 ml:100 μg purified nonhistone phosphoprotein, 5 μmoles of $MgCl_2$, 40 μmoles of Tris·HCl (pH 7.5), and 5 nmoles [$\gamma$-$^{32}$P]ATP (770–3650 mCi/mmole). After 15 minutes of incubation at 37°C, the reaction is stopped by adding an equal volume of 0.3 *M* EDTA (pH 7.5).

## B. Assay of Nonhistone Phosphatase Activity

An important consideration in attempting to purify nuclear phosphatases is the type of assay employed. First of all, $^{32}$P-labeled nonhistone protein is the obvious substrate of choice when searching for nonhistone phosphatases. Second, the release of $^{32}$P from this substrate into an acid-soluble form is *not* in itself a sufficient criterion for the detection of phosphatase activity, since any protease that cleaves off small $^{32}$P-containing peptides will be incorrectly detected as a phosphatase. For this reason, care has been taken to design an assay based on the selective detection of $^{32}$P released as inorganic phosphate.

$^{32}$P-labeled nonhistone protein (1–10,000 cpm) is incubated in a reaction mixture of 2.0 ml containing 80 $\mu$moles of Tris·HCl (pH 7.5) and various amounts of enzyme fractions to be tested for phosphatase activity. After incubation for 15 minutes at 37°C, the reaction is stopped by adding 0.5 ml of 4 *N* HCl–1 *N* $H_2SO_4$, followed quickly by 0.1 ml of 0.1 *M* silicotungstic acid in 0.1 *N* $H_2SO_4$. The protein precipitate is removed by centrifugation for 5 minutes at 1000 *g*. Two milliliters of the supernatant are removed and transferred to a glass-stoppered tube containing 0.5 ml of 5% ammonium molybdate in 4 *N* $H_2SO_4$. After thorough mixing, 2.5 ml of 1:1 isobutanol–benzene is added and the tube is stoppered and mixed for 15 seconds by inversion.

After centrifuging a few minutes at 500–1000 *g* to separate the layers cleanly, a 1.0-ml aliquot of the upper phase is taken and counted in a toluene-based liquid scintillation cocktail. The rationale for this assay is that only $^{32}P$ released in the form of inorganic phosphate can form a phosphomolybdate complex extractable with isobutanol–benzene.

## III. Application of Assay

Using the above procedure, nonhistone phosphatase activity has been found to be localized mainly in the soluble (nuclear sap) fraction of isolated

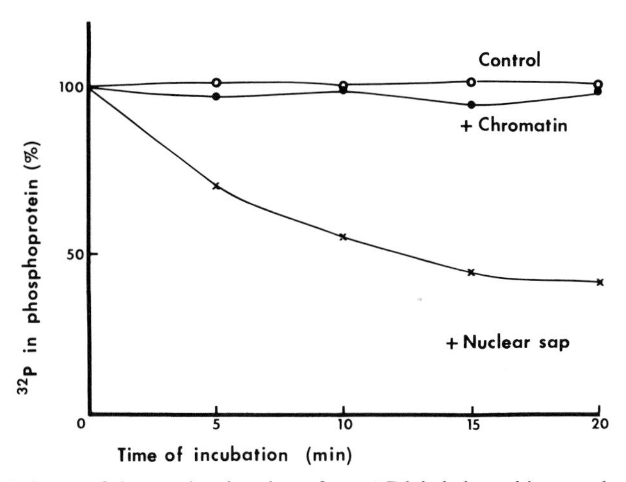


FIG. 1. Release of inorganic phosphate from $^{32}P$-labeled nonhistone phosphoprotein catalyzed by various nuclear fractions. Note that only the nuclear sap contains significant levels of nonhistone phosphatase activity.

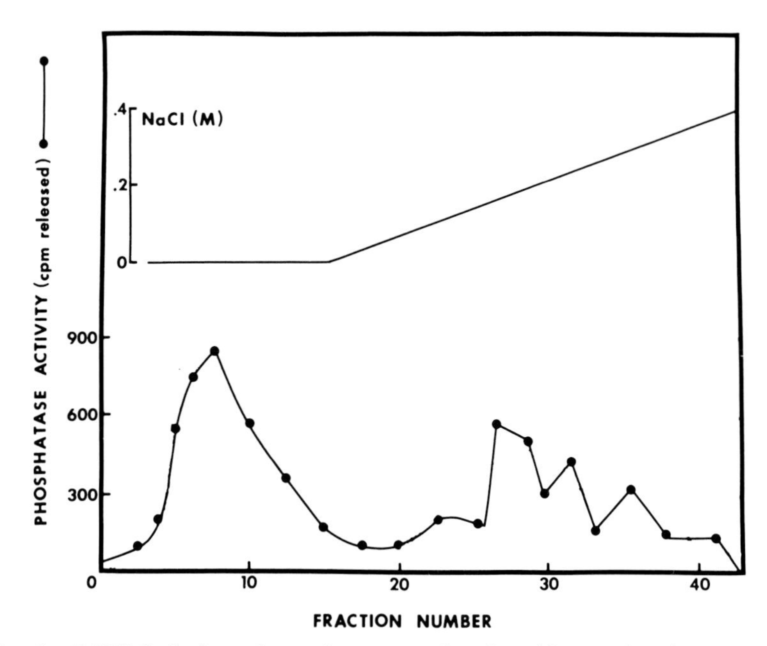


FIG. 2. DEAE-Sephadex column chromatography of nonhistone phosphatase activity present in calf thymus nuclear sap extract (0.14 *M* NaCl extract). Column size 1.0 × 20.0 cm, fraction size 2.0 ml.

nuclei (Fig. 1). Neither chromatin nor any of the major chromosomal protein fractions seem to contain significant levels of activity. Chromatography of nuclear sap fractions on DEAE-Sephadex results in the resolution of at least four major peaks of activity (Fig. 2). Experiments designed to characterize the substrate specificities of these enzyme fractions are currently under way.

## ACKNOWLEDGMENT

Studies on this subject in our laboratory have been supported by grants from the National Science Foundation (GB-23921 and BMS-23418).

## REFERENCES

1. Kleinsmith, L. J., *in* "Acidic Proteins of the Nucleus" (I. L. Cameron, and J. R. Jeter, Jr., eds.), p. 103. Academic Press, New York, 1974.
2. Stein, G. S., Spelsberg, T. C., and Kleinsmith, L. J., *Science* **183**, 817 (1974).
3. Kleinsmith, L. J. *J. Cell. Physiol.* **85**, 459 (1975).
4. Kleinsmith, L. J., *in* "Chromosomal Proteins and Their Role in the Regulation of Gene Expression" (G. S. Stein, and L. J. Kleinsmith, eds.), p. 45. Academic Press, New York, 1975.
5. Takeda, M., Yamamura, H., and Ohga, Y., *Biochem. Biophys. Res. Commun.* **42**, 103 (1971).
6. Ruddon, R. W., and Anderson, S., *Biochem. Biophys. Res. Commun.* **46**, 1499 (1972).

7. Kamiyama, M., Dastugue, B., and Kruh, J., *Biochem. Biophys. Res. Commun.* **44**, 1345 (1971).
8. Desjardins, P. R., Luie, P. F., Liew, C. C., and Gornall, A. G., *Can. J. Biochem.* **50**, 1249 (1972).
9. Thomson, J. A., Chiu, J.-F., and Hnilica, L. S., *Biochim. Biophys. Acta* **407**, 114 (1975).
10. Kish, V. M., and Kleinsmith, L. J., *J. Biol. Chem.* **249**, 750 (1974).
11. Kleinsmith, L. J., Stein, J., and Stein, G., *Proc. Natl. Acad. Sci. U.S.A.* **73**, 1174 (1976).

# Chapter 17

# *ADP-Ribosylation of Nuclear Proteins*

LLOYD A. STOCKEN

*Department of Biochemistry, University of Oxford, Oxford, England*

## I. Introduction

The formation of polyadenosine diphosphate ribose,

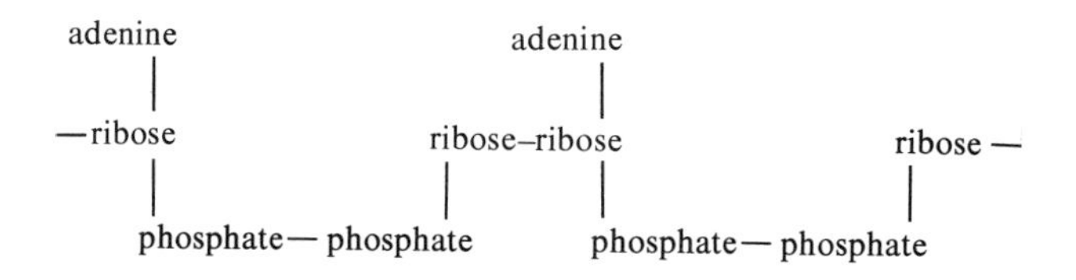


by nuclei was first suggested by Chambon *et al.* (*1*) after their earlier observations (*2*) that incorporation of the adenylate moiety of ATP into an acidinsoluble product was stimulated at least 1000-fold by nicotinamide mononucleotide. It was also concluded that the essential substrate was NAD.

Nuclei from a wide variety of tissues and species have been employed, and no activity has yet been found in higher plants or in prokaryotes. The first enzymic step is probably the attachment of the ADP-ribose moiety to a receptor protein followed by further additions in a glycosidic linkage between ribose units ($1' \rightarrow 2'$) to give derivatives of chain lengths varying from one to 20–30 units. The chain length may be limited by the presence in nuclei of a poly(ADP-ribose) glycohydrolase (*3*), and phosphodiesterase (*4*).

The different proteins utilized as receptors in nuclei have not yet been precisely defined, but histones are certainly implicated. It also appears that more than one receptor group is involved. Nishizuka *et al.* (*5*) were the first to show that most of the ADP-ribose residues could be rendered acid-soluble by a short treatment with dilute alkali, 0.1 *N* NaOH at 0°C for 30 minutes, or by 2 *M* hydroxylamine at pH 7.0. Since the glycosidic link is stable to alkali under these conditions (*1,6*), and since no release of radioactivity took place at pH 5.0 and 0°C, they proposed that a carboxyl group in the protein was a likely candidate as the ADP-ribose acceptor. This work has recently been extended by Adamietz and Hilz (*7*), who concluded that the ADP-ribosylation of nuclear proteins comprised two types of linkage, both of them susceptible to alkali, but only one split by neutral hydroxylamine. In the case of histone H1 it has been shown by the isolation of the modified protein from rat liver nuclei that a seryl residue is phosphorylated, and this is esterified in the ribose-1 position of ADP-ribose (*8*). ADP-ribose was quantitatively removed from the protein by alkali but not by neutral hydroxylamine.

More recently, Ueda *et al.* (*9*) have isolated histones from livers of rats injected with [$^{14}$C]ribose and [$^{3}$H]adenine as ADP-ribose precursors. They found radioactivity associated with histones H1, H2, and H3 when chromatographically separated on CM-cellulose. Subsequent treatment with $NH_2OH$ and phosphodiesterase led to the formation of $\psi$ADP-ribose, which led them to conclude that polymeric ADP-ribose was covalently attached to these histones. Incubation of a postnuclear extract of HeLa cells with isotopically labeled NAD leads to the production of labeled nonhistone proteins and most histones (*10*), and incubation of liver and thymus nuclei to labeled H1, H2b, and H3 (*11*).

Very strong circumstantial evidence indicates that the guanidino group of arginine is the receptor in the $\alpha$ polypeptide of DNA-dependent RNA polymerase in *Escherichia coli* following $T_4$ phage infection (*12,13*). On the other hand, the receptors in the chromatin-bound $Ca^{2+}$, $Mg^{2+}$-dependent endonuclease (*14*), or in the nuclear nonhistone proteins have not yet been established with certainty. Similarly, the receptors for cytoplasmic ribosylation involving elongation factor 2 or mitochondrial protein (*16*) are not yet clear.

This account is only concerned with the ribosylation of nuclear proteins, and it would be unsafe to assume that the methodology and the results obtained in extranuclear systems can be applied to chromatin or nuclei, nor should it be presumed that the products obtained following incubation of nuclei with NAD or its precursors *in vitro* exist *in vivo.*

Major complications have been introduced into the interpretation of the experimental results because of the different methods employed in the synthesis of the polymer and because of the different procedures used to isolate material purporting to be covalently attached to protein.

Practically all the evidence concerning the ADP-ribosylation of proteins has been obtained with nuclei or nuclear extracts incubated with $^{14}$C-labeled NAD *in vitro.*

## II. Incubation Conditions

Most authors have followed the method described by Nishzuka *et al.* (*6*) with only minor modifications. The optimum conditions appear to be, a medium consisting of 100 m*M* Tris HCl buffer (pH 8.0), 10 m*M* $MgCl_2$, 0.5–4 m*M* mercaptoethanol or dithiothreitol, 40–60 m*M* KCl, 80–200 $\mu M$ [$^{14}$C]adenine NAD incubated for 10 minutes at 37°C with a suitable quantity of enzyme. The enzyme preparation used has ranged from an amount of intact nuclei containing 1–2 mg protein (*17*) to 10 mg of protein (*18*) per milliliter of incubation mixture.

## III. Total Incorporation of [$^{14}$C]Adenine into Poly (ADP-Ribose)

The reaction is stopped by the addition of trichloroacetic acid TCA to give a final concentration of not less than 20%. This is to ensure the complete precipitation of all the histones. When either 5% TCA or 5% PCA is used to terminate the reaction and to wash the precipitate, histone H1, and some of the acid-soluble nonhistone proteins associated with it are otherwise discarded in the supernatant. At the same time this acid treatment, unless carried out very expeditiously and at low temperature, is liable to split the ribose-1-phosphate bond. The precipitate is collected on a Millipore filter (pore size 0.45 $\mu$m) and repeatedly washed with ice-cold 20% TCA (about 5 separate washes) until free from soluble counts. The material is washed with acetone and dried, and the filters are counted in 0.5% w/v 2:5 diphenyloxazole in a mixture 7:3 toluene and Triton X-100 (*19*).

## IV. Hydroxylamine-Labile Poly ADP-Ribose

If it is desired to distinguish the hydroxylamine-labile from nonlabile products, the nuclei are centrifuged down and repeatedly washed with isotope-free incubation medium. The nuclei are then extracted with 3% PCA, which takes 8% of the radioactivity into solution. The nuclear pellet is treated three times with 0.05 *M* Tris HCl (pH 8.0) and then with 0.4 *M* $NH_2OH$ (pH 7.5) for 30 minutes at 25°C in order to obtain the hydroxylamine-labile fraction. The soluble fraction contains 59% and the insoluble residue 33% of the radioactivity. If, however, the original nuclei are washed as before, incubated with 0.4 *M* hydroxylamine and then extracted with 3% PCA, 98% of the radioactivity now goes into solution (*17*). These authors have distinguished between the hydroxylamine-sensitive and -insensitive nucleus-bound radioactivity and have concluded that the $K_m$ for the former is 87 μ*M* and for the insensitive system is 0.65 m*M*.

However, as Hilz *et al.* (*20*) pointed out in a critical review of the problem, the extreme difficulty of separating the free polymer from nuclear protein by the usual CsCl gradients or high urea concentrations prevents a clear-cut assessment of how much of the polymer is free and how much is labile or nonlabile to hydroxylamine treatment.

## V. Determination of Chain Length

At the end of incubation 2 volumes of ethanol and 0.12 volume of 5 *M* acetate buffer (pH 5.0) are added and the precipitate washed successively with 66% ethanol–0.2 *M* acetate buffer (pH 5.0), ethanol, and ether. The precipitate is dissolved in 0.02 *M* Tris. HCl buffer (pH 7.4) and digested with Pronase (4 μg/ml) for 1 hour at 37°C. The reaction product is extracted with 0.8 volume of aqueous phenol (90% w/v) (pH 8.0) to remove protein, and then with ether. RNA, DNA, and the poly (ADP-ribose) are then precipitated with the ethanol/acetate buffer. The precipitate is dissolved in 20 ml of 0.1 *M* phosphate buffer (pH 6.8) and the solution is applied to a hydroxyapatite column and eluted with a linear gradient of phosphate buffer made from 120 ml of 0.1 *M* and 120 ml of 0.5 *M* phosphate. RNA and DNA are eluted at 0.15 and 0.25 *M* phosphate, and the polymeric ADP-ribose is eluted with increasing chain length according to increasing buffer strength. The fractions containing 10–400 pmoles of poly(ADP-ribose) as ADP-ribose units are digested with snake venom diesterase (45 units/ml) in 20 m*M* Tris · HCl (pH 7.4)–10 m*M* $MgCl_2$ for 30 minutes at 37°C; a further equal amount of the diesterase is then added, and the incubation is continued for 60 minutes. Samples can then be developed on thin-layer cellulose polyethyleneimine plates using 1 *M* acetic acid/0.3 *M* LiCl. The ratio of total radioactivity in (AMP + PR AMP) to AMP gives the chain length (*21*).

Sugimura *et al.* (*22*) obtained fractions containing chain lengths of 4.5, 12, 16, and 28 units.

## VI. Variations that Affect the End Product

Some progress has been made in the purification of the polymerase, but if a partially purified extract containing about 0.2 mg of protein (*23*) is used it is necessary to add DNA and histone in addition to the above-mentioned reagents. Omission of the lysine-rich histone reduces the incorporation of ADP-ribose by half, while histone in excess of a certain histone: DNA ratio ($\cong$ 1 : 1) inhibits the reaction. Purification of the enzyme seems to reduce the average chain length of the polymer similar to the action of DNase on whole nuclei. If the incubation is carried out at 25°C instead of 37°C (*17*), lower polymeric forms tend to be produced.

Yamada and Sugimura (*24*) have also shown that the lysine-rich histone and DNA are essential for increasing the number and length of the polymer chains. The enzyme is inhibited by chloromercuribenzoate (*1*), nicotinamide (*25*), and by thymine nucleoside or nucleotide (*26*).

The activity of the polymerase is markedly affected by DNase and this has been explained (*27*) in part by the inhibitory action of thymidine-containing oligo- and polynucleotides and in part (*28*) by the DNA requirement in the conformation of the enzyme bound to chromosomal RNA. A further observation by Hayaishi and Ueda (*27*) is that DNA contributes by inhibiting poly-ADP-ribose glycohydrolase.

## VII. Protein Receptors

The question which of the nuclear proteins are the ADP-ribose acceptors is not properly resolved. Nishizuka *et al.* (*5, 18*) showed that 75% of the radiolabeled poly (ADP-ribose) was associated with histones but with a considerable activity in the nonhistone fraction. On the other hand, Dietrich and Siebert (*29*) found only 34% in the histone fraction if the acidic proteins were first removed with sodium chloride.

It is evident that until the several histone and nonhistone protein receptors are isolated, characterized and quantified no firm conclusions can be made about the linkage(s) to the proteins or the physiological significance of the poly (ADP-ribose)–protein complexes.

Methods are now available for the separation of histones. The H1, P1 group is extracted from nuclei or chromatin with 5% PCA and the remainder of the histones with 250 m*M* HCl.

The H1, P1 group in 50 m*M* Tris · HCl buffer (pH 7.2) is separated on DEAE-cellulose by elution with the same buffer to give H1, followed by

0.5 *M* NaCl–50 m*M* Tris HCl (pH 7.2) (*30*) to give partially ADP-ribosylated P1 (*31*) H1 is further separated, using the buffer system devised by Balhorn *et al.* (*32*) on Amberlite IRC 50 or Bio-Rex 70 to give ADP-ribosylated H1a and H1b. The column is equilibrated with 3% NaCl–0.1 *M* Tris (pH 8.2), the H1 mixture is applied, and the H1a is eluted with the same buffer. H1b is obtained with 9% NaCl (*8*).

The nonhistone proteins are much less easy to separate and as yet, except for the nonhistone protein P1 associated with H1, no ADP-ribosylated nonhistone protein has been characterized. Separation methods for these proteins have been devised; for details of the methods in current use, the articles by Kleinsmith and Kish (*33*) and MacGillivray *et al.* (*34*) should be consulted, as well as the large-scale preparation described by Goodwin *et al.* (*35*).

## Addendum

Standard poly(ADP-ribose) polymerase (synthase) assay for comparison of enzyme preparations in different laboratories. This was agreed, since the above was written, at the Fourth Workshop on poly(ADP-ribose), August 2–4, 1976, Hamburg-Blankenese, Germany.

The assay measures the incorporation of radioactivity from $NAD^+$ into acid-insoluble material.

### A. Reagents

The DNA and histone H1 reagents are from Sigma Chemical Company Ltd.

The DNA is type 1: sodium salt "highly polymerized" from calf thymus; product No. D1501. The present batch held in the London office of Sigma is batch 26C 9540.

The histone H1 is Sigma histone type V-S: from calf thymus isolated as described by Johns (*36*); product No. H5880. The present batch held in the London office of Sigma is batch 85C 8080.

Both these products should be used, and the batch numbers should be specified.

### B. Assay Conditions

Total volume is 500 μl, containing the following components:
Tris-hydrochloride (pH 8.0), 100 m*M*
Mg $Cl_2$, 10 m*M*
Dithiothreitol, 1 m*M*
High molecular weight calf thymus DNA, 10 μg

Calf thymus histone H1 (lysine-rich, F1), 10 μg

NAD+: a double-reciprocal plot with NAD concentrations from 10 to 4000 μM should be established from which an apparent $V_{max}$ is calculated. An inferior alternative is to use 4.0 mM NAD.

Enzyme preparation

The assay temperature is 37°C. The reaction is terminated by adding trichloroacetic acid to 20% (w/v). The initial rate only should be used; because the assay is sometimes nonlinear, estimates should be made at 0, 1, 2, 5, and 10 minutes. From a plot of these 5 estimates, the initial rate should be calculated.

The pH of the assay solutions should be checked at 37°C after preparation.

*Definition:* One enzyme unit is defined as the enzyme activity which incorporates 1 μmole of radioactive ADP-ribose from NAD+ into acid-insoluble material in 1 minute at 37°C under the above-described conditions.

## Acknowledgment

I am indebted to Professor S. Shall for collating the several different methods used by different workers.

## References

1. Chambon, P., Weill, J. D., Doly, J., Strosser, M. T., and Mandel, P., *Biochem. Biophys. Res. Commun.* **25**, 638 (1966).
2. Chambon, P., Weill, J. D., and Mandel, P., *Biochem. Biophys. Res. Commun.* **11**, 39 (1963).
3. Miwa, M., Tanaka, M., Matsushima, T., and Sugimura, T., *J. Biol. Chem.* **249**, 3475 (1974).
4. Futai, M., Mizuno, D., and Sugimura, T., *J. Biol. Chem.* **243**, 6325 (1968).
5. Nishizuka, Y., Ueda, K., Yoshihara, K., Yamamura, H., Takeda, M., and Hayaishi, O., *Cold Spring Harbor Symp. Quant. Biol.* **34**, 781 (1969).
6. Nishizuka, Y., Ueda, K., Nakazawa, K., and Hayaishi, O., *J. Biol. Chem.* **242**, 3164 (1967).
7. Adamietz, P., and Hilz, H., *Biochem. Soc. Trans.* **3**, 1118 (1975).
8. Smith, J. S., and Stocken, L. A., *Biochem. Biophys. Res. Commun.* **54**, 297 (1973).
9. Ueda, K., Omachi, A., Kawaichi, M., and Hayaishi, O., *Proc. Natl. Acad. Sci. U.S.A.* **72**, 205 (1975).
10. Roberts, J. H., Stark, P., Giri, C. P., and Smulson, M., *Arch. Biochem. Biophys.* **171**, 305 (1975).
11. Stocken, L. A., Smith, J. S., and Ord, M. G., *in* "Poly(ADP-Ribose)—An International Symposium" (M. Harris, ed.), Fogarty Int. Cent. Proc. No. 26, p. 257. Nat. Inst. Health, Bethesda, Maryland (DHEW Publ. 74–477), 1974.
12. Goff, C. G., *J. Biol. Chem.* **249**, 6181 (1974).
13. Rohrer, H., Zillig, W., and Mailhammer, R., *Eur. J. Biochem.* **60**, 227 (1975).
14. Yoshihara, K., Tanigawa, Y., Burzio, L., and Koide, S. S., *Proc. Natl. Acad. Sci. U.S.A.* **72**, 289 (1975).
15. Maxwell, E. S., Robinson, E. A., and Henriksen, O., *J. Biochem.* (*Tokyo*) **77**, 9P (1975).

16. Kun, E., Zimber, P. H., Chang, A. C. Y., Puschendorf, B., and Grunicke, H., *Proc. Natl. Acad. Sci. U.S.A.* **72**, 1436 (1975).
17. Dietrich, L. S., and Siebert, G., *Hoppe-Seyler's Z. Physiol. Chem.* **354**, 1133 (1973).
18. Nishizuka, Y., Ueda, K., Honjo, T., and Hayaishi, O., *J. Biol. Chem.* **243**, 3765 (1968).
19. Brightwell, M. D., Leech, C. E., O'Farrell, M. K., Whish, W. J. D., and Shall, S., *Biochem. J.* **147**, 119 (1975).
20. Hilz, H., Adamietz, P., Bredehorst, R., and Leiber, U., *in* "Poly (ADP-Ribose)—An International Symposium" (M. Harris, ed.), Fogarty Int. Cent. Proc. No. 26, p. 47. Natl. Inst. Health, Bethesda, Maryland (DHEW Publ. 74–477), 1974.
21. Lehmann, A. R., Kirk-Bell, S., Shall, S., and Whish, W. J. D., *Exp. Cell Res.* **83**, 63 (1974).
22. Sugimura, T., Yoshimura, N., Miwa, M., Nagai, H., and Nagao, M., *Arch. Biochem. Biophys.* **147**, 660 (1971).
23. Yamada, M., Miwa, M., and Sugimura, T., *Arch. Biochem. Biophys.* **146**, 579 (1971).
24. Yamada, M., and Sugimura, T., *Biochemistry* **12**, 3303 (1973).
25. Fujimura, S., Hasegawa, S., Shimizu, Y., and Sugimura, T., *Biochim. Biophys. Acta* **145**, 247 (1967).
26. Preiss, J., Schlaeger, R., and Hilz, H., *FEBS Lett.* **19**, 244 (1971).
27. Hayaishi, O., and Ueda, K., *in* "Poly (ADP-Ribose)—An International Symposium" (M. Harris, ed.), Fogarty Int. Cent. Proc. No. 26, p. 69. Nat. Inst. Health, Bethesda, Maryland (DHEW Publ. 74–477), 1974.
28. Ueda, K., and Hayaishi, O., *in* "Poly (ADP-Ribose)—An International Symposium" (M. Harris, ed.), Fogarty Int. Cent. Proc. No. 26, p. 77. Natl. Inst. Health, Bethesda, Maryland (DHEW Publ. 74–447), 1974.
29. Dietrich, L. S., and Siebert, G., *Biochem. Biophys. Res. Commun.* **56**, 283 (1974).
30. Buckingham, R. H., and Stocken, L. A., *Biochem. J.* **117**, 157 (1970).
31. Ord, M. G., and Stocken, L. A., *Proc. FEBS Meet., 9th, 1974* Vol. 34, p. 113, (1975).
32. Balhorn, R., Rieke, W. O., and Chalkley, R., *Biochemistry* **10**, 3952 (1971).
33. Kleinsmith, L. J., and Kish, V. M., *in* "Methods in Enzymology" (B. W. O'Malley and J. G. Hardman, eds.), Vol. 40, p. 177. Academic Press, New York, 1975.
34. MacGillivray, A. J., Rickwood, D., Cameron, A., Carroll, D., Ingles, C. J., Krauze, R. J., and Paul, J., *in* "Methods in Enzymology" (B. W. O'Malley and J. G. Hardman, eds.), Vol. 40, p. 160, Academic Press, New York, 1975.
35. Goodwin, G. H., Nicolas, R. H., and Johns, E. W., *Biochim. Biophys. Acta* **405**, 280 (1975).
36. Johns, E. W., *Biochem. J.* **92**, 55 (1964).


## Note Added in Proof

Recent relevant papers:


Wong, N. C. W., Poirier, G. G., and Dixon, G. H., *Eur. J. Biochem.* **77**, 11 (1977).

Mullins, D. W. Jr., Giri, C. P., and Smulson, M. *Biochemistry* **16**, 506 (1977).

Tanuma, S-I., Enomoto, T., and Yamada, M-A., *Biochem. Biophys. Res. Commun.* **74**, 599 (1977).

Hilz, H. and Stone, O., *Rev. Physiol. Biochem. Pharmacol.* **76**,1 (1976).

Kristensen, T. and Holtlund, J., *Eur. J. Biochem.* **70**, 441 (1976).

Ord, M. G. and Stocken, L. A., *Biochem J.* **161**, 583 (1977).

# Chapter 18

# *Chromatin-Bound Proteases and Their Inhibitors*

DONALD B. CARTER, PEGGY H. EFIRD, AND CHI-BOM CHAE

*Department of Biochemistry,*
*University of North Carolina,*
*Chapel Hill, North Carolina*

## I. Introduction

In the late 1960s protein catabolism in calf thymus nuclei was studied by Furlan and Jericijo (*1,2*), Furlan *et al.* (*3*), Bartley and Chalkley (*4*), and Combard and Vendrely (*5*). Using histones and deoxyribonucleoprotein as substrates, Furlan and Jericijo (*1*) observed two pH optima, pH 4.4 and pH 7.8, for proteolytic activity in calf thymus nuclei. However, they showed that contamination of thymus nuclei with whole cells was the source of the pH 4.4 proteolytic activity which is characteristic of a cytoplasmic protease in thymocytes. These results tend to suggest that the low pH activities observed by the earlier investigators may also have been due to cytoplasmic contamination of nuclei during isolation of nuclei from whole cells. The neutral protease was found to be associated exclusively with nuclei.

In later work Furlan and Jericijo (*2*) showed that the chromatin-bound neutral protease was extractable in acid with the lysine-rich histones. Chromatography of the acid extract on a Sephadex G-100 column exhibited a neutral protease of 24,000 daltons. Purification of the neutral protease was achieved by salt extraction (2.5 *M* NaCl) of calf thymus chromatin and subsequent gel filtration on Sephadex G-75 (*3*). The purified protease showed a preference for nucleohistone as a substrate when compared to hemoglobin or bovine serum albumin. Combard and Vendrely (*5*) and Bartley and Chalkley (*4*) have shown that the lysine-rich histones (F1 and F2B) tend to be the first of the five major histone species digested in chromatin. Garrels *et al.* (*6*) have reported the presence of a neutral protease in rat liver chromatin isolated from whole tissue, and Chong *et al.* (*7*) have reported the isolation of a high molecular weight protease from the rat liver chromatin isolated from whole tissue. Kurecki and Toczko (*8,9*) have verified the existence of a small protease (16,000–18,000) associated with calf thymus chromatin.

That a protease does exist in rat liver chromatin which is active in the presence of 2 *M* NaCl–5 *M* urea was reported from this laboratory (*10–14*). These observations are significant for those investigators interested in studying the structure and reconstitution of chromatin, since chromatin can be easily dissociated in high salt and urea solutions. Several experiments have been published in which chromatin was dissociated and reconstituted in high salt and urea (*15–18*). Other investigators have dissociated chromatin in 2 *M* NaCl–5 *M* urea or 3 *M* NaCl in order to fractionate nonhistone proteins (*19–22*). In general, no attention has been given to the effects of proteolysis on the interpretation of dissociation or reassociation experiments involving labile chromatins. The methods discussed in this chapter are directed toward an understanding of how one may routinely assay a particular chromatin preparation for proteolytic activity, and block proteolytic activity in a variety of denaturing solvents.

# II. Methods

## A. Preparation of Chromatin

Our studies on chromatin degradation are largely confined to various rat tissues. Pure nuclei were prepared by centrifugation of crude nuclei through 2.3 *M* sucrose–3 m*M* $CaCl_2$–10 m*M* potassium phosphate (pH 6.5–6.8) and subsequent washing of the nuclei with 1% Triton X-100 in 0.25 *M* sucrose–3 m*M* $MgCl_2$–10 m*M* potassium phosphate (pH 6.5). Chromatin was prepared from the nuclei by standard procedures; that is, the nuclei were washed successively with 0.075 *M* NaCl–0.024 *M* EDTA (pH 7), 0.3 *M* NaCl or 0.05 *M* Tris (pH 7.9), 5 m*M* potassium phosphate (pH 6.5) or 0.01 *M* Tris (pH 7.9). The final chromatin was then swollen in ice-cold deionized water and sheared in a VirTis homogenizer (Model 23) at 30 V for 60 seconds. The concentration of the sheared chromatin was adjusted to $A_{260} = 20$ and small aliquots were frozen at −20°C. Chromatin-bound protease is active for at least one year in the case of rat liver chromatin.

We have also studied proteolysis of rat liver chromatin prepared by the method of Stein *et al.* (*23*) and Elgin and Bonner (*24*).

## B. SDS Polyacrylamide Gel Electrophoresis of Whole Chromatin

Sodium dodecyl sulfate (SDS) polyacrylamide gel electrophoresis is a convenient method to examine the proteolytic degradation of chromatin.

The integrity of chromosomal proteins during fractionation of chromosomal proteins and also chromatin reconstitution should be examined at least by comparing electrophoretic profiles of proteins with those of native chromatin.

We rarely see in the literature that the integrity of fractionated chromosomal proteins, especially nonhistone proteins, is demonstrated by comparison with those proteins present in native chromatin. This seems to be due either to the hindrance of DNA to the electrophoresis of nonhistone proteins or to the extra time and trouble involved in the removal of DNA. We described an SDS gel electrophoresis procedure which minimizes the interference by DNA of the electrophoresis of chromosomal proteins. The main difficulty we usually encounter during electrophoresis of the chromatin dissociated in SDS is the blocking of polyacrylamide gel pores by DNA, thus causing a streaking pattern of chromosomal proteins. This can be obviated by extensive shearing of DNA and use of stacking gel as described by Smith and Chae (*25*). Modifications of this original procedure were described in a later report (*10*).

## C. Electrophoretic Profiles of Chromatin Exposed to Various Dissociation Conditions

The dissociation of chromatin in various denaturing solvents for purposes of reconstitution, or fractionation of chromatin proteins has been reported (*16–18,20,26,27*). However, few investigators have examined the integrity of either chromosomal nonhistone proteins or histones in the presence of the commonly used denaturing solvents, such as high salt and urea, urea alone, salt alone, or guanidine·HCl. Figure 1 shows that considerable degradation of nonhistone proteins occurs under these conditions. Histones extracted from the treated chromatin by cold 0.4 $N\,H_2SO_4$ and electrophoresed on the acid–urea gels of Panyim and Chalkley (*28*) also show extensive degradation (*10,13*). Rat liver chromatin also undergoes similar degradation when dissociated in 2 *M* guanidine·HCl–5 *M* urea (pH 6.5) or salt in concentrations from 0.15 *M* NaCl to 2 *M* NaCl. It is also apparent from Fig. 1 that dissociation in 1% SDS or boiling chromatin for 2 minutes inactivates the proteolytic activity.

## D. Degradation of [$^3$H]Acetate-Labeled Histones by Chromatin

The release of acid-soluble labeled peptide fragments is a convenient method for assaying whether a particular chromatin preparation may contain proteolytic activity. Lysine-rich histones from calf thymus were prepared by the method of Johns (*29*). Labeling of a 1:1 mixture of F1 and F2B

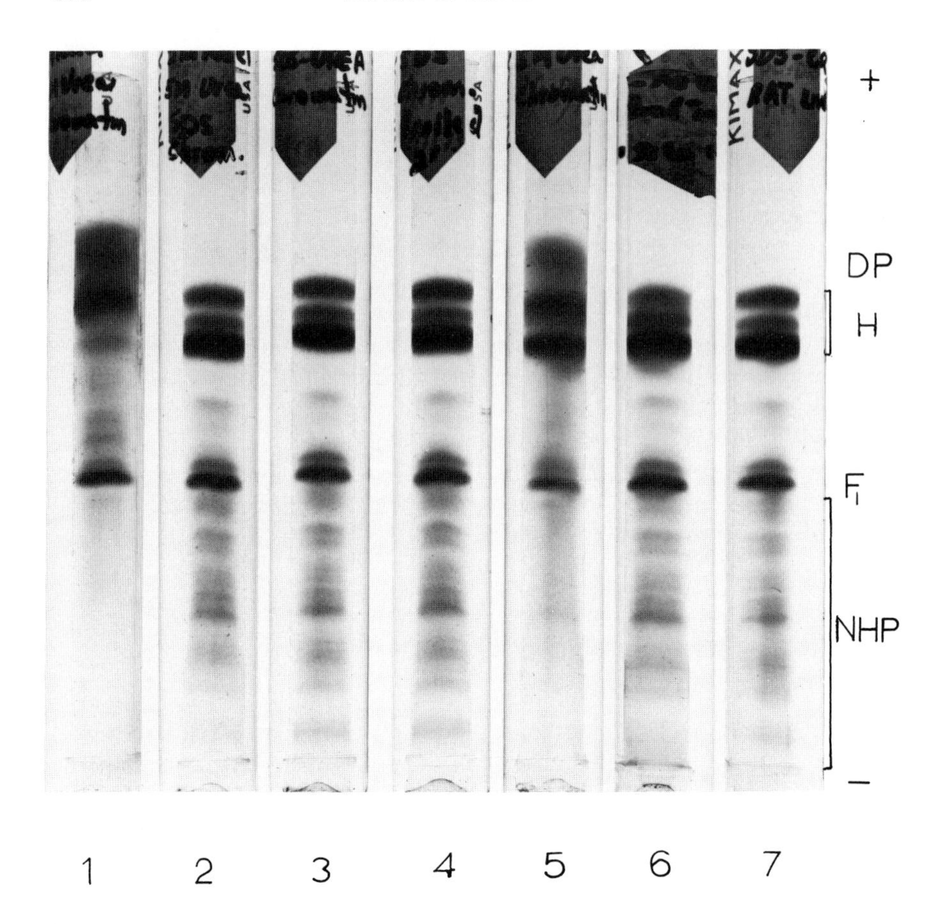


FIG. 1. Degradation of rat liver chromatin in the presence of 2 *M* NaCl–5 *M* urea or 5 *M* urea alone. Rat liver chromatin, prepared as described in Materials and Methods, was incubated in either 2 *M* NaCl–5 *M* urea–10 m*M* $KPO_4$ (pH 7.0) or 5 *M* urea–10 m*M* $KPO_4$ (pH 7.0) after treatment with 1% sodium dodecyl sulfate (SDS) or heating at 100°C for 2 minutes.

Chromatin incubated in salt and urea 16 hours at 4°C (1); chromatin incubated in salt and urea at 25°C but first made 1% in SDS (2); chromatin made 1% in SDS and incubated in urea at 25°C (3); chromatin heated to 100°C, 2 minutes before incubation at 25°C in salt and urea (4); chromatin incubated in urea alone at 4°C (5); chromatin heated and incubated in urea alone at 25°C (6); chromatin dissolved in 1% SDS as a control (7). Migration is from the bottom (−) to the top (+) in 7.5% SDS–polyacrylamide gels. DP refers to degraded protein, and NHP to nonhistone proteins. Histones are indicated by the nomenclature of Johns (*29*), and the letter H represents the histones F2A1, F2A2, F3 + F2B in order of increasing electrophoretic activity. Reprinted with permission from *Biochemistry* (*13*). Copyright by the American Chemical Society.

was carried out as follows: Two millicuries of tritium-labeled acetic anhydride (3.9 Ci/mmole) were incubated with 100 mg of F1-F2B mixture in 2 ml of 10 m*M* potassium phosphate (pH 7.0) for 2 hours at 4°C. This was followed by exhaustive dialysis against 2 *M* NaCl–5 *M* urea–10 m*M* potassium phosphate (pH 7.0) at 4°C. The labeled histones were then subjected to three rounds of precipitation with cold 25% TCA to remove nondialyzable acid-soluble peptide fragments. Label was incorporated into F2B three times more effectively than into F1 as determined by electrophoresis on acid-urea gels (*28*) and counting 1.5-mm gel slices. The gel slices were incubated at 37°C overnight in 5 ml of a scintillation cocktail consisting of 1 part $NH_4OH$, 5 parts NCS tissue solubilizer, and 50 parts toluene–POPOP–PPO. The toluene–POPOP–PPO cocktail contained 2.5 gm of PPO and 0.15 gm of POPOP dissolved in 500 ml of toluene. Typical recoveries of radioactivity applied range between 65 and 80%.

The proteolytic activity of various chromatins in 2 *M* NaCl–5 *M* urea or 5 *M* urea alone using the labeled histone substrates is carried out by the following protocol: Chromatin having a concentration of 0.3 mg to 1.0 mg DNA per milliliter in salt and urea or urea alone is incubated for various periods of time at 25°C or 4°C with the labeled [$^3$H]histones (F1 + F2B). The ratio of labeled histones added to the reaction mixture is 0.1 mg:1 mg DNA in chromatin. Total final volumes used in this laboratory are 1.1 ml. To determine the acid-soluble radioactive peptides released during an incubation period, an equal volume of cold 50% TCA is added to the reaction mixture; after 15 minutes in ice, the acid-insoluble material is removed by centrifugation for 20 minutes at 20,000 *g*. An aliquot of the supernatant is counted in the Triton X-100–toluene scintillation fluid. A zero time control is necessary to establish acid-soluble background. A reaction with boiled chromatin (2 minutes at 100°C inactivates proteolytic activity) is also run to detect any release of acid-soluble peptides due to processes other than proteolysis during incubation. The zero time control and control with boiled chromatin agree within 5% of each other in content of acid-soluble radioactivity. Acetylated histones [$^3$H](F1 + F2B) are progressively degraded by rat liver chromatin prepared from purified nuclei as shown in Fig. 2. Up to 30% of the added labeled histones are made acid-soluble for rat liver chromatin, 50% in the case of rabbit bone marrow chromatin. No degradation of label is detectable for chicken erythrocyte chromatin in 2 *M* NaCl–5 *M* urea. These results are consistent with SDS gels of rat liver chromatin incubated in either 2 *M* NaCl–5 *M* urea–10 m*M* $KPO_4$ (pH 7.0) or 5 *M* urea–10 m*M* $KPO_4$ (pH 7.0) for various periods of time as shown in Fig. 3. There is a progressive disappearance of high-molecular-weight nonhistone proteins (NHP) and an accumulation of degraded protein as a function of time.

The tissues and cells tested in our laboratory for chromatin-bound pro-

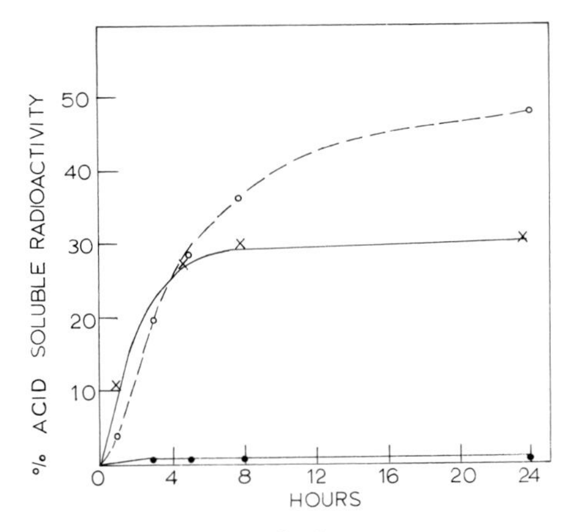


FIG. 2. Time course of degradation of [$^{3}$H](F1 + F2B) histones by chromatin in the presence of 2 *M* NaCl–5 *M* urea–10 m*M* $KPO_4$ (pH 7.0) at 25°C. Rabbit bone marrow (○), rat liver (X), and chicken erythrocyte (●). Reprinted with permission from *Biochemistry* **15**, 180(1976). Copyright by the American Chemical Society.

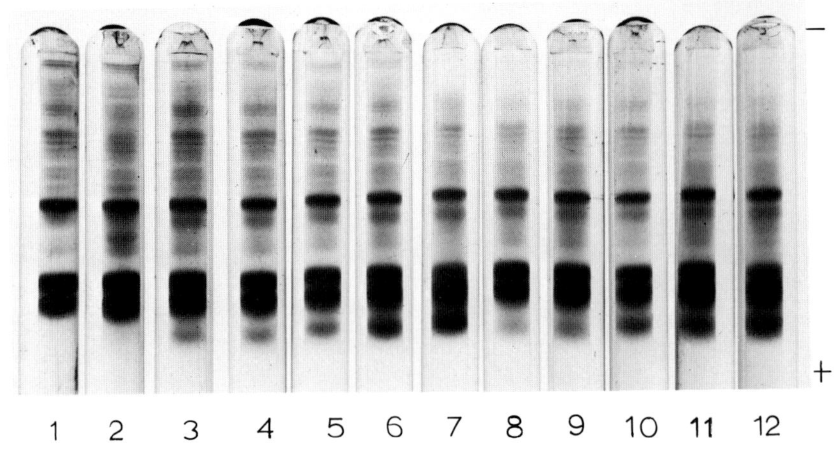


FIG. 3. Time course degradation of rat liver chromatin incubated in 2 *M* NaCl–5 *M* urea–10 m*M* $KPO_4$ (pH 7.0) and 5 *M* urea–10 m*M* $KPO_4$ (pH 7.0). Chromatin in 1% sodium dodecyl sulfate (SDS) as a control (1); chromatin heated to 100°C for 2 minutes and incubated in urea for 24 hours at 25°C (2); chromatin incubated in urea at 25°C for 1 hour (3), 2 hours (4), 4 hours (5), 8 hours (6), 24 hours (7); chromatin incubated in 2 *M* NaCl–5 *M* urea–10 m*M* $KPO_4$ (pH 7.0) at 25°C for 1 hour (8), 2 hours (9), 4 hours (10), 8 hours (11), and 24 hours (12). Migration is from top (−) to bottom (+) on 7.5% SDS polyacrylamide gels.

tease activity are liver, thymus, kidney, spleen, brain, testis, Morris hepatomas, Ehrlich ascites carcinoma cells, chicken reticulocytes and erythrocytes, rabbit bone marrow, mouse fibroblasts (3T3), 3T3 cells transformed by simian virus 40 (SV 3T3). Among these, chromatin isolated from erythrocyte, 3T3, SV3T3, and Ehrlich ascites carcinoma cells show low proteolytic activity in 2 *M* NaCl–5 *M* urea.

Bovine serum albumin was labeled with [$^3$H]acetic anhydride and prepared as described for the [$^3$H](F1 + F2B) substrate. When incubated with rat liver chromatin in 2 *M* NaCl–5 *M* urea or 5 *M* urea (pH 8), no detectable release of acid-soluble peptides was observed. One may conclude that even under denaturing conditions the protease or proteases bound to chromatin demonstrate some selectivity in cleavage.

## E. Inhibitors

Various inhibitors of chromatin-associated proteolytic activity have been utilized by different groups of investigators.

Diisopropylfluorophosphate (DFP), the serine site-specific reagent has been demonstrated to inhibit effectively calf thymus chromatin proteolytic activity. Furlan *et al.* (*2*) showed that 0.5 m*M* DFP reduced the initial activity of a crude preparation of calf thymus chromatin neutral protease by 70% in 0.033 *M* phosphate-citrate buffer (pH 7.8). Panyim *et al.* (*30*) also showed the effectiveness of DFP at 1 m*M* concentrations for prevention of histone hydrolysis during isolation from calf thymus nuclei. The substrate used in this study was heated (75°C, 20 minutes) in order to inactivate endogenous proteolytic activity. Carter and Chae (*13*) incubated whole rat liver chromatin prepared by the method of Huang and Huang (*16*) with 1 m*M* DFP in 10 m*M* Tris (pH 8.0) for 10 minutes in order to achieve complete blockage of chromatin proteolytic activity when chromatin was subsequently dissolved in 2 *M* NaCl–5 *M* urea–20 m*M* Tris (pH 8.0). Since DFP is a water-soluble compound at concentrations convenient for dilution into chromatin at 1 m*M* final concentration, it is useful for prevention of proteolysis under chromatin reconstitution conditions. The disadvantage of exposure of chromatin to organic solvents necessary for effective dissolution of other water-insoluble inhibitors is obviated by the solubility of DFP in aqueous buffers. However, in most cases the use of DFP is hazardous because of its extreme toxicity. Other less toxic inhibitors will be discussed.

Phenylmethanesulfonylfluoride (PMSF) has been used to block degradation of histones isolated from rat liver chromatin by Nooden *et al.* (*31*). In this case, the histones in 10 m*M* Tris (pH 7) were incubated with 0.1 m*M* PMSF–1% isopropyl alcohol. The inhibition in this case was apparently irreversible since dialysis of the incubated histones against 10 m*M* Tris (pH 8) and 24 hours of storage did not produce detectable proteolysis. Others have tried PMSF as an inhibitor in 5 *M* urea with various concentrations of salt without benefit of organic solvent (*26*). However, in our hands the protection of chromatin proteins (both nonhistone and histone), as assayed by SDS gel electrophoresis or labeled histone substrate, is only partial when PMSF is used without the presence of organic solvent. Isopropyl alcohol, dimethylsulfoxide, and *p*-dioxane are effective solvents for PMSF when used for the inhibition of whole chromatin proteolytic activity. Stock solutions of PMSF in either of the above solvents are made fresh to 0.1 *M* concentration. The final concentration in chromatin is 1 m*M* PMSF–1% organic solvent. Chromatin freshly prepared from Triton-washed nuclei by any of the previously mentioned methods is incubated for 20 minutes at room temperature with PMSF and organic solvent. The chromatin may then be dialyzed into 10 m*M* Tris (pH 8), deionized water or 2 *M* NaCl–5 *M* urea without subsequent proteolytic degradation. This treatment generally protects chromatin from degradation and removes the organic solvents from chromatin. A variation on the above protocol is to make chromatin 10% in *p*-dioxane before adding PMSF in *p*-dioxane and incubate for 20 minutes as before. This treatment provides consistent protection of chromatin to proteolysis after dialysis of chromatin into 2 *M* NaCl–5 *M* urea; 10% *p*-dioxane alone partially inhibits rat liver chromatin proteolytic activity reversibily by 35%. PMSF has been used to successfully inhibit proteolytic activity in chromatin isolated from purified nuclei of rat thymus, liver, calf thymus, and rabbit bone marrow (Fig. 4).

The trypsin and chymotrypsin substrate analog inhibitors, tosyllysine choromethylketone and tosylphenylalanine chloromethylketone, respectively, were found to be ineffective in the blockage of chromatin proteolytic activity. However, a chymotrypsin inhibitor, carbobenzoxyphenylalanine chloromethylketone (ZPCK) was found to be effective, but only when rat liver chromatin was adjusted to a high concentration of organic solvent, for example, 10% *p*-dioxane, prior to the addition of ZPCK to 1 m*M* final concentration. Even under these conditions the addition of the extremely water-insoluble ZPCK caused the chromatin mixture to become quite turbid. However, after incubation of the chromatin and ZPCK for 20 minutes and dialysis into 2 *M* NaCl–5 *M* urea, there was no proteolytic activity detectable by either SDS gel assay or release of acid-soluble peptide fragments from [$^3$H](F1 + F2B) substrate.

From the standpoint of interest in substrate specificity of the 25,000 dalton protease of rat liver chromatin (this will be discussed later), reversible inhitbitors will also be mentioned here. Carbobenzoxyphenylalanine (CBZ-Phe), at a 20 m*M* concentration of both D or L form, inhibits rat liver chromatin proteolytic activity in 2 *M* NaCl–5 *M* urea, as shown in Fig. 5. Although,

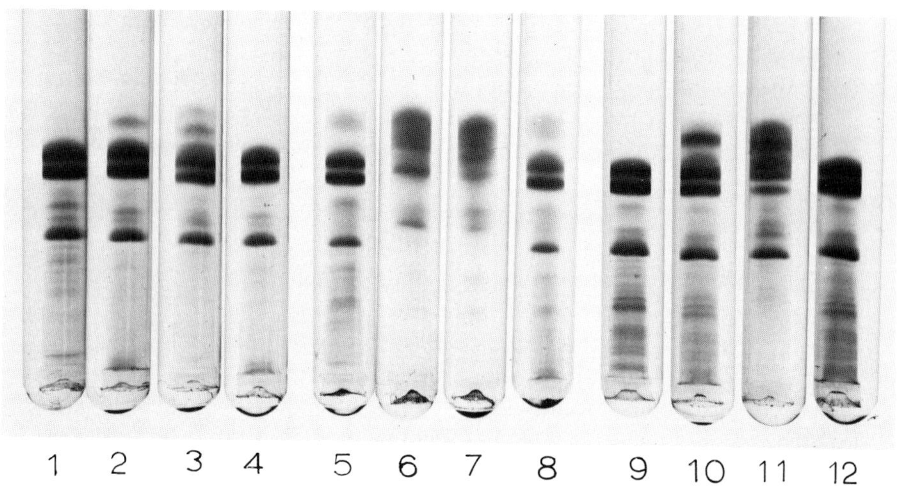


FIG. 4. The ability of 1 m*M* phenylmethanesulfonyl fluoride (PMSF) to inhibit the proteolytic activity of calf thymus, rabbit bone marrow, and rat liver chromatin in the presence of 2 *M* NaCl–5 *M* urea–10 m*M* Tris (pH 8.0). In each case chromatin was incubated at 25°C for 16 hours. Calf thymus control chromatin (1); calf thymus chromatin in 10% *p*-dioxane–salt and urea (2); salt and urea alone (no *p*-dioxane) (3); salt and urea–10% *p*-dioxane–1 m*M* PMSF (4); rabbit bone marrow control chromatin (5); rabbit bone marrow chromatin in 10% *p*-dioxane–salt and urea (6); salt and urea (7); salt and urea–10% *p*-dioxane–1 m*M* PMSF (8); rat liver control chromatin (9); rat liver chromatin in 10% *p*-dioxane–salt and urea (10); salt and urea (11); salt and urea–10% *p*-dioxane–1 m*M* PMSF (12). Migration is from bottom (−) to top (+) on 7.5% sodium dodecyl sulfate polyacrylamide gels. Reprinted with permission from *Biochemistry* **15** (180) 1976. Copyright by the American Chemical Society.

CBZ-L or D-Phe is water soluble at 0.1 *M* concentration at pH 8, there are disadvantages to using this compound. At pH levels somewhat lower than 8 the CBZ-Phe tends to precipitate from solution complexed with chromatin. Attempted acid extractions of histones from chromatin with 0.4 *N* $H_2SO_4$ in the presence of 20 m*M* CBZ-Phe yield very little F1 histone and none of the other histones, whereas acid extraction of histones from chromatin in salt and urea without CBZ-Phe presents no difficulties. At a concentration of 20 m*M*, CBZ-Phe also inhibits the proteolytic activities of rat testis, rat thymus, and calf thymus, but is not effective against rabbit bone marrow chromatin proteolytic activity in 2 *M* NaCl–5 *M* urea (pH 8.0).

Reversible inhibition of rat liver chromatin-bound protease is also achieved with *p*-nitrophenylacetate (PNPA). PNPA dissolved in dimethylsulfoxide (DMSO) and added to chromatin to give a final concentration of 10 m*M* PNPA in 10% DMSO effected complete inhibition of rat liver chromatin proteolytic activity in the presence of 2 *M* NaCl–5 *M* urea (pH 8.0). Under the same conditions *p*-nitrophenol did not produce any observable inhibition; hence, though some hydrolysis of PNPA takes place in 2 *M* NaCl–5 *M* urea (pH 8.0), it is the acetate form which provides inhibition. At 10% DMSO, we see negligible inhibition of rat liver chromatin proteolytic activity in 2 *M* NaCl–5 *M* urea.

Koehler and Lienhard (*32*) have reported the use of 2-phenylethaneboronic acid (PEBA) as a competitive inhibitor of the $\alpha$-chymotrypsin-catalyzed hydrolysis of methyl hippurate. Since PEBA is water-soluble at a broad range of pH around 7, it was tested on rat liver chromatin protease for its possible inhibitory properties. At a 20 m*M* concentration of PEBA, complete inhibition of chromatin proteolytic activity in the presence of 2 *M* NaCl–5 *M* urea was achieved. The extractability of histones under these conditions has not been tested.

Sodium bisulfite has been reported to be an inhibitor of endogenous chromatin proteolytic activity (*6, 28*) at a concentration of 50 m*M*. Generally, these results were obtained in solutions of low ionic strength. However, in the presence of 2 *M* NaCl–5 *M* urea (pH 7 and 8), 50 m*M* sodium bisulfite protects neither histones nor nonhistone chromatin proteins; at pH 6 there is partial protection. When sodium bisulfite is dissolved in water or in 10 m*M* Tris (pH 8) at a concentration of 50 m*M*, the pH of the resulting solution is 5.4 and 6.0, respectively. The results of Carter and Chae (*13*) on the protease activity of chromatin as a function of pH would suggest that the intrinsic pH of the chromatin solution is the determinant of proteolytic activity, and not the presence or the absence of sodium bisulfite. It is concluded that caution must be observed when sodium bisulfite is used with the intent of inhibiting endogenous chromatin proteolytic activity in the presence of salt and urea.

Table I summarizes the properties of neutral proteases associated with chromatin and compares the inhibitory capacity of the listed compounds. One finds a great deal of similarity between the chromatin-bound proteases of calf thymus chromatin and rat liver chromatin investigated by Carter and Chae (*13*).

The high-molecular-weight protease isolated from rat liver chromatin prepared from whole tissue by Chong *et al.* (*7*) has quite different inhibition properties as well as molecular properties. Thus it appears that rat liver chromatin may possess multiple forms of proteolytic activity.

## F. Labeling of Nuclei and Chromatin by [$^3$H]DFP

Several investigators have reported the presence in chromatin of proteases which are inhibited by diisopropylfluorophosphate (DFP), the active-site serine phosphorylating reagent. These results suggest that the degradation of chromosomal proteins under various conditions is largely due to serine-

TABLE I

COMPARISON OF PROPERTIES OF NEUTRAL PROTEASE FROM RAT LIVER AND CALF THYMUS CHROMATIN

| | Chong *et al.* (*7*) | This report | Furland and Jericijo (*1,2*) | Kurecki and Toczko (*9*) |
|---|---|---|---|---|
| Source | Rat liver | Rat liver | Calf thymus | Calf thymus |
| Method of preparation | Enzyme extracted from chromatin prepared from whole tissue | Enzyme extracted from chromatin prepared from purified nuclei | Enzyme extracted from chromatin prepared from purified nuclei | Enzyme extracted from chromatin prepared from whole tissue |
| pH optimum | 7.0 | 8—9 | 7.8 | 8.5 |
| Effect of: | | | | |
| β-Mercaptoethanol | High concentrations inhibit | 0.2 *M* not inhibitory | — | — |
| Iodoacetamide | — | 1 m*M* not inhibitory | 1 m*M* not inhibitory | 0.2–5 m*M* not inhibitory |
| Sodium bisulfite | 10 m*M* inhibits | 50 m*M*, pH 6 partially inhibits<br>50 m*M*, pH 7 or 8 not inhibitory | 1 m*M* not inhibitory | 20 m*M* not inhibitory |
| $Hg^{2+}$ | Low concentrations inhibit | 1 m*M* not inhibitory | 1 m*M* not inhibitory | — |
| NaCl | > 1 *M* NaCl inhibits | Increasing NaCl increases activity to 2 *M* NaCl | Increasing NaCl increases activity to 1 *M* NaCl | — |
| Urea | 2 *M* Urea partially inhibits | Increasing urea increases activity to 5 *M* urea | — | — |
| PMSF[a] | 1 m*M* inhibits | 1 m*M* inhibits | — | — |
| DFP[a] | 1 m*M* inhibits | 1 m*M* inhibits | 1 m*M* inhibits | — |
| Molecular weight | 100,000 subunit | 25,000 (binding of DFP-$^3$H) | 24,000 | 16,000–18,000 |

[a]PMSF, phenylmethanesulfonylfluoride; DFP diisopropylfluorophosphate.

protease(s). DFP is highly selective in reacting with the serine in the active site of various proteases (*33*), and other enzymes. Therefore, studies on the labeling patterns of chromatin with radioactive DFP provide useful information on the possible multiplicity of chromatin-bound protease.

### 1. Affinity Labeling with [$^3$H]DFP

Nuclei prepared by sedimentation through 2.3 *M* sucrose as described previously are washed twice in 0.25 *M* sucrose–3 m*M* $MgCl_2$–10 m*M* $KPO_4$ (pH 6.5)–1% Triton X-100 (3 v/w). The nuclear pellet is lysed in distilled water and adjusted to 10 m*M* Tris (pH 8) and a concentration of 0.75 mg/ml DNA. Chromatin is also adjusted to 0.75 mg DNA per milliliter and [$^3$H] DFP (3.9 Ci/mmole) in propylene glycol is added to either nuclei or chromatin in a ratio of 67 μCi/mg DNA. The mixture is incubated for 15 hours at 4°C or 25°C, and excess label is removed by exhaustive dialysis against 1% SDS–0.01 *M* sodium phosphate–0.1% β-mercaptoethanol (pH 7.0). This treatment for rat liver chromatin generally results in a label incorporation of 4000–6000 cpm per 50 μl of chromatin. Results for nuclei are more variable, in terms of incorporation, yielding from 2000 cpm to 6000 cpm per 50 μl of nuclei. Ideally, between 2000 and 4000 cpm are layered on the 7.5% SDS polyacrylamide gel system already discussed. The gels are electrophoresed as described before, stained and destained in Coomassie Blue, sliced into 1.5-mm slices and counted as described previously.

### 2. Labeling of Rat Liver Nuclei and Chromatin Prepared by Different Methods by [$^3$H]DFP

Rat liver nuclei and chromatin prepared by the procedures given previously were labeled by reaction with [$^3$H]DFP in order to compare the DFP-binding proteins of chromatin prepared by different methods. Rat liver nuclei labeled with [$^3$H]DFP show three peaks of radioactivity (A of Fig. 5) corresponding to molecular weights of approximately 70,000, 60,000, and 25,000. Chromatin prepared from purified nuclei by the method of Smith and Chae (*25*) and Huang and Hunag (*16*) (B and C of Fig. 5) indicates qualitatively the same pattern of labeling that occurs with whole nuclei.

The relative amounts of radioactivity in the three proteins were somewhat variable between different preparations of nuclei and chromatin. The most typical labeling patterns of the several experiments are shown in Fig. 5. In general the smallest molecular weight DFP-binding protein contained more radioactive DFP than two other proteins in the case of chromatin. Chromatin prepared by the method of Stein *et al.* (*23*) shows a labeling pattern qualitatively similar to whole nuclei differing only in the relative quantity

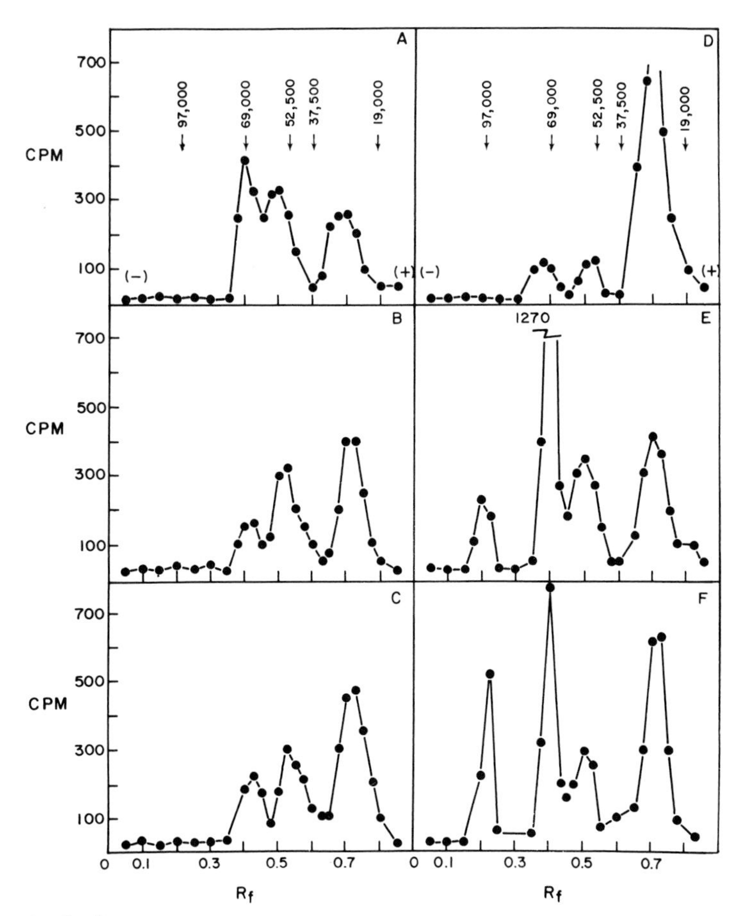


FIG. 5. [³H]Diisopropyl fluorophosphate labeling of chromatin prepared by different methods. Proteins of rat liver nuclei and chromatin prepared and labeled with [³H]DFP as described in Materials and Methods were electrophoresed on 7.5% polyacrylamide sodium dodecyl sulfate gels. The gels were sliced into 1.5-mm fractions and counted for radioactivity. The abscissa gives the ratio of the slice number counted to the slice number corresponding to the position of the tracking dye, bromophenol blue. Purified nuclei (A); chromatin prepared by the method of Smith and Chae (*25*) (B), Huang and Huang (*16*) (C), Stein *et al.* (*23*) (D), Elgin and Bonner (*24*), clear supernatant (E), and Elgin and Bonner (*24*), translucent supernatant (F). The ordinates are in counts per minute above background. The position of molecular weight markers used in the same gel systems are indicated in panels A and D. Reprinted with permission from *Biochemistry* **15**, 2603 (1976). Copyright by the American Chemical Society.

of high- and low-molecular-weight proteins (D of Fig. 5). Rat liver chromatin prepared from whole tissue by the method of Elgin and Bonner (*24*), shows a different labeling pattern (E and F of Fig. 5).

### 3. [$^3$H]DFP LABELING OF CHROMATIN FROM DIFFERENT TISSUES

Chromatin from various tissues was prepared by the method of Huang and Huang (*16*) and labeled with [$^3$H]DFP (Fig. 6). There are qualitative as well as quantitative differences in the DFP-binding proteins. However, all tissues examined thus far contain a DFP-binding protein of molecular weight 25,000 (by SDS gel electrophoresis), which quantitatively is the predominant DFP-binding protein in the chromatins. Evidence was presented which

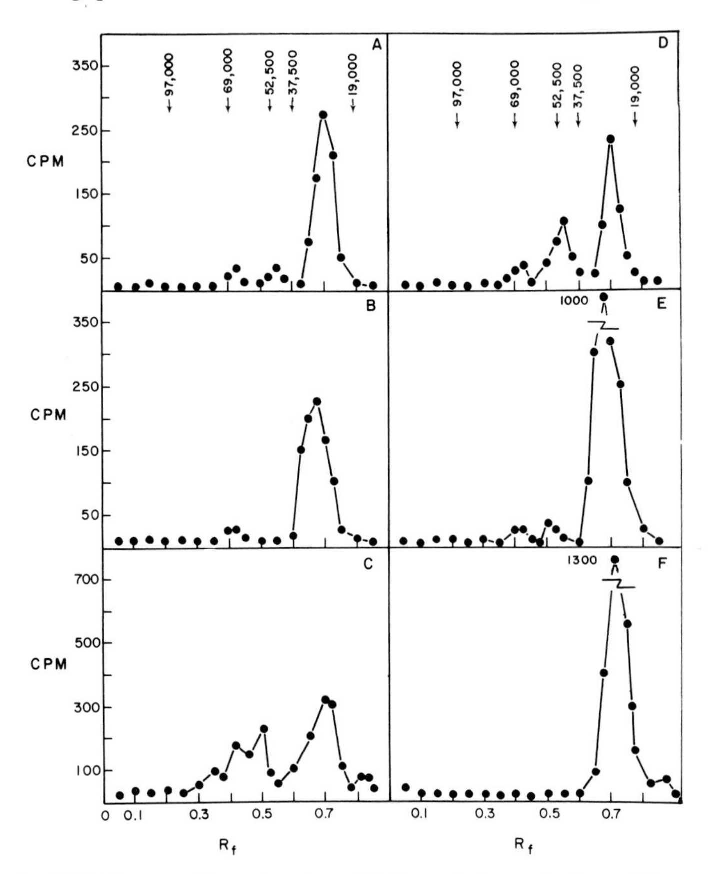


FIG. 6. [$^3$H]Diisopropylfluorophosphate labeling of chromatin from different tissues. Chromatin prepared by the method of Huang and Huang (*16*) from various tissues was labeled with [$^3$H]DFP as described in Materials and Methods and electrophoresed on 7.5% polyacrylamide sodium dodecyl sulfate gels. The gels were sliced into 1.5-mm fractions. The abscissa gives the ratio of the slice number counted to the slice number corresponding to the position of the tracking dye. Rat chromatins from the following tissues are displayed; spleen (A), kidney (B), Morris hepatoma 9121 (C), testis (D), lung (E), and thymus (F). Reprinted with permission from *Biochemistry* **15**, 2603 (1976). Copyright by the American Chemical Society.

suggests that the low-molecular weight DFP-binding protein is a protease active in 2 *M* NaCl–5 *M* urea (pH 8.0) (*14*). The previous results indicate that the chromatin-bound protease active in NaCl and urea is a rather ubiquitous constituent of a variety of chromatins (*13*). However, there are interesting differences in the amounts of high-molecular weight DFP-binding proteins present in the chromatin of various tissues. Although liver and testis appear very similar in the distribution of DFP-binding proteins, spleen, lung, thymus, and kidney chromatin appear to contain a relatively small quantity of the high-molecular-weight DFP-binding proteins. It is not certain at this time whether these differences are due to the possible variations in the amount of the three DFP-binding proteins or to the possible difference in the state of binding of the proteins in chromatins from different tissues.

Among the three DFP-binding proteins present in rat liver chromatin, only the 25,000-dalton DFP-binding protein is selectively extracted by 0.25 *N* HCl, and the 25,000-dalton DFP-binding protein appears to be converted to a 20,000-dalton form during acid extraction (*14*). The acid-extract obtained from chromatin is proteolytically active in 2 *M* NaCl–5 *M* urea, and the proteolysis can be blocked by DFP. When the 0.25 *N* HCl acid-extract of rat liver chromatin is labeled with [$^{3}$H]DFP, the label is exclusively in a 20,000-dalton protein as determined by SDS gel electrophoresis. The radioactively labeled DFP-binding protein and the enzyme activity in 0.25 *N* HCl extract have the same elution volume during chromatography on a Sephadex G-75 column (*14*).

These results suggest that the 25,000-dalton DFP-binding protein which is common for all tissues and cells is a protease which has a serine at its active site.

## III. Concluding Remarks

Work in this laboratory is currently being directed to fractionation and isolation of the three DFP-binding proteins in rat liver nuclei. Whether the two high-molecular weight DFP-binding proteins of rat liver nuclei are proteases or are perhaps precursors of proteases is being investigated. It is hoped that this information will aid in the assignment of a biological role for the protease activity endogenous to chromatin.

### Acknowledgment

The work done in our laboratory was supported by grants from the University of North (VF 336), American Cancer Society (IN15-0), USPHS General Research Support Award (5 S01-FR-05406), National Institute of General Medical Sciences, N.I.H. (GM 21846), and a grant (HD 05277) awarded to Dr. J. Logan Irvin from the National Institute of Child Health and Human Development, N.I.H.

## References

1. Furlan, M., and Jericijo, M., *Biochim. Biophys. Acta* **147**, 135 (1967).
2. Furlan, M., and Jericijo, M., *Biochim. Biophys. Acta* **147**, 145 (1967).
3. Furlan, M., Jericijo, M., and Suhar, A., *Biochim. Biophys. Acta* **167**, 154 (1968).
4. Bartley, J., and Chalkley, R., *J. Biol. Chem.* **246**, 4286 (1970).
5. Combard, A., and Vendrely, R., *Biochem. J.* **118**, 875 (1970).
6. Garrels, J. I., Elgin, S. C. R., and Bonner, J., *Biochem. Biophys. Res. Commun.* **46**, 545 (1972).
7. Chong, M. T., Garrard, W. T., and Bonner, J., *Biochemistry* **13**, 5178 (1974).
8. Kurecki, T., and Toczko, K., *Bull. Acad. Pol. Sci.* **20**, 543 (1972).
9. Kurecki, T., and Toczko, K., *Acta Biochim. Pol.* **21**, 225 (1974).
10. Chae, C.-B., and Carter, D. B., *Biochem. Biophys. Res. Commun.* **57**, 740 (1974).
11. Chae, C.-B., *Biochemistry* **14**, 900 (1975).
12. Chae, C.-B., Gadski, R. A., Carter, D. B., and Efird, P. H., *Biochem. Biophys. Res. Commun.* **67**, 1459 (1975).
13. Carter, D. B., and Chae, C.-B., *Biochemistry* **15**, 180 (1976).
14. Carter, D. B., Efird, P. H., and Chae, C.-B., *Biochemistry*, **15**, 2603 (1976).
15. Bekhor, I., King, G. M., and Bonner, J., *J. Mol. Biol.* **39**, 351 (1969).
16. Huang, R. C. C., and Huang, P. C., *J. Mol. Biol.* **39**, 365 (1969).
17. Gilmour, R. S., and Paul, J., *J. Mol. Biol.* **40**, 137 (1969).
18. Stein, G., Ghaudhuri, S., and Baserga, R., *J. Biol. Chem.* **247**, 3918 (1972).
19. Gilmour, R. S., and Paul, J., *FEBS Lett.* **9**, 242 (1970).
20. MacGillivray, A. J., Cameron, A., Krauze, R. J., Rickwood, D., and Paul, J., *Biochim. Biophys. Acta* **277**, 384 (1972).
21. Richter, K. H., and Sekeris, C. E., *Arch. Biochem. Biophys.* **148**, 44 (1972).
22. Van Den Broek, H. W. J., Nooden, L. D., Sevall, J. S., and Bonner, J., *Biochemistry* **12**, 229 (1973).
23. Stein, G. S., Mans, R. J., Gabbay, E. J., Stein, J. L., Davis, J., and Adawadkar, P. D., *Biochemistry* **14**, 1859 (1975).
24. Elgin, S. C. R., and Bonner, J., *Biochemistry* **9**, 4440 (1969).
25. Smith, M. C., and Chae, C.-B., *Biochim. Biophys. Acta* **317**, 10 (1973).
26. Bekhor, I., Lapeyre, J.-N., and Jung, K., *Arch. Biochem. Biophys.* **161**, 1 (1974).
27. Ipelsberg, T. C., Hnilica, L. S., and Ansevin, A. T., *Biochim. Biophys. Acta* **228**, 550 (1971).
28. Panyim, S., and Chalkley, R., *Arch. Biochem. Biophys.* **130**, 337 (1969).
29. Johns, E. W., *Biochem. J.* **92**, 55 (1964).
30. Panyim, S., and Chalkley, R., *Biochim. Biophys. Acta* **160**, 252 (1968).
31. Noodén, L. D., Van Den Broek, H. W. J., and Sevall, J. S., *FEBS Lett.* **29**, 326 (1973).
32. Koehler, K. A., and Lienhard, G. E., *Biochemistry* **10**, 2477 (1971).
33. Cohen, J. A., Oosterbaan, R. A., and Berends, J., *in* "Methods in Enzymology" (C. H. W. Hirs, ed.), Vol. 11, p. 686. Academic Press, New York, (1967).

# Chapter 19

## *Histone Hydrolase*

WOON KI PAIK AND SANGDUK KIM

*Fels Research Institute and Department of Biochemistry,*
*Temple University School of Medicine,*
*Philadelphia, Pennsylvania*

### I. Introduction

Histones are synthesized in the cytoplasm and are transported into the nucleus, where they exist in conjunction with DNA at an approximate ratio of unity (*1*). The role that histones play in controlling genetic expression is not yet clear. The cell apparently maintains a vital regulatory process in controlling the amount of histone available in order to limit its binding to DNA in a constant ratio, thus it would seem that any excess histones must be disposed of in some way. One of the mechanisms to maintain the homeostasis is the presence of proteolytic enzymes which are highly specific toward histones. Neutral histone hydrolase has been observed in various cellular fractions; chromatin (*2–4*), microsomes (*5*), or cytosol (*6, 7*).

### II. Assay Method

#### A. Principle

Histones are hydrolyzed by the enzyme and the material, which is then rendered soluble in 5% trichloroacetic acid, is determined by allowing it to react with ninhydrin. Since the optimum pH for histone hydrolase activity varies with the tissue, one must first determine this parameter. The following assay condition has been used for the enzyme from rat kidney.

#### B. Reagents

Histone suspension (3 mg per milliliter of water; histone type II-A of Sigma Chemical Co. is routinely used)

Tris-HCl buffer (pH 9.0), 0.5 *M*
Trichloroacetic acid, 10%
NaOH, 1 *N*
Sodium citrate buffer 0.2 M, (pH 5.0)
Ninhydrin, 4%, in ethyl Cellosolve (ethylene glycol monoethyl ether; Fisher Chemical Co.)
$SnCl_2$ suspension in water, 50 mg/ml; prepare immediately before use

## C. Procedure

The tissue was homogenized in 0.25 *M* sucrose–6 m*M* $CaCl_2$ or in cold water to yield a 20% homogenate with an electrically driven Teflon-glass homogenizer. The homogenate was passed through a double layer of cheesecloth. One-tenth or 0.2 ml of tissue homogenate or a fractionated enzyme preparation, 0.2 ml of histone suspension, 0.1 ml of Tris-HCl buffer in a total volume of 0.5 ml were incubated at 37°C for 20 minutes. The reaction was terminated by the addition of 0.5 ml of 10% trichloroacetic acid solution. For a control, the enzyme suspension was added after the reaction was stopped by trichloroacetic acid. The mixture was transferred into a Sorvall centrifuge tube and was centrifuged at 39,000 *g* for 10 minutes. A portion of the clear supernatant (usually 0.1 × 0.2 ml) was transferred to a Coleman spectrophotometer cuvette (19 × 105 mm), and a predetermined amount of NaOH solution was added to bring the pH of the solution to about 5. The volume was adjusted to 1.0 ml with water, and the ninhydrin color was developed as follows.

Thirty milliliters of sodium citrate buffer and an equal volume of ninhydrin solution was first mixed thoroughly, and 1 ml of $SnCl_2$ solution was added. The solution was completely mixed. One milliliter of this ninhydrin solution was added into the above Coleman cuvette and the cuvette was capped with a rubber stopper with a capillary tube. The cuvette was heated in a boiling water bath for 5 minutes, and was then cooled for 3 minutes. Five milliliters of water was added into the cuvette, and absorbancy at 580 nm was read in a Coleman spectrophotometer. Duplicate determinations at two enzyme concentrations have been determined and the value was corrected for the control.

## D. Definition of Specific Activity

The enzyme activity is expressed as specific activity, which corresponds to $A_{580}$/20 minutes per milligram of enzyme protein. One unit of $A_{580}$ corresponds to 0.24 μmole of leucine. The enzyme activity represents an increase of ninhydrin color in 5% trichloroacetic acid-soluble fraction after allowing the histone to react with the enzyme preparation. Enzyme protein concentration was determined by the method of Lowry *et al.* (*8*).

## III. Properties

### A. Specificity

The enzyme is highly specific toward histones. Table I lists specificity of histone hydrolase obtained from various sources. With calf thymus chromatin-associated histone hydrolase, the susceptibility of various histones to hydrolysis is greatly dependent on whether the histones are conjugated with DNA or not (*3*). In the nucleohistone complex, the lysine-rich (F1) and arginine-rich (F3) histones are degraded, and the rest of the histones are resistant. When freed, only lysine-rich histone becomes resistant to the hydrolytic activity, and the rest of the histones are rapidly degraded.

### B. Optimum pH

Optimum pH for histone hydrolase varies depending on the sources. However, the enzyme activities listed in Table I indicate that the enzyme in general may be characterized as a neutral protease.

### C. Subcellular Distribution

The neutral histone hydrolase has been found in chromatin isolated from both calf thymus (*2,3*) and rat liver nuclei (*4*), in the microsomal fraction of rat kidney (*5*), in the cytosol fraction of tadpole liver (*6*) and regenerating limb of the adult newt *Diemictylus viridescens* (7).

### D. Inhibitors

Chromatin-bound enzyme is completely inhibited by sodium bisulfite at concentration of 0.05 *M* (*4*). Serine-specific protease inhibitors, such as phenylmethanesulfonyl fluoride and diisopropylfluorophosphate, and the alkylating reagent, carbobenzoxyphenylalanine chloromethylketone inhibit the enzyme reversibly at 1 m*M* concentration (*9*). Rat liver chromatin-bound protease is still active in the presence of 2 *N* NaCl–5 *M* urea (pH 6–8), which has been most often used for dissociation and reconstitution of chromatin (*9*).

TABLE I

SPECIFICITY OF HISTONE HYDROLASE

| Organisms and organs: | Rat liver | Rat kidney | Tadpole liver | Newt limb |
|---|---|---|---|---|
| Subcellular localization: | Chromatin | Microsome | Cytosol | Cytosol |
| Optimum pH: | 8.2 | 8–9 | 6–8 | 6–7 |
| Substrate | | | | |
| Histone (crude) | 100[a] | 100 | 100 | 100 |
| F1 histone | | 157 | 180 | 24 |
| F2 histone | | 128 | 227 | |
| F3 histone | | 44 | 72 | 17 |
| Polylysine | 320 | 43 | 290 | 74 |
| Protamine | 9 | 129 | 150 | 4 |
| Casein | 18 | | | |
| Hemoglobin (denatured) | 0 | | | |
| Ribonuclease | 5 | 44 | | |
| Lysozyme | 2 | 32 | | |
| Polyarginine | | 6 | | |
| Albumin | | 0 | | |
| Globulin | | 0 | | |
| Polyleucine | | 0 | | |
| Polyglutamic acid | | 0 | | |
| Reference no. | (*4*) | (*5*) | (*6*) | (*7*) |

[a] Percentage activity.

## E. Enzyme Purification

Furlan *et al.* (*10*) purified histone hydrolase from calf thymus nuclei 220-fold with 64% yield. Approximately 700-fold purification was also achieved from rat liver chromatin (*11*). It has a molecular weight of 200,000.

# IV. Biological Significance

Biological significance of histone hydrolase is not clear at present. Since nuclear histones are turning over at an extremely slow rate (*12*), it does not seem probable that histone hydrolase removes histones as a means of gene activation. However, the cytosolic histone hydrolase of regenerating limb of the adult newt *D. viridescens* increased about 2-fold during the wound healing period (*7*). A similar increase in hydrolase activity was observed in liver of tadpoles during thyroxine-induced metamorphosis (*13*). These two observations suggest that histone hydrolase might play an important role in maintaining the balance between histone synthesis and the amount of histone influx to the nucleus.

It should be noted here that recent evidence indicates that so-called chromatin-bound histone hydrolase might represent contamination with a cytosolic enzyme (*14, 15*).

## REFERENCES


*1.* Hnilica, L. S., "The Structure and Biological Function of Histones," p. 57. CRC Press, Cleveland, Ohio (1972).
*2.* Furlan, M., and Jericijo, M., *Biochim. Biophys. Acta* **147**, 135 (1967).
*3.* Bartley, J., and Chalkley, R., *J. Biol. Chem.* **245**, 4286 (1970).
*4.* Garrels, J. I., Elgin, S. C. R., and Bonner, J., *Biochem. Biophys. Res. Commun.* **46**, 545 (1972).
*5.* Paik, W. K., and Lee, H. W., *Biochem. Biophys. Res. Commun.* **38**, 333 (1970).
*6.* Paik, W. K., and Lee, H. W., *Experientia* **27**, 630 (1971).
*7.* Procaccini, D. J., Procaccini, R. L., and Pease, J. B., *Oncology* **29**, 265 (1974).
*8.* Lowry, O. H., Rosebrough, M. J., Farr, A. L., and Randall, R. J., *J. Biol. Chem.* **193**, 265 (1951).
*9.* Carter, D. B., and Chae, C. B., *Biochemistry* **15**, 180 (1976).
*10.* Furlan, M., Jericijo, M., and Suhar, A., *Biochim, Biophys. Acta* **167**, 154 (1968).
*11.* Chong, M. T., Garrard, W. T., and Bonner, J., *Biochemistry* **13**, 5128 (1974).
*12.* Hnilica, L. S., "The Structure and Biological Function of Histones," p. 64. CRC Press, Cleveland, Ohio (1972).
*13.* Paik, W. K., and Kim, S., unpublished results.
*14.* Raydt, G., and Heinrich, P. S., *Hoppe-Seyler's Z. Physiol. Chem.* **356**, 267 (1975).
*15.* Destree, O. H. J., D'Adelhart-Toorop, H. A., and Charles, R., *Biochim. Biophys. Acta* **378**, 450 (1975).

# Part C. Histone Messenger RNAs

## Chapter 20

# *Translation of HeLa Cell Histone Messenger RNAs in Cell-Free Protein Synthesizing Systems from Rabbit Reticulocytes, HeLa Cells, and Wheat Germ*

DIETER GALLWITZ, EBO BOS, AND HANS STAHL

*Physiologisch-Chemisches Institut I,*
*Philipps-Universität Marburg,*
*Marburg/Lahn, West Germany*

## I. Introduction

*In vitro* protein synthesizing systems from eukaryotic cells have been successfully used in studies to elucidate the basic mechanism of protein synthesis and to identify specific mRNAs by their translational products. Moreover, *in vitro* systems which accurately translate mammalian mRNAs are extremely useful to investigate the regulation of mRNA translation and to quantitate biologically active mRNAs from whole cells or from cellular subfractions by their translational capacity.

The synthesis of histones in cultured mammalian cells is coupled with DNA replication during the S phase of the cell cycle (*1–4*). There is evidence that the regulation of histone synthesis is exerted at the transcriptional and translational level (*3–10*).

To investigate the regulatory mechanisms involved in histone biosynthesis, various *in vitro* RNA and protein-synthesizing systems should prove to be of outmost importance. In this article we describe the translation of HeLa cell histone mRNAs in several *in vitro* systems and a method for titrating the amount of translatable histone mRNAs in a rabbit reticulocyte lysate. The methods described can also be used to identify histone mRNAs from other sources.

## II. Cell Growth and Synchronization

HeLa S3 cells are grown in suspension culture (2 to 4 $\times$ $10^5$ cells/ml) in Eagle's minimum essential medium containing 5% calf serum. Since, during their life cycle, cells in mid S phase contain the largest amount of histone mRNAs, the cells are synchronized and allowed to enter S phase prior to isolation of the mRNAs. Synchronization of logarithmically growing cells is achieved by interrupting DNA replication either with thymidine (*5, 11*) or hydroxyurea (*7, 12*).

### A. Reagents

Thymidine, 0.2 *M* (sterilized)
Hydroxyurea, 0.3 *M* (sterilized)

### B. Procedure

Cells are diluted with fresh medium to a density of 1.5 to 2.0 $\times$ $10^5$ cells/ml. Thymidine or hydroxyurea is added to a final concentration of 2 m*M* or 1.5 m*M*, respectively. Fifteen hours later, cells are pelleted at 800 *g* for 5 minutes at 37°C and resuspended in fresh medium. After 10 hours a second block can be performed as described. The block is reversed 15 hours later.

The synchronization procedure has been successfully used with 8-liter-suspensions of cells grown in 10-liter-flasks on magnetic stirrers.

## III. Isolation of Histone mRNAs

Translatable histone mRNAs have been isolated from HeLa cell polyribosomes (*13, 14*) and the postmitochondrial supernatant (*8, 9*). The mRNAs coding for the individual 5 histones have approximate molecular weights of 1.5 to 2.3 $\times$ $10^5$ (*5, 15*) and can be purified in different steps including sucrose gradient centrifugation, poly(U)-Sepharose chromatography, salt precipitation, and preparative polyacrylamide gel electrophoresis. Since histone mRNAs lack poly(A)-sequences at their 3'-OH ends (*16*), poly(U)-Sepharose purifies these mRNAs from most other cellular messengers.

### A. Reagents

All buffers are sterilized.
Lysis buffer: 10 m*M* Tris-HCl (pH 7.5)–1.5 m*M* $MgCl_2$-5 m*M* 2-mercaptoethanol

Wash buffer: 10 m*M* Tris-HCl (pH 7.5)–0.15 *M* NaCl
Detergent buffer: 10 m*M* Tris-HCl (pH 7.5)–2 m*M* $(CH_3COO)_2Mg$–0.15 *M* KCl–0.1% Triton X-100
Polyribosome extraction buffer: 30 m*M* Tris-HCl (pH 7.6)–0.1 *M* NaCl–10 m*M* EDTA–0.5% SDS
Acetate buffer: 3 *M* $CH_3COONa$–5 m*M* EDTA (pH 6.0)
Gradient solutions: 15% and 30% sucrose in 20 m*M* Tris-HCl (pH 7.6)–1 m*M* EDTA–0.1 *M* NaCl–0.25% SDS
Poly(U)-equilibration buffer: 10 m*M* Tris-HCl (pH 7.6)–0.5 *M* NaCl–0.1% SDS
Poly(U)-elution buffer: 10 m*M* Tris-HCl (pH 7.6)–0.1% SDS
Phenol, water-saturated
Proteinase K
Ethanol, absolute, p.A.
Chloroform, p.A.

## B. Procedure

Three to four hours after reversal of a double block with either thymidine or hydroxyurea (Section II) cells are pelleted in a refrigerated centrifuge and washed twice with cold wash buffer. $10^9$ cells are lysed in 50 ml of lysis buffer, or alternatively in detergent buffer, in a Dounce homogenizer with a tightly fitting pestle. The lysate is centrifuged for 15 minutes at 20,000 *g* and, after removal of the lipid layer, the resulting supernatant is made 0.5% in sodium dodecyl sulfate (SDS) with a 10% SDS solution. Proteinase K (0.2 mg/ml) is added and the cytoplasmic fraction is incubated for 30 minutes at 37°C. An equal volume of water-saturated phenol is added, and the mixture is shaken vigorously for 15 minutes at room temperature. Phases are separated by centrifugation at 20°C for 10 minutes at 10,000 *g*. The aqueous phase is reextracted once with phenol–chloroform, 1:1 (v/v), and once with chloroform, and the RNA is precipitated with 2 volumes of ethanol at −20°C overnight. After pelleting, the RNA is smeared over the inside of the glass tube with a sterile glass rod and washed twice with 5–10 ml of acetate buffer to remove tRNA and 5 S rRNA. The RNA is then dissolved in glass-distilled water and precipitated with 2 volumes of ethanol.

A similar RNA preparation can be obtained from polyribosomes which are pelleted from the postmitochondrial supernatant by centrifugation for 90 minutes at 120,000 *g*. Polyribosomes are suspended in polyribosome extraction buffer (25 ml/$10^9$ cells), and RNA is extracted as described above.

The RNA prepared either from polyribosomes or the postmitochondrial supernatant is dissolved to a concentration of 100–150 $A_{260}$ units/ml in 10 m*M* Tris-HCl (pH 7.6) containing 0.25% SDS. About 150 $A_{260}$ units are loaded on a 35 ml 15–30% sucrose gradient and centrifuged at 20°C for 36–38 hours at 23,000 rpm in a Beckman SW-27 rotor. The absorbance at 260 nm is monitored in a spectrophotometer adapted with a flow cell. The RNA sedimenting between 5 S and 15 S is pooled and precipitated with 2 volumes of ethanol. The precipitated RNA is dissolved in poly(U) equilibration buffer (about 50 $A_{260}$ units/ml) and passed at room temperature over a poly(U)–Sepharose column (0.8 cm × 5 cm) equilibrated with the same buffer. The nonbound RNA fraction containing more than 90% of the histone mRNAs is diluted with an equal volume of glass-distilled, sterile water and the RNA is precipitated with ethanol. The precipitated RNA is centrifuged again through sucrose gradients as described above and the RNA sedimenting between 5 S and 15 S is collected. After ethanol precipitation, the RNA is dissolved in water (20–50 $A_{260}$/ml) and stored in aliquots at −80°C or in liquid nitrogen.

As analyzed by polyacrylamide gel electrophoresis in 6% gels the preparations give very reproducible profiles of a heterogeneous RNA population. The gel profiles of RNA derived from polyribosomes and the postmitochondrial supernatant are identical.

When translated in a wheat germ system, the only products formed under the direction of this RNA preparation are the five histones H1, H2A, H2B, H3, and H4 (Fig. 1C).

## IV. Preparation of Rabbit Reticulocyte Lysate

The lysate is prepared with slight modifications according to published procedures (*17, 18*).

### A. Reagents

Phenylhydrazine hydrochloride, 2.5% (w/v), neutralized to pH 7.0 with NaOH. The solution is stored in aliquots at −20°C.
Sodium pentobarbital, 50 mg/ml
Heparin (Liquemin®, Roche), 5000 USP units/ml
Wash buffer: 0.14 *M* NaCl–5 m*M* KCl–5 m*M* $MgCl_2$
Lysing solution: 2 m*M* $MgCl_2$–5 m*M* 2-mercaptoethanol–50 μ*M* hemin

### B. Procedure

Rabbits weighing 2.5–3 kg are made anemic by five daily subcutaneous injections of 0.3 ml of phenylhydrazine per kilogram of body weight. The rabbits are bled on day 7 under pentobarbital anesthesia. Pentobarbital (1 ml/kg) and heparin (1000 units/kg) are mixed and injected into the marginal vein of the ear. The blood is collected from the open chest cavity after cutting the right heart chamber. About 100–120 ml of blood obtained from each rabbit are poured through 4 layers of gauze. All operations are performed at 2°–4°C. The blood cells are pelleted for 10 minutes at 6000 *g*; they are washed four times with 5 volumes of wash buffer and suspended in about 2 volumes of lysing solution. The suspension is stirred for 3 minutes in an ice bath, and cell debris is removed by centrifugation at 15,000 *g* for 15 minutes. The dark red supernatant is stored in aliquots in liquid nitrogen.

## V. Preparation of a Protein-Synthesizing System from HeLa Cells

A cell-free extract from HeLa cells is prepared in principle according to published procedures (*19–21*). A 30,000 *g* extract of HeLa cells, which were lysed with hypotonic medium, is preincubated under conditions of protein synthesis, chromatographed on Sephadex G-25, and used as such for translation experiments. For maximal activity the system has to be supplemented with tRNA (*20*) and initiation factors (*22*).

### A. Reagents

All buffers are sterilized.

Wash buffer: 10 m*M* Tris-HCl (pH 7.5)–0.15 *M* NaCl

Lysis buffer: 10 m*M* Tris-HCl (pH 7.5)–1.5 m*M* $MgCl_2$–5 m*M* 2-mercaptoethanol

Concentrated buffer: 0.222 *M* Tris-HCl (pH 7.5)–0.95 *M* KCl–18 m*M* $(CH_3COO)_2$ Mg–20 m*M* 2-mercaptoethanol

Column buffer: 30 m*M* Tris-HCl (pH 7.5)–95 m*M* KCl–3 m*M* $(CH_3COO)_2Mg$–6 m*M* 2-mercaptoethanol

Sephadex G-25 fine, autoclaved

Energy mixture: 10 m*M* ATP–2 m*M* GTP–0.15 *M* creatine phosphate–0.5 mg/ml of creatine kinase–0.5 m*M* 20 unlabeled amino acids. Before adding creatine kinase to the solution containing all components, the pH is adjusted to pH 6–7 with KOH.

## B. Procedure

HeLa cells are harvested by centrifugation, washed twice with wash buffer, and suspended in lysis buffer at a concentration of 6 to 8 $\times 10^7$ cells/ml. After 10 minutes in an ice bath, the cells are lysed with 10–15 strokes in a Dounce homogenizer. Isotonicity is restored by adding 0.1 volume of concentrated buffer. The cell homogenate is centrifuged for 20 minutes at 30,000 *g*, the lipid layer is removed, and the upper two-thirds of the supernatant are aspirated with a Pasteur pipette. One-tenth volume of the energy mixture is added, and the extract is incubated at 35°C for 30 minutes. The incubated extract is cooled in an ice bath before centrifugation for 10 minutes at 30,000 *g*. The supernatant (3–4 ml) is passed at a flow rate of 2.5 ml/minute over a column of Sephadex G-25 fine (2 cm $\times$ 15 cm) equilibrated with column buffer. The opalescent fractions are pooled, aspirated with a 2-ml pipette, and dropped into liquid nitrogen. The resulting beads have a volume of about 40 $\mu$l and are stored in liquid nitrogen.

# VI. Preparation of a Protein-Synthesizing System from Wheat Germ

The cell-free system is prepared according to procedures described (*23–25*). The 30,000 *g* extract is used without preincubation.

## A. Reagents

Untreated, fresh wheat germ

Extraction buffer: 20 m*M* HEPES (pH 7.6) (adjusted with KOH)–0.1 *M* KCl–1 m*M* $(CH_3COO)_2Mg$–2 m*M* $CaCl_2$–6 m*M* 2-mercaptoethanol

Column buffer: 20 m*M* HEPES (pH 7.6)–0.125 *M* KCl–5 m*M* $Mg(Ac)_2$–6 m*M* 2-mercaptoethanol

Sephadex G-25 fine, autoclaved

## B. Procedure

All steps are performed at 2°–4° C. Four grams of wheat germ are ground with 4 gm of powdered glass (prepared by crashing Pasteur pipettes) for 45–60 seconds in a cooled mortar. Eight milliliters of extraction buffer are added and gentle grinding is continued for about 20–30 seconds. The paste is transferred to a centrifuge tube using a glass rod and centrifuged for 15 minutes at 30,000 *g*. After removing the lipid layer the supernatant (3–4 ml) is passed over a column of Sephadex G-25 fine (2 cm × 15 cm) equilibrated with column buffer. The flow rate is 3 ml/minute. Fractions of 1 ml are collected, and fractions having less than 80–90 $A_{260}$ units are discarded. The other fractions are pooled, centrifuged for 15 minutes at 30,000 *g*, and stored as small beads in liquid nitrogen (Section V).

To prevent inactivation of the cell-free system, all steps should be performed as fast as possible.

## VII. Preparation of Unfractionated tRNA

Unfractionated tRNA is required for maximal activity of the cell-free system from HeLa cells. It is conveniently prepared from the ribosome-free supernatant fraction derived from rabbit reticulocytes (Section IV).

### A. Reagents

Phenol, water-saturated
Chloroform, p.A.
Tris-HCl, 0.1 *M* (pH 8.9)
Potassium acetate, 1 *M* and 10 m*M* (pH 7.0)
Ethanol, absolute, p.A.

### B. Procedure

To the postribosomal supernatant from reticulocytes, an equal volume of phenol is added; the mixture is vigorously shaken in Erlenmeyer flasks for 15 minutes at room temperature. After centrifugation at 10,000 *g* for 10 minutes, the aqueous phase is recovered and reextracted once with phenol–chloroform, 1:1 (v/v), and once with chloroform. One-tenth volume of a 1 *M* potassium acetate solution is added to the aqueous phase, and the RNA is precipitated with 2 volumes of ethanol at −20°C overnight. The RNA is reprecipitated once and separated by Sephadex G-200 chromatography at 4°C according to Delihas and Staehelin (*26*). tRNA elutes as the second peak from the Sephadex column, which is equilibrated with column buffer. The ethanol-precipitated tRNA is dried under vacuum and dissolved in 0.1 *M* Tris-HCl (pH 8.9) at a concentration of about 10 mg/ml. Deacylation of tRNA is performed at 35°C for 20 minutes, 0.1 volume of 1 *M* potassium acetate is added, and the RNA is precipitated with ethanol. The precipitated tRNA is dried, dissolved in sterilized glass-distilled water at a concentration of 100 $A_{260}$ units/ml and stored in aliquots at $-80\,°C$.

## VIII. Preparation of Reticulocyte Initiation Factors

For maximal activity the HeLa cell-free protein synthesizing system (Section V) has to be supplemented with initiation factors. A crude initiation factor preparation is obtained from rabbit reticulocyte polyribosomes washed with 0.5 *M* KCl.

### A. Reagents

All solutions are sterilized.
Lysing solution: 2 m*M* $MgCl_2$–0.5 m*M* dithiothreitol (DTT)
Sucrose solution: 1.5 *M* sucrose–0.15 *M* KCl–2 m*M* $MgCl_2$
Polyribosome suspension buffer: 5 m*M* Tris-HCl (pH 7.6)–0.25 *M* sucrose–0.5 m*M* DTT–5 m*M* $MgCl_2$
Dilution buffer: 10.5 m*M* Tris-HCl (pH 7.6)–0.5 m*M* DTT–0.55 m*M* $MgCl_2$
Equilibration buffer: 10 m*M* Tris-HCl (pH 7.6)–0.5 m*M* DTT–50 m*M* KCl–1 m*M* $MgCl_2$
Elution buffer: 10 m*M* Tris-HCl (pH 7.6)–0.5 m*M* DTT–0.35 *M* KCl–1 m*M* $MgCl_2$
KCl, 4 *M*

### B. Procedure

Reticulocytes are prepared as described in Section IV. Lysis is achieved by suspending the packed blood cells in about 2 volumes of lysing solution and stirring the suspension for 5 minutes in an ice bath. Two-tenths volumes of sucrose solution are added, and cell debris is removed by centrifugation at 30,000 *g* for 15 minutes. Polyribosomes are pelleted from the supernatant by centrifugation for 2 hours at 120,000 *g* and suspended in polyribosome suspension buffer at a concentration of 150–250 $A_{260}$ units/ml. In an ice bath, 4 *M* KCl is added dropwise under stirring to give a final concentration of 0.5 *M*. Stirring is continued for 10 minutes, and the extract is centrifuged for 2 hours at 240,000 *g*. Nine volumes of dilution buffer are added to the supernatant and the extract is passed over a DEAE-cellulose column equilibrated with equilibration buffer (3 cm × 5 cm column for an extract from 4000 $A_{260}$ units of polyribosomes). The proteins adsorbed to DEAE-cellulose are eluted from the column with elution buffer and the peak fractions (3.5–4.5 $A_{260}$ units/ml) are collected and stored in aliquots in liquid nitrogen.

## IX. Translation and Quantitation of Histone mRNAs in a Rabbit Reticulocyte Lysate

For translating histone mRNAs the reticulocyte lysate has certain advantages over other cell-free systems since there is no endogenous histone synthesis in this system, more than 90% of the endogenous product represents globin which can be separated from histones on CM-cellulose, and, furthermore, the system has very good initiation properties. A method of titrating the amount of biologically active histone mRNA by its translational capacity in the reticulocyte lysate has been described (*7, 9*).

### A. Histone mRNA Translation

#### 1. Reagents

Components for the assay mixture
RNase A, 1 mg/ml
EDTA, 0.1 *M*
HCl, 0.5 *N*, 0.25 *N*, 0.1 *N*
Acetone, p.A.

#### 2. Translation

Translation of histone mRNAs and identification of the products are routinely performed in assay mixtures with a volume of 0.3 ml containing: 10 m*M* Tris-HCl (pH 7.5)–0.1 *M* $NH_4Cl$–2 m*M* $MgCl_2$–1 m*M* DTT–1 m*M* ATP–0.2 m*M* GTP–15 m*M* creatine phosphate–20 μg of creatine kinase–25 μCi of L-[$^3H$]lysine (specific activity about 40 Ci/mmole)–50 μ*M* 19 unlabeled amino acids–0.15 ml reticulocyte lysate (Section IV) and varying amounts of histone mRNA. Incubations are performed in closed glass tubes at 26°C for 45–60 minutes. EDTA and RNase are added at the end of the incubation to give final concentrations of 10 m*M* and 10 μg/ml, respectively. Incubations are continued for 15 minutes at 26°C. All following steps are performed at 0–4°C. The incubation mixtures are transferred to an ice bath and an equal volume of cold 0.5 *N* HCl is added dropwise under stirring.

The volume is then adjusted to 1 ml with 0.25 *N* HCl, and acid-soluble proteins are extracted for 30 minutes. After centrifugation at 20,000 *g* for 10 minutes, the acid extract is dialyzed for 4–6 hours against 0.1 *N* HCl, and the proteins are then precipitated with 9 volumes of acetone at −20°C overnight.

## B. CM-Cellulose Chromatography of the Translational Products

### 1. Reagents

Tris–urea buffer: 0.1 *M* Tris-HCl (pH 7.6)–6 *M* urea
HCl–urea solution: 0.25 *N* HCl–6 *M* urea
HCl, 0.1 *N*
Acetone p.A.

### 2. Procedure

The proteins precipitated with acetone are centrifuged for 10 minutes at 15,000 *g*, dried under vacuum, and dissolved in 1 ml of Tris–urea buffer. They are slowly passed through a 0.5 cm × 5 cm column of CM-cellulose equilibrated with Tris–urea buffer. Globin passes through the column (CM-1 fraction) whereas histones are quantitatively adsorbed to the ion exchanger and eluted with HCl–urea solution. Fraction of 1.8 ml are collected and 50 μl aliquots of each fraction are pipetted directly into scintillation vials containing 0.4 ml of water. Six milliliters of Bray's scintillation fluid (*27*) are added, and radioactivity is measured. The radioactive fractions of the HCl–urea wash (CM-2 fraction) are pooled, dialyzed for 4 hours against 0.1 *N* HCl, and precipitated with 9 volumes of acetone for at least 20 hours at −20°C. The precipitated proteins are centrifuged at 20,000 *g* for 15 minutes, dried under vacuum, and used for further analysis of the *in vitro* products (i.e., polyacrylamide gel electrophoresis, column chromatography, tryptic digestion). A gel electrophoretic profile of the *in vitro* products formed under the direction of HeLa cell histone mRNAs is shown in Fig. 1A.

## C. Quantitation of Histone mRNA in the Reticulocyte Lysate

To quantitate biologically active histone mRNA from cells at different stages of the cell cycle or from subcellular fractions, for instance, the mRNAs are isolated as outlined in Section III. The amount of mRNA is calculated from the radioactivity incorporated into protein according to the equation $(A - B)/a$. $A$ and $B$ are (cpm in CM-2 × 100)/(cpm in CM-1 × 100)

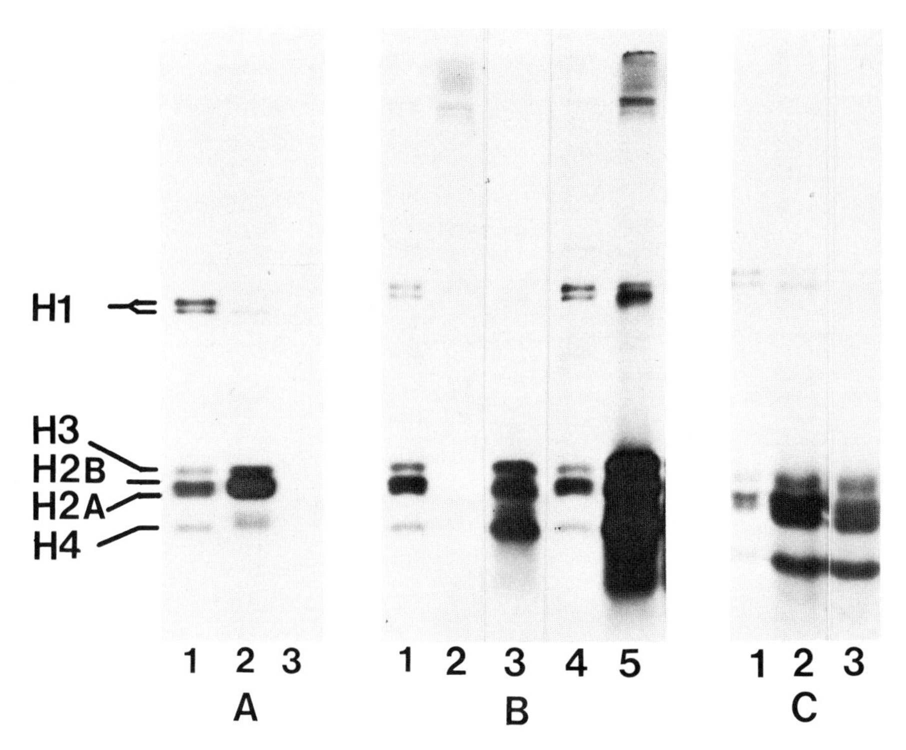


FIG. 1. Sodium dodecyl sulfate (SDS) gel electrophoresis of cell-free products synthesized under the direction of HeLa cell histone mRNA in: (A) reticulocyte lysate (slot 1, marker histones; slot 2, CM-2 fraction of histone mRNA directed products; slot 3, CM-2 fraction from a control without added mRNA); (B) HeLa cell-free system (slots 1 and 4, marker histones; slot 2, cell-free product without added mRNA; slots 3 and 5, histone mRNA-directed products); (C) wheat germ system (slot 1, marker histones; slots 2 and 3, histone mRNA directed products). Radioactivity was detected by fluorography. To show synthesis of histones H1 the films on (B), slot 5, and (C), slot 2, were exposed for longer periods of time.

from lysates incubated in the presence and in the absence of histone mRNA, respectively; $a$ is $A_{260}$ units of polyribosomal or whole cytoplasmic poly(A)-(−)-RNA from which histone mRNA is recovered. For these studies polyribosomal or whole cytoplasmic RNA is passed over poly(U)–Sepharose before sucrose gradient centrifugation and the amount of RNA from which the 5–15 S RNA is recovered represents $a$ in the equation above. As shown in Fig. 2, a linear relationship between the concentration of mRNA added to the incubation mixture and lysine incorporation $(A - B)$ is obtained. It can be calculated from Fig. 2 that the relative amount of histone mRNA in hydroxyurea-blocked cells is 15% of that in S-phase cells $(A - B/a)$.

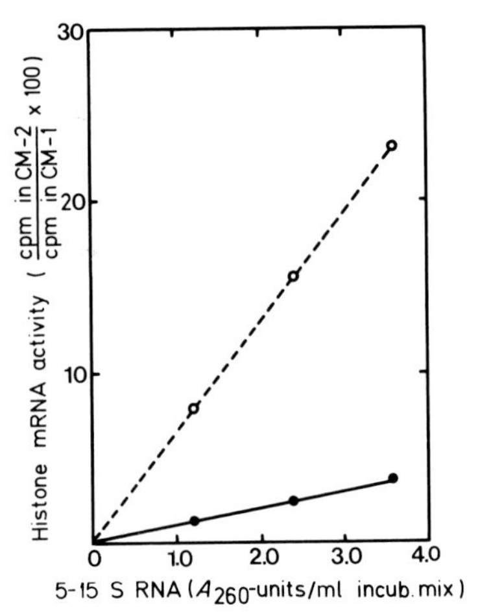


FIG. 2. Quantitation of translatable HeLa cell histone mRNA derived from polyribosomes of S-phase cells (○——○) and hydroxyurea-blocked cells (●——●).

## X. Translation of Histone mRNAs in a Cell-Free System from HeLa Cells

For maximal activity and translation of exogenously added mRNA, the preincubated system has to be supplemented with unfractionated tRNA (Section VII) and initiation factors (Section VIII). The amount of the components necessary for optimizing the system varies from one extract to another and should be determined after preparation.

Histone mRNAs from HeLa cells have been translated in cell-free extracts prepared from S phase and from hydroxyurea-blocked S-phase cells (7). As shown in Fig. 1B, all histones are synthesized in the HeLa cell-free extract as judged by comparing the electrophoretic mobility of the $^{3}$H-labeled *in vitro* products with authentic HeLa cell histones labeled with [$^{14}$C]lysine *in vivo*.. The two bands of histone H1 are seen only after exposing the film for longer periods of time. The radioactive band moving in the SDS-gel between histones H2A and H4 (Fig. 1B, slot 3) is usually very prominent when no histone H1 is detectable. It might, therefore, represent a degradation product of histones H1 or a defined unfinished H1-polypeptide.

### A. Reagents

Components for the assay mixture:
Trichloroacetic acid, 20% (w/v)

RNase A, 1 mg/ml
EDTA, 0.1 *M*

## B. Procedure

Each 50-μl assay mixture contains: 30 m*M* Tris-HCl (pH 7.5)–95 m*M* KCl–3 m*M* $(CH_3COO)_2Mg$–6 m*M* 2-mercaptoethanol–1 m*M* ATP–0.2 m*M* GTP–15 m*M* creatine phosphate–3 μg of creatine kinase–5 μCi of L-[$^3$H]lysine (specific activity about 40 Ci/mmole)–50 μ*M* 19 unlabeled amino acids 0.2–0.5 $A_{260}$ unit of preincubated S-30 extract (Section V), 0.1 $A_{260}$ units of unfractionated tRNA (Section VII), crude reticulocyte initiation factors (15–40 μg of protein in 0.35 *M* KCl; Section VII), 0.1–0.15 $A_{260}$ unit of histone mRNA (Section III). Incubations are performed in open glass tubes at 35°C for 30 minutes. Incorporation of labeled amino acids is linear for 30–45 minutes. Protein synthesis of a large number of samples is conveniently measured by pipetting aliquots of the incubation mixtures onto Whatman No. 3 MM filter paper disks previously soaked in 20% trichloroacetic acid. The filters are air-dried and transferred to 20% trichloroacetic acid (the high concentration of trichloroacetic acid is necessary to precipitate all histones). After heating at 90°C for 20 minutes, the filters are washed twice with fresh trichloroacetic acid, once with ethanol, ethanol–ether, 1:1 (v/v), and ether. The dried papers are counted in a toluene-based liquid scintillation fluid.

For analysis of the products on polyacrylamide gels, at the end of the incubations EDTA and RNase are added to give final concentrations of 10 m*M* and 20 μg/ml, respectively. Incubations are continued for 15 minutes at 35°C, and the samples are mixed with sample buffer and loaded directly onto slab gels (Section XII), as shown in Fig. 1B.

# XI. Translation of Histone mRNAs in a Cell-Free System from Wheat Germ

Histone mRNAs from HeLa cells are efficiently translated in the wheat germ system. As in all other cell-free systems, histones H1 are poorly synthesised in the wheat germ extract as well (Fig. 1C). This might mean that the mRNA preparations obtained contain only small amounts of H1 messengers or that histones H1 are readily degraded by proteases present in the cell-free extracts.

## A. Reagents

Components for the assay mixture:
Trichloroacetic acid, 20% (w/v)
RNase A, 1 mg/ml
EDTA, 0.1 *M*

## B. Procedure

Each 50-μl assay mixture contains: 20 m*M* HEPES (pH 7.6) (adjusted with KOH)–2 m*M* $(CH_3COO)_2Mg$–85 m*M* KCl–6 m*M* 2-mercaptoethanol–1 m*M* ATP–0.2 m*M* GTP–15 m*M* creatine phosphate–3 μg of creatine kinase–5 μCi of L-[$^3$H]lysine (specific activity about 40 Ci/mmole)–50 μ*M* of 19 unlabeled amino acids–1.5 mg of polyethylene glycol (MW 6000)–0.1 m*M* spermine, and 4–5 $A_{260}$ units of wheat germ extract, as well as varying amounts of histone mRNA. An aliquot of the wheat germ extract is thawed immediately before use. Storage, even in an ice bath, results in a quick loss of activity. Incubations are performed in open glass tubes at 30°C for 20 minutes. Protein synthesis is measured and histones are identified as described in Section X.

# XII. Identification of *in Vitro* Synthesized Histones by Polyacrylamide Gel Electrophoresis

For analysis of *in vitro* synthesized products, two methods of polyacrylamide gel electrophoresis can be used, e.g., electrophoresis in acid–urea gels according to Panyim and Chalkley (*28*) and in SDS gels according to the procedure of Laemmli (*29*). Slab gels with 10 cm × 15 cm × 0.13 cm (20 ml) for the separating gel and 2 cm × 15 cm × 0.13 cm (4 ml) for the spacer gel are used. Either the whole incubation mixture or acid-extracted proteins are loaded onto the gels. The radioactive proteins are detected by fluorography (*30*).

## A. SDS Gels

### 1. Reagents

Acrylamide solution: 30% (w/v) acrylamide–0.8% (w/v) *N*,*N*′-bismethylene acrylamide
Tris-HCl buffer: 1.5 *M* (pH 8.8) and 0.5 *M* (pH 6.8)
SDS solution, 10% (w/v)

Ammonium persulfate, 10% (w/v), freshly prepared before use

*N,N,N′,N′*-Tetramethylethylenediamine (TEMED)

Electrophoresis buffer, concentrated: 0.77 *M* glycine–0.1 *M* Tris–0.4% SDS, distilled water to 1 liter. Before use 1 volume of the buffer is diluted with 3 volumes of water.

Sample buffer: 0.1 *M* Tris-HCl (pH 6.8)–2% SDS–2% (v/v) 2-mercaptoethanol–20% (v/v) glycerol

#### 2. Preparation of Slab Gels

The slab gels are prepared and electrophoresis is performed according to Studier (*31*). Ten milliliters of acrylamide solution, 5 ml of 1.5 *M* Tris-buffer (pH 8.8), 0.2 ml of SDS solution, 0.2 ml of ammonium persulfate solution, and 4.6 ml of water are mixed; 10 μl of TEMED is added, and the solution is polymerized under a 1-cm layer of water between two glass plates which are separated by 6 mm-wide Lucite spacers placed at the side and bottom edges. To make the chamber liquid-tight, 2% agarose is polymerized around the outside edges.

After 1 hour the 7.5% spacer gel (1 ml of acrylamide, 1 ml of 0.5 *M* Tris-HCl (pH 6.8), 40 μl of SDS, 2 ml of water, and 4 μl of TEMED) is polymerized for 1 hour on top of the 15% separating gel. Thirteen sample wells (6 mm wide) are formed with a Lucite comb.

#### 3. Sample Preparation and Electrophoresis

To the incubation mixtures of wheat germ or HeLa cell extracts, 1 volume of sample buffer is added and the solution is heated for 1 minute in a boiling water bath. Proteins of the CM-2 fraction (Section IX) are dissolved in half-strength sample buffer by heating for 1 minute at 100°C. Aliquots of not more than 50 μl are placed into the slots, and immediately thereafter electrophoresis is started and performed at 50 V constant for 13–14 hours. Gels are stained with 0.1% Amido Black in 40% methanol–10% acetic acid (v/v) for at least 2 hours and destained with 40% methanol–10% acetic acid.

### B. Acid-Urea Gels

#### 1. Reagents

Acrylamide solution 1: 60% (w/v) acrylamide–0.4% (w/v) *N,N′*-bismethylene acrylamide

Acrylamide solution 2: 40% (w/v) acrylamide–0.3% (w/v) *N,N′*-bismethylene acrylamide

Urea solution: 4 *M* urea–0.2% (w/v) ammonium persulfate, freshly prepared before use

TEMED–acetic acid: 4% (v/v) *N,N,N′,N′*-tetramethylethylenediamine–43.2% (v/v) acetic acid
Acetic acid, 0.9 *N* (deaerated)
Sample buffer: 6 *M* urea–1% 2-mercaptoethanol

### 2. Preparation of Slab Gels

For the 15% separating gel 6 ml of acrylamide solution 1, 15 ml of urea solution, and 3 ml of TEMED–acetic acid are mixed, degassed for 1 minute, and polymerized under a 1-cm layer of water for at least 2 hours. A 7.5% gel is used as spacer (2 ml of acrylamide solution 2, 1 ml of TEMED-acetic acid, 5 ml of urea solution). The slab gels are formed as described in Section XII,A,2.

### 3. Sample Preparation and Electrophoresis

Preelectrophoresis is performed for at least 6 hours at 50 V constant. Samples are dissolved in sample buffer by incubating for 1 hour at room temperature. Aliquots of not more than 25 μl are applied, and electrophoresis is performed at 100 V constant for 7–8 hours, at room temperature.

## C. Fluorography

The radioactive proteins separated on polyacrylamide slab gels are detected by fluorography according to Laskey and Mills (*30*).

### 1. Reagents

Dimethylsulfoxide (DMSO)
DMSO–PPO: 20% (w/v) 2,5 diphenyloxazole (PPO) in DMSO
Kodak RP Royal "X-Omat" medical X-ray film (13 cm × 18 cm)

### 2. Procedure

Stained or unstained slab gels are soaked in 20 volumes of DMSO for 30 minutes and then for another 30 minutes in fresh DMSO. The gels are immersed in 4 volumes of PPO–DMSO for 3 hours, and finally in water for 1 hour. The gels are then transferred to a piece of Whatman No. 3 MM filter paper on a porous sheat of Teflon and dried under vacuum in a chamber sealed with Saran wrap. The dried gels are exposed to RP Royal X-Omat film at −70°C.

### Acknowledgments

We thank Mrs. Renate Seidel for expert technical assistance. This work was supported by a grant from the Deutsche Forschungsgemeinschaft to Dieter Gallwitz.

## REFERENCES


1. Spalding, J., Kajiwara, K., and Mueller, G. C., *Proc. Natl. Acad. Sci. U.S.A.* **56**, 1535 (1966).
2. Robbins, E., and Borun, T. W., *Proc. Natl. Acad. Sci. U.S.A.* **57**, 409 (1967).
3. Borun, T. W., Scharff, M., and Robbins, E., *Proc. Natl. Acad. Sci. U.S.A.* **58**, 1977 (1967).
4. Gallwitz, D., and Mueller, G. C., *J. Biol. Chem.* **244**, 5947 (1969).
5. Breindl, M., and Gallwitz, D., *Eur. J. Biochem.* **32**, 381 (1973).
6. Perry, R. P., and Kelley, D. E., *J. Mol. Biol.* **79**, 681 (1973).
7. Breindl, M., and Gallwitz, D., *Eur. J. Biochem.* **45**, 91 (1974).
8. Borun, T. W., Gabrielli, F., Ajiro, K., Zweidler, A., and Baglioni, C., *Cell* **4**, 59 (1975).
9. Gallwitz, D., *Nature (London)* **257**, 247 (1975).
10. Stein, G. S., Park, W. D., Thrall, C. L., Mans, R. J., and Stein, J. L., *Biochem. Biophys. Res. Commun.* **63**, 945 (1975).
11. Galavazi, G., Schenk, H., and Bootsma, D., *Exp. Cell Res.* **41**, 428 (1966).
12. Adams, R. L. P., and Lindsay, J. G., *J. Biol. Chem.* **242**, 1314 (1967).
13. Gallwitz, D., and Breindl, M., *Biochem. Biophys. Res. Commun.* **47**, 1106 (1972).
14. Jacobs-Lorena, M., Baglioni, C., and Borun, T. W., *Proc. Natl. Acad. Sci. U.S.A.* **69**, 2095 (1972).
15. Bos, E. S., Roskam, W., and Gallwitz, D. *Biochem. Biophys. Res. Commun* **73**, 404 (1976).
16. Adesnik, M., and Darnell, J. E., *J. Mol. Biol.* **67**, 397 (1972).
17. Adamson, S. D., Godchoux, W., and Herbert, E., *Arch. Biochem. Biophys.* **125**, 671 (1968).
18. Gilbert, J. M., and Anderson, W. F., *J. Biol. Chem.* **245**, 2342 (1970).
19. Mathews, M. B., and Korner, A., *Eur. J. Biochem.* **17**, 328 (1970).
20. Aviv, H., Boime, I., and Leder, P., *Proc. Natl. Acad. Sci. U.S.A.* **68**, 2303 (1971).
21. McDowell, M. J., Joklik, W. K., Villa-Komaroff, L., and Lodish, H. F., *Proc. Natl. Acad. Sci. U.S.A.* **69**, 2649 (1972).
22. Metafora, S., Terada, M., Dow, L. W., Marks, P. A., and Bank, A., *Proc. Natl. Acad. Sci. U.S.A.* **69**, 1299 (1972).
23. Roberts, B. E., and Paterson, B. M., *Proc. Natl. Acad. Sci. U.S.A.* **70**, 2330 (1973).
24. Marcus, A., Efron, D., and Weeks, D. P., *Methods Enzymol.* **30**, Part F, 749 (1974).
25. Marcu, K., and Dudock, B., *Nucleic Acids Res.* **1**, 1385 (1974).
26. Delihas, M., and Staehelin, M., *Biochim. Biophys. Acta* **119**, 385 (1966).
27. Bray, G. A. A., *Anal. Biochem.* **1**, 279 (1960).
28. Panyim, S., and Chalkley, R., *Biochemistry* **8**, 3972 (1969).
29. Laemmli, U. K., *Nature (London)* **227**, 680 (1970).
30. Laskey, R. A., and Mills, A. D., *Eur. J. Biochem.* **56**, 335 (1975).
31. Studier, F. W., *J. Mol. Biol.* **79**, 237 (1973).

# Chapter 21

# *Isolation of Histone Messenger RNA and Its Translation in Vitro*

LEE A. WEBER, TIMOTHY NILSEN, AND CORRADO BAGLIONI

*Department of Biological Sciences,*
*State University of New York at Albany,*
*Albany, New York*

## I. Introduction

The synthesis of histones in embryos of marine invertebrates and animal cells in culture appears to be regulated both at the level of transcription and translation of messenger RNA (mRNA). Histone mRNA is present in the cytoplasm of sea urchin eggs (*1,2*), but is not translated until after fertilization. Very little protein synthesis takes place prior to fertilization, however, and histones synthesis may be under the control of a general regulatory mechanism which prevents the translation of all mRNAs in unfertilized eggs. After fertilization, histone mRNA is actively synthesized (*3*), and both newly synthesized and maternal histone mRNA are translated. A different type of regulation of histone synthesis occurs in *Spisula solidissima* eggs, since no histone mRNA is found among the maternal mRNAs and active histone synthesis upon fertilization seems uniquely dependent on *de novo* transcription of histone templates (*4*).

In HeLa cells, histones are synthesized during the S phase of the cell cycle (*5,6*). A block in DNA synthesis by different inhibitors results in a rapid decline of histone synthesis without a detectable effect on the synthesis of other cellular proteins (*7*). Histone mRNA is synthesized and is present on polyribosomes only during the S phase, and rapidly disappears upon a block in DNA synthesis (*8–11*). The lifetime of histone mRNA during S phase is at least a few hours (*12,13*), but histone synthesis stops within 30 minutes after a block in DNA synthesis (*8,10*). Therefore, the coordination of DNA replication and histone synthesis cannot be explained solely by a transcriptional control mechanism, and regulation of histone synthesis must also operate at the translational level (*8–10, 12, 14*). While the molecular basis of this translational control is not yet understood, an insight into this problem has become possible only after the successful isolation, translation and identification of histone mRNA (*15–17*).

Messenger RNA coding for histones has several unique properties which facilitate its isolation and identification. Histone mRNA is composed of three size classes of molecules which correspond to the three size classes of histones. The mRNAs sediment on sucrose gradients from approximately 7.5 S to 12 S, with a major peak at 9 S (*16*). Histone mRNA is much smaller in size than the mRNA coding for the majority of proteins of animal cells which has an average sedimentation coefficient of about 20 S (*18, 19*). Centrifugation in sucrose gradients or preparative gel electrophoresis can therefore be used to separate histone mRNA from the bulk of cellular mRNA as well as from ribosomal and transfer RNA. Furthermore, histone mRNA lacks the 3′-poly(A) sequence found in most eukaryotic mRNAs. The poly(A)-containing mRNA can be selectively removed during RNA isolation (*20, 21*) and contamination of histone mRNA with other cellular templates is considerably reduced. Finally, histones are soluble in acid and possess characteristic electrophoretic and chromatographic properties which simplify the analysis of the products of *in vitro* translation (*15, 22*).

This article will discuss: (a) the isolation of histone mRNA from HeLa cells and marine eggs and embryos; (b) its translation in cell-free protein-synthesizing systems obtained from mouse ascites, rabbit reticulocytes, and wheat germ; and (c) the analysis of the products of *in vitro* translation. The procedures to be discussed can be used to isolate and translate any mRNA present in discrete amounts in different cell types and tissues. The techniques involved are quite straightforward, and success in the isolation and translation of mRNA depends on the care with which RNA and protein synthesizing systems are prepared.

A very simple procedure for the isolation of HeLa cell histone mRNA labeled to high specific activity with radioactive precursors will also be described. Because of its small size and the absence of poly(A), labeled histone mRNA of high radiochemical purity can be easily isolated. The labeled histone mRNA can then be used as a marker in the isolation of unlabeled mRNA to be used in translation experiments. Radioactive histone mRNA can also be used to study the entry of mRNA into initiation complexes (*23*). The association of histone mRNA with ribosomal subunits in cell extracts can be followed by sucrose gradient analysis and provides a direct assay for the ability of a cell-free system to initiate protein synthesis (*23*).

# II. Isolation of Histone Messenger RNA

## A. General Considerations

The most critical step in cell-free translation experiments is the preparation of structurally intact mRNA free from contaminants which inhibit protein synthesis. It must be stressed that methods designed for the study of RNA metabolism are not always suitable for the preparation of mRNA for translation experiments. Since the isolation of mRNA requires considerably more "handling," more precautions must be taken to avoid contamination with traces of RNase and metal ions. While all the precautions taken in our laboratory might not be absolutely necessary, we have learned by experience that extra caution minimizes the unexplained failure of some RNA preparations to be translated.

All glassware that is to come in contact with RNA is washed carefully by gloved hands, using a fresh detergent solution. The glassware is then rinsed thoroughly with tap water and rinsed several times with glass-distilled water, covered with aluminum foil, and heat treated at 250°C for 6 hours. Glassware that has come in contact with cell homogenates or cell extracts is cleaned with sulfuric acid prior to washing with detergent and heat treatment. The use of cleaning solutions containing dichromate is strictly avoided! All aqueous stock solutions used in the preparation of mRNA are made up using autoclaved glass-distilled water. The solutions are filtered through prerinsed 0.45 $\mu$m nitrocellulose filters (Millipore) and stored frozen. Ethanol used for precipitating RNA is filtered through acid-hardened filter paper to remove particles of dust. Pipettes are always rinsed with buffer prior to immersion in a solution that is to come in contact with RNA. Vessels containing RNA solutions are kept covered with parafilm to prevent contamination by dust and mold.

The phenol used to deproteinize RNA is often contaminated with oxidation products and metal ions, which catalyze hydrolysis of RNA and also inhibit cell free protein synthesis. While some commercial preparations may be satisfactory, phenol is always redistilled. Phenol must be free of preservatives or else it cannot be safely redistilled.

## B. Isolation of Histone mRNA from HeLa Cells

HeLa cells are enriched in histone mRNA content by synchronizing cultures in the S phase of the cell cycle. The cells are collected when histone synthesis reaches its peak in mid S. This can easily be monitored by incubating aliquots of a culture with [$^3$H]thymidine. The peak in DNA synthesis corresponds well with the peak in histone synthesis (*8*). Polyribosomes are prepared from the cells collected and the RNA is purified by phenol extraction. The histone mRNA is separated from other RNAs by sucrose gradient centrifugation and prepared for translation by repeated ethanol precipitation.

### 1. Reagents

Thymidine stock: 0.1 m*M* thymidine in sterile 0.85% NaCl.
Cycloheximide stock: 10 mg/ml in water.
RSB: 10 m*M* NaCl–10 m*M* Tris (pH 7.5)–1.5 m*M* $Mg(OAc)_2$
SDS buffer: 0.1 *M* NaCl–10 m*M* Tris (pH 7.5)–1 m*M* EDTA–0.5% sodium dodecyl sulfate (SDS)
Redistilled phenol: saturated with water and stored in the dark at 4°C.
Salt solution: 5 *M* NaCl–5 *M* Li CL–4 *M* $K(OAc)_2$ (pH 5.1)

### 2. Growth and Synchrony of Cells

HeLa cells are maintained in suspension culture at 37°C in Joklik's modified minimal essential medium (MEM) [Grand Island Biological Co. (Gibco)] supplemented with 7% horse serum. The cell concentration is maintained between 2 and 4 $\times$ $10^5$ cells/ml. In order to synchronize the cells at the beginning of S phase, a double thymidine block is used (*24*). DNA synthesis is inhibited by the addition of thymidine to a final concentration of 1 m*M*. After 12 hours the cells are collected by centrifugation and resuspended in one-tenth of the original volume of warm spinner salts solution (Gibco). The cells are centrifuged again and resuspended in normal medium. After 9 hours DNA synthesis is again inhibited for 12 hours by the addition of 1 m*M* thymidine. This second thymidine block is terminated in the same manner as the first block, and the cells are resuspended in fresh medium again at a concentration of about 3 $\times$ $10^5$ cells/ml. Over 95% of the cells are in S phase after release from the second thymidine block. In order to collect cells when histone synthesis is at its peak, DNA synthesis is monitored by incubating 1 ml aliquots of cell cultures with 0.1 $\mu$Ci of [$^{14}$C]thymidine for 15 minutes. The cells are then chilled on ice, collected by centrifugation, washed once with spinner salts solution, and dissolved in 0.5 ml of 0.2 *N* KOH. Two milliliters of 10% trichloroacetic acid (TCA) are added, and the precipitate is collected on glass-fiber filters for counting. DNA and histone synthesis is maximal 4.5–5 hours after release from the thymidine block.

In order to save time and culture media, a single thymidine block can be used to synchronize the cells in S. The cells are released from the block in fresh medium at a density of 1.5–2 $\times$ $10^6$ cells/ml. DNA and histone synthesis is at a maximum at 3 hours after release from a single thymidine block. This procedure seems to be less toxic to cells than a double block. Cells partially synchronized with a single thymidine block incorporate [$^3$H]uridine into histone mRNA as well as, if not better than, cells synchronized by a double block.

### 3. Preparation of RNA

Histone mRNA must remain bound to polysomes during cell collection and homogenization. Protein synthesis must be rapidly stopped to prevent the release of histone mRNA from polysomes which occurs when cell cultures are slowly cooled below 37°C (unpublished observations). Small cultures can be rapidly chilled by the addition of 4–5 volumes of cold spinners salts solution. When working with several liters of material it is more practical to add cycloheximide to a concentration of 10 μg/ml 10 minutes before harvesting the cells. The cells can then be collected by centrifugation at room temperature and washed afterward in cold spinner salts solution. The subsequent handling of the cells is carried out at 0°C.

The cells are resuspended at $2 \times 10^7$/ml in RSB, and allowed to swell for 10 minutes before homogenization in a Dounce homogenizer. The homogenate is centrifuged for 10 minutes at 27,000 *g*, and the supernatant is layered over 2 ml of 30% sucrose made up in RSB. The polysomes are isolated by 2 hours of centrifugation at 49,000 rpm in the Ti 50 rotor. The polysome pellet is rinsed several times with RSB and dissolved in SDS buffer by vortexing at room temperature. The preparation is quite stable at this point and can be frozen until polysomes from at least $4 \times 10^9$ cells have been accumulated.

Deproteinization of RNA with phenol at low temperature at neutral pH results in the retention of most poly(A)-containing RNA in the phenol–water interphase (*20, 21*). The polysome preparation is thus mixed with an equal volume of water-saturated phenol and extracted by either vortexing in centrifuge tubes intermittently kept at 0°C or by stirring with a magnetic stirrer in a cold room. The emulsion is centrifuged at 10,000 *g* for 10 minutes, and the organic phase and interphase are discarded. The aqueous phase is extracted with phenol a second time as before, except that a volume of chloroform–isoamyl alcohol (24:1) equal to the volume of phenol is added just prior to centrifugation. This results in a compact interphase and allows removal of the aqueous phase without contamination with interphase material. The RNA is then precipitated from the aqueous phase by adjusting the NaCl concentration to 0.25 *M* and adding 2.1 volumes of 95% ethanol. Precipitation occurs overnight at −20°C.

The RNA precipitated is recovered by a 10-minute centrifugation at 10,000 *g*, washed with 70% ethanol and dried under vacuum. The yield from $4 \times 10^9$ cells is around 100 mg of RNA (2000 $A_{260}$ units). The RNA is dissolved in SDS buffer at a concentration of 4 mg/ml and up to 12 mg of RNA

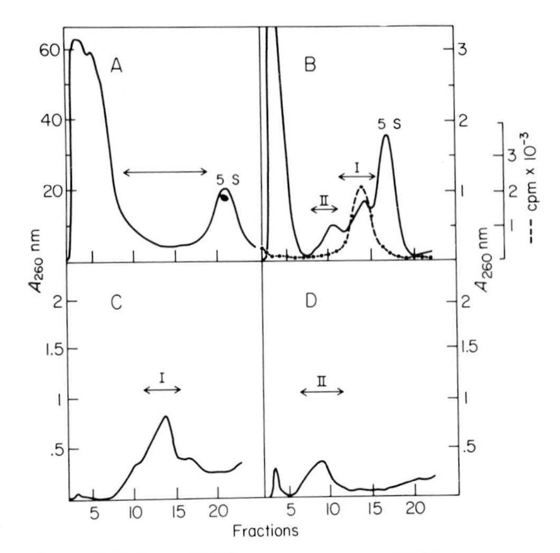


FIG. 1. Preparation of HeLa cell histone messenger RNA by repeated centrifugation on sucrose gradients. The RNA is prepared and centrifuged as described in the text. (A) Absorbance profile of the initial preparative gradient. Fractions under the arrow are pooled, precipitated with ethanol and recentrifuged on an identical gradient. (B) Typical fractionation achieved by a third consecutive preparative gradient. Marker [$^3$H]histone-mRNA is centrifuged in a parallel gradient for comparison. Peaks designated I and II are recovered and precipitated. (C) Further fractionation of peak I by 36 hours of centrifugation on a fourth gradient. The fraction indicated by the arrow codes almost exclusively for all histones except H1 when translated *in vitro*. (D) Further fractionation of peak II, as described for peak 1. The fraction indicated by the arrow codes for histones including H1 and for a number of other polypeptides when translated *in vitro*.

are layered over one 35-ml 5 to 20% sucrose gradient made in SDS buffer. The gradients are centrifuged for 24 hours at 24,000 rpm in the SW 27 rotor with a temperature setting of 20°C. The gradients are analyzed through a recording spectrophotometer and 1.5-ml fractions are collected (Fig. 1A). The fractions corresponding to the material sedimenting between the 4–5 S RNA peak and the 18 S rRNA at the bottom of the gradient are pooled, adjusted to contain 0.25 NaCl and precipitated with 2.1 volumes of 95% ethanol. The RNA from several gradients is collected by centrifugation, redissolved in SDS buffer, and fractionated under conditions identical to those used for the first gradient. The $A_{260}$ profile of the second gradient is similar to the first one, and the fractions containing RNA sedimenting between 5 S and 18 S are again pooled and precipitated with ethanol. The RNA recovered is fractionated on a 17-ml 5 to 20% sucrose gradient as described above. Two components are now visible between the 5 S and the 18 S RNA (Fig. 1B). The lighter peak corresponds with radioactive marker histone mRNA, which has been prepared as will be described in Section II,C.

The fractions corresponding to the two peaks are pooled, and the RNA is precipitated with ethanol and fractionated again on a fourth 17-ml sucrose gradient (Fig. 1C and D). The RNA is precipitated from each peak with ethanol at $-70°C$. The ethanolic solution forms a gel, which is allowed to thaw at $-20°C$ before centrifugation. This procedure allows more quantitative recovery of small quantities of RNA.

Prior to translating RNA in a cell-free system, the preparation must be freed from SDS, EDTA, and salts. This is accomplished by reprecipitating the RNA from ethanol three additional times. The RNA precipitated is treated as described above and redissolved at 0°C in 0.1–0.2 ml of water. In the first two precipitations LiCl is added to 0.25 *M* and 2.1 volumes of 95% ethanol are added. In the last precipitation $K(OAc)_2$ is added to 0.25 *M* in order to obtain the RNA as the potassium salt. Finally, the RNA is dissolved in water at about 1 mg/ml and stored at $-70°C$ in small aliquots. The mRNA preparations are stable indefinitely when stored in this way and do not lose translational activity upon repeated freezing and thawing.

## C. Preparation of Labeled Histone mRNA from HeLa Cells

HeLa cells synchronized by a single thymidine block as described in Section II,B,1 are resuspended in 50 ml of fresh medium at a concentration of $2 \times 10^6$ cells/ml. After 30 minutes 2.5 mCi of [$^3$H]uridine are added and the cells are incubated for 2 hours. Ten minutes before collecting the cells by centrifugation, cycloheximide is added at 10 μg/ml. The cells are washed and homogenized as described in Section II,B,1, except for the addition of 50 μg/ml of dextran sulfate (a nuclease inhibitor) to the buffer used to resuspend the cells before homogenization. The preparation of polysomes and the isolation of RNA by cold phenol extraction are exactly as described in Section II,B,2. The RNA is fractionated by 24 hours of centrifugation on a single 17-ml 5 to 20% sucrose gradient. Fractions containing 0.7 ml are collected, and 5-μl aliquots of each are spotted onto Whatman No. 3 MM filter paper disks. These are rinsed with cold 5% TCA, dried and counted in a toluene scintillant. A discrete peak of labeled material sedimenting at 9–10 S is observed (Fig. 2). The fractions corresponding to this peak are pooled and the RNA is precipitated with ethanol after addition of 50 μg of carrier 18 S rRNA prepared from rabbit reticulocytes. Following a second sucrose gradient centrifugation (Fig. 2), the fractions corresponding to the 9–10 S peak are collected and precipitated with ethanol after addition of 50 μg of *Escherichia coli* tRNA as a carrier. The RNA is reprecipitated three times as described in Section II,B,2 to remove SDS and salts.

The purity of [$^3$H]RNA preparations can be assessed by electrophoretic analysis on polyacrylamide gels run according to Loening (*25*). A good

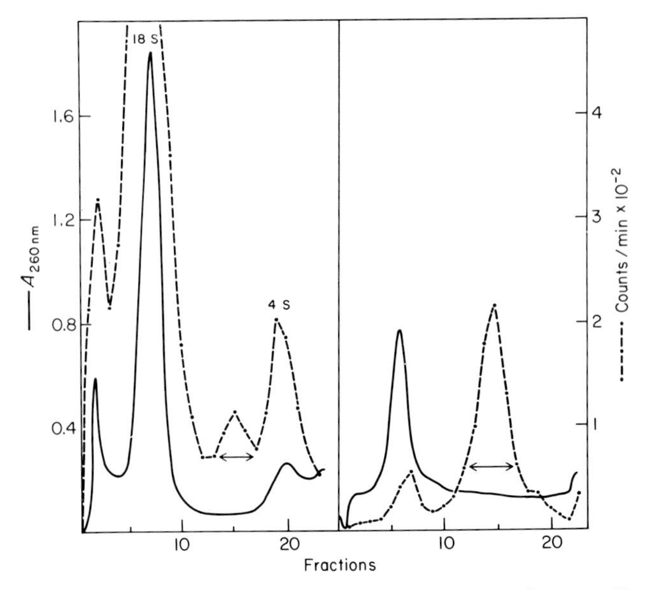


FIG. 2. Preparation of $^3$H-labeled histone mRNA by sucrose gradient centrifugation. Left panel: the initial preparative gradient. The fractions indicated by the arrows are precipitated. The analysis is shown in the right panel. The fractions indicated by the arrow are precipitated and used in the experiment shown in Fig. 3.

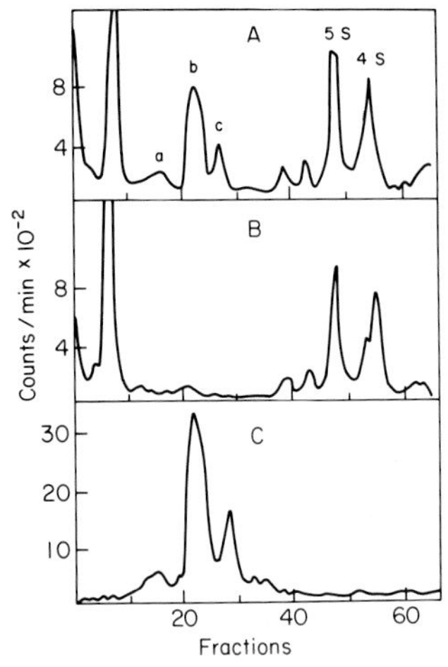


FIG. 3. Electrophoretic analysis of $^3$H-labeled histone mRNA. The gels were prepared and run as described in the text. (A) Total polyribosomal RNA from HeLa cells 1 hour into S phase, labeled with 10 $\mu$Ci/ml of [$^3$H]uridine for 30 minutes. (B) Total polyribosomal RNA from HeLa cells labeled as in (A), except that arabinosylcytosine is added to 25 $\mu$g/ml 30 minutes prior to the addition of [$^3$H]uridine. (C) $^3$H-Labeled histone mRNA prepared by centrifugation on two successive sucrose gradients as shown in Fig. 2.

resolution of histone mRNA is obtained by using 10 cm-long gels of 4.5% acrylamide with a 1.5 cm top spacer gel of 3% acrylamide. A small aliquot of RNA is electrophoresed for 4 hours at 5 mA per cylindrical gel (0.6 cm diameter). The gels are extruded from tubes, frozen, and cut in 0.75 mm slices with a gel slicer. The slices are counted after addition of Protosol (New England Nuclear) and a toluene scintillant.

Figure 3C shows a gel analysis of histone mRNA prepared as described above. The total [$^3$H]uridine-labeled polysomal RNA from cells actively synthesizing histones and from cells in which histone and DNA synthesis has been inhibited by arabinosylcytosine have been run in parallel gels (Fig. 3A and B). The preparation of purified histone mRNA (Fig. 3C) is predominantly composed of the three classes of RNA designated *a*, *b*, and *c*, which are present only on polysomes of cells actively synthesizing histones.

The biological activity of a preparation of labeled mRNA and the relative template content is assessed by its ability to bind to ribosomes in cell extracts during the course of cell-free protein synthesis. We routinely use rabbit reticulocyte lysates for this assay because of their high initiation activity (*26*). The reactions are assembled as described in Section III,B,2 for protein synthesis, except for the addition of 0.1 m*M* sparsomycin. This inhibitor of polypeptide chain elongation allows each mRNA molecule to bind to only a single ribosome and prevents ribosome runoff (*27*). A few thousand counts per minute of [$^3$H]mRNA are incubated 2 minutes with 100 $\mu$l of lysate, diluted afterward with buffer, and analyzed on 15 to 30% sucrose gradients according to Darnbrough *et al.* (*27*). The gradients are scanned with a recording spectrophotometer, and the fractions collected are counted directly after addition of a water-miscible scintillant. In a control reaction kept at 0°C, histone mRNA sediments at about 20 S (Fig. 4A). The increase in sedimentation coefficient from 9–10 S to 20 S is a consequence of protein binding to the RNA. After the 2-minute incubation a large amount of mRNA sediments with the 80 S monosome peak (Fig. 4B and C). The percentage of labeled mRNA which binds to ribosomes in this assay provides a measurement of the amount of active mRNA in the preparation. The maximum binding observed is 75–80%, and this cannot be improved by increasing the ratio of lysate to mRNA.

Much of the material that fails to bind to ribosomes is not mRNA. This is illustrated by the experiment shown in Fig. 5. Labeled histone mRNA is incubated with HeLa cell extract under conditions for protein synthesis in the presence of 0.1 m*M* sparsomycin (*23*). The results are very similar to those of Fig. 4. In an incubation kept at 0°C, the mRNA sediments at 20 S; after 2 minutes at 30°C, 70% of the mRNA binds to 80 S ribosomes. The bound mRNA as well as the unbound mRNA (20 S peak) are precipitated from gradient fractions adjusted to 1% SDS and 0.2 *M* NaCl with 2 volumes

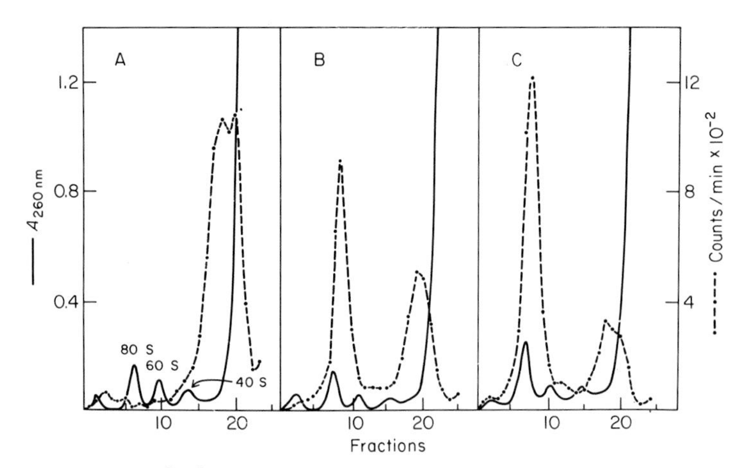


FIG. 4. Binding of [$^3$H]histone-mRNA to ribosomes in the reticulocyte cell-free system (A and C) and in the HeLa cell-free system (B), in the presence of 0.1 m*M* sparsomycin. (A) Control incubation kept at 0°C. (B) Two-minute incubation at 30°C in the HeLa cell-free system; for details of this cell-free system see Weber *et al.* (*23*). (C) Two-minute incubation at 30°C in the reticulocyte cell-free system.

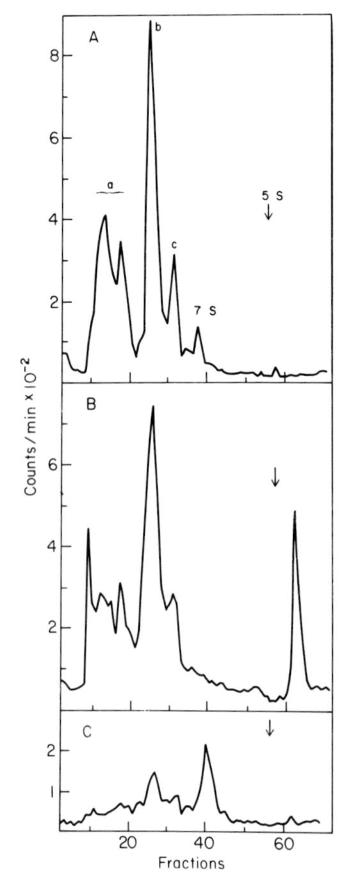


FIG. 5. Determination of the mRNA content of a preparation of [$^3$H]histone mRNA. The RNA is incubated in the HeLa cell-free system as shown in Fig. 3B. The RNA bound to ribosomes (80 S peak) and the unbound RNA (20 S peak) are recovered by ethanol precipitation and analyzed by electrophoresis as described in the text. (A) Electrophoretic analysis of RNA incubated with extract at 0°C in a control reaction. The RNA is recovered in the 20 S peak (see Fig. 4A). (B) RNA bound to the 80 S peak in a 2-minute incubation at 30°C. (C) RNA which fails to bind to ribosomes and remains in the 20 S peak.

of ethanol. The RNA is then analyzed by electrophoresis on polyacrylamide gels as described above. The preparation of histone mRNA used in this experiment is different from that shown in the gel analysis of Fig. 3, and is contaminated with some mRNA of higher molecular weight than histone mRNA (Fig. 5). The three histone peaks are present predominantly in the RNA fraction which binds to 80 S ribosomes. The peak migrating slightly faster than 5 S RNA marker (Fig. 5B) may be a fragment of mRNA protected from nuclease degradation by its association with ribosomes. The unbound RNA (Fig. 5C) contains a 7 S peak, which is noticeably absent in the gel of bound RNA (Fig. 5B). This may be the 7 S RNA described by Jacobs-Lorena and Baglioni (*28*) in reticulocytes, which is not a template for protein synthesis. The unbound RNA may also contain some heterodisperse material, which probably corresponds to fragments of higher molecular weight RNA, and a small amount of unbound histone mRNA.

## D. Isolation of Histone mRNA from Eggs and Embryos

The same basic procedure described for the isolation of histone mRNA from HeLa cells can be used for eggs and embryos of marine invertebrates, provided that steps are taken to counteract the higher levels of RNase activity present in this material. While HeLa cell RNase activity is hardly detectable below 8°C (*29*), the hydrolytic enzymes of cold blooded organisms seem to be active at 0°C. However, when histone mRNA is isolated for translation experiments, the use of RNase inhibitors which may copurify with RNA (i.e., heparin or dextran sulfate) should be avoided. Addition of bentonite which physically removes RNase by absorption is therefore preferred.

### 1. Collection and Fractionation of Eggs

Shedding of eggs can be induced in some species of sea urchin by electric shock or by the injection of a KCl solution. In other species and in the surf clam *Spisula solidissima*, eggs can be collected only by dissection; see Costello and Henley (*30*) for a discussion of this topic. All subsequent procedures are carried out at 0°C. Eggs are washed twice in Millipore-filtered sea water and twice in calcium-magnesium-free sea water by low speed centrifugation. After an initial washing in TNM [20 m*M* Tris-HCl (pH 7.6)–0.2 *M* NaCl–5 m*M* $Mg(OAc)_2$], the eggs are homogenized in 3 volumes of TNM containing 1 mg/ml of bentonite. The homogenate is centrifuged 10 minutes at 27,000 *g* to yield a postmitochondrial supernatant. Histone mRNA is found in this fraction. To obtain the highest yield of mRNA it is best to directly phenol-extract the postribosomal supernatant without attempting further fractionation. This is accomplished by adding

SDS and neutralized EDTA to a final concentration, of 1% and 10 m*M*, respectively, briefly warming the extract to room temperature to dissolve the SDS, and adding an equal volume of water-saturated phenol. Deproteinization and isolation of histone mRNA by repeated sucrose gradient centrifugation is carried out as described for HeLa cells (Section II,B,2).

After fertilization, embryos become increasingly difficult to homogenize as cell size decreases with each cleavage. The detergent Triton X-100 is therefore included (0.1%) in the buffer used to homogenize embryos. The detergent aids in breaking down the cells, but it introduces a potential problem by releasing hydrolytic enzymes from lysosomes. Histone mRNA becomes associated with polysomes in embryos and, depending upon the purpose of the experiment, it may be convenient to prepare polysomes by centrifuging the postribosomal supernatant for 1 hour at 150,000 *g*. The RNA can be extracted from the polysome pellet as described above (Section II,B,2). Messenger RNA not associated with polysomes can be extracted from the postribosomal supernatant by the same procedure used for eggs.

## III. Translation of Histone Messenger RNA *in Vitro*

### A. Ascites Cell-Free System

Histone mRNA was first successfully translated *in vitro* using a cell-free system prepared from mouse ascites cells (*15*). This protein-synthesizing system is easy to prepare and has a low level of endogenous activity.

REAGENTS

TBS: 0.14 *M* NaCl–35 m*M* Tris/HCl (pH 7.6).

Swelling buffer: 15 m*M* KCl–1.5 m*M* Mg $(OAc)_2$–6 m*M* 2-mercaptoethanol

Extract buffer (10 ×): 0.2 *M* HEPES (pH 7.5)–1.2 *M* KCl–50 m*M* $Mg(OAc)_2$–60 m*M* 2-mercaptoethanol

Preincubation energy mix: 20 mg/ml creatine phosphate–2 mg/ml creatine phosphokinase–2 m*M* GTP–10 m*M* ATP.

Column buffer: 0.12 *M* KCl–20 m*M* *N*-2-hydroxyethylpiperazine-*N*′-2-ethanesulfonic acid (HEPES)-KOH (pH7.6)–5 m*M* $Mg(OAc)_2$–1 m*M* DTT.

The preparation of the ascites cell extract has been previously discussed (*31*). The ascitic tumors are maintained by injection of 0.2 ml of ascitic fluid into the peritoneal cavity of Swiss mice. After 7 days the ascitic fluid is removed from the peritoneal cavity with a pipette and filtered through gauze into centrifuge tubes containing cold TBS. All subsequent steps are performed at 0°C, except where noted. The cells are washed five times with TBS by centrifugation at 80 *g* for 5 minutes. A tightly packed cell pellet is then obtained by centrifugation at 2000 *g* in a graduated conical centrifuge tube. The pellet is resuspended in 3 volumes of swelling buffer, and, after standing for 10 minutes, the cells are broken with 30–50 strokes of a tight-fitting Dounce homogenizer. Isotonicity is restored by the addition of 1/10 volume of 10 × extract buffer, and the homogenate is centrifuged 20 minutes at 27,000 *g*. The supernatant is filtered through gauze to remove the lipid layer, and 1/10 volume of preincubation energy mix is added. Preincubation is for 40 minutes at 37°C. The extract is clarified by centrifugation and passed through a 2.5 × 30 cm column of Sephadex G-25 equilibrated with column buffer. The opalescent excluded fractions eluted with the same buffer are pooled and rapidly frozen in small aliquots in liquid $N_2$. The aliquots are stored at −70°C and remain active for several years.

## B. Rabbit Reticulocyte Lysate

The reticulocyte lysate is the most active cell-free protein synthesizing system available in terms of the total amount of protein synthesized. However, the vast majority of the protein synthesized, even in the presence of added mRNA, is globin. The addition of histone mRNA, as well as of other mRNAs, to the reticulocyte cell-free system does not result in a stimulation of protein synthesis over that obtained in the absence of added mRNA. The amount of histone synthesized can only be determined by a quantitative analysis of the translation products. Fortunately, histones can easily be separated from globin by chromatography. The reticulocyte cell-free system has thus been used to quantitate the amount of histone mRNA present in HeLa cell polysomes during the cell cycle (*32*).

### 1. Reagents

Saline: 0.14 *M* NaCl–1.5 m*M* $Mg(OAc)_2$–5 m*M* KCl

Phenylhydrazine: 2.5% solution in 10 m*M* 2-mercaptoethanol adjusted to pH 7 (store frozen in the dark; do not refreeze)

### 2. Procedure

The rabbits (about 3 kg) are made anemic by 6 consecutive daily subcutaneous injections of 1 ml of phenylhydrazine (during the winter months a large number of rabbits may succumb to respiratory infections). After 1 day of recovery the rabbits are bled by cardiac puncture using an 18-gauge needle and a 100-ml syringe previously rinsed with heparin solution. The blood is mixed with about 3 volumes of ice cold saline containing 0.001%

heparin. Reticulocytes are obtained by washing the blood cells 4 times with saline and removing after each centrifugation the top layer of white blood cells with an aspirator. A compact cell pellet is obtained by a final centrifugation at 5000 *g* and the cells are lysed by addition of an equal volume of 6 m*M* 2-mercaptoethanol. The lysate is gently stirred on ice with a glass rod for 5 minutes and centrifuged at 27,000 *g* for 15 minutes. Small aliquots of the supernatant are frozen in liquid $N_2$ and stored at $-70°C$. The lysate is stable for at least one year.

## C. Wheat Germ Extract

The best cell-free system presently available is prepared from ungerminated wheat embryos. Wheat germ extracts translate most cellular and viral mRNAs with higher efficiency (picomoles of amino acid incorporated per microgram of mRNA) than other unfractionated mRNA-dependent cell-free systems. Endogenous protein synthesis is almost negligible even without preincubation of the extract. This facilitates analysis of translation products. Moreover, large amounts of extract can be prepared from an inexpensive source of material which is available at most health food stores.

### 1. Reagents

Grinding buffer: 0.1 *M* KCl–1 m*M* $Mg(OAc)_2$–2 m*M* $CaCl_2$–6 m*M* 2-mercaptoethanol–20 m*M* HEPES-KOH (pH 7.6)

Sterile sand: boiled in 0.1 *M* EDTA (pH 7.0), rinsed with water, and treated at 400°C for 12 hours

Column buffer: same composition as for ascites column buffer

### 2. Procedure

Wheat germ extract is prepared as described by Roberts and Patterson (*33*), except that the preincubation step is omitted. All procedures are carried out at 0°–4°C. Six grams of dry wheat germ (stored in a desiccator at 4°C) are mixed with an equal weight of sand in a chilled mortar. A paste is made by gradually adding 28 ml of grinding buffer while homogenizing with a pestle. Three minutes of vigorous grinding are sufficient to homogenize the wheat germ. The homogenate is transferred to centrifuge tubes, and the bulk of the sand is allowed to settle by gravity for 2–3 minutes. The supernatant is centrifuged at 30,000 *g* for 10 minutes. The supernatant (S30) is carefully removed with a pipette, avoiding the pellicle of lipid, and is immediately passed through a column of Sephadex G-25 exactly as described (Section III,B) for ascites. The most turbid excluded fractions are pooled, and small aliquots are frozen in liquid $N_2$. The extracts retain activity for several years when stored at $-70°C$. The activity of extracts prepared from wheat germ obtained from different suppliers, or even from different batches of the same supplier, may vary widely. We suggest trying different samples of wheat germ in order to determine the relative activity and stability of the extracts.

## D. Translation of Histone mRNA

### 1. Reagents

Ascites master mix: 240 m*M* KCl–100 m*M* DTT–80 m*M* HEPES-KOH (pH 7.6)–10 m*M* ATP–2 m*M* GTP–20 mg/ml creatine phosphate–6.7 mg/ml creatine phosphokinase

Reticulocyte master mix: 250 m*M* KCl–7.5 m*M* $Mg(OAc)_2$–5 m*M* DTT–110 m*M* HEPES-KOH (pH 7.4)–1 m*M* ATP–1 m*M* GTP–75 m*M* creatine phosphate–10 mg/ml creatine phosphokinase

Wheat germ energy mix: 80 m*M* HEPES-KOH (pH 7.1)–13 m*M* DTT–6 m*M* ATP–1.25 m*M* GTP–75 m*M* creatine phosphate–500 $\mu$g/ml creatine phosphokinase

Hemin stock solution: 1 m*M*, dissolve hemin in 1/100 volume of 1 *N* KOH, add 0.6 volume of 20 m*M* HEPES buffer, and 0.39 volume of water

Wheat germ salts: 0.64 *M* KCl–15 m*M* $Mg(OAc)_2$

Unlabeled amino acid stock: mixture of 19 amino acids lacking lysine; 0.5 m*M* each in 1 m*M* DTT

### 2. Procedure

The final concentrations of components present in each of the three cell-free protein synthesizing systems described above are listed in Table I. The salt concentrations listed for ascites and wheat germ have been determined to be optimal for the translation of histone mRNA. The salts listed for the reticulocyte lysate are optimal for endogenous protein synthesis. Since reticulocyte lysates are not passed through Sephadex, they do not contain defined concentrations of salts. Some variation exists between individual lysates and optimal $K^+$ and $Mg^{2+}$ concentrations should be determined empirically. Reactions are carried out in 6 $\times$ 50 mm glass tubes. Incubation volumes as small as 10 $\mu$l can be used if the tubes are capped with parafilm. The appropriate amount of labeled amino acid is dried in a reaction tube along with the correct amount of unlabeled amino acid mixture lacking the labeled amino acid. We normally use [$^3$H]lysine as the labeled moiety when translating histone mRNA and add 5–25 $\mu$Ci per incubation of 50 $\mu$l (specific activity 20 Ci/mmole). The amount of cell extract, master mix, and other components added for each cell-free system are indicated in Table I. After the reactions are assembled, protein synthesis is carried out at the indicated temperature.

TABLE I

COMPOSITION OF THE INCUBATIONS FOR *in Vitro* TRANSLATION OF MESSENGER RNA[a]

| Component (final concentration) | Ascites | Reticulocyte | Wheat germ |
|---|---|---|---|
| KCl ($\mu$moles/ml) | 96 | 50[b] | 100 |
| $Mg(OAc)_2$ ($\mu$moles/ml) | 3 | 1.5[b] | 3 |
| 2-Mercaptoethanol ($\mu$moles/ml) | — | 4.2 | — |
| Dithiothreitol ($\mu$moles/ml) | 10 | 1 | 2.6 |
| HEPES-KOH ($\mu$moles/ml) | 20 (pH 7.6) | 22 (pH 7.4) | 22 (pH 7.3) |
| ATP ($\mu$moles/ml) | 1 | 0.2 | 1.2 |
| GTP ($\mu$moles/ml) | 0.2 | 0.2 | 0.25 |
| Unlabeled amino acids[c] ($\mu$moles/ml) | 0.05 | 0.05 | 0.05 |
| Creatine phosphate ($\mu$moles/ml) | 8 | 15 | 15 |
| Creatine phosphokinase (mg/ml) | 0.67 | 1 | 0.1 |
| Cell extract (% final vol) | 60% | 70% | 30% |

[a]The ascites cell free system is assembled by mixing 6 parts of cell extract, 1 part of ascites master mix, and 4 parts of RNA solution or water. The reticulocyte system contains 7 parts lysate, 2 parts reticulocyte master mix, and 1 part RNA solution or water. One-twentieth volume of hemin solution is added in addition. The wheat germ system is assembled by mixing 3 parts wheat germ extract, 2 parts wheat germ energy mix, 1 part 10 × salt solution, and 4 parts RNA solution or water. The incubation temperature is 30°C for ascites and reticulocyte systems and 24°C for wheat germ.

[b]Concentration of added salts; actual composition in the reaction is not defined.

[c]Concentration of each amino acid.

Protein synthesis is determined by removing 5-$\mu$l aliquots with disposable capillary pipettes before starting the reaction (time 0) and after 60 minutes or longer. Control incubations containing water in place of RNA are included to measure endogenous protein synthesis. Incorporation is measured by spotting the aliquots onto 2.4 cm Whatman No. 3 *MM* filter paper disks. In the case of ascites and wheat germ the filters are allowed to dry and are dropped into a beaker containing 5% TCA. This is heated until it begins to boil then allowed to cool. The filters are rinsed three times with 5% TCA, twice with 95% ethanol, dried under a heat lamp, and counted in a toluene-based scintillation cocktail. For the highly colored reticulocyte lysate, the filters are dropped immediately into 5% TCA in acetone. The heme group is extracted by this treatment and color quenching during counting is avoided. After soaking in TCA-acetone for 30 minutes the disks are transferred into aqueous 5% TCA and processed as described above.

The protein-synthesizing activity of each cell-free extract should be tested before attempting to translate a valuable preparation of RNA. Ascites and wheat germ extracts are tested by measuring stimulation of protein synthesis by a standard mRNA preparation, known to have template activity. Rabbit globin mRNA is usually the template of choice. The simplest, most foolproof method of preparing active globin mRNA is by direct oligo (dT)-cellulose chromatography of reticulocyte polysomes as described by Pemberton *et al.* (*34*). We find that oligo(dT) can leach off commercial preparations of oligo(dT)-cellulose and can strongly inhibit protein synthesis. Therefore, it is best to chromatograph nearly saturating amounts of polyribosomal RNA on small columns. Ten milligrams of polyribosomal RNA is passed through a column containing 0.5 gm of oligo(dT)-cellulose (Collaborative Research, Inc.) to yield approximately 200–300 μg of mRNA fraction.

In a good ascites extract, protein synthesis is stimulated at least 10- to 20-fold by saturating amounts of globin mRNA. However, the endogenous activity of ascites extracts is quite variable from preparation to preparation. The most active extracts often have a relatively high endogenous activity, whereas the extracts with low endogenous activity give high stimulation (defined as cpm + RNA/cpm − RNA with time 0 values subtracted from both) but relatively less incorporation per microgram of mRNA. The wheat germ extracts have consistently low endogenous activity (less than 1000 cpm per 5-μl aliquot). An active wheat germ extract incorporates 40,000–100,000 cpm of [$^3$H]lysine per 5-μl aliquot in the presence of 20 μg of globin mRNA per milliliter.

Reticulocyte lysates must be active in initiation of protein synthesis if they are to be used in translating exogenous mRNA. Lysates are tested for initiation by following the kinetics of protein synthesis. Active lysates incorporate labeled amino acids into protein for nearly 1 hour at a linear rate. A useful control is an incubation in the absence of added hemin. Protein synthesis usually stops within 10 minutes in this case, because in the absence of hemin the ability to initiate new chains is rapidly lost (*26*). Not all the lysates, however, show hemin dependency for initiation.

Preliminary experiments must be carried out to determine the concentration of added mRNA that gives optimal translation in the ascites and wheat germ cell-free systems. This is determined by measuring the stimulation of amino acid incorporation as a function of mRNA concentration. This is accomplished by first assembling a reaction containing the highest concentration of mRNA to be tested. One half of the content is then serially diluted with an equal volume of reaction containing water instead of mRNA. The input of histone mRNA which gives maximal stimulation varies with individual extracts and mRNA preparations. The optimal concentration of histone mRNA is usually around 0.1 mg/ml. Higher concentrations may be inhibitory.

The addition of exogenous mRNA may inhibit total protein synthesis in the reticulocyte cell-free system. However, the amount of specific protein synthesized in response to the added mRNA increases with the mRNA input. We have not determined the optimum concentration of histone mRNA for translation in the reticulocyte cell-free system, but 70 μg/ml does not appear to be saturating.

## IV. Analysis of Translation Products

### A. General Considerations

The proteins synthesized *in vitro* can be analyzed in a variety of methods designed to specifically identify histones or in general any product of translation. Electrophoresis on slabs of polyacrylamide gels in the presence of SDS is the most direct and sensitive method to identify the products of translation on the basis of relative electrophoretic mobility. Most proteins migrate according to molecular weight on this type of gel (*35*). This analysis provides a direct assessment of the purity of a preparation of mRNA. If pure histone mRNA is translated *in vitro*, only histones should be observed among the proteins resolved by gel electrophoresis. Histones can be specifically identified by their migration in polyacrylamide gels run in acidic buffers containing urea. Most other proteins do not enter into these gels. Therefore, this analysis cannot be used to examine contamination of histone mRNA preparations with other templates. Finally, definitive proof of the faithful translation of mRNA into histones may require the analysis of the peptides obtained by digestion with trypsin or other proteolytic enzymes. Conclusive identification of the mRNA is obtained if the peptides are identical to those obtained from histones synthesized *in vivo* by the same cells from which the histone mRNA has been prepared.

### B. Electrophoretic Analysis

#### 1. Polyacrylamide Gels Containing SDS

The products of translation in the ascites or wheat germ cell-free system are analyzed on polyacrylamide gels according to Laemmli (*36*). The gels are cast between glass plates separated by silicon rubber strips and sealed with vacuum grease. The separating gels contain 15% acrylamide–0.15% *N,N*′-methylene-bisacrylamide–0.1% SDS–0.375 *M* Tris-HCl (pH 8.8). This gel is polymerized by the addition of ammonium persulfate and TEMED to final concentrations of 0.05% and 0.033%, respectively. Stacking gels are cast with a slot-former and contain 5% acrylamide–0.133% *N,N'*-methylene-bisacrylamide–0.125 *M* Tris (pH 6.8)–0.1% SDS. This is polymerized with 0.1% ammonium persulfate and 0.05% TEMED. Samples from a cell-free incubation are mixed with an equal volume of buffer containing 0.125 *M* Tris-HCl (pH 6.8)–2% SDS–10% mercaptoethanol–20% glycerol–0.002% bromophenol blue and are immersed in boiling water for 2 minutes. The samples are applied to the slots in the spacer gel and overlaid with running buffer which contains 0.192 *M* glycine–0.025 *M* Tris–0.1% SDS. Electrophoresis is carried out at about 10 mA for 2–3 hours and terminated when the tracking dye reaches the bottom of the gel. The gels are stained overnight in 0.2% Coomassie Blue–50% methanol–7% acetic acid. Destaining is carried out in 5% methanol–7% acetic acid.

The most complete analysis of the translation products is carried out by autoradiography of the dried gels. This is made possible with $^{3}H$-labeled proteins by a procedure described by Laskey and Mills (*37*), which allows quantitation of both $^{3}H$ and $^{14}C$ in polyacrylamide gels by autoradiography. Details of this technique can be found in the original publication.

### 2. Polyacrylamide Gels at Acid pH

Histones are separated by electrophoresis in 15% acrylamide–0.1% methylene-bisacrylamide–2.5 *M* urea–10% glycerol–0.9 *N* acetic acid (pH 2.7) (*38*). The gels are cast in 0.6 cm (i.d.) glass tubes 14 cm long. The gels are preelectrophoresed overnight at 1.5 mA/gel in 0.9 acetic acid. The samples to be analyzed are dialyzed overnight against 0.9 *N* acetic acid–2.5 *M* urea–0.4% 2-mercaptoethanol. The samples are applied to the gels and run at 2 mA/gel toward the cathode. The length of the run is monitored with cytochrome *c*, which migrates approximately in the position of H3 histone. The gels are stained in 0.5% Amido Schwartz–40% methanol–7% acetic acid and destained in the same solvent. The gels are sliced for counting as described in Section II,C. The inclusion of marker histones labeled with a different isotope is recommended in order to precisely localize the position of histones. In gels which are analyzed by autoradiography the position of marker histones is established by including with the samples analyzed a preparation of unlabled histones and staining the gels.

An improved procedure for the separation of histones has been devised by Zweidler (*39*) and used by Borun *et al.* (*22*) for the analysis of the translation products of histone mRNA in the ascites cell-free system. At the end of the incubation the sample is dialyzed against 0.9 *N* acetic acid and centrifuged at 27,000 *g* for 10 minutes. The supernatant is precipitated with 7 volumes of acetone after adding 0.1 mg of carrier histones. The precipitate is dissolved in 8 *M* urea–10% 2-mercaptoethanol–0.001% Pyronine Y and applied to 11 cm-long cylindrical gels containing 12% acrylamide–0.008% bisacrylamide–8 *M* urea–6 m*M* Triton X-100–5% acetic acid. The gels are prerun for 4 hours at 200 V and then 0.1 ml of 0.5 *M* cysteamine is added for 30 minutes at 100 V. The samples are electrophoresed for 12.5 hours at 100 V, stained and destained as described above, and counted by slicing.

In our experience, treatment of reaction mixtures with DNase, precipitation of the proteins and dialysis are not required when analyzing the products of translation by electrophoresis on gels containing SDS. However, the samples must be dialyzed against 0.9 *N* acetic acid for electrophoretic analysis on acidic gels containing urea.

## C. Chromatographic Analysis

This procedure for the isolation of histones is used in order to remove hemoglobin when histone mRNA is translated in the reticulocyte cell-free system. Globin represents at least 95% of the protein synthesized by reticulocyte lysates and interferes with the electrophoretic analysis because its molecular weight is very close to that of some histones. Breindl and Gallwitz (*17*) have described a convenient procedure for the fractionation of reticulocyte lysates by chromatography on carboxymethylcellulose. The reaction mixtures are digested with 20 $\mu$g/ml of RNase A for 20 minutes at 26°C. After cooling on ice, an equal volume of 0.5 *N* HCl is added dropwise with continuous stirring. The samples are allowed to stand on ice for 60 minutes and then centrifuged at 27,000 *g* for 15 minutes. The supernatant is removed, dialyzed against 0.1 *N* HCl, and precipitated with 10 volumes of acetone. The precipitate is dried, redissolved in 0.2 ml of Tris-HCl (pH 7.7) – 6 *M* urea, and passed over a CM-cellulose column (0.8 × 8 cm) equilibrated with the same buffer. The column is washed with buffer and the proteins absorbed to the column are eluted with 0.25 *N* HCl–6 *M* urea. This fraction is dialyzed against 0.1 *N* HCl, precipitated with 10 volumes of acetone, and dried. Histones present in this fraction are then separated by electrophoresis on polyacrylamide gels as described above.

## D. Peptide Analysis

The translation products of histone mRNA can be directly digested with trypsin or can be fractionated beforehand by chromatography. Trypsin is used because it reproducibly cleaves proteins at lysine and arginine residues. In the case of histones, this leads to the formation of numerous peptides which cannot be well resolved. However, if a labeled amino acid present in only a few residues per histone molecule is used in the translation assay, only few labeled peptides are obtained by tryptic digestion. We

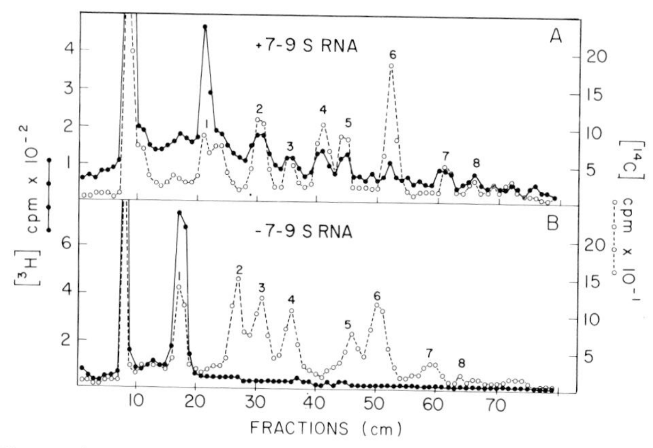


FIG. 6. Electrophoretic analysis of a tryptic digest of the translation product of HeLa cell histone mRNA in the ascites cell-free system. (A) Analysis of the tryptic digest of a reaction containing 20 $\mu l$ of histone mRNA (7–9 S RNA) solution in 200 $\mu l$ final volume. Thirty microcuries of [$^3$H]phenylalanine (50 Ci/mmole) are used to label the translation product, which is digested with trypsin after addition of marker histones uniformly labeled with [$^{14}$C]phenylalanine in intact HeLa cells. (B) An incubation run as in (A), except that histone mRNA is omitted. The numbers indicate specific peptides, except for 1, which corresponds to free phenylalanine. For details see Jacobs-Lorena *et al.* (*15*).

have found that [$^3$H]phenylalanine can conveniently be used in the translation of histone mRNA and that corresponding phenylalanine-labeled tryptic peptides are easily separated by one-dimensional high-voltage paper electrophoresis (*15*).

The samples to be digested with trypsin are dialyzed against water after addition of marker histones labeled *in vivo* with [$^{14}$C]phenylalanine (the ratio of $^3$H to $^{14}$C should be kept close to 10:1 for convenience in counting). The pH is adjusted to 8.5 with 0.5 *M* trimethylamine using phenolphthalein as an indicator. Five microliters of a solution containing 1 mg/ml of trypsin (TCPK-treated, Worthington) in 1 m*M* HCl is added to a 50-$\mu l$ reaction, which is then incubated for 2 hours at 37°C. Another 5-$\mu l$ aliquot of trypsin is added, and the incubation continued for an additional 2 hours. The pH is maintained constant by the addition of trimethylamine when necessary. At the end of the digestion the samples are added directly (spotted and dried) to a 120 cm-long Whatman No. 3 MM paper. The tryptic peptides are separated by electrophoresis at 4.5 kV for 5 hours in a Varsol cooled tank, using the pH 3.5 pyridine–acetate buffer (0.5% pyridine–5% acetic acid) as described by Sanger *et al.* (*40*). The paper is dried and 1-cm strips are cut, transferred to vials, and counted after addition of 1 ml of 0.1 *N* NaOH for 2 hours, neutralization with 0.1 ml of 1 *N* acetic acid, and addition of 10 ml of a water-miscible scintillant (Fig. 6).

## References


1. Gross, K. W., Jacobs-Lorena, M., Baglioni, C., and Gross, P. *Proc. Natl. Acad. Sci. U.S.A.* **70**, 2614 (1973).
2. Skoultchi, A., and Gross, P. R. *Proc. Natl. Acad. Sci. U.S.A.* **70**, 2840 (1973).
3. Kedes, L. H., and Gross, P. R. *Nature (London)* **223**, 1335 (1969).
4. Gabrielli, F., and Baglioni, C. *Dev. Biol.* **43**, 254 (1975).
5. Robbins, E., and Borun, T. W. *Proc. Natl. Acad. Sci. U.S.A.* **57**, 409 (1967).
6. Spalding, J., Kajiwara, K., and Mueller, G. C. *Proc. Natl. Acad. Sci. U.S.A.* **57**, 409 (1966).
7. Gabrielli, F., and Baglioni, C., *Eur. J. Biochem.* **42**, 121 (1974).
8. Borun, T. W., Scharff, M. D., and Robbins, E., *Proc. Natl. Acad. Sci. U.S.A.* **58**, 1977 (1967).
9. Gallwitz, D., and Mueller, G. C., *J. Biol. Chem.* **244**, 5947 (1969a).
10. Gallwitz, D., and Mueller, G. C, *Science* **163**, 1351 (1969b).
11. Stein, J., Thrall, C., Park, W., Mans, R., and Stein, G. S., *Science* **189**, 557 (1975).
12. Liberti, P., Festucci, A., and Baglioni, C., *Mol. Biol. Rep.* **1**, 61 (1973).
13. Perry, R. P., and Kelley, D. E., *J. Mol. Biol.* **79**, 681 (1973).
14. Butler, W. B., and Mueller, G. C., *Biochim. Biophys. Acta* **294**, 481 (1973).
15. Jacobs-Lorena, M., Baglioni, C., and Borun, T. W., *Proc. Natl. Acad. Sci. U.S.A.* **69**, 1425 (1972).
16. Gross, K. W., Ruderman, J. V., Jacobs-Lorena, M., Baglioni, C., and Gross, P. R., *Nature New Biol. (London)* **241**, 272 (1972).
17. Breindl, M., and Gallwitz, D., *Eur. J. Biochem.* **32**, 381 (1973).
18. Hirsch, M., Spradling, A., and Penman, S., *Cell* **1**, 31 (1973).
19. Milcarek, C., Price, R., and Penman, S., *Cell* **3**, 1 (1974).
20. Perry, R. P., LaTorre, J., Kelley, D. E., and Greenberg, J. R., *Biochim. Biophys. Acta* **262**, 220 (1972).
21. Brawerman, G., *Methods Cell Biol.* **7**, 1 (1973).
22. Borun, T. W., Gabrielli, F., Adjiro, K., Zweidler, A., and Baglioni, C. *Cell* **4**, 59 (1975).
23. Weber, L. A., Feman, E. R., and Baglioni, C., *Biochemistry* **14**, 5315 (1975).
24. Stein, G. S., and Borun, T. W., *J. Cell Biol.* **52**, 292 (1972).
25. Loening, U. K., *Biochem. J.* **113**, 131 (1969).
26. Zucker, W. V., and Schulman, H. M., *Proc. Natl. Acad. Sci. U.S.A.* **59**, 582 (1968).
27. Darnbrough, C. H., Legon, S., Hunt, T., and Jackson, R. J., *J. Mol. Biol.* **76**, 37 (1973).
28. Jacobs-Lorena, M., and Baglioni, C., *Proc. Natl. Acad. Sci. U.S.A.* **69**, 1425 (1972).
29. Penman, S., Greenberg, H., and Willems, M., *in* "Fundamental Techniques in Virology" (K. Habel and N. Salzman, eds.), Vol. 1, p. 49 Academic Press, New York, 1969.
30. Costello, D. P., and Henley, C., "Methods for Obtaining and Handling Marine Eggs and Embryos." Marine Biological Laboratory, Woods Hole, Massachusetts (1971).
31. Villa-Komaroff, L., McDowell, M., Baltimore, D., and Lodish, H., *Methods Enzymol.* **30**, Part F, 709 (1974).
32. Breindl, M., and Gallwitz, D., *Eur. J. Biochem.* **45**, 91 (1974).
33. Roberts, B. E., and Patterson, B. M., *Proc. Natl. Acad. Sci. U.S.A.* **70** 2330 (1973).
34. Pemberton, R. E., Liberti, P., and Baglioni, C., *Anal. Biochem.* **66**, 18 (1975).
35. Weber, K., and Osborn, M., *J. Biol. Chem.* **244**, 4406 (1969).
36. Laemmli, U. K., *Nature (London)* **227**, 680 (1970).
37. Laskey, R., and Mills, A., *Eur. J. Biochem.* **56**, 335 (1975).
38. Panyim, S., and Chalkley, R., *Arch. Biochem. Biophys.* **130**, 337 (1969).
39. Zweidler, A., *J. Cell Biol.* **59**, 378a (1973).
40. Sanger, F., Brownlee, G. G., and Barrell, B. G., *J. Mol. Biol.* **13**, 373 (1965).

# Chapter 22

# *In Vitro* Synthesis of Single-Stranded DNA Complementary to Histone Messenger RNAs

C. L. THRALL,[1] A. LICHTLER, J. L. STEIN, AND G. S. STEIN

*Department of Biochemistry and Molecular Biology,*
*University of Florida,*
*Gainesville, Florida*

## I. Introduction

The discovery and demonstration of RNA-dependent DNA polymerase (reverse transcriptase) (1–3) has provided a very useful approach for studying specific messenger RNAs. The enzyme makes possible the synthesis of high-resolution probes, complementary DNA (cDNA) molecules, for the detection and quantitation of specific RNA sequences by nucleic acid hybridization techniques. Complementary DNAs have been prepared to several eukaryotic mRNAs including those coding for globin from mouse (*4*), rabbit (*5–7*), birds (*5*), and human reticulocytes (*8*); silk worm chorion proteins (*7*); myosin (*9*); ovalbumin (*10*); and HeLa cell histone (*11*). These cDNAs have been used to assay quantitatively and qualitatively the presence of defined RNA sequences in cell fractions, as well as in *in vitro* chromatin transcripts (*4–8, 12–21*). Over the past several years, such cDNA probes have been instrumental in gaining insight into mechanisms operative in the regulation of gene expression at the transcriptional and post-transcriptional levels in a broad spectrum of biological systems.

Because RNA-dependent DNA polymerase requires a primer, the presence of 150–200 AMP residues at the 3′-OH termini of most eukaryotic mRNAs has facilitated transcription by the enzyme. Exogenous oligo(dT) readily anneals to these terminal poly(A) sequences, thus providing an effective primer for transcription of cDNA.

[1] *Present address:* Department of Molecular Medicine, The Mayo Clinic, Rochester, Minnesota.

In this article, a procedure is described for synthesis of a single-stranded DNA complementary to histone messenger RNAs. The histone mRNAs when isolated from the polyribosomes of S-phase HeLa cells lack poly(A) at their 3′-OH termini. However, AMP residues can be added to these messages, and as such they become effective templates for transcription by RNA-dependent DNA polymerase. Characterization and specificity of the histone cDNA probe are considered.

## II. Isolation of Histone mRNAs

### A. Background

Histone mRNAs of HeLa $S_3$ cells were identified by Borun *et al.* (*22*) and by Gallwitz and Mueller (*23*) in the late 1960s. Initial criteria for histone mRNAs included electrophoretic mobility on sodium dodecyl sulfate (SDS) polyacrylamide gels and ability of the light polysomes, with which these mRNAs are associated, to complete synthesis of histone polypeptides *in vitro*. Other properties of the messages reported in these early studies were association with polyribosomes of S-phase cells but not with polyribosomes of $G_1$ cells, and disappearance of the messages from S-phase polyribosomes following inhibition of DNA synthesis. Confirmation that the 7–11 S mRNAs of S-phase HeLa cells contain histone mRNAs was provided in 1972 by results from cell-free translation studies (*24,25*). Histone mRNAs have subsequently been isolated from other tissue culture cells (e.g., L cells) and from invertebrates (e.g., sea urchins and *Drosophila*). Considerable attention has been focused on histone mRNAs due to the important role of histones in the structural and functional properties of the eukaryotic genome.

### B. mRNA Isolation Procedure

#### 1. General Considerations

Perhaps the most important precaution which must be adhered to when isolating and fractionating RNAs is avoiding ribonuclease activity. Gloves should be worn during procedures that might allow glassware and solutions to come in contact with fingers. We routinely treat glassware for 15 minutes with a 0.1% aqueous suspension of diethylpyrocarbonate (produced by vigorous agitation and used immediately after preparation) and then autoclave at 121 °C for 30 minutes. Alternatively, glassware is sterilized in an oven at 200°C for 6 hours or 400°C for 2 hours. Dry heat sterilization can be conveniently achieved in a self-cleaning, commercial electric range on the "self-cleaning" cycle. Plastics are sterilized by immersion in 0.1% diethylpyrocarbonate for 15 minutes and dried in an oven at a temperature below the melting point of the material. Care should be taken because diethylpyrocarbonate reacts with some plastics. Whenever possible, solutions are treated with 0.005–0.01% diethylpyrocarbonate and then autoclaved at 121 °C for 30 minutes. This treatment destroys diethylpyrocarbonate, which could otherwise ethylate the RNA. Tris readily reacts with diethylpyrocarbonate, so the pH of Tris buffers should be checked after use of diethylpyrocarbonate. While other laboratories employ bentonite as a ribonuclease inhibitor, we have encountered an enhancement of ribonuclease activity with some lots of bentonite. A frequent source of trouble during organic extraction of RNA is oxidized phenol. Addition of 8-hydroxyquinoline to a concentration of 0.1% immediately upon opening a fresh bottle of phenol effectively prevents generation of oxidized products. Phenol can also be purified by distillation, prior to use for nucleic acid extraction. However, extreme caution must be exercised with this procedure. For a general discussion of RNA extraction, fractionation, and handling, see Chapters 20 and 21 in this volume.

## 2. Synchronization of HeLa $S_3$ Cells

The procedure that we utilize for obtaining S-phase HeLa $S_3$ cells involves two cycles of treatment with 2 m*M* thymidine. All aspects of synchronization are carried out at 37°C, employing sterile techniques. Synchronization of tissue culture cells with excess thymidine was initially developed by Bootsma *et al.* (*26*) and by Xeros (*27*). Details of the protocol we use for cell synchronization have been reported (*28*).

Thymidine is added to exponentially growing HeLa $S_3$ cells in suspension culture (cells are grown in Joklik-modified Eagle's minimal essential medium supplemented with 7% calf serum) to a final concentration of 2 m*M*. At 12–16 hours the cells are harvested, the medium is carefully removed from the cell pellet, and the cells are resuspended in fresh medium. Nine hours later thymidine is again added to a final concentration of 2 m*M* for 12–16 hours. The cells are released from the second thymidine block by centrifugation and resuspension in fresh growth medium. Cells are harvested 3 hours following release of the second thymidine block at which time 98% of the cells are in the S-phase of the cell cycle as determined by autoradiographic assessment of thymidine-labeled nuclei (Fig. 1). Synchronization of the cells can also be monitored by pulse labeling with thymidine and determining TCA precipitable radioactivity (Fig. 1).

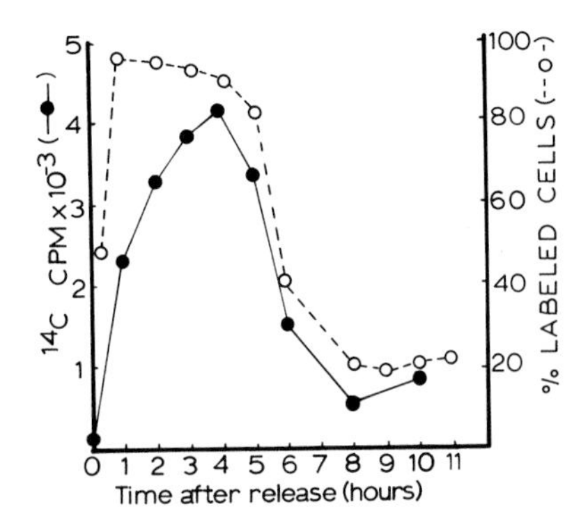


FIG. 1. Cell synchrony following release from two cycles of 2 m$M$ thymidine block. Cells were pulse-labeled with [$^{14}$C]thymidine for 15 minutes at various times following release from thymidine block, and trichloroacetic acid-precipitable radioactivity was assayed (●). Data are not corrected for fluctuations in the thymidine pool. Cells were similarly pulse-labeled with [$^{3}$H]thymidine, and the percentage of radioactively labeled nuclei was determined autoradiographically (○).

## 3. Messenger RNA Isolation Procedure

Cells are harvested by centrifugation at 1000 *g* for 5 minutes at 37 °C. The pellets are carefully drained of residual culture medium, and the cells are immediately resuspended in cold (4 °C) Earle's balanced salt solution. The cells are washed three times in cold balanced salt solution (4 °C) followed each time by centrifugation at 1000 *g* for 5 minutes. It is essential that the balanced salt solution is completely removed following the last washing, to facilitate hypotonic lysis of cells in the next step.

Cells are resuspended in cold 10 m$M$ KCl–10 m$M$ Tris–1.5 m$M$ $MgCl_2$ (pH 7.2) (RSB) at a concentration of $4 \times 10^7$ cells/ml and stirred at 4 °C for 20 minutes while osmotic swelling is periodically monitored by phase-contrast microscopy. The cells are then transferred to a Dounce tissue homogenizer and lysed by 12–15 strokes with a tight-fitting ("B") pestle. The lysate should be examined by phase-contrast microscopy to ascertain whether greater than 95% of the cells are lysed and the nuclei are free of significant amounts of cytoplasmic material. If cell lysis appears inadequate, addition of 0.1 volume of RSB and several additional strokes with the "B" pestle should remedy the situation. The lysate is centrifuged at 27,000 *g* for 15 minutes at 4 °C and the postmitochondrial supernatant is centrifuged at 100,000 *g* for 90 minutes at 4 °C to pellet the polyribosomes.

The polyribosomes are resuspended at room temperature in 0.1 $M$ NaCl–10 m$M$ sodium acetate–1 m$M$ EDTA–1% SDS (pH 5.4) by several strokes with a loose-fitting ("A") pestle in a Dounce homogenizer or by stirring vigorously with a sterile metal spatula. The low pH and presence of EDTA and SDS have the effect of minimizing ribonuclease activity. Deproteinization of the polyribosomes is continued by adding 1 volume of phenol [buffered three times with 1 volume of 0.1 *M* NaCl–10 m*M* sodium acetate–1 m*M* EDTA (pH 5.4)], shaking for 3 minutes, adding 1 volume of chloroform–isoamyl alcohol (24:1 v/v), shaking for 3 minutes and then centrifuging the milky white suspension at 3000 *g* for 30 minutes. Organic extractions and associated centrifugations have been carried out at room temperature or at 4°C; it appears that phases separate better and interfaces are smaller at low temperature. The aqueous (upper) layer containing the RNA is removed and set aside. To rescue RNA trapped in the remaining interphase (middle layer) and organic phase (lower layer), 0.5 volume of 0.1 *M* NaCl–10 m*M* sodium acetate–1 m*M* EDTA (pH 5.4) is added. After shaking for 3 minutes and centrifugation at 3000 *g* for 10 minutes, the aqueous phase is combined with the aqueous phase from the previous organic extraction. The pooled aqueous fraction is again extracted with 1 volume of phenol and 1 volume of chloroform–isoamyl alcohol. Additional extractions with chloroform–isoamyl alcohol are executed until no visible protein interphase is observed. Two volumes of ethanol are added to the aqueous fraction and the polyribosomal RNA is precipitated at −40°C for 12 hours.

Polyribosomal RNA is then fractionated in a 5–30% sucrose gradient, utilizing a Beckman SW-27 swinging-bucket rotor or a Beckman Ti-14 zonal rotor depending upon the quantity of RNA to be separated. In both cases, ethanol-precipitated RNA is prepared for sucrose gradient fractionation by centrifuging at 10,000 *g* for 30 minutes, discarding the ethanol supernatants, evaporating the RNA pellets to dryness under a gentle stream of nitrogen and resuspending the RNA in 0.1 *M* NaCl–10 m*M* sodium acetate–1 m*M* EDTA–0.1% SDS–2% sucrose (pH 5.4).

We routinely fractionate up to 100 mg of polysomal RNA in a single zonal gradient run. Into the rotor is pumped a 200-ml cushion of 30% sucrose–0.1 *M* NaCl–10 m*M* sodium acetate–1 m*M* EDTA–0.1% SDS (pH 5.4), a 400-ml 5–30% sucrose gradient containing 0.1 *M* NaCl–10 m*M* sodium acetate–1 m*M* EDTA–0.1% SDS (pH 5.4), the RNA sample in 20 ml, and a 50-ml overlay of 0.1 *M* NaCl–10 m*M* sodium acetate–1 m*M* EDTA–0.1% SDS (pH 5.4). Centrifugation is at 32,000 rpm for 16 hours at 22°C. Upon completion of the run, the gradient is pumped out through an absorbance monitor and 20-ml fractions are collected. The optical density profile of a typical gradient is shown in Fig. 2A. Regions A and B of the gradient contain 18 S and 28 S ribosomal RNAs, respectively, and region D contains 4 S and 5 S RNAs. Histone mRNAs are found in region C. The identity of these RNA species is confirmed by polyacrylamide gel electrophoresis in the presence of standard markers. Two volumes of ethanol are added to the material sedimenting in region C of the gradient, and the RNAs are allowed to

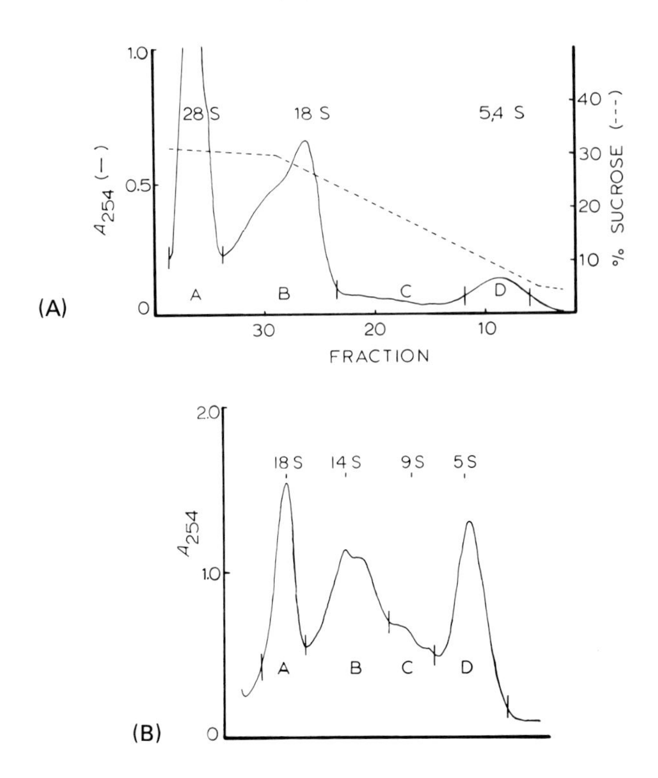


FIG. 2. (A) Absorbance profile of polysomal RNA fractionated by centrifugation in a 5 to 30% linear sucrose gradient in a Beckman Ti-14 zonal rotor. Polysomal RNA was isolated from S-phase HeLa $S_3$ cells synchronized by 2 m*M* thymidine block. The indicated sedimentation values were confirmed by electrophoretic fractionation in sodium dodecyl sulfate polyacrylamide gels. (B) Absorbance profile of 4–18 S RNA fractionated by centrifugation in a 5–30% linear sucrose gradient in a Beckman SW-27 rotor. Centrifugation was at 26,000 rpm for 22 hours at 22°C.

precipitate at −20°C for at least 12 hours. Further resolution of the histone mRNAs is then achieved by repeated centrifugation in 5 to 30% sucrose gradients in an SW-27 rotor using approximately 1 mg of RNA per gradient. A typical optical density profile of such fractionations of 4–18 S RNAs is shown in Fig. 2B. Histone mRNAs are found in region C of this gradient, and the indicated sedimentation values are confirmed by polyacrylamide gel electrophoresis in the presence of standards. Fractions from initial gradients are pooled on the basis of the pattern of the optical density scan rather than by calculated sedimentation values. For convenience, the histone mRNA preparation is referred to as 7–11 S RNA. Figure 3 shows the extent to which the 4–18 S RNA fraction can be resolved on the first SW-27 gradient on a good day. *In vitro* translation data indicate that the

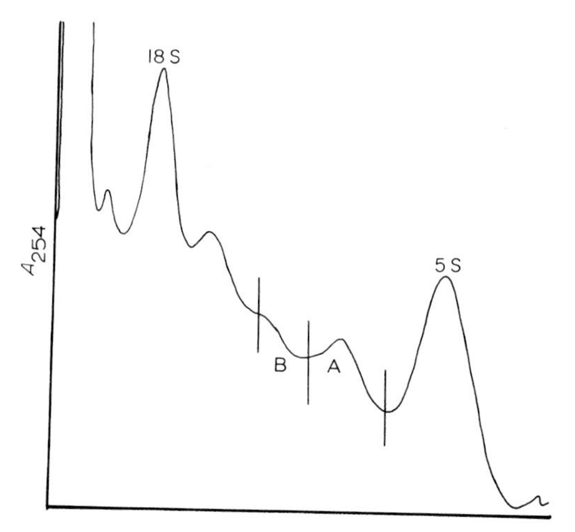


FIG. 3. Absorbance profile of 4–18 S RNA fractionated by centrifugation in a 5 to 30% linear sucrose gradient in a Beckman SW-27 rotor. Centrifugation was at 26,000 rpm for 22 hours at 22°C. Cell-free translation of RNAs that sediment in regions A and B of the gradient is shown in Fig. 8.

mRNAs for H4, H2A, H2B, and H3 are in region A of the gradient and H1 mRNA is in region B of the gradient.

For isolation of histone mRNAs from 1–2 liters of S-phase HeLa $S_3$ cells we initially fractionate total polysomal RNA in a 5–30%, 38-ml sucrose gradient in an SW-27 rotor using 1 mg of RNA per gradient, and subsequently fractionate the 4–18 S fraction in a 5 to 30% sucrose gradient in an SW-27.1 rotor. Figure 4 shows the resolution of the 4–18 S RNA fraction in the first SW-27.1 gradient. Histone mRNAs are found in region H.

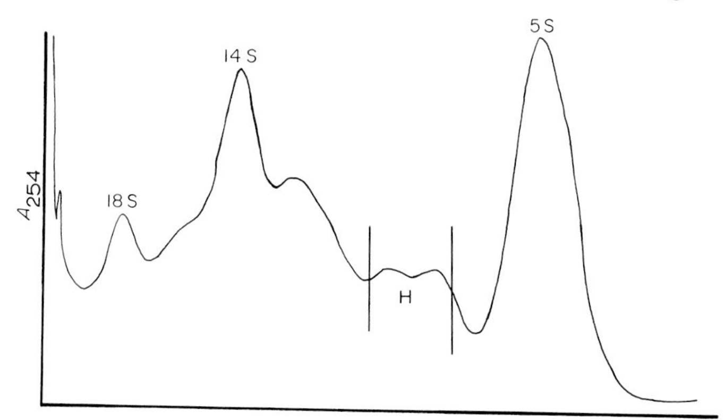


FIG. 4. Absorbance profile of 4–18 S RNA fractionated by sedimentation in a 5 to 30% linear sucrose gradient in a Beckman SW-27.1 rotor. Centrifugation was at 26,000 rpm for 29 hours at 22°C. Histone mRNAs sediment in region H of the gradient.

Following sucrose gradient fractionation, the 7–11 S RNAs are ethanol precipitated, evaporated to dryness under nitrogen, and then subjected to nitrocellulose filtration and/or oligo(dT)-cellulose chromatography to remove poly (A)-containing RNA species. The procedure which we utilize for nitrocellulose filtration is that of Lee *et al.* (*29*) and is carried out at 4°C. RNAs are suspended in 0.5 *M* KCl–1 m*M* $MgCl_2$–10 m*M* Tris (pH 7.6) at a concentration of 10 μg/ml and filtered by gravity through a 25 mm, 0.45 μm Millipore nitrocellulose filter which has been extensively washed with the buffer. Under these high-salt conditions, poly(A)-containing RNAs adhere to nitrocellulose filters and poly(A) minus RNAs (histone mRNAs) pass through. After extensively washing the filters with buffer, bound material [poly(A) plus RNA] is eluted with 0.5% SDS–0.1 *M* Tris (pH 9.0). The technique which we utilize for oligo (dT)-cellulose chromatography is that of Aviv and Leder (*30*) and is carried out at room temperature. Oligo(dT)-cellulose is equilibrated with 0.5 *M* KCl–10 m*M* Tris (pH 7.5), poured into a Pasteur pipette column and then packed by gravity flow. The column is extensively washed with buffer until the buffer eluted from the column has zero absorbance at 260 nm. The RNA in 0.5 *M* KCl–10 m*M* Tris (pH 7.5) is applied to the column and chromatographed at a flow rate of 3–6 ml/hour. The poly (A) minus RNA is collected, and elution of the column with high-salt buffer continues until the absorbance of the eluted material at 260 nm returns to zero. Poly (A) plus RNA is then eluted with 10 m*M* Tris (pH 7.5). By simply assessing absorbance at 260 nm, we do not detect the presence of significant amounts of poly (A) plus material in our 7–11 S RNA preparations by either nitrocellulose filter fractionation or by oligo (dT)-cellulose chromatography. However, the presence of poly (A) plus RNAs prior to nitrocellulose filtration or oligo (dT)-cellulose chromatography is apparent since before these procedures are carried out the 7–11 S RNAs are effective as templates for $dT_{10}$-primed reverse transcription. This activity is completely lost when the procedures for elimination of poly(A) plus material are executed.

The size distribution of the histone mRNA preparation on a 2.7% SDS-polyacrylamide gel is shown in Fig. 5; an apparent heterogeneity is evident with four defined peaks. When a "broad cut" of sucrose gradient-fractionated, $^{32}P$-labeled, polysomal RNA (5–18 S) is electrophoresed on 6% poly acrylamide slab gels in Loening E buffer (*31*) at 20°C (Fig. 6) by the method of Kedes *et al.* (*32*), the typical heterogeneous pattern for histone mRNAs (*33*) is observed. RNAs from these slab gels can be readily extracted by cutting out RNA bands located using an autoradiograph according to the method of de Wachter and Fiers (*34*) and electrophoresing the RNA into a dialysis bag, as described by Grunstein and Schedl (*35*).

To assay for the presence of histone mRNAs in RNA fractions from

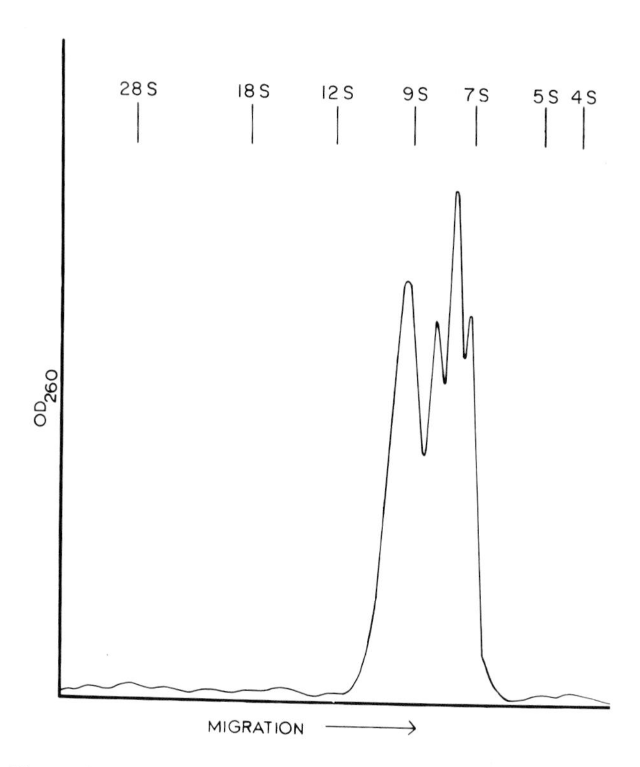


FIG. 5. Electrophoresis of histone mRNAs in a 2.7% sodium dodecyl sulfate polyacrylamide gel.

sucrose gradients and from preparative polyacrylamide gels, we have translated the RNAs in a cell-free protein-synthesizing system derived from wheat germ. The wheat germ lysate is prepared essentially as described originally by Roberts and Paterson (*36*). Details of translation of histone mRNAs in the wheat germ system are described in the legend to Fig. 7 and in Chapters 20 and 21 in this volume.

In our laboratory the activity of the wheat germ isolate was assayed using poly A plus HeLa mRNA prepared from total polysomal RNA by chromatography on oligo(dT)-cellulose (*30*). Incorporation of [$^3$H]lysine into hot acid-resistant (5% TCA), trichloroacetic acid-precipitable material was linear for 70 minutes (Fig. 7A). Maximal stimulation was obtained with 5 $\mu$g of RNA per 50-$\mu$l reaction. Aliquots of the reaction mixture to be analyzed by polyacrylamide gel electrophoresis were mixed with unlabeled HeLa histones and adjusted to contain one-half concentration of Laemmli stacking gel buffer plus 2% SDS and 10% $\beta$-mercaptoethanol for Laemmli

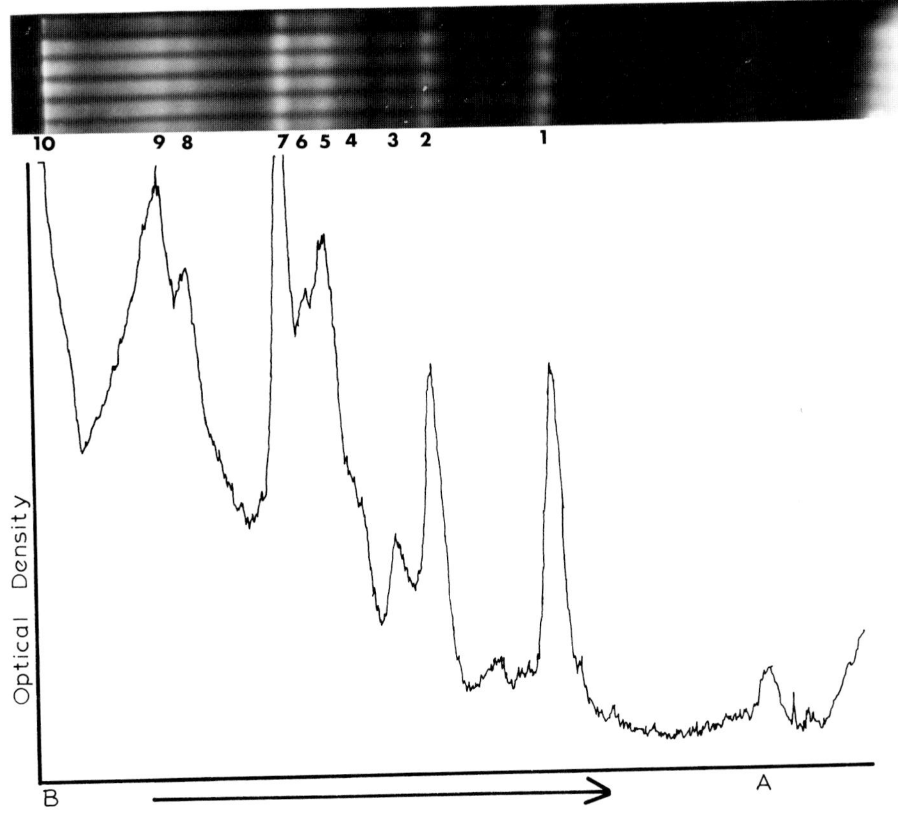


FIG. 6. Electrophoretic fractionation of 5–18 S RNAs from S-phase HeLa $S_3$ cells. $^{32}P$-labeled RNA, initially fractionated on a linear 5 to 30% sucrose gradient, was fractionated electrophoretically in a 6% polyacrylamide slab gel (30 cm × 15 cm × 3 mm) containing a 3% stacking gel. Details of the electrophoretic procedure have been reported (*32*). Electrophoresis was carried out using Loening buffer E (*31*) at 68 V, 60 mA (constant current) for 37 hours at 20°C. The buffer was continuously circulated between the upper and lower chambers of the electrophoresis apparatus throughout the procedure. Autoradiography was carried out as follows: The gel was carefully removed from the electrophoresis apparatus, wrapped in Saran Wrap, and placed in contact with Kodak X-ray film at 4°C for 8.5 hours in total darkness. The film was developed in an automated X-ray processer and printed on Kodak Polycontrast paper. A densitometer tracing of region A-B is also shown. Translation in a wheat germ system of RNAs extracted from this gel showed that RNAs from regions 2 and 3 code for histone H4, and regions 8 and 9 code for histone H1. The RNAs in regions 4, 5, 6, and 7 are not completely resolved for one another, but code for histones H3, H2B, and H2A. The RNAs from region 1 and region 10 (containing 18 S RNA) did not direct protein synthesis in the wheat germ system. 5 S RNA has been run off the gel.

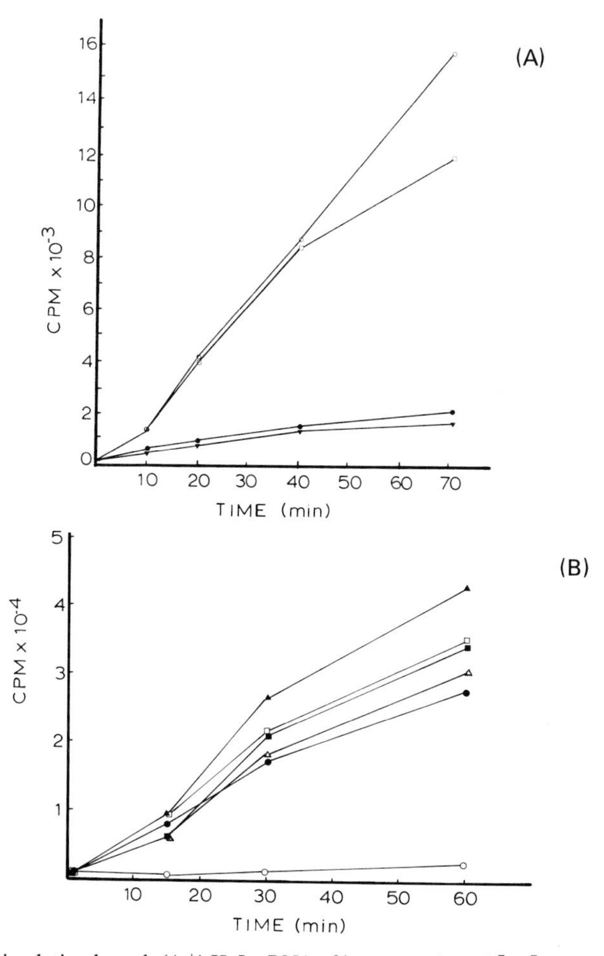


FIG. 7. (A) Stimulation by poly(A$^+$) HeLa RNA of incorporation of [$^3$H]lysine into trichloroacetic acid-precipitable material in a wheat germ lysate. Poly(A$^+$) HeLa RNA was isolated from HeLa polysomal RNA according to the method of Aviv and Leder (*30*), without the 0.1 *M* KCl wash. Wheat germ S30 extract was isolated according to Robert and Paterson (*36*), except that the wheat germ was initially ground dry for 30 seconds with a mortar and pestle containing crushed pipettes instead of sand. Reaction mixtures of 50 μl contained: 100 m*M* KCl, 3 m*M* $Mg(OAc)_2$, 2.6 m*M* dithiothreitol, 22 m*M* HEPES-KOH (pH 7.3), 1.2 m*M* ATP, 0.25 m*M* GTP, 0.5 m*M* amino acids except lysine, 15 m*M* creatine phosphate, 0.1 mg/ml creatine phosphokinase, 15 μl wheat germ S30 extract, the indicated amounts of added RNA, and 10 μCi of [$^3$H]lysine (38 Ci/ml, New England Nuclear Corporation). Incubations were carried out at 23°C. Aliquots (5 μl) were removed at the indicated times and spotted on 25 mm Whatman 3 MM disks presoaked in 20% TCA. The filters were dried, heated in 5% TCA to boiling, cooled, washed 3 times with 5% TCA and 3 times with 95% ETOH, and then dried and counted. Amounts of RNA: no mRNA (▲), 20 μg (●), 1 μg (□), and 5 μg (○). (B) Stimulation of incorporation of [$^3$H]lysine into TCA-precipitable material by histone mRNAs. Incubations were carried out as described in Fig. 7A. 1 μg (●), 2 μg (□), 5 μg (▲) and 10 μg (■) of RNA from region A of Fig. 3; 5 μg of RNA from region B of Fig. 3 (△).

gels (*37*) or were dialyzed against 0.9 *M* acetic acid–2.5 *M* urea–0.4% β-mercaptoethanol for acetic acid–urea gels (*38*).

In Fig. 8 are shown the results of translation of the RNAs isolated from regions A and B of the sucrose gradient shown in Fig. 3. It should be noted that while this RNA preparation was not subjected to nitrocellulose filter fractionation or oligo(dT)-cellulose chromatography to remove poly (A)-containing RNA sequences, such procedures are routinely employed for our preparations of histone mRNAs which are used as templates for transcription of cDNA. The lower-molecular-weight RNAs (7–9 S from region A) predominantly translate polypeptides that coelectrophorese with histones H4, H2A, H2B, and H3 in SDS polyacrylamide gels where fractionation is based on molecular weight (Fig. 8A). Confirmation that the *in vitro* synthesized polypeptides are histones is provided by coelectrophoresis with histones H4, H2A, H2B, and H3 in acetic acid–urea polyacrylamide gels where migration is dictated by charge as well as molecular weight (Fig. 8C). The higher molecular weight RNAs (9–14 S from region B) contain mRNAs for histone H1 as evidenced by coelectrophoresis of polypeptides synthesized *in vitro* with H1 histone marker, in SDS (Fig. 8B) as well as in acetic acid–urea (Fig. 8D) polyacrylamide gels.

## III. Synthesis and Characterization of Histone cDNA

### A. Synthesis of Histone cDNA

AMP residues are enzymically added to the 3′-OH termini of histone mRNAs with an ATP-polynucleotidyl exotransferase [poly(A) polymerase] isolated from maize seedlings. Isolation and characterization of the enzyme as well as conditions for addition of AMP residues to RNA molecules have been reported by Mans and Huff (*39*). Polyadenylation is carried out in enzyme excess, and by altering the conditions of the reaction the number of AMP residues added to RNA molecules can be varied. We have found that a poly(A) tail of 20–35 AMPs is optimal for $dT_{10}$-primed reverse transcription of histone mRNAs.

The reaction mixture for preparation of histone cDNA from polyadenylated histone mRNA templates is composed of the following in 0.5 or 1 ml: 50 m*M* Tris (pH 8.3), 20 m*M* KCl, 10 m*M* $MgCl_2$, 80 μ*M* dATP, 80 μ*M* dTTP, 15 μ*M* dithiothreitol, 10 μg/ml polyadenylated histone mRNA, 100 μg/ml actinomycin D, 1.6 μg/ml $dT_{10}$, 300 μCi/ml [$^3$H]dCTP, 200 μCi/ml [$^3$H]dGTP, and Rous sarcoma virus or avian myeloblastosis virus reverse transcriptase. After incubation at 37°C for 2 hours 100 μg of sonicated, denatured *E. coli* DNA are added as carrier. SDS is then added to 1%, and the reaction mixture is incubated for 10 minutes at 37°C. The reaction is deproteinized by extraction with phenol and chloroform–isoamyl alcohol as described previously for RNA isolation. The aqueous extract is brought to 0.25 *M* NaOH and incubated for 18 hours at 37°C to hydrolyze the RNA. The hydrolyzate is neutralized with HCl, concentrated to 0.5 ml by evaporation and chromatographed on Sephadex G-50 in 1 m*M* HEPES

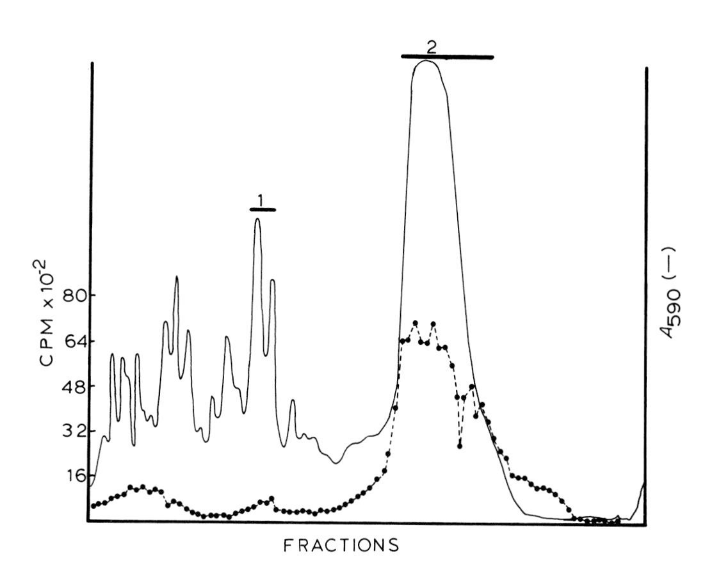


FIG. 8. Electrophoretic analysis of *in vitro* translation products of RNA from regions A and B of Fig. 3. A 20-μl aliquot of a translation reaction mixture which contained a total of 5 μg of RNA (Fig. 7B) was mixed with 20 μg of acid-extracted (0.4 *N* $H_2SO_4$) HeLa histones. Sodium dodecyl sulfate (SDS) was added to a final concentration of 2%, β-mercaptoethanol to 10%, and Tris (pH 6.8) to 0.0625 *M*. The samples were heated in boiling water for 2 minutes. Sucrose was added to a final concentration of 15%, and the samples were loaded onto a 10 cm × 15 cm × 0.3 cm, 12.5% polyacrylamide slab gel. The polyacrylamide gel electrophoresis system was that described by Laemmli (*37*). Electrophoresis was carried out for 7 hours at 30 mA. Following completion of electrophoresis the gels were fixed overnight in 12.5% trichloroacetic acid–40% ethanol, stained for 6 hours at 37°C in 0.25% Coomassie Blue–40% ethanol–7% acetic acid and then destained in 10% ethanol–7% acetic acid. For acetic acid–urea–polyacrylamide gel electrophoresis analysis, 15 μl of translation mixtures were combined with 40 μg of HeLa histones and dialyzed against 2.5 *M* urea–0.9 *M* HoAc–0.4% β-mercaptoethanol. The samples were then electrophoresed for 4 hours at 2 mA/gel in 6 mm × 9 cm acetic acid–urea–polyacrylamide gels according to Panyim and Chalkley (*38*). Gels were scanned at 590 nm, sliced into 1 mm fractions, solubilized in 30% $H_2O_2$, and then assayed for radioactivity in a liquid scintillation counter. (A) SDS polyacrylamide gel electrophoresis of proteins translated from RNAs isolated from region A of the sucrose gradient shown in Fig. 3. Region 1, $H_1$; region 2, $H_3$, $H_{2B}$, $H_{2A}$, $H_4$. Figure 8B–E on pp. 250 and 251.

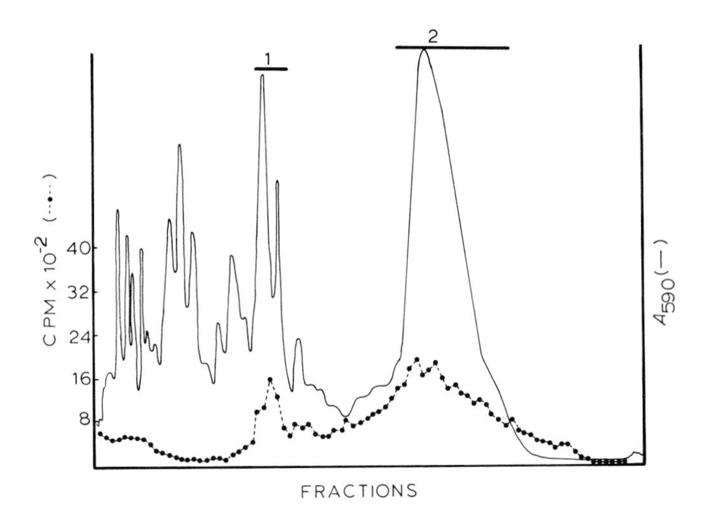


FIG. 8B. SDS polyacrylamide gel electrophoresis of proteins translated from RNAs isolated from region B of the sucrose gradient shown in Fig. 3. Region 1, $H_1$ ; region 2, $H_3$ , $H_{2B}$ , $H_{2A}$ , $H_4$. See page 249 for Fig. 8A.

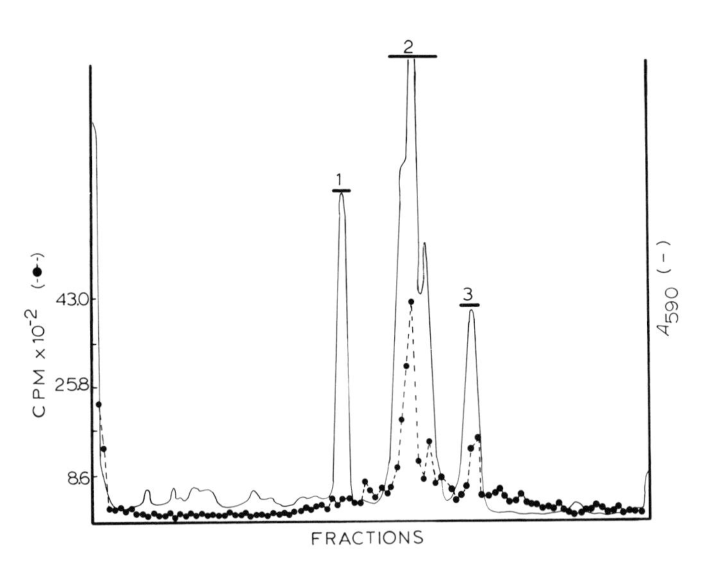


FIG. 8C. Acetic acid–urea polyacrylamide gel electrophoresis of proteins translated from RNAs isolated from region A of the sucrose gradient shown in Fig. 3. Region 1, $H_1$: region 2, $H_3$, $H_{2B}$, $H_{2A}$; region 3, $H_4$.

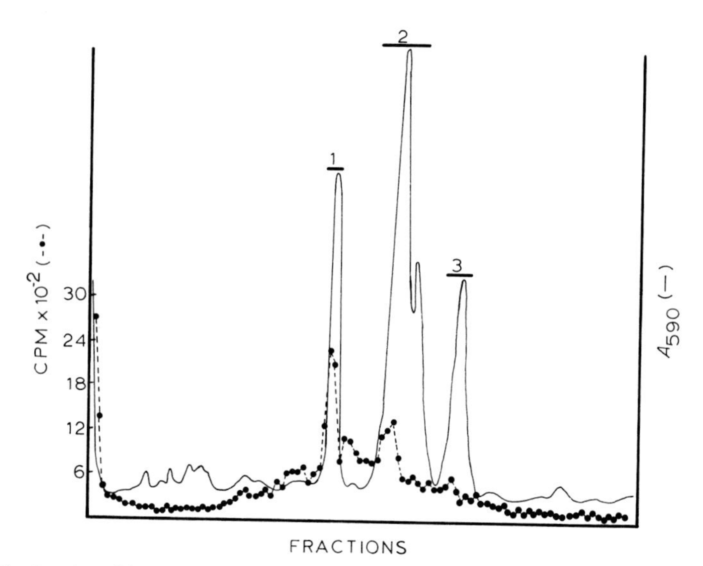


FIG. 8D. Acetic acid–urea polyacrylamide gel electrophoresis of proteins translated from RNAs isolated from region B of the sucrose gradient shown in Fig. 3. Region 1, $H_1$; region 2, $H_{2A}$, $H_{2B}$, $H_3$; region 3, $H_4$.

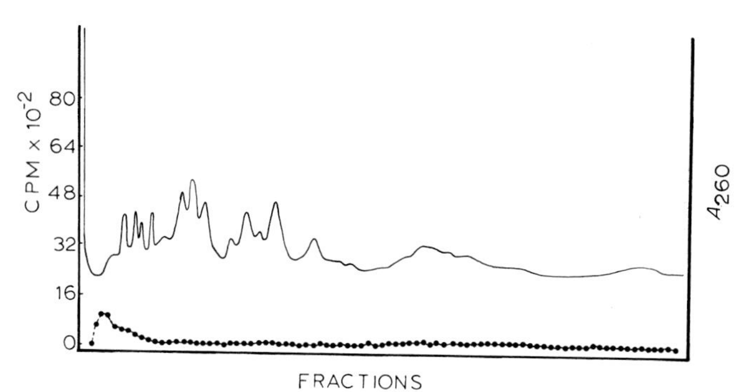


FIG. 8E. SDS polyacrylamide gel electrophoresis of an aliquot of an *in vitro* translation mixture which was incubated in the absence of added messenger RNAs. The aliquot electrophoresed was identical to those shown in Figs. 8A and 8B. The apparent variations in the absorbance profiles are due to the fact that the gels shown in Figs. 8A and 8B were scanned at 590 nm whereas the gel shown in Fig. 8E was scanned at 260 nm.

buffer (pH 7.0). The excluded fractions contain a complementary DNA separated from unincorporated nucleotides. The specific activity of the cDNA probe we prepare is generally in the range of 25,000–30,000 dpm/ng.

The optimal ratio of $dT_{10}$ to RNA template has been reported for globin mRNA (*5*) as 2 $dT_{10}$ per mRNA molecule. In the preparation of histone cDNA as described above, a ratio of 5 $dT_{10}$ per mRNA molecule ($10^5$ daltons) was employed. Reverse transcription was carried out in the presence of actinomycin D to ensure synthesis of a single-stranded cDNA molecule. In more recent histone cDNA preparations we have been using all the deoxynucleoside triphosphates at a concentration of 80 $\mu M$. Other workers have reported that using deoxynucleoside triphosphate concentrations in the range of 100–200 $\mu M$ (*7*) or carrying out reverse transcription in the presence of sodium pyrophosphate (*40*) facilitates synthesis of mostly complete cDNA copies of large mRNAs.

## B. Properties of cDNA

The mean sedimentation coefficient of the cDNA in alkaline sucrose is 6.1 S, which corresponds to a size of approximately 400 nucleotides (*41,42*). The kinetics of hybridization of the cDNA probe to histone mRNA isolated from polysomes of S-phase HeLa $S_3$ cells reflects the purity of the template RNA and cDNA (Fig. 9). The reaction proceeds with a $Cr_0t_{1/2}$ of $1.7 \times 10^{-2}$ (from double-reciprocal plot of the data) (*21*). Since the molar concentration of the nucleotide sequences in solution determines the rate of hybridization, comparison of the $Cr_0t_{1/2}$ of the histone mRNA with that

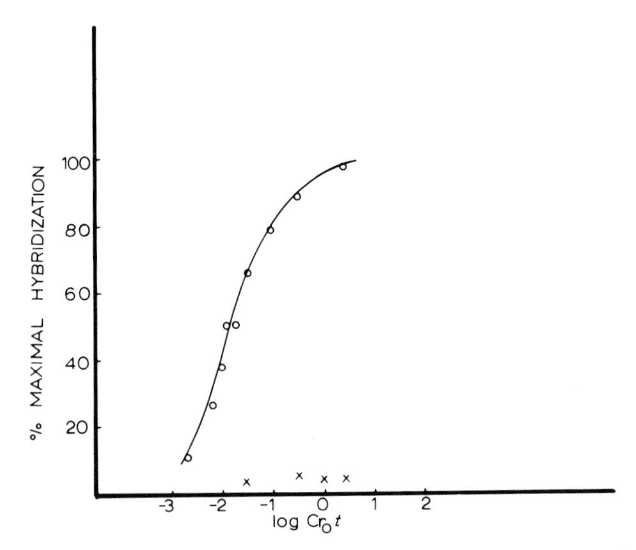


FIG. 9. Kinetics of annealing of histone cDNA to histone messenger RNAs isolated from the polyribosomes of S-phase HeLa $S_3$ cells. Four-hundredths ng [$^3$H]cDNA (27,000 dpm/ng) was annealed at 52°C in a volume of 15 $\mu$l containing 50% formamide, 0.5 *M* NaCl, 25 m*M* HEPES (pH 7.0), and 1 m*M* EDTA with either 0.03 or 0.19 $\mu$g of histone mRNA in the presence of 3.75 $\mu$g of *Escherichia coli* RNA as carrier (○) or with 3.75 $\mu$g of *E. coli* RNA under identical conditions (x). No background values have been subtracted. $Cr_0t$ = ribonucleotides $\times$ seconds/liter.

of a kinetic standard such as globin mRNA [complexity of 1200 bases (*43*), and $Cr_0t_{1/2}$ in similar conditions of $3.8 \times 10^{-3}$ (*44*)] yields a calculated sequence complexity of approximately 5400 bases which is two times greater than that expected for the total complexity of the five histone messages. This is, however, within the range of variation found for the rate of RNA–DNA hybridization (*45–48*). When cDNA and histone mRNA are hybridized in the absence of formamide at 75°C (data not shown), the $Cr_0t_{1/2}$ is $5 \times 10^{-3}$ From the $Cr_0t_{1/2}$ of the globin mRNA in these conditions ($Cr_0t_{1/2} = 2 \times 10^{-3}$) (*17,49*) the sequence complexity of histone mRNA is estimated to be 3000 bases. When the probe is annealed with *E. coli* RNA in either of the above conditions, no significant level of hybrid formation is detected. The low level of $S_1$ nuclease-resistant, TCA-precipitable radioactivity may be accounted for by a limited amount (3%) of [$^3$H]cDNA which is not digested by the enzyme in the incubation conditions used. Thermal denaturation curves of the histone mRNA–cDNA hybrids exhibit a single transition with a $T_m$ of 65°C in 50% formamide–0.5 *M* NaCl–25 m*M* HEPES (pH 7.0)–1 m*M* EDTA and 95°C in 0.5 *M* NaCl–25 m*M* HEPES (pH 7.0)–1 m*M* EDTA. These $T_m$ values are consistent with a reported base composition of histone mRNA of 54% GC (*50*).

Additional evidence for specificity of the histone cDNA is its ability to form hybrids with total polysomal RNA isolated from intact S-phase HeLa cells ($Cr_0t_{1/2} = 1.8$) and its lack of hybrid formation with $G_1$ polysomal RNA (*20*). From these findings it is reasonable to conclude that ribosomal (5 S, 18 S, and 28 S) and transfer RNA complementary sequences are not present in the histone cDNA. Furthermore, polysomal RNA isolated from HeLa cells in which histone and DNA synthesis have been blocked by cytosine arabinoside does not form hybrids with the histone cDNA. These results are consistent with data from several laboratories which indicate that histone mRNA is not present on the polyribosomes of HeLa cells treated with inhibitors of DNA synthesis (*22,23,51–54*) and additionally rule out the possibility that the cDNA contains detectable amounts of sequences complementary of other S-phase specific nonpolyadenylated RNAs which have been reported to be insensitive to cytosine arabinoside (*51*).

## IV. Applications of Histone cDNA

Histone cDNA has been effectively utilized for studying the regulation of histone gene expression during the cell cycle of continuously dividing

HeLa $S_3$ cells as well as after stimulation of nondividing WI-38 human diploid fibroblasts to proliferate. The probe has been used to quantitatively assess histone mRNA sequences in various intracellular compartments—nuclei, polyribosomes, postpolysomal supernatant—during the cell cycle in these cells (*20,55*). Results from such studies suggest the possibility that regulation of histone gene expression resides, at least in part, at the transcriptional level. Histone cDNA has also been utilized to quantitatively evaluate the transcription of histone mRNA sequences from various native and reconstituted chromatin preparations (*21,55–57*). Results from the latter studies provide additional support for the possibility of transcriptional control of histone gene readout during the cell cycle in HeLa and WI-38 cells. Furthermore, these studies suggest an important role for a component of the S-phase nonhistone chromosomal proteins in rendering histone genes transcribable during the period of the cell cycle when DNA replication occurs (*21,55–57*). Undoubtedly, histone cDNA will offer the opportunity for further examining the mechanism by which histone gene expression is controlled.

## REFERENCES


*1.* Baltimore, D., *Nature* (*London*) **226**, 1209 (1970).

*2.* Temin, H., and Mizutani, S., *Nature* (*London*) **226**, 1211 (1970).

*3.* Spiegelman, S., Burny, A., Das, M. R., Keydar, J., Schlom, J., Travnicek, M., and Watson, K. *Nature* (*London*) **227**, 563 (1970).

*4.* Paul, J., Gilmour, R., Affara, H., Birnie, G., Harrison, P., Hell, A., Humphries, S., Windass, J., and Young, B., *Cold Spring Harbor Symp. Quant. Biol.* **38**, 885 (1973).

*5.* Verna, I., Temple, G., Fan, H., and Baltimore, D., *Nature* (*London*), *New Biol.* **253**, 163 (1972).

*6.* Ross, J., Aviv, E., Scolnick, E., and Leder, P. *Proc. Natl. Acad. Sci. U.S.A.* **69**, 264 (1972).

*7.* Efstratiadis, A., Maniatis, T., Kafatos, F., Jeffrey, A., and Vorunakis, J., *Cell* **4**, 367 (1975).

*8.* Kacian, D., Spiegelman, S., Bank, A., Terada, N., Metafora, S., Dow, L., and Marks, P., *Nature* (*London*), *New Biol.* **235**, 167 (1972).

*9.* Robbins, J., and Heywood, S., *Biochem. Biophys. Res. Commun.* **68**, 918 (1970).

*10.* Monahan, J., Harris, S., Woo, S., Robberson, D., and O'Malley, B., *Biochemistry* **15**, 223 (1976).

*11.* Thrall, C., Park, W., Rashba, H., Stein, J., Mans, R., and Stein, G., *Biochem. Biophys. Res. Commun.* **61**, 1443 (1974).

*12.* Harrison, P., Hell, A., Birnie, G., and Paul, J., *Nature* (*London*) **239**, 219 (1972).

*13.* Gilmour, R. A., and Paul, J., *Proc. Natl. Acad. Sci. U.S.A.* **70**, 3440 (1973).

*14.* Axel, R., Cedar, H., and Felsenfeld, G., *Proc. Natl. Acad. Sci. U.S.A.* **70**, 2029 (1973).

*15.* Sullivan, D., Palacios, R., Stavnezer, J., Taylor, J., Faras, A., Kiely, M., Morris, N., Bishop, J. M., and Schimke, R. T., *J. Biol. Chem.* **248**, 7530 (1973).

*16.* Harris, S., Means, A., Mitchell, W., and O'Malley, B. W., *Proc. Natl. Acad. Sci. U.S.A.* **70**, 3776 (1973).

*17.* Barrett, T., Maryanka, D., Hamlyn, P., and Gould, H., *Proc. Natl. Acad. Sci. U.S.A.* **71**, 5057 (1974).

18. Stavnezer, J., Huang, R. C. C., Stavnezer, E., and Bishop, J. M., *J. Mol. Biol.* **88**, 43 (1974).
19. Stein, G., Park, W., Thrall, C., Mans, R., and Stein, J., *Biochem. Biophys. Res. Commun.* **63**, 945 (1975).
20. Stein, J., Thrall, C., Park, W., Mans, R., and Stein, G., *Science* **189**, 557 (1975).
21. Stein, G., Park, W., Thrall, C., Mans, R., and Stein, J., *Nature* (*London*) **257**, 746 (1975).
22. Borun, T., Scharff, M., and Robbins, E., *Proc. Natl. Acad. Sci. U.S.A.* **58**, 1977 (1967).
23. Gallwitz, D., and Mueller, G., *J. Biol. Chem.* **244**, 5947 (1969).
24. Jacobs-Lorena, M., Baglioni, C., and Borun, T., *Proc. Natl. Acad. Sci. U.S.A.* **69**, 2095 (1972).
25. Gallwitz, D., and Brindl, M., *Biochem. Biophys. Res. Commun.* **47**, 1106 (1972).
26. Bootsma, D., Bidke, L., and Vos, O., *Exp. Cell Res.* **33**, 301 (1964).
27. Xeros, N., *Nature* (*London*) **194**, 683 (1962).
28. Stein, G. S., and Borun, T. W., *J. Cell Biol.* **52**, 292 (1972).
29. Lee, S. U., Medecki, J., and Brawerman, G., *Proc. Natl. Acad. Sci. U.S.A.* **68**, 1331 (1971).
30. Aviv, H., and Leder, P., *Proc. Natl. Acad. Sci. U.S.A.* **69**, 1408 (1972).
31. Loening, U. E., *Biochem. J.* **113**, 131 (1969).
32. Kedes, L. H., Cohn, R. H., Lowry, J. C., Chang, A. C. Y., and Cohen, S. N., *Cell* **6**, 359 (1975).
33. Grunstein, M., Levy, S., Schedl, P., and Kedes, L., *Cold Spring Harbor Symp. Quant. Biol.* **38**, 717 (1973).
34. de Wachter, R., and Fiers, W., *in* "Methods in Enzymology" (L. Grossman and K., Moldave, eds.), Vol. 21, p. 167. Academic Press, New York,1971.
35. Grunstein, M., and Schedl, P., *J. Mol. Biol.* **104**, 323 (1976).
36. Roberts, B., and Paterson, B., *Proc. Natl. Acad. Sci. U.S.A.* **70**, 2330 (1973).
37. Laemmli, U., *Nature* (*London*) **227**, 680 (1970).
38. Panyim, S., and Chalkley, R., *Biochemistry* **8**, 3972 (1969).
39. Mans, R. J., and Huff, N., *J. Biol. Chem.* **250**, 3672 (1975).
40. Kacian, D. L., and Myers, J. C., *Proc. Natl. Acad. Sci. U.S.A.* **73**, 2191 (1976).
41. McEwen, C. R., *Anal. Biochem.* **20**, 114 (1967).
42. Studier, F. W., *J. Mol. Biol.* **11**, 373 (1965).
43. Labrie, F., *Nature* (*London*) **221**, 1217 (1969).
44. Young, B. D., Harrison, P. R., Gilmour, R. S., Birnie, G. D., Hell, A., Humphries, S., and Paul, J., *J. Mol. Biol.* **84**, 555 (1974).
45. Birnstiel, M. L., Sells, B. H., and Purdom, I. F., *J. Mol. Biol.* **63**, 21 (1972).
46. Straus, N. A., and Bonner, T. I., *Biochim. Biophys. Acta* **227**, 87 (1972).
47. Bishop, J. O., *Biochem. J.* **113**, 805 (1969).
48. Bishop, J. O., *Biochem. J.* **126**, 171 (1972).
49. Gulati, S. C., Kacian, D. L., and Spiegelman, S., *Proc. Natl. Acad. Sci. U.S.A.* **71**, 1035 (1974).
50. Adesnik, M., and Darnell, J., *J. Mol. Biol.* **67**, 397 (1972).
51. Borun, T. W., Gabrielli, F., Ajiro, K., Zwiedler, A., and Baglioni, C., *Cell* **4**, 59 (1975).
52. Breinde, M., and Gallwitz, D., *Eur. J. Biochem.* **32**, 381 (1973).
53. Burler, W. B., and Mueller, G. C., *Biochim. Biophys. Acta* **294**, 481 (1973).
54. Wilson, M. C., Melli, M., and Birnstiel, M. L., *Biochem. Biophys. Res. Commun.* **61**, 404 (1974).
55. Jansing, R. L., Stein, J. L., and Stein, G. S., *Proc. Natl. Acad. Sci. U.S.A.* **74**, 173 (1976).
56. Park, W. D., Stein, J. L., and Stein, G. S., *Biochemistry* **15**, 3296 (1976).
57. Stein, J. L., Reed, K., and Stein, G. S., *Biochemistry* **15**, 3291 (1976).

# Chapter 23

# *Immunological Methods for the Isolation of Histone H5 mRNA from Chicken Reticulocytes*

A. C. SCOTT[1] AND J. R. E. WELLS

*Department of Biochemistry, University of Adelaide, Adelaide, South Australia*

## I. Introduction

The ability to purify specific mRNA species has had a major impact on the study of cell biology at the molecular level. However, the total number of mRNAs that have been isolated is relatively small. The reason for this lies in the techniques available for isolation, which generally depend upon the mRNA either being a major RNA species in the cell or having unusual properties such as a distinctive size (*1,2*) high G + C content (*3*), or poly(A) content (*4,5*). The number of mRNAs which meet these criteria is quite limited, and many of the potentially most interesting ones, such as those coding for control proteins, may not meet them.

The isolation of a minor mRNA species in pure form is a difficult task. We were interested in studying the genes and the mRNA coding for H5 histone. This protein is unique among histones in being synthesized in cells that have ceased to divide (*6,7*), and it may be involved in the progressive termination of macromolecular synthesis in these cells as maturation of the cell line proceeds (*7,8*). The H5 mRNA is a minor mRNA species in avian red cells, in which the major mRNAs present are those coding for globin. Since H5 mRNA was likely to be of similar size to these major globin mRNAs (*9,10*) size-dependent fractionation on sucrose gradients was inadequate and although chromatography on oligo(dT)-cellulose was useful [H5 mRNA lacks a 3′-poly(A) tract], only partial purification could be achieved by these means.

This prompted us to employ immunological methods, since the exquisite

[1] *Present address:* Forensic Biology Laboratory, Institute of Medical and Veterinary Science, Adelaide, South Australia.

resolution of which antibodies are capable should make this purification possible. The use of antibodies to precipitate specific polysomes is not new (*11–14*). However, early attempts suffered from lack of specificity, low yields, and partial degradation of the mRNA during its isolation. The methodology relied upon a direct precipitin reaction, whch required high concentrations of antibodies and of polysomes to be effective. Unfortunately such conditions promoted nonspecificity and ribonuclease problems. Recent modifications of these techniques which have involved indirect precipitation of the antibody–polysome complexes have made the isolation of minor mRNA species a distinct, if technically difficult, possibility (*14–17*). This article is intended to describe these methods as they apply to the isolation of H5 mRNA, and to discuss some of the general problems associated with these techniques of mRNA isolation.

The exact conditions employed in these procedures are determined by the individual antibody preparation used. It is therefore impossible to give figures that are universally applicable. We have included figures from our own work with one particular set of reagents and described how these figures were obtained. In the interest of brevity, we have omitted details that have been published elsewhere.

## II. Immunoadsorption

The immunoadsorption reaction depends upon the specific, monovalent binding of small amounts of highly purified antibodies to nascent peptides on polysomes. These antibody–polysome complexes are then precipitated by the addition of insolubilized antigen. The specificity and yield of the reaction depend upon two factors: (a) the purity and specificity of the reagents employed; and (b) the conditions of ionic strength, temperature, and incubation time at which the reaction is carried out.

### A. Reagents

The reagents employed in the immunoadsorption reaction are an antibody, an immunoadsorbent, and a solution of polysomes.

#### 1. Antibodies

The antibody must react specifically with nascent H5 peptides, and contain a minimum of contaminating antibodies, in order to reduce nonspecific adsorption. The antibody must also be capable of reacting rapidly, at low temperatures, in order to minimize ribonuclease activity. Suitable antibody preparations are thus quite rare, and it may be necessary to test a large number before finding a suitable one.

Anti-H5 was raised in rabbits by the injection of electrophoretically pure H5 (*9*) and the clarified serum was subjected to two precipitations with 45% saturated ammonium sulfate to prepare the $\gamma$-globulin fraction. This was then dialyzed against phosphosaline, [0.015 *M* NaCl–0.01 *M* sodium phosphate (pH 7.2)], and all subsequent operations were performed using sterile equipment, solutions, and techniques. The $\gamma$-globulin was next purified by affinity chromatography to prepare the purified anti-H5. Two affinity columns (2 × 6 cm) were prepared by coupling proteins to CNBr-activated Sepharose (*9, 18*). One column contained purified globin bound to Sepharose, and the other contained purified H5 bound to Sepharose. Since globin is the major product of the cells, it was imperative to remove any anti-globin activity in the antibody preparation. The $\gamma$-globulin was thus chromatographed on the globin-Sepharose column in phosphosaline, and any antiglobin or nonspecifically binding $\gamma$-globulin was thus removed. The unbound effluent was then chromatographed on the H5-Sepharose column by the method of Shapiro *et al.* (*16*), and the bound anti-H5 was subsequently eluted with 0.1 *M* glycine (pH 2.8). The resulting antibody preparation, while highly specific to H5 and of high titer, was still found to contain small amounts of ribonuclease activity on occasions. The ribonuclease was detected simply by incubating 18 S rRNA with the antibody, treating with formamide to expose any hidden breaks, and analyzing the RNA on sucrose gradients or polyacrylamide gels in formamide (*10*). The ribonuclease was removed by chromatography through a 3 cm-diameter column containing 5 cm of CM-cellulose over 5 cm of DEAE-cellulose, equilibrated with phosphosaline (*14*).

### 2. Polysomes

Polysomes were prepared, as described, from washed chicken reticulocytes (*9*) and suspended in immunoadsorption buffer [0.025 *M* Tris-HCl (pH 7.6)–0.025 *M* NaCl–0.005 *M* $MgCl_2$–1 $\mu$g/ml Trichodermin–0.1 mg/ml herparin] to the required concentration. The polysomes were normally freshly prepared before use but could be stored frozen in liquid nitrogen for several months. If frozen polysome suspensions were used, they were centrifuged at 20,000 *g* for 15 minutes before use, to remove any aggregates.

### 3. Immunoadsorbent

The immunoadsorbent employed in this reaction was prepared by glutaraldehyde cross-linking of purified total chicken histone (*9*). Since the only histone which the chicken reticulocyte synthesizes is H5 (*7*), the other histones present merely act as an inert carrier. Theoretically, one should use purified H5 to prepare the matrix, in order to reduce the size of the immunoprecipitate, as this has a marked effect on the degree of nonspecific adsorption. In practice, however, the difference is not great since total chicken histone contains 25% H5 (*6*) and the amount of immunoadsorbent required is small.

## B. Optimization of the Reaction

The importance of optimizing the reaction conditions cannot be overestimated, as small changes can have major effects on the specificity and efficiency of the reaction. Since all antibody preparations will vary, only an outline of the procedures can be given, and optimal conditions must be determined for each batch of antibodies. The three major variables are discussed below.

### 1. The Amount of Antibody and Antigen Matrix to Add

The first variable investigated was the amount of antigen matrix required to precipitate a given amount of anti-H5. This must give maximum yield without compromising specificity. It was estimated by carrying out a pre-

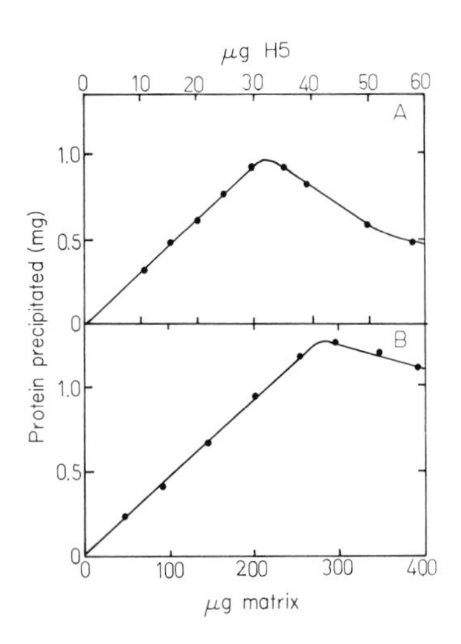


FIG. 1. Precipitin curves for the precipitation of anti-H5 by purified chicken histone H5 (A) and an insoluble matrix of total chicken histone (B). Increasing amounts of precipitant were incubated with a constant amount of anti-H5 (1.5 mg) for 2 hours. The precipitate was then collected by centrifugation (5000 *g*, 15 minutes), and washed three times with saline by suspension and centrifugation. The amount of protein precipitated was determined either by optical density measurements (*15*), or by scintillation counting when an isotopically labeled protein was used.

cipitin reaction. It is important to perform this experiment with immunoadsorbent, not with soluble antigen, as much of the antigen in the matrix is not available for reaction with antibody (*10*). For example, as shown in Fig. 1A, 1 mg of purified anti-H5 would precipitate 31 μg of purified H5. In contrast, as shown in Fig. 1B, this required 285 μg of total histone matrix, even though this is 25% H5 (*6*). This experiment resulted in an estimate of the amount of antigen matrix required to precipitate a given amount of antibody.

### 2. Duration of Incubation of the Reagents

The second variable investigated was the time required for the immune reaction at 0°C, which is the temperature of choice for minimizing ribonuclease activity. It must be sufficient to give a reasonable yield while minimizing ribonuclease activity. Figure 2 shows the reaction of an anti-H5 preparation with histone matrix, as measured by removal of antibody from the supernatant. At 0°C, the reaction was essentially complete within 60 minutes. Such efficient preparations are quite rare.

### 3. Concentration of Polysomes

If the concentration is too high, this interferes with the reaction.

It must be stressed that these figures served only as a guide to the optimal conditions, as the reaction of an antibody with a nascent peptide may well be quite different from that with antigen matrix. Thus it was necessary to

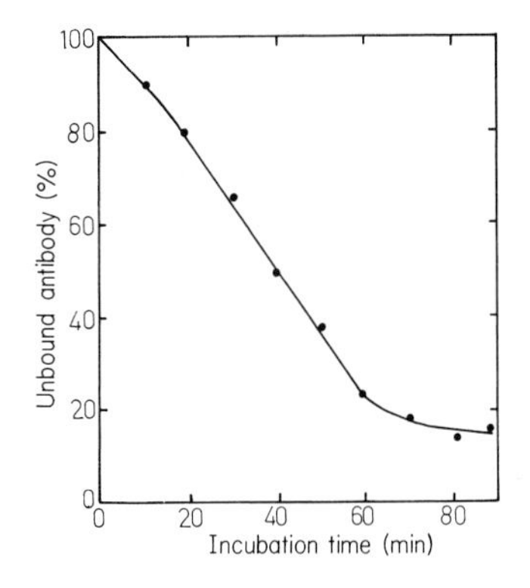


Fig. 2, Time course of the reaction of purified anti-H5 with total histone matrix. Purified anti-H5 (3.5 mg) was incubated with 1 mg of total histone matrix in 3.5 ml of phosphate-buffered saline at 0°C. After the required time, the precipitate was removed by centrifugation (10,000 *g*, 5 minutes), and the optical density at 280 nm of the supernatant was measured. This was used to calculate the percentage of unbound (unprecipitated) antibody.

carry out the entire immunoadsorption reaction varying the polysome concentration, the amount of anti-H5 added, the time of incubation, and the amount of immunoadsorbent added.

## 4. Procedure

Polysomes at a fixed concentration (5–50 $A_{260}$ units/ml) were incubated with stirring at 0°C, with varying amounts of anti-H5 (10–100 μg per 10 $A_{260}$ units of polysomes) for the estimated time (i.e., about 60 minutes from Fig. 2). The appropriate amount of antigen matrix required to precipitate the added antibody (i.e., about 285 μg per 1.0 mg of anti-H5 as determined above) was then added, and stirring was continued at 0°C for a further 60 minutes (by reference to Fig. 2). The matrix was then precipitated and very extensively washed, as described (*9, 14*) to remove nonspecifically bound polysomes. The RNA was then extracted from the matrix using phenol–sodium dodecyl sulfate (SDS) at pH 5.3 (*19, 20*), as this was found to be the most efficient method. The amount of RNA extracted was then estimated by optical density, after removal of any remaining traces of phenol.

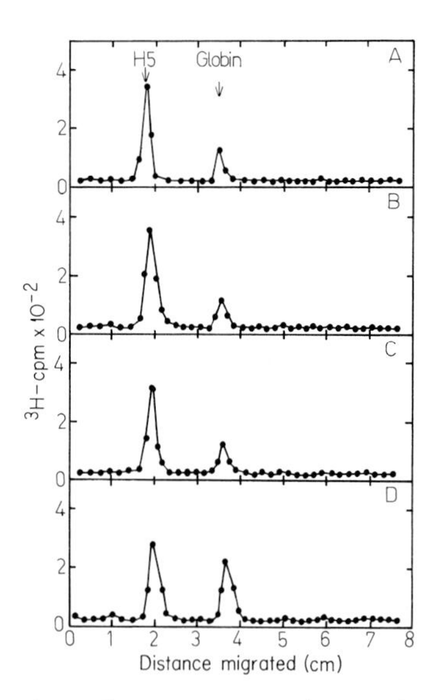


FIG. 3. Effect of increasing polysome concentration on the purity of H5 mRNA prepared by immunoadsorption. H5 mRNA was prepared as described in the text from polysomes at concentrations of (A) 10$A_{260}$/ml; (B) 15 $A_{260}$/ml; (C) 20 $A_{260}$/ml; (D) 25 $A_{260}$/ml. Each mRNA preparation was then translated at nonsaturating concentration in a wheat embryo cell–free system, and the products were analyzed on polyacrylamide gels containing sodium dodecyl sulfate and urea (*9*). Electrophoresis was from left to right in all cases, and the position of the two major products is marked.

By varying each parameter, one can thus estimate the conditions for optimal yield of total RNA. These figures, however, need not equate with the maximum yield or purity of H5 mRNA. In fact the degree of purity of H5 mRNA was greatest at concentrations slightly lower than those giving optimal yield of total RNA. Since purity was the essential requirement, the conditions were finally optimized by assaying the RNA prepared, by translation. This was done as follows.

The total phenol-extracted RNA was disaggregated by warming briefly in formamide and then fractionated, after suitable dilution, on sucrose–SDS gradients (*20*). The 8–14 S RNA, which contains the H5 mRNA and globin mRNA, was translated in the wheat embryo cell-free translation system, and the translation products were analyzed on SDS–urea gels (*9*). For example, as shown in Fig. 3, it was found that the optimum conditions for H5 mRNA purity were at less than 20 $A_{260}$ units of polysome per milliliter. This was comparable to a figure of 30 $A_{260}$ units per milliliter for optimal yield of total RNA.

In this way, one can derive a series of estimates for the yield and purity of H5 mRNA at increasing values of the variable parameters. One can then determine the optimal conditions, which in our case were those giving maximum specificity, i.e., minimum contaminating globin mRNA, and these are illustrated in Fig. 4.

## C. Summary of Immunoadsorption Technique

Obviously this technique is capable of producing H5 mRNA from reticulocytes, of reasonable purity (see Fig. 3). There are several drawbacks to the technique, however. First, antibody preparations that can be successfully employed in this technique are rare. Second, determining the optimum conditions under which the reaction is carried out is a long and tedious process. The most telling criticism of the technique, however, is that in our hands the yield of mRNA is extremely low. The preparation of even 1 μg of mRNA is a major task requiring several experiments. This in turn means that further purification procedures such as oligo (dT)-cellulose chromatography are virtually impossible to monitor. One way of improving the yield is by the use of drugs such as Trichodermin (an inhibitor of peptide termination) or cycloheximide (at appropriate concentrations an inhibitor of peptide elongation) during polysome preparation (*9*). This effect is probably due to the prevention of the rapid runoff of chicken polysomes during their isolation. Despite such modifications, the yield of H5 mRNA was too low to be of any long-term practical value. The yield is probably diminished markedly by the difficulty found in washing the immunoadsorbent to free it of nonspecifically bound polysomes. This could perhaps be improved by

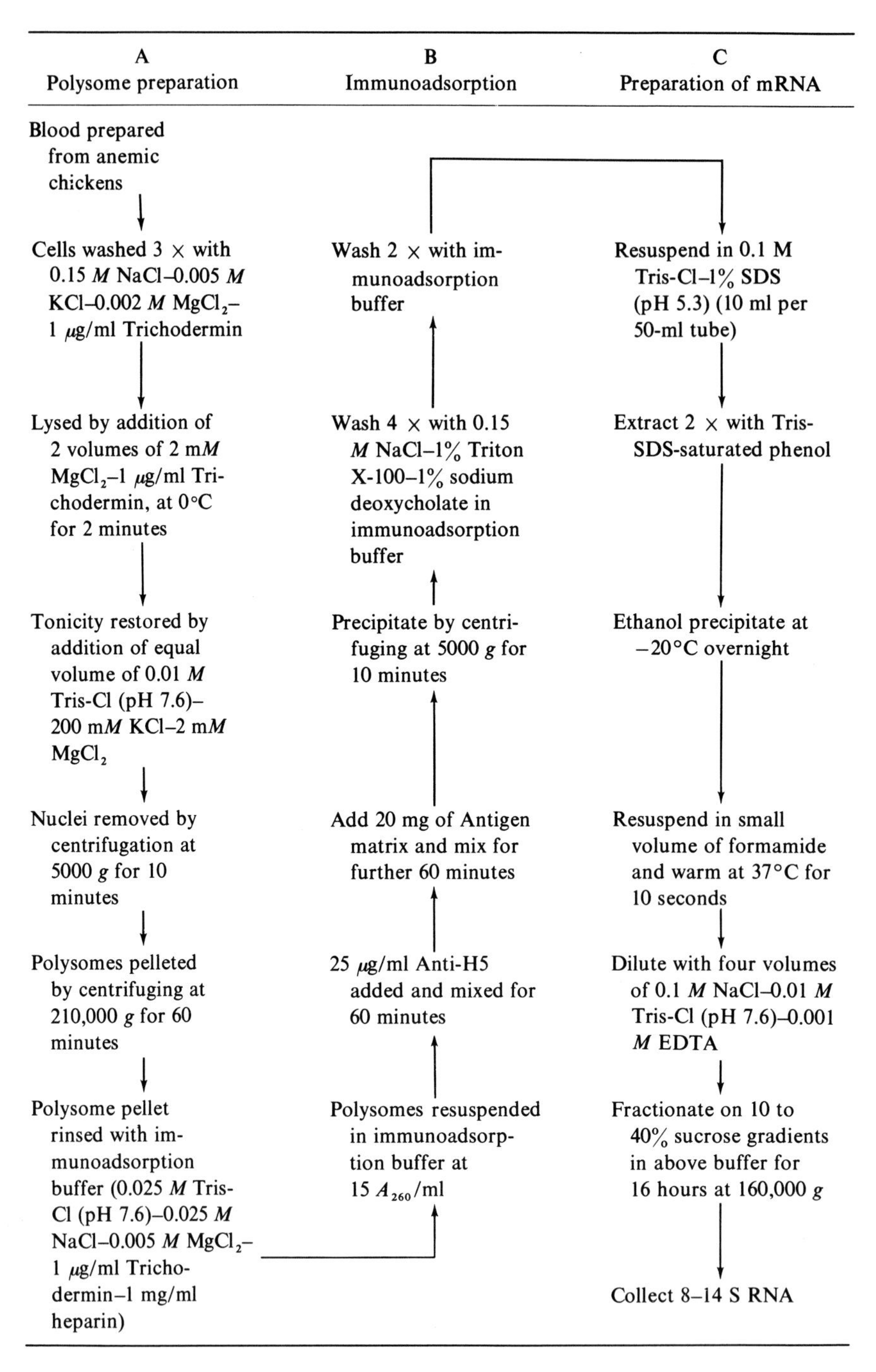


FIG. 4. Flow diagram of the method for preparing H5 mRNA by immunoadsorption. All operations were performed at 0°C unless otherwise stated.

using purified H5 for the immunoadsorbent; but a more attractive alternative was the use of indirect immunoprecipitation.

## III. Indirect Immunoprecipitation

The indirect immunoprecipitation reaction differs from immunoadsorption only in the method employed in precipitating the antibody–polysome complexes. In this case it is carried out by using an antibody to the anti-H5. The effect of this change was quite large, however. Not only was the technique simpler than immunoadsorption, but it resulted in higher yield and purity of H5 mRNA and was thus preferred. The technique required only the further isolation of a suitable antibody to the rabbit anti-H5.

### A. Reagents

The preparation of anti-H5 and polysomes was unchanged from the immunoadsorption reaction with the exception that the polysomes were suspended in immunoprecipitation buffer [0.025 *M* Tris-HCl (pH 7.6)–0.15 *M* NaCl–0.005 *M* $MgCl_2$–0.1 mg/ml heparin–1 $\mu$g/ml Trichodermin]. The higher salt concentration results in a lower level of nonspecific adsorption (*17*).

Antibody to anti-H5 [referred to as anti-(anti-H5)] was raised in goats by injection of purified rabbit $\gamma$-globulin. The further purification of the antibody was analogous to anti-H5 in that ammonium sulfate-fractionated-$\gamma$-globulin was chromatographed on globin-Sepharose to remove antiglobin activity and then on a column of rabbit $\gamma$-globulin-Sepharose to purify the antibody. The preparation was finally rendered ribonuclease-free by chromatography on a combined CM-cellulose, DEAE-cellulose column.

### B. Optimization of the Reaction

The optimization of the indirect immunoprecipitation reaction is just as important as for the immunoadsorption reaction. The methods are similar in many respects and fall into three categories.

#### 1. Estimation of the Amounts of Reagents Required and Incubation Time

The amount of precipitant required, in this case anti-(anti-H5), to precipitate a fixed amount of anti-H5 was first determined by carrying out a precipitin reaction between these two reagents. The time course of this reaction at 0°C must also be determined. Figure 5 shows the reaction of our two reagents under these conditions. The reaction in this case takes approximately 80 minutes to reach completion. The minimal requirements of a suitable antibody preparation are the same as for the immunoadsorption reaction, i.e., it must react specifically and rapidly with its antigen at 0°C.

### 2. Optimization of the Yield of Total RNA

This was optimized by carrying out the indirect immunoprecipitation reaction at varying concentrations of reagents and times of incubation, then extracting and quantitating the amount of RNA precipitated. In brief, the procedure was as follows.

Polysomes were suspended in immunoprecipitation buffer at 0°C to the desired concentration (5–50 $A_{260}$ units/ml), and a measured amount of purified anti-H5 was added (10–100 μg per 10 $A_{260}$ units of polysomes). This was stirred for a fixed time (in our case about 60 minutes from Fig. 2), and then sufficient anti-(anti-H5) was added to quantitatively precipitate the anti-H5 (in our case, about 28 mg/mg; anti-H5 calculated as in Section I above). The stirring was continued at 0°C while a precipitate formed (about 75 minutes, from Fig. 5). The entire mixture was then layered over discontinuous sucrose gradients consisting of 3 ml of 15% sucrose over 6 ml of 30% sucrose, both in immunoprecipitation buffer plus 1% Triton X-100 and 1% sodium deoxycholate. The immune precipitate was then collected after centrifugation at 20,000 *g* for 20 minutes. This was resuspended in immuno-

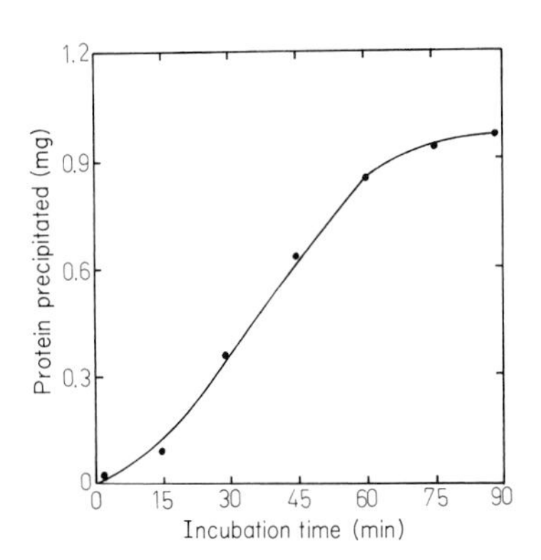


Fig. 5. Time course of the reaction of purified anti-H5 with anti-(anti-H5). Anti-(anti-H5) (2 mg) was incubated with 60 μg of anti-H5 in a total volume of 2 ml of phosphate-buffered saline at 0°C. After the required interval, any precipitate was collected by centrifugation (10,000 *g*, 5 minutes) and washed three times by suspension and recentrifugation with phosphate-buffered saline. The precipitate was then dissolved in 0.1 *N* NaOH and the $OD_{280}$ was recorded.

precipitation buffer and centrifuged again through identical sucrose gradients. The RNA was then extracted from the precipitate with phenol-SDS (pH 5.3), and the amount was estimated.

The great advantage of this technique lies in the washing procedure, which was adapted from Shapiro *et al.* (*16*). This was found to be highly efficient in removing nonspecifically bound polysomes as well as being rapidly performed.

### 3. Optimization of Specificity of the Reaction

The reaction was optimized for specificity as described for the immunoadsorption reaction. Thus the RNA was disaggregated with formamide, fractionated on sucrose gradients, the 8–14 S RNA translated in the wheat embryo cell-free system, and the translation products were analyzed on polyacrylamide gels.

It should be remembered, however, that this reaction is not a simple antigen–antibody reaction, and the conditions can only be derived by experimentation. For example, Fig. 6 shows the effect of polysome concentration on the precipitation of anti-H5–polysome complexes by anti-(anti-H5) at 0°C, after different incubation times. For example, at 15 $A_{260}$ units of polysomes per milliliter increasing incubation time from 75 to 120 minutes has negligible effect upon the RNA yield. However, at 25 $A_{260}$/ml this change in incubation time results in a 50% higher yield of RNA. It can be seen that the polysome concentration affects the rate at which the reaction occurs. This is useful information as it means that by extending the incubation time of

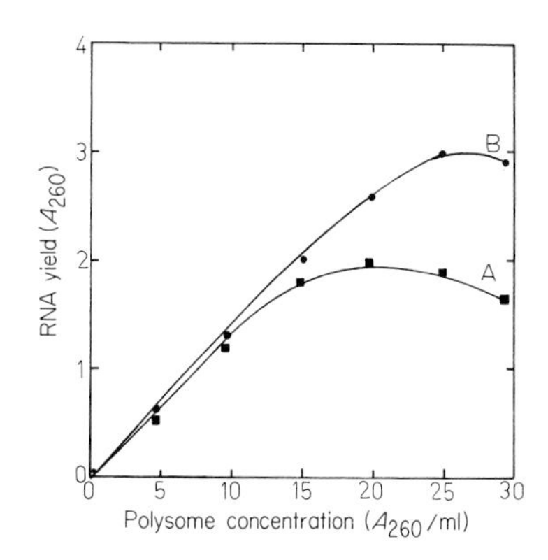


FIG. 6. Effect of increasing polysome concentration on the yield of RNA prepared by indirect immunoprecipitation. Four hundred $A_{260}$ units of polysomes at varying concentrations subjected to indirect immunoprecipitation as described in the text, and H5-synthesizing polysomes were prepared. The RNA was extracted from these, and the amount was measured in $A_{260}$ units. The anti-(anti-H5) was incubated with the other reagents for 75 minutes (curve A) prior to collecting the precipitate, or for 120 minutes (curve B).

these reagents to 120 minutes, we can handle larger amounts of material with a higher yield of H5 mRNA.

## C. Further Purification of H5 mRNA

The higher yields of H5 mRNA produced by the indirect immunoprecipitation reaction made the further purification of the mRNA possible. This was performed by more conventional techniques.

The 8–14 S RNA collected from the sucrose gradient fractionation described above was chromatographed twice through a 0.5 × 2 cm column of oligo (dT)-cellulose in 0.3 *M* NaCl–0.01 *M* Tris-HCl (pH 7.4). The unbound RNA was then ethanol-precipitated, disaggregated with formamide, and fractionated on a sucrose gradient (*20*). The H5 mRNA was then collected as a 10–12 S species. This mRNA coded for the synthesis of more than 90% H5 in the wheat embryo system, as estimated by polyacrylamide gel electrophoresis (Fig. 7). The level of globin mRNA contamination was too low to be accurately measured in this manner.

The purification procedure was designed to remove most of the globin mRNA which binds to oligo (dT)-cellulose under these conditions owing to its poly (A) tract. The procedure is not selective enough to be feasible

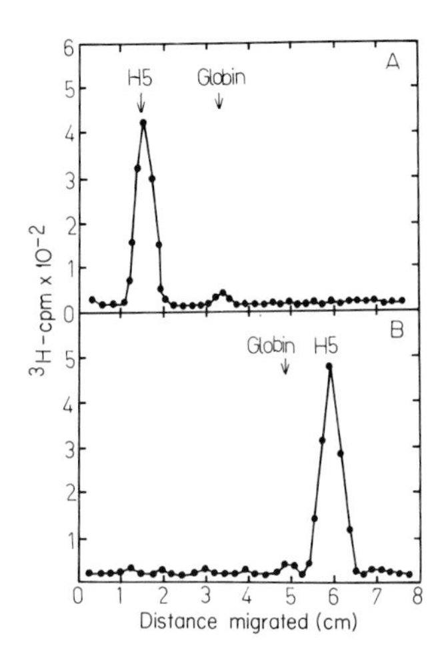


FIG. 7. Analysis of the products of *in vitro* translation of H5 mRNA prepared by indirect immunoprecipitation and oligo (dT)-cellulose chromatography. H5 mRNA was prepared as described in the text, and 0.3 μg was translated in a wheat embryo cell-free system. The [$^3$H] lysine-labeled *in vitro* products were analyzed by electrophoresis on (A) sodium dodecyl sulfate-urea polyacrylamide gels (*9*), and (B) low pH urea polyacrylamide gels. Electrophoresis was from left to right in both cases. The position of $^{14}$C-labeled standards is indicated.

with total reticulocyte polysomal RNA, as a preparative method for H5 mRNA, but is useful in further purification of the immunologically prepared mRNA. The final sucrose gradient may seem unnecessary but was found to be essential for efficient translation of the mRNA.

Figure 8 shows a flow diagram of a typical H5 mRNA preparative experiment. The figures shown are from our own work and will vary from one antibody preparation to another.

### D. Summary of the Indirect Immunoprecipitation Technique

Using this technique, we have been able to prepare microgram quantities of H5 mRNA, which is, by translation and hybridization analysis (*20*), over 90% pure. Thus the technique is both more efficient and more specific, and for these reasons it has supplanted immunoadsorption in our laboratory. Much of the credit for the improvement is due to the efficient and rapid washing procedure for the immune precipitate. The precipitate formed is a diffuse one as opposed to the bulky immunoadsorbent and is thus more readily washed.

## IV. Conclusions

Schimke and his collaborators have developed much of the methodology involved in the use of antibodies to isolate a specific mRNA species. The principle is straightforward enough, but in practice we have found it to be an art as much as a science. The precise conditions to employ depend absolutely upon the individual antibody preparation, and we can only give general techniques for determining these conditions. They will also vary depending upon the mRNA being isolated. For example, since H5 mRNA is a minor mRNA species, the amount of anti-H5 antibody required is also small. The principles are exactly the same, however, irrespective of the antibody used or mRNA sought.

While both immunoadsorption and indirect immunoprecipitation were described, the latter is simpler and, in our hands, more efficient and is thus recommended.

Although most of the techniques described (e.g., precipitin reactions, sucrose gradient fractionation) are well known, it would be incorrect to say that these procedures are routine methods for mRNA preparation, in their present form. Suitable antibodies are uncommon, and the procedure is long

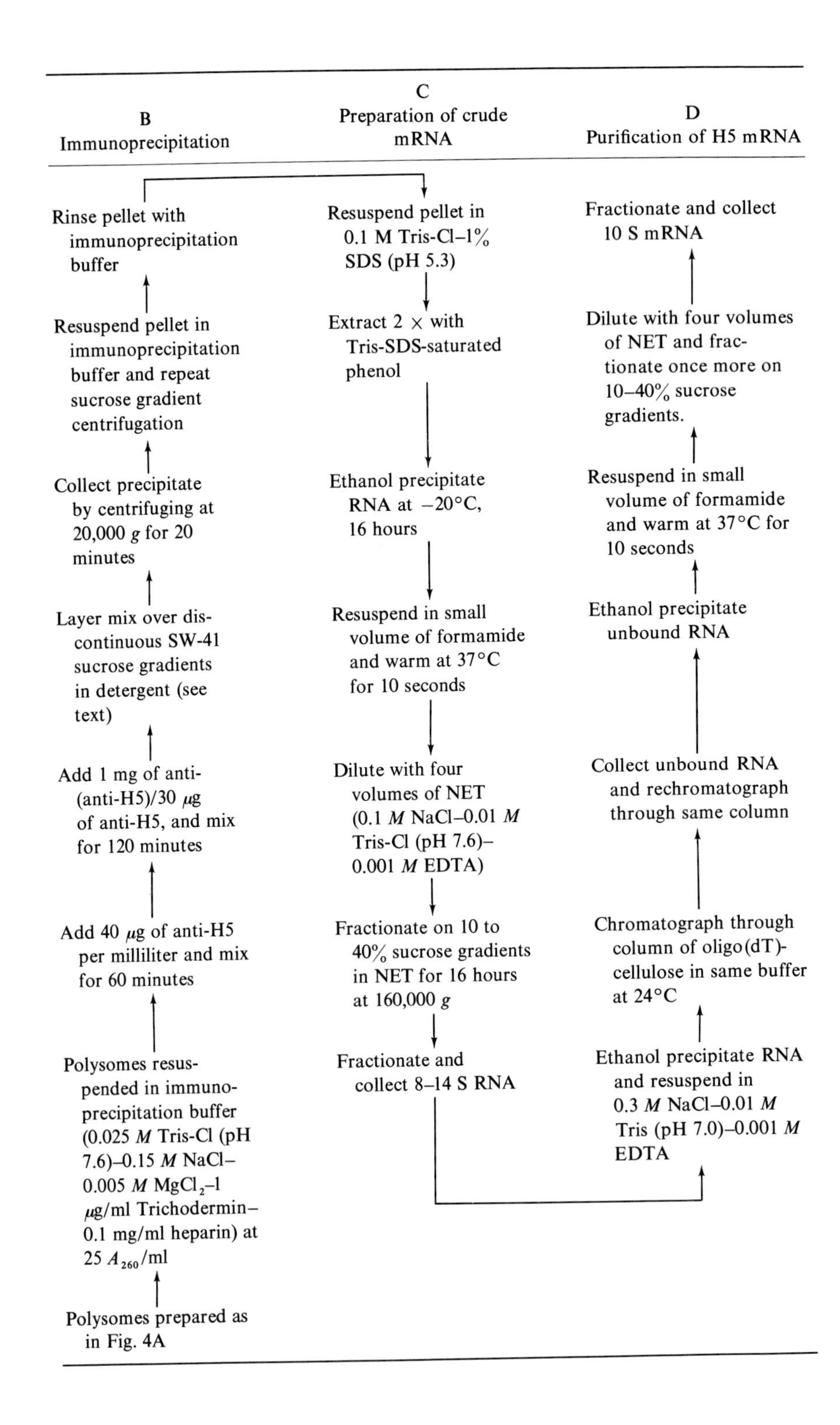

B
Immunoprecipitation
C
Preparation of crude mRNA
D
Purification of H5 mRNA
Polysomes prepared as in Fig. 4A
Polysomes resuspended in immunoprecipitation buffer (0.025 M Tris-Cl (pH 7.6)–0.15 M NaCl–0.005 M $MgCl_2$–1 μg/ml Trichodermin–0.1 mg/ml heparin) at 25 $A_{260}$/ml
Add 40 μg of anti-H5 per milliliter and mix for 60 minutes
Add 1 mg of anti-(anti-H5)/30 μg of anti-H5, and mix for 120 minutes
Layer mix over discontinuous SW-41 sucrose gradients in detergent (see text)
Collect precipitate by centrifuging at 20,000 g for 20 minutes
Resuspend pellet in immunoprecipitation buffer and repeat sucrose gradient centrifugation
Rinse pellet with immunoprecipitation buffer
Resuspend pellet in 0.1 M Tris-Cl–1% SDS (pH 5.3)
Extract 2 × with Tris-SDS-saturated phenol
Ethanol precipitate RNA at –20°C, 16 hours
Resuspend in small volume of formamide and warm at 37°C for 10 seconds
Dilute with four volumes of NET (0.1 M NaCl–0.01 M Tris-Cl (pH 7.6)–0.001 M EDTA)
Fractionate on 10 to 40% sucrose gradients in NET for 16 hours at 160,000 g
Fractionate and collect 8–14 S RNA
Ethanol precipitate RNA and resuspend in 0.3 M NaCl–0.01 M Tris (pH 7.0)–0.001 M EDTA
Chromatograph through column of oligo(dT)-cellulose in same buffer at 24°C
Collect unbound RNA and rechromatograph through same column
Ethanol precipitate unbound RNA
Resuspend in small volume of formamide and warm at 37°C for 10 seconds
Dilute with four volumes of NET and fractionate once more on 10–40% sucrose gradients.
Fractionate and collect 10 S mRNA

and complicated. As a result, stringent precautions against ribonucleases degradation, such as carrying out the reaction at 0° C, and using heparin as a ribonuclease inhibitor, are employed. Owing to its technical difficulty, immunological methods are unlikely to supplant other mRNA isolation techniques where these are appropriate. The great virtue of such immunological techniques, lies in their ability to effect the purification of minor mRNA species such as H5 mRNA, which are otherwise impossible. The state of the art is still young and further study may well find ways of simplifying these procedures. One might anticipate an improvement in the specificity of the polysome–antibody reaction with the use of antibody specific for N-terminal fragments represented in nascent polypeptides (*21*).

The present work in our laboratory is involved in using the H5 mRNA as a reagent for investigating gene control. For example, we can prepare polyadenylated H5 mRNA *in vitro*, and from it, produce H5 cDNA. This cDNA has been used to calculate the gene reiteration frequency of H5 in the chicken genome (*20*). We are now investigating the possible linkage relationship of the gene for this unusual histone and the other histone genes.

## REFERENCES


*1.* Lebleu, B., Marbaix, G., Huez, G., Temmerman, J., Burny, A., and Chantrenne, H., *Eur. J. Biochem.* **19**, 264 (1971).
*2.* Jacobs-Lorena, M., and Baglioni, C., *Proc. Natl. Acad. Sci. U.S.A.* **69**, 1425 (1972).
*3.* Suzuki, Y., and Brown, D. D., *J. Mol. Biol.* **63**, 409 (1972).
*4.* Swan, D., Aviv, H., and Leder, P., *Proc. Natl. Acad. Sci. U.S.A.* **69**, 1967 (1972).
*5.* Aviv, H., and Leder, P., *Proc. Natl. Acad. Sci. U.S.A.* **69**, 1408 (1972).
*6.* Appels, R. (1971). Ph.D. Thesis, University of Adelaide.
*7.* Appels, R., and Wells, J. R. E., *J. Mol. Biol.* **70**, 425 (1972).
*8.* Johns, E. W., *in* "Histones and Nucleohistones" (D. M. P. Phillips, ed.), p. 2. Plenum, New York, 1971.
*9.* Scott, A. C., and Wells, J. R. E., *Biochem. Biophys. Res. Commun.* **64**, 448 (1975).
*10.* Scott, A. C. (1976). Ph.D. Thesis, University of Adelaide.
*11.* Hartlief, R., and Koningsberger, V., *Biochim. Biophys. Acta* **166**, 512 (1968).
*12.* Holme, G., Delovitch, T. L., Boyd, S. L., and Sehon, A. H., *Biochim. Biophys. Acta* **274**, 104 (1971).
*13.* Delovitch, T. L., Davis, B. K., Holme, G., and Sehon, A. H., *J. Mol. Biol.* **69**, 373 (1972).
*14.* Palacios, R., Sullivan, D., Summers, N. M., Kiely, M. L. and Schimke, R. T., *J. Biol. Chem.* **248**, 540 (1973).
*15.* Schechter, I., *Biochemistry* **13**, 1875 (1974).
*16.* Shapiro, D. J., Taylor, J. M., McKnight, G. S., Palacios, R., Gonzalez, C., Kiely, M. L., and Schimke, R. T., *J. Biol. Chem.* **249**, 3665 (1974).


FIG. 8. Flow diagram of the method for preparing H5 mRNA by immunoadsorption. All operations were performed at 0°C unless otherwise stated.

17. Sidorova, E. V., Trudolyubova, M. G., and Lerman, M. I., *Mol. Biol. Rep.* **1**, 401 (1974).
18. March, S. C., Parikh, I., and Cuatrecasas, P., *Anal. Biochem.* **60**, 149 (1974).
19. Thrall, C. L., Park, W. D., Rashba, H. W., Stein, J. L., Mans, R. J., and Stein, G. S., *Biochem. Biophys. Res. Commun.* **61**, 1443 (1974).
20. Scott, A. C., and Wells, J. R. E., *Nature* (*London*) **259**, 635 (1976).
21. Young, N. S., Curd, J. G., Eastlake, A., Furie, D., and Schechter, A. N., *Proc. Natl. Acad. Sci. U.S.A.* **72**, 4759 (1975).

# Chapter 24

# *Enrichment and Purification of Sea Urchin Histone Genes*

ERIC S. WEINBERG AND G. CHRISTIAN OVERTON[1]

*Department of Biology,*
*Johns Hopkins University,*
*Baltimore, Maryland*

## I. Introduction

The purification of specific reiterated genes has contributed much to the understanding of genome organization in higher organisms. The ribosomal genes of *Xenopus laevis* were first purified by CsCl density gradient centrifugation (*1,2*). Equilibrium density gradient centrifugation has been especially useful in purifying other tandemly repeated genes such as 5 S ribosomal DNA (*3*), tRNA genes (*4*), and sea urchin histone genes (*5*). These procedures, using standard cesium salt gradients as well as the binding of actinomycin and heavy metal ions to DNA [reviewed in Brown and Stern (*6*)], depend on differences in base composition and sequence for the purification of particular genes.

Recent advances in purification of eukaryotic gene sequences via plasmid and phage vectors and cloning procedures have made it possible to isolate, in a pure state, both reiterated and unique genes (*7–9*). Restriction enzyme-produced fragments of histone genes from the sea urchins *Strongylocentrotus purpuratus*, *Lytechinus pictus* (*10*) and *Psammechinus miliaris* (*11*) have been isolated in this manner. We have followed the methods of Kedes *et al.* (*10*) with success. The reader is advised to refer to this publication for methods dealing with this approach to the isolation of histone genes.

[1] *Present address:* The Wistar Institute, Philadelphia, Pennsylvania.

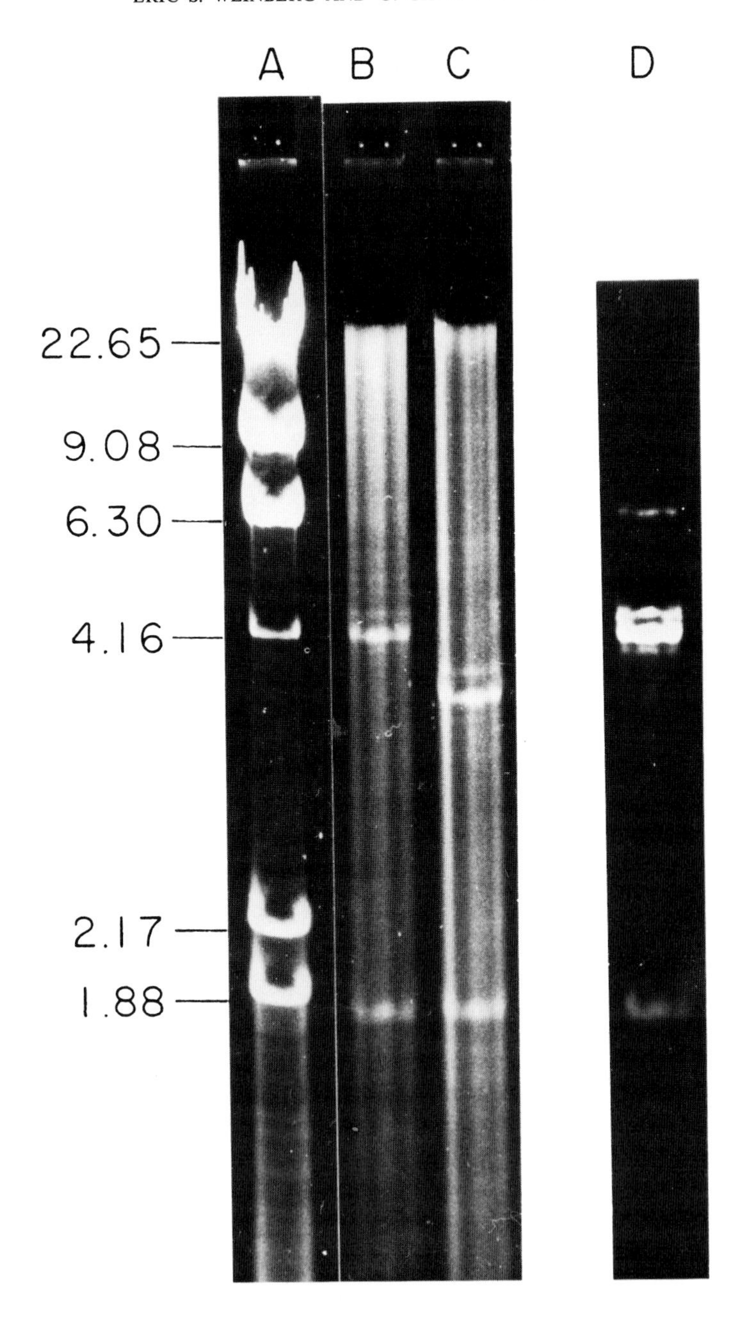
A
B
C
D
22.65
9.08
6.30
4.16
2.17
1.88

## II. Advantages of Physical Methods in Gene Purification

There are circumstances in which it is advantageous to use physical techniques to isolate reiterated genes. Tandemly repeated sequences often can be isolated as a satellite in various types of density gradients. Restriction enzyme digestion of this DNA yields information that may be difficult to obtain from cloning procedures alone. Centrifugation to equilibrium in actinomycin-CsCl and Hg-$Cs_2SO_4$ gradients (*5*) yields sufficiently pure histone DNA to do restriction enzyme mapping (*11–14*). The histone genes of *S. purpuratus* have been shown to consist of two Eco RI-produced fragments of approximately 1.8 and 4.3 kb both by digestion of gradient-enriched histone satellite DNA (referred to here as hDNA) (*12, 14*) and by analysis of cloned fragments (*10, 14*). Digestion of the hDNA with Hin dIII yields a fragment of approximately 6.1 kb, the sum of the sizes of the two RI fragments. Partial digests of the hDNA with Hin dIII (*12*) have shown that the 6.1 kb unit is the predominant repeating sequence. This information would be more difficult to obtain by analysis of cloned fragments alone.

The question of how the two Eco RI fragments are oriented with respect to each other is best answered using the purified hDNA. N. Kunkel in our laboratory found an HpaI site in the large Eco RI fragment, about 0.64 kb from one of the Eco RI sites (Fig. 1, lanes B, C). Similarly it has been shown that the small RI fragment contains a Hin dIII site about 0.1–0.15 kb from one of the RI sites (*12, 14*). Depending on the orientation of the Eco RI fragments, the two alternatives illustrated in Fig. 2 are possible. Sequential digestion of hDNA with HpaI and Hin dIII gives a fragment of 5.30 kb (Fig. 3), indicating that alternative A is correct. Now other restriction enzyme sites may be mapped relative to this scheme by using hDNA or the two cloned Eco RI fragments.

Another advantage of studying the hDNA is that it can be used to demonstrate the presence of heterogeneity of repeat length. Investigation of this property with individually cloned fragments would be laborious for rare length classes. Using Eco RI-digested *S. purpuratus* hDNA, substantial length heterogeneity of the larger Eco RI fragment can be demonstrated by hybridization with [$^3$H]histone mRNA (Fig. 1, lane D).

---

FIG. 1. Agarose gel electrophoresis of histone DNA (about 50-fold enriched) digested with restriction enzymes. (A) Hin dIII $\lambda$DNA standard (molecular weights given in the margin are kilobase pairs, as determined by electron microscopy by Dana Carroll). (B) Histone DNA digested with Eco RI (New England Biolabs). (C) Histone DNA digested with Eco RI and HpaI (New England Biolabs). (D) Hybridization of [$^3$H]histone mRNA ($2 \times 10^5$ cpm) to the DNA of lane (B), which had been transferred to a Millipore strip by the Southern technique, (*15*) and visualized by fluorography (*16*).

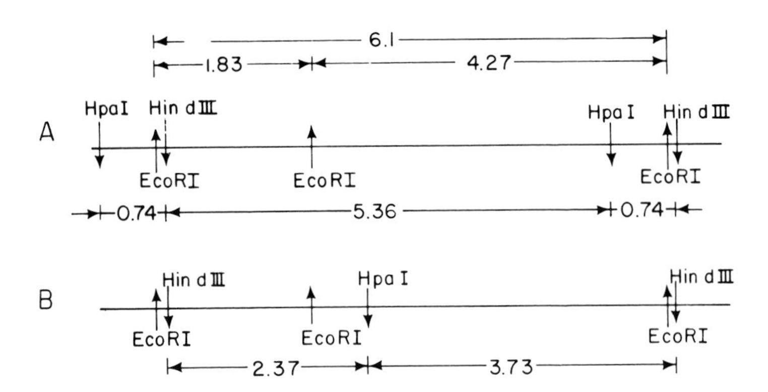


FIG. 2. Possible arrangements of the Eco RI histone fragments. Alternative A is correct. Numbers refer to kilobase pairs.

There may be cases in which a high degree of purity is required before cloning. A protocol of density gradient, R-loop (*26*), and restriction enzyme-agarose gel methods might allow some sequences to be isolated in virtually pure form, without the use of cloning procedures. Presented here are the physical methods we have used to purify the *S. purpuratus* histone genes.

## III. Methods of Purification and Assay of Histone DNA

### A. Preparation of [$^3$H]Histone mRNA

The assay for purification of histone DNA involves hybridization with labeled histone mRNA. The methods we use to obtain [$^3$H]histone mRNA of high specific radioactivity are modifications of previously published procedures (*17–19*).

Eggs and sperm are obtained from excised gonads of *S. purpuratus* (Pacific Biomarine Co.). The eggs are washed several times with seawater, fertilized with sperm, and allowed to develop to the 4–8-cell stage. The cultures are usually started with 1–2 ml of packed eggs suspended in 100–200 ml of seawater. After two or three cleavages, 5 mCi of [$^3$H]uridine (20,000–30,000 mCi/mmole, Amersham/Searle) is added and development allowed to proceed to early blastula (12 hours after fertilization).

Embryos are washed three times with an ice-cold homogenization buffer [0.4 *M* KCl–0.01 *M* $MgCl_2$–0.05 *M* Tris (pH 7.8)]. The embryos are then homogenized in 3–4 volumes of this buffer in a tight-fitting Dounce tissue grinder. After centrifugation at 10,000 rpm for 10 minutes in the Sorvall

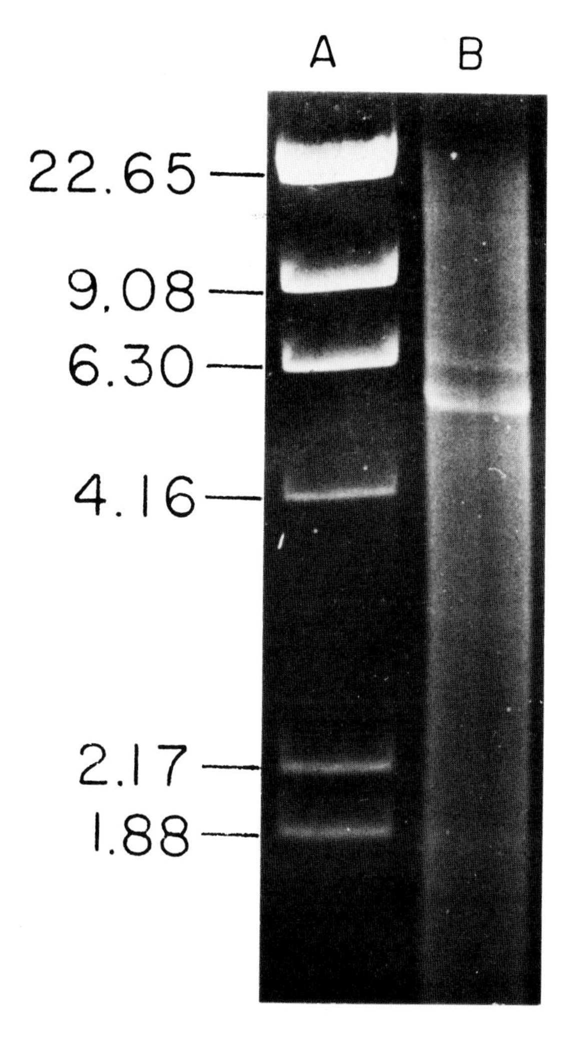


FIG. 3. Agarose gel electrophoresis of histone DNA digested sequentially with Hin dIII and HpaI. (A) λDNA Hin dIII fragments (kilobase values given in margin) (B) Hin dIII (Miles)–HpaI-digested DNA. The presence of a 6.1 kilobase pairs fragment indicates an incomplete digest with one of the enzymes.

HB-4 rotor to remove nuclei, mitochondria, and unbroken cells, the supernatant is layered over 30% sucrose cushions (3 ml) in SW-50.1 tubes, and centrifuged for 2 hours at 45,000 rpm at 4°C. The resulting glassy polysomal pellet may be rinsed with a small volume of the buffer and dissolved in 0.1% sodium lauryl sulfate (SLS). At this point it is no longer necessary to keep the preparation cold.

The polysomal solution is extracted twice with water-saturated phenol and once with chloroform–isoamyl alcohol (24:1). The aqueous phase is made 0.3 *M* in NaCl, the RNA is precipitated at −20°C after the addition of 2 volumes of 95% ethanol, then centrifuged at 10,000 rpm for 30 minutes in the Sorvall HB-4 rotor and immediately dissolved in 0.1% SLS. About 1 ml of RNA (1 mg/ml) is layered over a 15 to 30% sucrose gradient, which is centrifuged for 22 hours at 24,000 rpm in the SW-27 rotor at 20°C. A typical absorbance profile and distribution of radioactivity of such a gradient can be seen in Fig. 4. Fractions containing the 9 S peak of [$^3$H]RNA are pooled, precipitated with ethanol, and redissolved in 0.1% SLS. The RNA can be further fractionated on acrylamide gels (*12, 14, 20*) to obtain specific histone mRNAs, or it can be used directly in hybridization reactions to assay for the presence of histone DNA.

## B. RNA–DNA Hybridization Procedure

The filter hybridization procedure of Gillespie and Spiegelman (*21*) as adapted by Birnstiel *et al.* (*22*) is used. Millipore HAWP filters, 13 mm in diameter, are soaked in 2 × SSC [0.15 *M* NaCl–0.015 *M* sodium citrate (pH 7.0)] and placed into homemade plastic holders. Aliquots of DNA solutions (e.g., fractions of density gradients) are brought to 1 ml with 0.1 × SSC and are denatured by the addition of an equal volume of 1 *M* NaOH. The solution is neutralized 15 minutes later by addition of about 2.5 volumes of a solution made of 1 part 1 *M* Tris (pH 8.0), 2 parts 3 *M* NaCl, and 1 part 1.27 *N* HCl and is then immediately dripped through the filters. The filters are then washed with 5 ml of 2 × SSC, dried for 2 hours in a vacuum oven at 80°C, and numbered with a soft lead pencil.

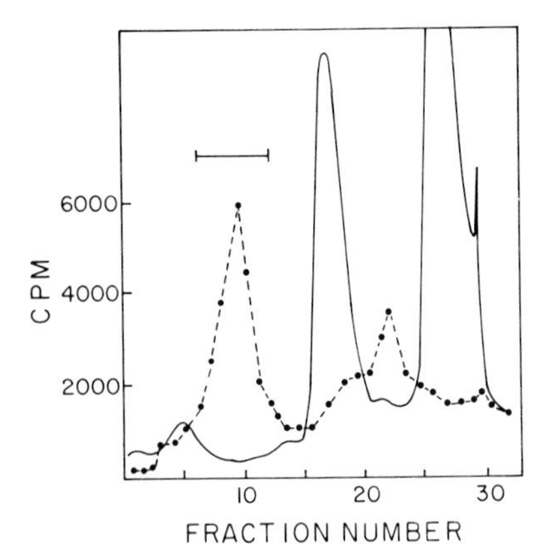


FIG. 4. Sucrose gradient of sea urchin blastula $^3$H-labeled polysomal RNA. Exact details are given in the text. For scintillation counting 10 μl of each 1.2-ml fraction was used. —$A_{260nm}$, 2 units full scale: ---, cpm. Top of gradient is to left.

The filters are presoaked in 2 × SSC, 0.1% SLS and stacked in a glass vial with an inside diameter of 14 mm, containing the [$^3$H]histone mRNA in 2 × SSC, 0.1% SLS. The solution is brought to a volume sufficient to cover the stack of filters, and mineral oil is overlayed to prevent evaporation during hybridization. The vial is kept at 55°–65°C for at least 12 hours. The filters are then washed in batchwise fashion (*23*) with 6 × SSC and several changes of 2 × SSC and are finally treated with RNase A [Sigma, pancreatic, preboiled in 2% sodium acetate (pH 6.0)] at 10 μg/ml in 2 × SSC. The filters are then dried and counted in a toluene fluor in a scintillation counter.

## C. Preparation of Sea Urchin DNA

We have found the procedures worked out in Stafford's laboratory (*24,25*) to be very reliable and have used them with slight modification. Sperm are obtained by soaking the testes of 4–5 male *S. purpuratus* in ice-cold 0.5 *M* NaCl–0.03 *M* KCl. The suspension is strained through 4 layers of cheesecloth, and the sperm are washed with this solution after repeated centrifugations in the Sorvall HB-4 rotor at 5000 rpm for 5 minutes. The sperm (5–7 ml) can be used immediately or stored in liquid nitrogen after suspension in an equal volume of glycerol.

A tube may be removed from liquid nitrogen and pulverized with a hammer. Five hundred milliliters of a lysing buffer [0.1 *M* Tris (pH 8.0)–0.01 *M* EDTA–0.01 *M* NaCl–0.5% Sarcosyl] is autoclaved and cooled to 60°C. Then 50 mg of proteinase K (EM Laboratories, Inc.) is added, and the solution is mixed with the pulverized sperm. The suspension is incubated at 60°C for 2 hours, with occasional stirring, and then shaken for 4 hours at room temperature on a New Brunswick rotary shaker at 120 rpm.

The suspension is extracted with phenol–chloroform–isoamyl alcohol (25:24:1) presaturated with buffer. The phenol (Baker, crystalline) is twice distilled before use. After centrifugation at 5000 rpm in the Sorvall GSA rotor to separate phases, the aqueous phase is reextracted 2 or 3 times. Since the viscosity of the DNA solution is very high, it is sometimes useful to add additional buffer prior to the extraction. The interface is excluded after the first extraction.

The DNA is dialyzed into a shearing buffer [1 *M* NaCl–30 m*M* Tris (pH 8.5)–10 m*M* EDTA] and the concentration is adjusted to 250 μg/ml; 300 ml of a DNA solution is sheared in a 600–ml Sorvall Omni-mix vessel at 4000 rpm for 8 hours at 0°C. The vessel is occasionally removed and shaken to ensure even mixing. The final molecular weight, determined by velocity sedimentation in the Model E centrifuge, is about 20 × $10^6$ for double-stranded DNA.

## D. Density Gradient Purification

Actinomycin-CsCl and Hg-$Cs_2SO_4$ gradients are both useful in purifying histone DNA from the sea urchin, *P. miliaris* (*5*). We have adapted these methods for the enrichment of hDNA from *S. purpuratus*. We have used a succession of Hg-$Cs_2SO_4$ and actinomycin-CsCl gradients. Each type of gradient, when initially used to purify hDNA, gives a 4–10-fold enrichment. Our practice has been to use one or two rounds of purification in Hg-$Cs_2SO_4$ followed by two or three rounds in actinomycin-CsCl. These procedures have yielded up to 100-fold purification of hDNA. We estimate that, at this purification, the hDNA is approximately 25% of the sample (*12*).

### 1. Hg-$Cs_2SO_4$ Gradients

DNA, purified and sheared as described above, is dialyzed into a borate buffer [2.5 m*M* $Na_2B_4O_7$ (pH 9.2)–0.1 *M* $Na_2SO_4$–0.5% Sarcosyl]. The Sarcosyl is added to guard against any nuclease activity and does not seem to affect the banding of the DNA in the gradient. The DNA is kept at a concentration of approximately 250 $\mu$g/ml.

For each sheared DNA preparation, a series of different Hg concentrations are tested to determine the optimum separation of the hDNA. Typical Hg:nucleotide ratios for the test runs are 0.25, 0.275, 0.30, 0.325. The refractive index of the solution to be centrifuged is also tested, with optima found to be in the range of 1.3760–1.3780. For the purification series shown in Fig. 5, a Hg:nucleotide ratio of 0.325 was used, and for the series in Fig. 6, the ratio was 0.275. The refractive index of the solutions was adjusted to 1.3765 and 1.3780 respectively.

A typical purification series is illustrated in Fig. 6. For each centrifuge tube (20 tubes were run), 4.5 ml of a DNA solution (220 $\mu$g/ml) and 1.675 ml of the borate buffer are mixed, then 0.825 ml of a freshly made solution of 1 m*M* $HgCl_2$ is added dropwise. The solution is left in the dark for 20–30 minutes to equilibrate, 7 ml of a saturated $Cs_2SO_4$ solution is added, and the refractive index is adjusted to 1.3780. Before use, the $Cs_2SO_4$ (Gallard Schlessinger, Analar) solution is filtered twice through Whatman No. 5 paper, Millipore HA (0.45 $\mu$m) and Millipore GS (0.22 $\mu$m) to remove impurities, and concentrated to saturation by boiling. The 14 ml of solution are added to each polycarbonate tube and centrifuged in the 60 Ti rotor at 28,000 rpm or the Type 30 rotor at 25,000 rpm for 3 days at 20°C. The gradients are collected in 0.6-ml fractions with an Isco gradient fractionator. For hybridization, 50 $\mu$l of each fraction is denatured, loaded onto filters, and hybridized as described above. As determined from the hybridization results of one gradient (Fig. 6A) or the hybridization of the initial test gradient, the first 8 fractions of each of 20 tubes are combined and recentrifuged in a second round of Hg–$Cs_2SO_4$. The total volume (85 ml) is split

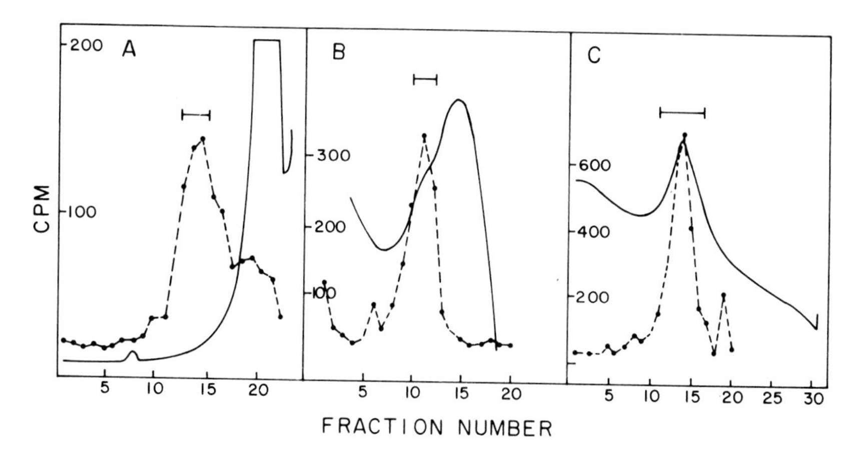


FIG. 5. Sequential purification of histone DNA by one round of Hg-$Cs_2SO_4$ (A) and three rounds of actinomycin-CsCl gradients, the first (B) and third (C) of which are shown. Conditions are described in the text; 50 $\mu l$ of each fraction of the first round and 25 $\mu l$ of subsequent round were used for hybridization. ——, $A_{260nm}$, full scale, 2 units; -----, cpm of [$^3H$]histone mRNA hybridized. Top of gradients are to left.

equally into 4 polycarbonate tubes and centrifuged in the 60 Ti rotor at 28,000 rpm for 48–72 hours at 20°C. The gradients are collected and aliquots are hybridized as described above (Fig. 6B).

## 2. ACTINOMYCIN-CsCl GRADIENTS

The peak fractions from one (Fig. 5) or two (Fig. 6) rounds of Hg–$Cs_2SO_4$ gradients are pooled and dialyzed against 25 m$M$ $Na_2B_4O_7$–2.0 m$M$ EDTA

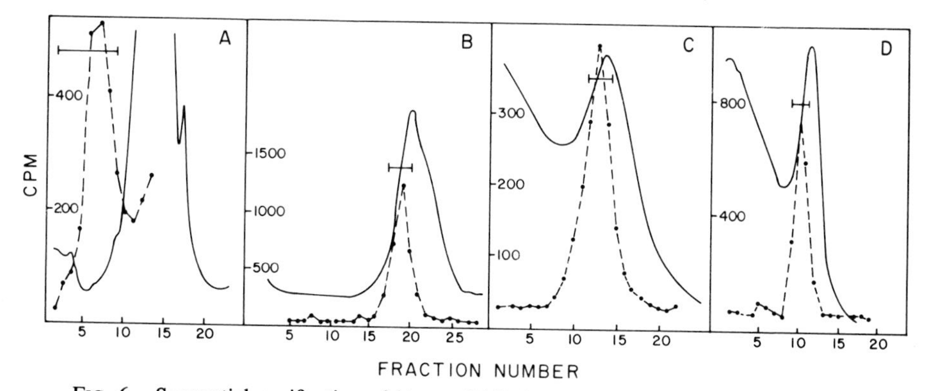


FIG. 6. Sequential purification of histone DNA by two rounds of Hg-$Cs_2SO_4$ (A, B) and two rounds of actinomycin–CsCl gradients (C, D). Exact details are given in text. Of each fraction, 50 $\mu l$ of the first round and 25 $\mu l$ of subsequent rounds were used for hybridization. ——, $A_{260nm}$, full scale, 2 units; -----, cpm of [$^3H$] histone mRNA hybridized. Top of gradients are to left.

(pH 9.2). If necessary, the DNA can be concentrated at this stage with *sec*-butanol (*24*) or by rapid evaporation from a fan-blown dialysis tube.

The actinomycin–CsCl gradients are run at 4°C (*5*). All solutions to be used in the preparation of the gradient are kept at this temperature. Again, using the purification series illustrated in Fig. 6 as an example, the DNA is brought to 16.5 ml with a borate buffer [25 m*M* $Na_2B_4O_7$ (pH 9.2)–2 m*M* EDTA]. With constant stirring, 1.815 ml of actinomycin D solution (1 mg/ml, Merck) is added dropwise to the DNA. The final concentration of actinomycin is then 0.1 mg/ml. The preparation is left in the dark for 30 minutes, then 17.4 gm of solid CsCl (Merck, optical grade) are added and the refractive index is adjusted to 1.3893. The solution is added to a polycarbonate tube, which is centrifuged in the 60 Ti rotor at 28,000 rpm for 48–72 hours at 4°C. The gradients are collected and assayed as described above (Fig. 6C). The peak fractions are pooled for the next round of purification.

It is necessary to remove the actinomycin between gradient rounds. The DNA solution is dialyzed against 0.01 *M* Tris (pH 8.5)–1 m*M* EDTA, and the actinomycin is removed (as the DNA is concentrated) by shaking with *sec*-butanol. The DNA is again dialyzed against the 25 m*M* borate buffer. 13.5 ml of the DNA solution is mixed with 1.5 ml of actinomycin D (1 mg/ml) to which 14 gm of solid CsCl are added. The refractive index is again adjusted to 1.3893, and the solution is centrifuged in the 60 Ti rotor at 33,000 rpm for 48 hours at 4°C. The gradients are collected and hybridized once again (Fig. 6D).

The DNA is then pooled, dialyzed against 0.01 *M* Tris (pH 8.5) 1 m*M* EDTA, extracted with *sec*-butanol, and redialyzed into this buffer. If another round of actinomycin-CsCl gradient centrifugation is desired, the procedure described above may be repeated.

### 3. Estimation of Enrichment

The calculation of the degree of enrichment at each stage may be made by comparing the hybridization of [$^3$H]histone mRNA to the purified DNA (per microgram) with the hybridization to unfractionated DNA (per microgram) (*5*). As long as the RNA is far from depletion in the reaction vessel, this comparison will yield the enrichment factor. The data and calculations made for the purification shown in Fig. 5 are illustrated in Table I.

An estimation of the degree of enrichment may also be made by noting the size distribution of the DNA after restriction enzyme digestion. Since the enzyme Hin dIII cuts the histone genes once per repeating unit into fragments of 6.1 kb, the amount of DNA of this size class after enzyme digestion will monitor the degree of purification of the hDNA. Agarose gels of digested DNA of different degrees of purity are shown in Fig. 7. Such agarose

TABLE I
CALCULATION OF ENRICHMENT OF HISTONE DNA[a]

| DNA | Amount of sample (mg)[b] | μg DNA/filter[c] | [$^3$H]RNA[d] hybridized (cpm) | Enrichment[e] |
|---|---|---|---|---|
| *Micrococcus lysodeikticus* (background filter) | — | 8.0 | 8 | — |
| Total *Strongylocentrotus purpuratus* | 59.7 | 4.52 | 83 | 1 |
| Hg-$Cs_2SO_4$ | 1.3 | 1.0 | 102 | 5.7 |
| First actinomycin-CsCl | 0.140 | 0.37 | 96 | 14.5 |
| Second actinomycin-CsCl | 0.0314 | 0.27 | 264 | 58 |

[a] Filters were prepared with the above amounts of untreated or enriched DNA and hybridized with 40,000 cmp of [$^3$H]histone mRNA in 6 × SSC at 65°C for 17 hours. In the purification of the DNA, only 70% of the material obtained in the Hg-$Cs_2SO_4$ gradient step was applied to the actinomycin-CsCl gradient.

[b] Total amount of sample indicates the initial amount of DNA and the amount recovered after each of the indicated purification steps.

[c] The amounts indicated were mixed with sufficient *M. lysodeikticus* DNA so that a total of 8 μg of DNA were placed on each filter.

[d] Average of 3 filters, 8 count background already subtracted from sea urchin DNA filters.

[e] Enrichment was calculated by determining the ratio of counts per minute hybridized (after subtraction of background) per microgram of enriched DNA to that of unfractionated DNA.

gels can also be used as a preparative tool. Extraction of the DNA in the 6.1 kb band yields a preparation greatly enriched in hDNA.

## 4. LIMITATIONS OF THE METHOD

In general, the purification of hDNA by physical techniques is well suited to restriction enzyme analysis and agarose gel electrophoresis. However, it may present certain limitations for other experimental approaches. Although hDNA purified by physical procedures may be used for surveying the amount of length heterogeneity, it is less suitable for electron microscopy. One problem, which may be overcome with suitable precautions, is the tendency for $Hg^{2+}$ ions to produce nicks in the individual strands of a DNA duplex, especially in the presence of light. For this reason we recommend that the DNA be exposed to as little light as possible during the gradient steps and that the $Hg^{2+}$ be removed quickly with a chelating agent such as EDTA after the final round of Hg–$Cs_2SO_4$ purification. In this way, single-stranded hDNA of 6–8 × $10^6$ molecular weight has been obtained.

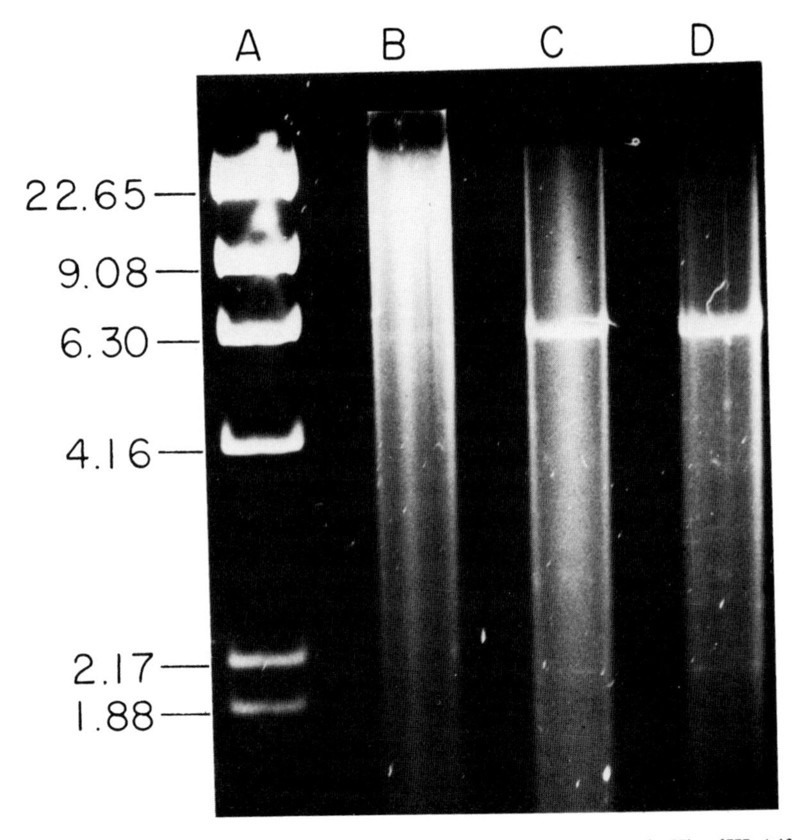


FIG. 7. Agarose gel electrophoresis of histone DNA digested with Hin dIII (gift of D. Carroll). (A) $\lambda$DNA Hin dIII standard. (B) Histone DNA from one round of Hg-$Cs_2SO_4$ gradients (7-fold enriched). (C) Histone DNA after an additional two rounds of actinomycin–CsCl gradients (30-fold enriched). (D) Histone DNA after 3 rounds of actinomycin CsCl gradients (70-fold enriched).

A more serious problem is that gradient methods, alone, do not yield a pure sample of histone DNA. If the histone genes comprise 0.25–0.45% of the total DNA (*12*), then the data of Table I indicate that, in the final sample of only 15–19% purity, nearly 95% of the initial histone DNA was lost. There is also the possibility that these fractionation methods may preferentially select for purification of certain classes of histone genes, although our preliminary experiments seem to indicate that the purified sample is representative of the total histone gene population. Although the methods used alone have obvious drawbacks, in combination with restriction enzyme-agarose gel electrophoresis and R-loop formation (see below), they are useful in producing a high degree of purity and a reasonable yield of hDNA.

## E. R-Loop Formation

Since an RNA–DNA hybrid is more stable than the corresponding DNA duplex (*22*), it is possible to form a DNA–RNA triplex, in which a segment of the DNA duplex is displaced by hybridization of the RNA to the complementary DNA strand (*26*). This procedure has been used to analyze the organization of sea urchin histone genes by electron microscopy (*27*). We have adapted the procedures Wellauer and Dawid (*28*) used to isolate the *Drosophila* ribosomal genes, to further purify the histone genes of *S. purpuratus*.

DNA previously enriched by actinomycin–CsCl and/or Hg–$Cs_2SO_4$ gradients is dialyzed against a Tricine buffer [33 m*M* Tricine (Calbiochem) (pH 8.5)–0.99 M NaCl–3.3 *M* EDTA]. The DNA should be previously sheared (see above) to achieve the best results. A typical experiment consists of mixing 35 μg of 15-fold-enriched histone DNA in 450 μl of Tricine buffer with 28 μg of unlabeled histone mRNA (prepared essentially as described for the [$^3$H]mRNA) in 50 μl of Tricine buffer; 1.66 ml of 100% formamide (Fisher) is then added, bringing the final formamide concentration to 70% and NaCl concentration to 0.3 *M*. The solution is mixed well, overlaid with mineral oil, and left at 51°C for 12 hours.

The solution is dialyzed against two changes of 10 m*M* Tris (pH 8.0)–1 m*M* EDTA at 4°C. The volume is adjusted to 2.67 ml by addition of Tris–EDTA buffer, and 2.0 ml of saturated $Cs_2SO_4$ (see above) is added. The refractive index is adjusted to 1.3702 at 25°C, and the solution is spun in a polycarbonate tube at 29,000 rpm in the Type 50 rotor for 48 hours at 25°C. The gradient is then collected with an Isco gradient fractionator and the histone DNA is assayed by denaturing an aliquot of each fraction and hybridizing to [$^3$H] histone mRNA. Figure 8 shows the great enrichment of histone DNA in the central region of the gradient.

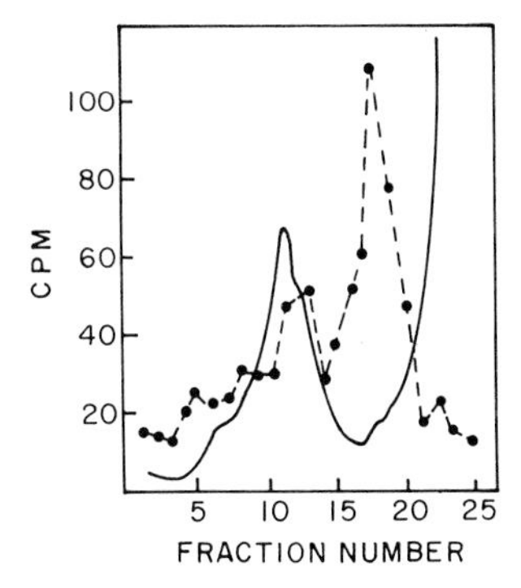


FIG. 8. $Cs_2SO_4$ gradient separation of R-looped histone DNA. See text for exact details. Top of gradient is to left.

## Acknowledgment

This work was suported by NIH grant GM2215 to E. W. and by fellowships from the NSF and NIH training grant GM00716 to G. C. Overton. This work is in partial fulfillment of the doctoral degree requirements of G. C. Overton.

## References

1. Birnstiel, M. L., Wallace, H., Serlin, J., and Fischberg, M., *Natl. Cancer Inst., Monogr.* **23**, 431 (1966).
2. Wallace, H., and Birnstiel, M. L., *Biochim. Biophys. Acta* **114**, 296 (1966).
3. Brown, D. D., Wensink, P. C., and Jordan, E., *Proc. Natl. Acad. Sci. U.S.A.* **68**, 3175 (1971).
4. Clarkson, S. G., and Kurer, V., *Cell* **8**, 183 (1976).
5. Birnstiel, M. L., Telford, J., Weinberg, E. S., and Stafford, D., *Proc. Natl. Acad. Sci. U.S.A.* **71**, 2900 (1974).
6. Brown, D. D., and Stern, R., *Annu. Rev. Biochem.* **43**, 667 (1974).
7. Morrow, J. F., Cohen, S. N., Chang, A. C. Y., Boyer, H. W., Goodman, H. M., and Helling, R. B., *Proc. Natl. Acad. Sci. U.S.A.* **71**, 1743 (1974).
8. Wensink, P. C., Finnegan, D. J., Donelson, J. E., and Hogness, D. S., *Cell* **3**, 315 (1974).
9. Glover, D. M., White, R. L., Finnegan, D. J., and Hogness, D. S., *Cell* **5**, 149 (1975).
10. Kedes, L. H., Chang, A. C. Y., Houseman, D., and Cohen, S. N., *Nature (London)* **255**, 533 (1975).
11. Schaffner, W., Gross, K., Telford, J., and Birnstiel, M., *Cell* **8**, 471 (1976).
12. Weinberg, E. S., Overton, G. C., Shutt, R. H., and Reeder, R. H., *Proc. Natl. Acad. Sci. U.S.A.* **72**, 4815 (1975).
13. Birnstiel, M. L., Gross, K., Schaffner, W., and Telford, J., *Fed. Enr. Biochem. Soc. Meet., 10th, 1975* Vol. 38, p. 3 (1975).
14. Kedes, L. H., Cohen, R. H., Lowry, J. C., Chang, A. C. Y., and Cohen, S. N., *Cell* **6**, 359 (1975).
15. Southern, E., *J. Mol. Biol.* **98**, 503 (1975).
16. Laskey, R., and Mills, A., *Eur. J. Biochem.* **56**, 335 (1975).
17. Rinaldi, A. M., and Monroy, A., *Dev. Biol.* **19**, 73 (1969).
18. Kedes, L. H., and Birnstiel, M. L., *Nature (London), New Biol.* **230**, 165 (1971).
19. Weinberg, E. S., Birnstiel, M. L., Purdom, I. M., and Williamson, K., *Nature (London)* **240**, 225 (1972).
20. Gross, K., Probst, E., Schaffner, W., and Birnstiel, M. L., *Cell* **8**, 455 (1976).
21. Gillespie, D., and Spiegelman, S., *J. Mol. Biol.* **12**, 829 (1965).
22. Birnstiel, M. L., Sells, B. H., and Purdom, I. F., *J. Mol. Biol.* **63**, 21 (1972).
23. Birnstiel, M. L., Speirs, J., Purdom, I., Jones, K., and Loening, U. E., *Nature (London)* **219**, 454 (1968).
24. Stafford, D. W., and Bieber, D., *Biochim. Biophys. Acta* **378**, 18 (1975).
25. Joseph, D. R., and Stafford, D. W., *Biochim. Biophys. Acta* **418**, 167 (1976).
26. White, R. L., and Hogness, D. S., *Cell* **10**, 177 (1977).
27. Holmes, D. S., Cohn, R. H., Kedes, L. H., and Davidson, N., *Biochem* **16**, 1504 (1977).
28. Wellauer, P. K. and Dawid, I. B., *Cell* **10**, 193 (1977).

# Part D. Chromatin Transcription and Characterization of Transcripts

# Chapter 25

## *Visualization of Genes by Electron Microscopy*

BARBARA A. HAMKALO

*Departments of Molecular Biology and Biochemistry*
*and of Developmental and Cell Biology,*
*University of California, Irvine,*
*Irvine, California*

### I. Introduction

The analysis of the structure of transcriptionally active or replicating genomes should provide information on nucleoprotein rearrangements which are a consequence of the activation of the genome. A simple procedure is described in detail for the preparation of cellular or nuclear contents for electron microscopy which allows the ultrastructural analysis of individual genes from single cells. The technique was developed by Miller and Beatty (*1,2*) and has been applied successfully to many systems by several investigators (*3–22*). This chapter details the technique and describes problems that may be encountered and the ways in which they may be solved.

### II. Technical Details

The procedure is outlined in Fig. 1. With the exception of the isolation step (A), the remainder of the procedure has been followed successfully for organisms from *Escherichia coli* (*3*) to human cells in culture (*4*).

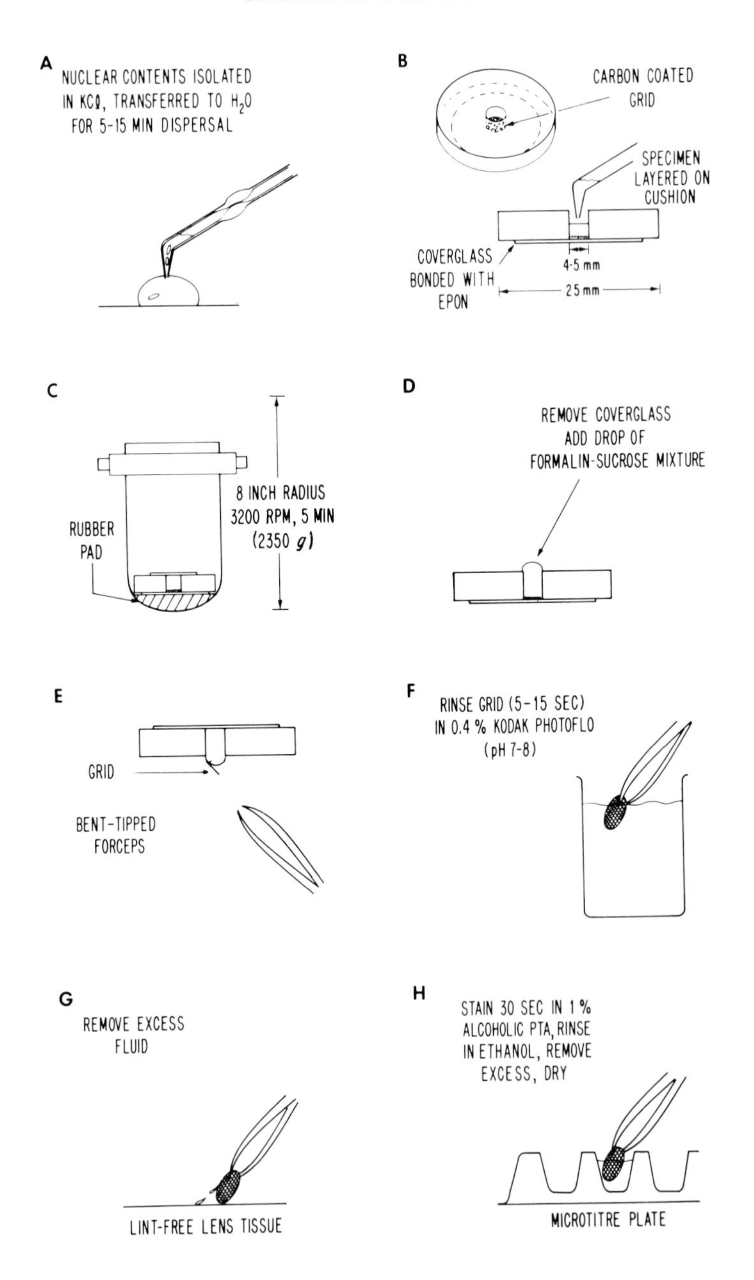


FIG. 1. Schematic representation of the Miller procedure for the preparation of material for electron microscopy (*4*).

## A. Carbon Films

A thin, strong support film with uniform, fine grain provides the optimum background for viewing either prokaryotic or eukaryotic genomes. This is especially true if one contrasts the preparation by positive staining rather than by heavy-metal shadowing. The benefit of the former over the latter is the short amount of time required to stain versus that to shadow a preparation.

We obtain good films by depositing carbon on a glycerine-coated glass slide in the following manner:

1. Make up a 50% solution of glycerine in water; filter through an 0.2 $\mu$m disposable filter to remove particulate material and place the filtered solution in a Coplin jar.

2. Wipe a precleaned glass slide with a Kimwipe, pass the slide through a bunsen burner flame, dip it into the glycerine solution, and stand the slide on its short edge against a supporting surface, such as a test tube rack, on bibulous paper to drain. Heating the slide causes the glycerine to drain rapidly, leaving a thin, uniform coating. Drainage occurs from the sides of the slide to the middle, and from top to bottom. Interference colors should be visible on the slide after draining. Areas with less glycerine will adsorb more carbon and vice versa.

3. Slides are placed in a vacuum evaporator on a sheet of white paper. The evaporator must be equipped with a glow discharge unit as plasma cleaning of the surface of the glycerine film is essential for a uniform carbon film which does not shatter. The evaporator is pumped down to about 0.1 Torr using the mechanical pump and the slides are glow-discharged for approximately 1 minute. Immediately after glowing, the evaporator is pumped down to 5 to 7 $\times 10^{-4}$ Torr, and carbon is evaporated slowly in order to avoid sparking. The thickness of the film is estimated by the color of the paper on which slides are placed; pale beige corresponds to a thin film, which is usually quite strong.

4. After evaporation, slides are scored with a razor blade into squares slightly larger than the diameter of a grid; the film is floated off onto a clean water surface in a dish with a dark background. The dark background improves visibility of even thin film when the water surface is illuminated by an overhead desk lamp.

5. Grids which have been rinsed in acetone and air-dried are placed on bibulous paper. A square of film is packed up from the water surface from beneath with a wire loop attached to a wooden stick handle. The film is lowered onto the grid while holding the edge of the grid immobile with a straight wire. After all films have been picked up, they are air-dried in a covered container, such as a Petri dish or cardboard box. Grids placed in the

refrigerator to dry do so more slowly than at room temperature, and films appear to be less brittle.

6. One grid is usually checked in the electron microscope for strength of film and evenness of grain. Coverage of each grid can be assessed with reflected light and a dissecting microscope. Carbon-coated grids can be stored in a closed container for over a month without apparent deterioration in strength.

It is also possible to utilize plastic-coated grids on which a small amount of carbon has been evaporated (*20*) although these films are usually thicker than carbon-glycerine films.

A hydrophilic support film is essential for extensive attachment of material to grids and for uniform spreading. Grids can be rendered hydrophilic either by glow-discharging them in a vacuum evaporator shortly before use or by dipping the grid in 95% ethanol for 30 seconds to a few minutes immediately before placing them, in the microcentrifugation chamber. The former technique has the liability of making films brittle.

## B. Solutions

All solutions are made up in deionized glass-distilled water kept in a glass container. The lysing medium, dispersing medium, and sucrose–formalin are adjusted to approximately pH 9.0 with pH 10 standard pH meter borate buffer. Stock staining solutions are made up to 4% in water and stored in brown bottles; the stains are diluted to 1% in 95% ethanol in a well of a disposable microtiter plate immediately before use. Sucrose–formalin and stock stains are filtered through a 0.2 $\mu$m disposable filter unit. The pH of the sucrose–formalin solution is checked periodically, and the solution is refiltered every few weeks. The Photoflo solution, which is made up fresh daily, is adjusted to pH 7 with borate buffer.

## C. Procedure

The centrifugation chamber, diagrammed in Fig. 1B, consists of a slice of a Plexiglas rod with a hole drilled through the center sufficiently large to accommodate an electron microscope grid (4 mm in diameter). A No. 2 round glass coverslip is glued to one surface; the coverglass serves as a table for the grid and allows light-microscopic viewing of the material in the chamber and on the grid. The dimensions of the chamber and its precise construction are not critical, provided that the hole is large enough for a grid and the chamber fits into the swinging bucket of a centrifuge. If one uses an 100-ml bucket, a convenient chamber size is 25 mm in diameter and 8 mm deep.

The depth of the chamber can be varied over a wide range of sizes to fit the requirements of a given problem.

A chamber is filled with sucrose–formalin and checked for the presence of bubbles, which, if under the grid, will break the carbon film upon centrifugation. A hydrophilic grid is rinsed with a few drops of the sucrose–formalin solution and dropped into the chamber with the carbon-coated side face up. The level of sucrose–formalin is lowered by pipetting out a portion from the chamber. The depth of cushion remaining varies with the material, but 1/3 to 1/2 is used routinely. A cell or nuclear lysate is layered over the cushion, and a No. 1 round coverslip is placed over the hole in the chamber and blotted with bibulous paper to adhere to the chamber. The chamber is then carefully placed in a swinging bucket, using long forceps. Material is centrifuged through the sucrose–formalin cushion by a 5-minute centrifugation at 2350 *g* (C). The time and speed of centrifugation can be modified according to the preparation. After centrifugation, the chamber is removed from the centrifuge, the cover slip is removed from the chamber, and a drop of sucrose—formalin is added to the chamber to aid in removing the grid (E). When the chamber is inverted, the grid floats to the top and is caught in the drop of sucrose–formalin. The grid is removed and the sucrose–formalin and material which does not adhere tightly to the grid are rinsed off in an aqueous solution containing 0.4% Kodak Photoflo 200 solution. The presence of Photoflo, a detergent, reduces stretching of fibers and aggregation as the grid surface dries. The excess Photoflo is removed from the grid by running the edge along a piece of Ross lint-free lens tissue, followed by air drying the grid completely. If one blots the grid at this point clumping of material results. Bent-tipped forceps are helpful in drying the grid since solutions do not get caught between the forceps' arms. The dried grid is then stained by immersing it in a 1% staining solution for 30 seconds, followed by a thorough rinse in 95% ethanol to remove excess stain prior to air drying. If one utilizes two stains in sequence (e.g., phosphotungstic acid then uranyl acetate), the sequence is 30 seconds stain; ethanol rinse; 30 seconds stain; ethanol rinse; dry. Once stained, preparations are stable for as long as the carbon films remain intact.

## III. Cell and Nuclear Lysis

In order to retain nucleoprotein complexes in an arrangement that reflects their interactions *in vivo*, it is necessary to lyse cells and nuclei gently. Ideally, this step is performed rapidly, since our experience indicates increased difficulty in the dispersal of chromatin with time after cells are killed, perhaps owing to extensive denaturation and cross-linking of nuclear proteins. The following procedures have been used successfully in a variety of cell types.

## A. Prokaryotic Cells

Logarithmically growing bacterial cells are rapidly chilled by pouring over an equal volume of crushed ice; cells are pelleted and resuspended in 1/5 volume of 1.0 *M* sucrose (pH 8.5) to effect plasmolysis. The sucrose concentration then is lowered to 0.25 *M* by addition of water. Plasmolyzed cells are treated with T4 lysozyme at a final concentration of 50 $\mu$g/ml for 3 minutes at 4°C. Lysis occurs upon dilution of spheroplasts into cold $H_2O$ (pH 8.5). The degree of dilution varies with the system, but with cells at $10^8$/ml, we use a 100-fold dilution. T4 Lysozyme can be replaced by egg white lysozyme, but it may be necessary to extend the time of lysozyme treatment.

## B. Eukaryotic Cells

Because of their large size (0.5–1.0 mm diameter), developing amphibian oocytes can be opened by manual dissection using 2 pairs of jeweler's forceps; nuclei released from these cells are also opened by manual dissection (*4*). Unfortunately, few eukaryotic cells are of sufficient demensions to be handled in this manner. Mechanical lysis by homogenization has been used, but with variable results. As a result, lysis is normally performed by either detergent solubilization or osmotic shock of membranes.

Metaphase cells which do not have a nuclear envelope are readily lysed by addition of the nonionic detergents Triton X-100 or NP-40 (pH 9) to a final concentration of 0.25% (*18*). These detergents usually do not lyse nuclear membranes, although there are exceptions, such as early sea urchin embryos, where either detergent will also lyse nuclei. Miller and Bakken (*4*) found that dilution of HeLa cells into cold Joy (Proctor and Gamble) to a final concentration of 0.33% results in lysis of cells and nuclei and liberation of chromatin, which retains nascent ribonucleoprotein fibers. There appears to be some variation in the formula of this detergent because we have noted that lower concentrations (0.05%) remove histones from chromatin, while other groups observe intact histone–DNA complexes even in the presence of higher concentrations of the detergent.

Hypotonic shock of eukaryotic cells usually lyses both plasma and nuclear membranes. Cells are concentrated in a small volume of cold buffer and diluted 100 to 400-fold into cold distilled $H_2O$ (pH 9–10). Lysis is more efficient if cells are precooled; this may be due to a reduction of transport processes. In addition, lysis is improved at elevated pH levels which may weaken the nuclear envelope. Olins *et al.* (*23*) find that swelling isolated nuclei in 0.2 *M* KCl prior to hypotonic shock improves lysis and dispersal of nuclear contents.

## IV. Dispersal of Genomes

Although nuclear lysis is critical to the success of this technique, perhaps the most important aspect of the preparation is the dispersal of highly folded genomes characteristic of both prokaryotes and eukaryotes. The contents of logarithmically growing bacterial cells are usually readily dispersed upon osmotic shock; eukaryotic genomes, however, appear to be more refractile to dispersal. Since we do not know all the factors involved in maintaining both euchromatin and heterochromatin in a folded form in the nucleus, it is impossible to detail a dispersal procedure that can be applied to all cells at all times. The following suggestions are based on their success in improving dispersal in at least some systems. The time of dispersal varies from 5 minutes to an hour or more. Each system is somewhat different, and problems of nuclease activity during dispersal vary considerably. It is possible, however, to incorporate nuclease inhibitors, such as RNA, into dispersal medium to reduce these activities and to disperse material at ice temperature.

### A. Low Ionic Strength

Transfer of amphibian oocyte nuclear contents from physiological saline to distilled water (pH 9) results in extensive dispersal of nucleoli (*1,2*). This observation implies that the low ionic strength of dispersal medium promotes chromatin dispersal. Recently, Rattner *et al.* (*18*) noted that lysis of mitotic L cells by hypotonic shock results in liberation of metaphase chromosomes, which disappear rapidly when viewed in the phase contrast microscope and are unwound into extended nucleoprotein fibrils when viewed in the electron microscope.

An interesting observation made by O. L. Miller, Jr., that amphibian oocyte nuclear contents which are isolated and kept in saline (0.1 *M* KCl) for a few minutes are refractile to dispersal upon transfer to pH 9 water, implies that there is rapid nucleoprotein aggregation or cross-linking when the material is removed from the cell. As a result, we attempt to lyse cells or nuclei and transfer lysates to the dispersing medium as rapidly as possible.

## B. $H_1$ Removal

Oudet *et al.* (*24*) found that the $H_1$-depleted chromatin from several sources is readily visualized as discrete fibers rather than as a compact mass. Foe *et al.* (*20*) described conditions for extensive dispersal of *Oncopeltus fasciatus* chromatin by the addition of yeast RNA (100 $\mu$g/ml) to the dispersal medium. Since Ilyin *et al.* (*25*) showed that there is rapid removal and transfer of $H_1$ from chromatin to tRNA, it has been suggested that the tRNA is improving chromatin dispersal by $H_1$ removal (*20*).

## C. Formalin

Although formalin is used in the Miller procedure as a fixative because of its ability to cross-link proteins to nucleic acids, Rattner *et al.* (*18*) noted that the addition of buffered formalin (10%; pH 9) to condensed metaphase chromosomes results in extensive dispersal of the material. Although the mechanism by which formalin induces such dispersal is unknown, it may result from altered interactions of folding proteins with DNA or chromatin.

## D. Mercaptoethanol

The presence of 0.5% mercaptoethanol in the dispersal medium aids in spreading of some chromatins. It is likely that this compound acts by reducing disulfide cross-links, which may form among sulfhydryl-containing nuclear proteins when nuclei are lysed and exposed to the air.

## E. DNase

An approach to chromatin dispersal which has not been extensively investigated to date, but which seems to have promise, is the brief DNase digestion of either nuclei or nuclear lysates in order to cleave the chromatin into large fragments, which then should be more readily spread out from the mass of the compact interphase nucleus. Either nonspecific nucleases, such as micrococcal nuclease (to which nuclei are permeable) or DNase I or II, or restriction endonucleases can be adapted to the procedure. Noll *et al.* (*26*) introduced micrococcal nuclease treatment of nuclei prior to lysis and extended digestion in the preparation of chromatin subunits, and Reeder *et al.* (*27*) showed that it is possible to restrict amphibian oocyte nucleolar chromatin while preserving nascent RNP complexes. Since conditions for optimum activity of these enzymes may not be those suitable to maintain nucleoprotein complexes, one must find conditions to satisfy both the enzyme and the chromatin.

## V. The Analysis

Electron microscopic analysis of well-dispersed genomes provides information about the structure of both transcriptionally active and inactive genes. Figures 2 and 3 illustrate a number of such observations. These micrographs represent a selected area of a grid which is, in most cases, exceptional because of the level of transcriptive activity. Figure 2 shows an active structural operon (Fig. 2A) (*3*) and an active ribosomal RNA-coding region (Fig. 2B) (*28*) from *Escherichia coli*. The identification of these two types of genes is straightforward, based upon predictions drawn from biochemical and genetic analyses of the structure of such regions.

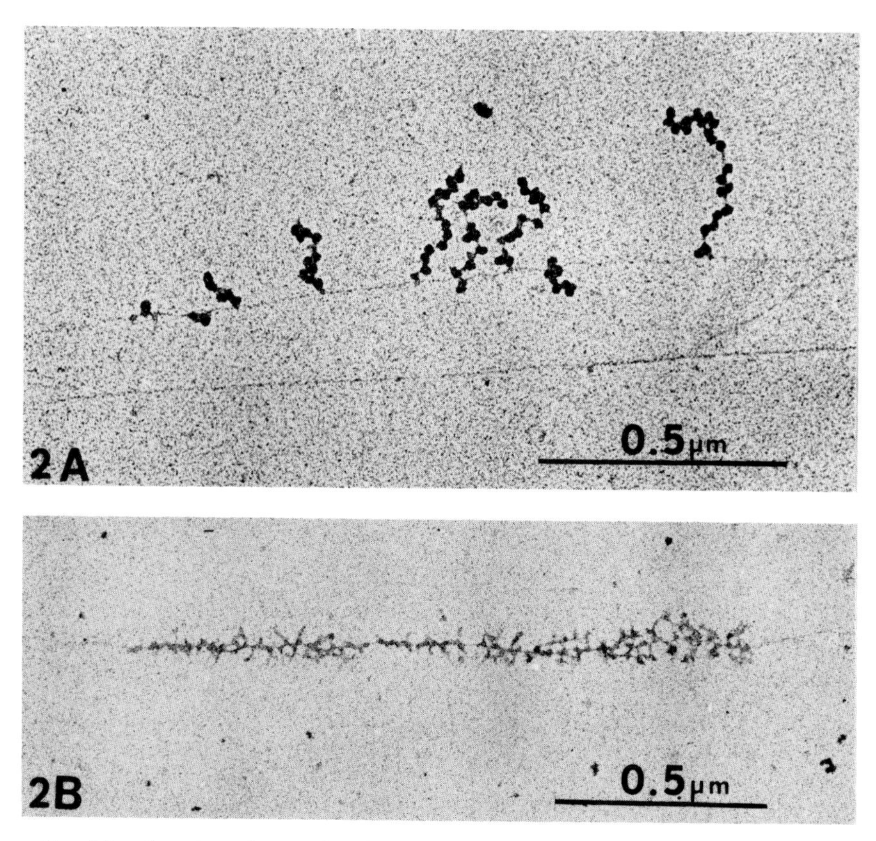


FIG. 2A. A region of an *Escherichia coli* genome illustrating coupled transcription and translation of a structural operon (*3*). Copyright 1970 by the Association for the Advancement of Science. B: An active ribosomal RNA-coding region from *E. coli* (*28*).

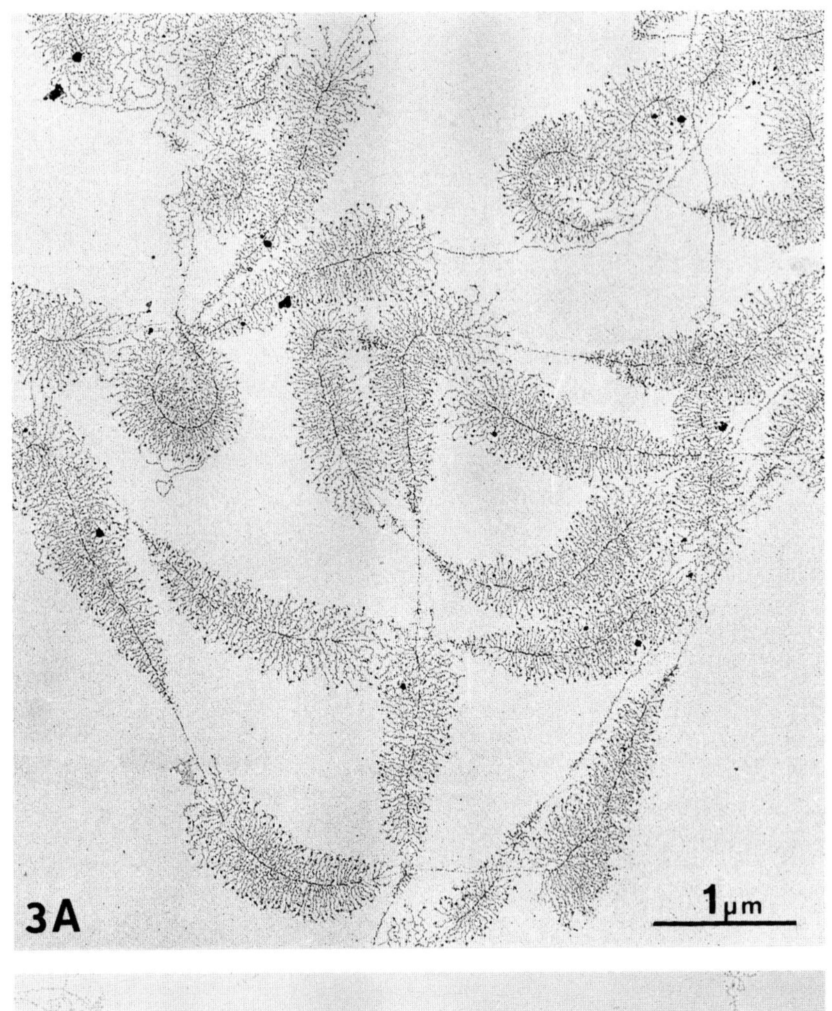
3A
1μm

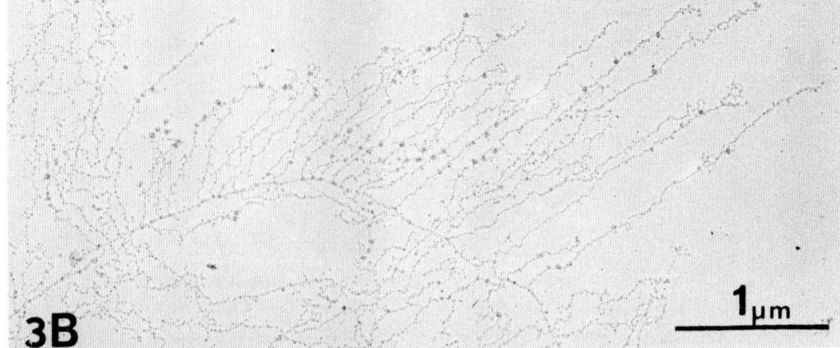
3B
1μm

Figure 3A illustrates active ribosomal precursor RNA genes from a developing amphibian oocyte, and Fig. 3B shows a region active in nonribosomal RNA synthesis from the cellular blastoderm stage of a *Drosophila melanogaster* embryo. The redundancy, linkage, and typically high level of transcription of ribosomal precursor genes, combined with the predicted length of such genes based on the molecular weights of the precursor RNAs provide a basis on which to identify these genes in preparations from a vast number of organisms.

With the exception of oocytes during the lampbrush stage, one normally sees few nonribosomal genome segments active in transcription in a typical preparation. This is consistent with estimates from molecular hybridization of the small fraction of the genome which is active. However, since the ultrastructural analyses have been confined to regions bearing several closely spaced nascent RNPs, the percentage of the genome transcribed is underestimated, since it has been calculated for sea urchin embryos that the average frequency of nonrepetitive transcripts is one per cell (*29*).

At present, only general statements can be made about the activity of nonribosomal genes from electron microscopic studies of such transcribing units (Fig. 3B). It is possible to assess the level of activity of a given segment (i.e., the number of nascent chains per unit length), which is a reflection of the frequency of transcriptional initiation. In addition, when it is possible to extrapolate to an approximate initiation site, one can calculate the approximate amount of shortening of an RNA within an RNP. Detailed analyses of this sort have been carried out in Laird's laboratory for several insect systems (*20,22,30*). The development of techniques for the identification of specific nascent RNAs in these preparations would provide an exceedingly powerful tool for the direct analysis of gene regulation at the level of individual loci.

When the technique outlined above was applied to the analysis of chromatin structure, Olins and Olins (*10, 11*) and Woodcock (*12*) noted the existence of a regular beaded repeat along the length of the interphase chromatin fiber. These early observations of so-called nu-bodies or nucleosomes serve as a critical piece of data for the existence of such nucleohistone complexes as a characteristic feature of eukaryotic chromosomes. Recently, Rattner *et al.* (*18*) observed an identical conformation of chromatin in dispersed mitotic chromosomes (Fig. 4).

Obviously, the statements made from electron microscopic studies are

FIG. 3A. Electron micrograph of a portion of a dispersed extrachromosomal nucleolar core from an oocyte of the amphibian *Notophthalmus viridescens* (*1*). Copyright 1969 by the Association for the Advancement of Science. B: Electron micrograph of a transcriptionally active nonribosomal RNA region for the embryonic cellular blastoderm stage of *Drosophila melanogaster* (*19*). Copyright 1976 by the MIT press.

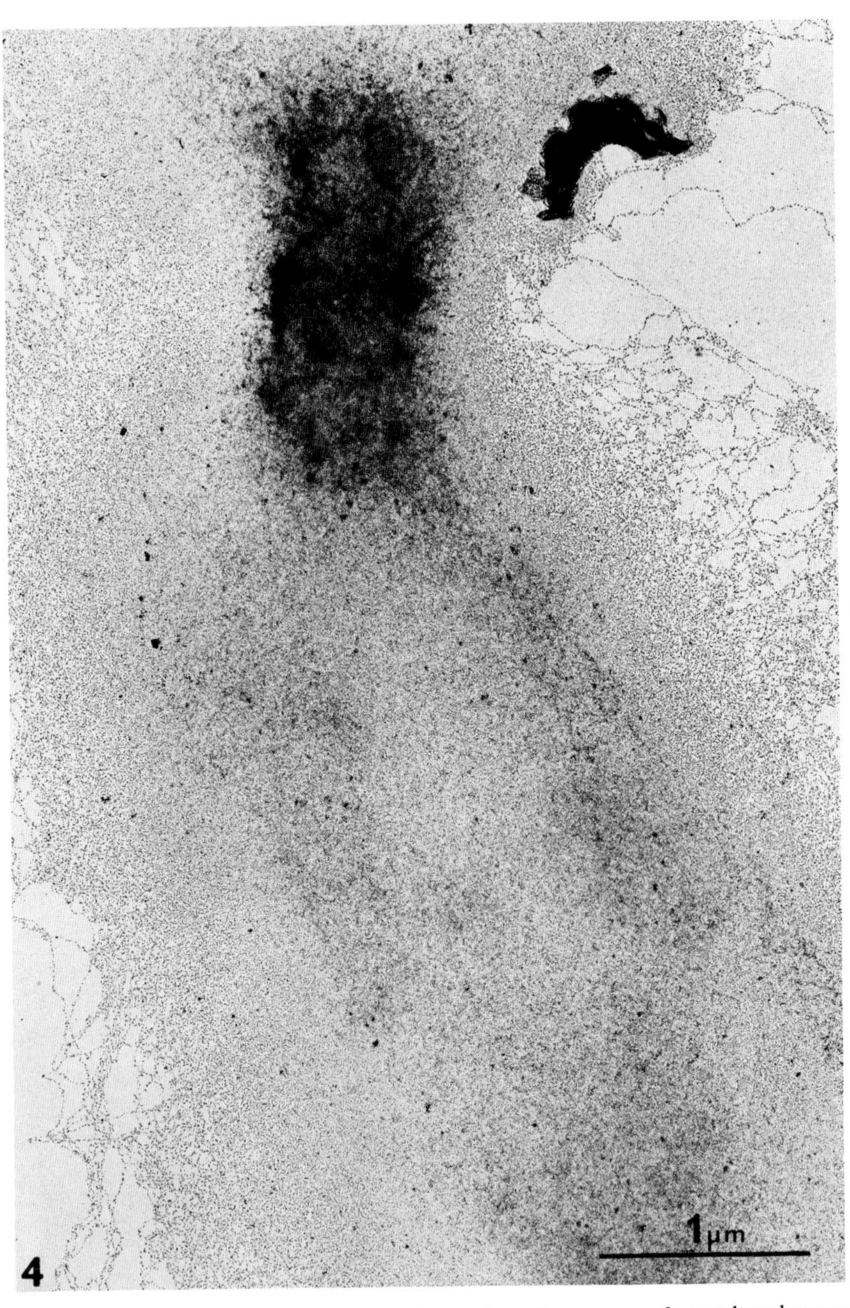


FIG. 4. Micrograph of partially dispersed metaphase chromosomes from cultured mouse L cells (*18*). Copyright 1975 by Springer-Verlag.

most convincing when coupled with information derived from independent lines of investigation, such as biochemical, biophysical, and genetic analyses.

The Miller technique can be coupled with other approaches in order to extract even more information from electron microscopic analysis of genomes. In order to aid in identification of components of a preparation, the preparation can be treated enzymically prior to deposition on the grid, or the grid with attached sample can be floated on a drop of enzyme solution. If digestion is done on the grid, however, samples should be deposited on gold, rather than copper, grids. In addition, it is possible to combine autoradiography with the electron microscopic preparative procedure in order to localize the site of synthesis of specific macromolecules. In this case, material is deposited on the grid as usual, and standard high resolution autoradiography is performed (*31*). The identification of specific molecules can be achieved by combining the usual preparative technique with immunoelectron microscopy. Several groups are pursuing this line of investigation with the hopes of identifying both specific proteins associated with the genome as well as specific nascent transcripts (*32,33,34*).

While a great deal of information has been provided by the application of the technique detailed in this article, it is likely that its combination with other experimental approaches will provide a powerful approach to the detailed understanding of chromosome structure as related to function.

## REFERENCES

*1.* Miller, O. L., Jr., and Beatty, B. R., *Science* **164**, 955 (1969).
*2.* Miller, O. L., Jr., and Beatty, B. R., *J. Cell. Physiol.* **74** Suppl. 1, 225 (1969).
*3.* Miller, O. L., Jr., Hamkalo, B. A., and Thomas, C. A., Jr., *Science* **169**, 392 (1970).
*4.* Miller, O. L., Jr., and Bakken, A. H., *Acta Endocrinol.* (*Copenhagen*), Suppl. **168**, 155 (1972).
*5.* Hamkalo, B. A., Miller, O. L., Jr., and Bakken, A. H., *Cold Spring Harbor Symp. Quant. Biol.* **38**, 915 (1973).
*6.* Scheer, U., Trendelenberg, M. F., Franke, W. W., and Herth, W., *Exp. Cell Res.* **80**, 175 (1973).
*7.* Derksen, J., Trendelenberg, M. F., Scheer, U., and Franke, W. W., *Exp. Cell Res.* **80**, 476 (1973).
*8.* Trendelenberg, M. F., Scheer, U., and Franke, W. W., *Nature* (*London*), *New Biol.* **245**, 167 (1973).
*9.* Trendelenberg, M. F., Spring, H., Scheer, U., and Franke, W. W., *Proc. Natl. Acad. Sci. U.S.A.* **71**, 3626 (1974).
*10.* Olins, A. L., and Olins, D. E., *J. Cell Biol.* **59**, 252a (1973).
*11.* Olins, A. L., and Olins, D. E., *Science* **183**, 330 (1974).
*12.* Woodcock, C. L. F., *J. Cell Biol.* **59**, 368a (1973).
*13.* Kierzenbaum, A. L., and Tres, L. L., *J. Cell Biol.* **63**, 923 (1974).
*14.* Trendelenberg, M. F., *Chromosoma* **48**, 119 (1974).
*15.* Spring, H., Trendelenberg, M. F., Scheer, U., Franke, W. W., and Herth, W., *Cytobiologie* **1**, 65 (1974).

16. Angelier, N., and Lacroix, J. C., *Chromosoma* **51**, 323 (1975).
17. Glatzer, K. H., *Chromosoma* **53**, 371 (1975).
18. Rattner, J. B., Branch, A., and Hamkalo, B. A., *Chromosoma* **52**, 329 (1975).
19. McKnight, S. L., and Miller, O. L., Jr., *Cell* **8**, 305 (1976).
20. Foe, V. E., Wilkinson, L. E., and Laird, C. D., *Cell* **9**, 131 (1976).
21. Chooi, W. Y., and Laird, C. D., *J. Mol. Biol.* **100**, 493 (1976).
22. Laird, C. D., Wilkinson, L. E., Foe, V. E., and Chooi, W. Y., *Chromosoma* **58**, 169 (1976).
23. Olins, A. L., Carlson, R. D., and Olins, D. E., *J. Cell Biol.* **64**, 528 (1975).
24. Oudet, P., Gross-Bellard, M., and Chambon, P., *Cell* **4**, 281 (1975).
25. Ilyin, Y. V., Varshavsky, A. Ya., Mickelsaar, U. N., and Georgier, G. P., *Eur. J. Biochem.* **22**, 235 (1971).
26. Noll, M., Thomas, J. O., and Kornberg, R. D., *Science* **187**, 1203 (1975).
27. Reeder, R. H., Higashinakagawa, T., and Miller, O. L., *Cell* **8**, 449 (1976).
28. Miller, O. L., Jr., and Hamkalo, B. A., *Int. Rev. Cytol.* **33**, 1 (1972).
29. Hough, B. R., Smith, M. J., Britten, R. J., and Davidson, E. H., *Cell* **5**, 291 (1975).
30. Laird, C. D., and Chooi, W. Y., *Chromosoma* **58**, 193 (1976).
31. Caro, L. G., *Methods Cell Physiol.* **1**, 327 (1964).
32. Chooi, W. Y., Swift, H. H., and Stoffler, G., *J. Cell Biol.* **67**, *68a (1975).*
33. Chooi, W. Y., *Handbook of Genet.* **5**, 219 (1976).
34. McKnight, S. L., Bustin, M., and Miller, O. L., Jr., *Cold Spring Harbor Symp. Quant. Biol.* 42, (1977).

# Chapter 26

# *Techniques of in Vitro RNA Synthesis with Isolated Nucleoli*

TOSHIO ONISHI AND MASAMI MURAMATSU[1]

*Department of Biochemistry,*
*Tokushima University School of Medicine,*
*Tokushima, Japan*

## I. Introduction

The nucleolus of a cell contains ribosomal RNA cistrons (rDNA) coding for 45 S preribosomal RNA (45 S pre-rRNA) together with a high concentration of DNA-dependent RNA polymerase (*1*). rDNA in the nucleolus is most likely bound with nuclear proteins, such as histones and acidic proteins, to form a specific nucleoprotein complex where the transcription and processing of 45 S pre-rRNA is adequately controlled according to the need of the cell. Recently, we have demonstrated that the nucleolus contains an excess amount of RNA polymerase I (or A), only a part of which is bound with the template, transcribing rDNA (*2*). Just what kind of factors are responsible for the regulation of transcription on the nucleolar template remains to be determined.

The fact that isolated nucleoli could synthesize RNA in an *in vitro* system was first demonstrated by Ro *et al.* (*3*). This system, similar to lysed nuclei of Weiss (*4*), required all four nucleoside triphosphates and was inhibited by a low concentration of actinomycin D. On the basis of nucleotide composition and competitive hybridization, the product was found to be predominantly pre-rRNA (*5–7*). Isolated nucleoli also carry out methylation of pre-rRNA at specific sites using *S*-adenosylmethionine (*8–13*). Processing of 45 S pre-rRNA, which involves certain specific cleavages and possibly trimmings of this precursor together with the maturation of preribosomal

[1]*Present address:* Department of Biochemistry, Cancer Institute, Toshima-ku, Tokyo, Japan.

ribonucleoprotein particles may occur, if partially, in isolated nucleoli (*14–16*).

Thus, isolated nucleoli appear to provide an excellent system for studying *in vitro* transcription of rDNA as well as posttransciptional modification and processing of pre-rRNA and preribosomes. However, the high RNase activity present in the nucleolus (*17–20*) has been hampering the analysis of the product, since the newly synthesized RNA is usually susceptible to this nucleolytic attack. Recently, Grummt and Lindgkeit (*21*) reported that isolated rat liver nucleoli could synthesize 45 S RNA or even larger molecules under appropriate conditions. We have repeated their experiments and found that isolated rat liver nucleoli as well as mouse hepatoma nucleoli could indeed synthesize RNA molecules as large as 45 S.

In this chapter we shall describe critical conditions for synthesizing 45 S RNA with isolated nucleoli of these cells together with some characteristics of the system and the products (*21a,21b*).

## II. Procedures for *in Vitro* RNA Synthesis with Isolated Nucleoli

### A. Isolation of Nucleoli

Nucleoli with ability to synthesize 45 S pre-rRNA may be isolated from rat liver as well as ascites tumor cells, such as Ehrlich ascites tumor and Novikoff ascites hepatoma (*12,13,21–22*). We routinely isolate nuclei from either rat liver or mouse ascites hepatoma, MH 134 by our magnesium (*23*) or detergent (*24*) procedure, respectively, except that the concentration of sodium deoxycholate is reduced to 0.12% since this ionic detergent is rather inhibitory to RNA polymerase activity. Nuclei are then disrupted by sonication, and nucleoli are purified by sedimentation through 0.88 *M* sucrose–0.05 m*M* $MgCl_2$ as described previously (*23,24*). Details of the procedure are described (*25*). The nucleolar pellet is suspended in 10 m*M* Tris-HCl (pH 7.4)–60% glycerol (v/v)–0.1 m*M* EDTA, and stored at −80°C until use. Little loss of RNA-synthesizing capacity was noted over a period of several months during storage. However, several times freezing and thawing resulted in a considerable reduction of newly synthesized RNA species with the sedimentation coefficient of 45 S. Addition of 1 m*M* $Mg^{2+}$ during continuous storage at −80°C also caused a decrease in the synthesis of 45 S RNA without any loss of total incorporation of nucleoside triphosphates into acid-insoluble material. These effects are thought to be due to the nicks that occur on the rDNA template in the nucleoli under those circumstances.

## B. Incubation Conditions

### 1. Incubation Mixture

The standard reaction mixture contains, in 0.15 ml: 33 m*M* Tris-HCl (pH 7.9) (at 37°C), 6.6 m*M* dithiothreitol, 5 m*M* $MgCl_2$, 50 m*M* KCl, 0.66 m*M* each of ATP, GTP, and CTP (when labeled UTP is employed), 0.012–0.024 m*M* [$^{14}$C]UTP, an appropriate amount of RNase inhibitor purified from rat liver cytoplasmic fraction (see below) and nucleoli containing 5 $\mu$g DNA. When [$\alpha$-$^{32}$P]CTP was used to label RNA, 0.013–0.026 m*M* concentration of this nucleotide was added together with 0.66 m*M* cold UTP.

For the assay of overall incorporation of the labeled precursor into acid-insoluble material, 1 ml of 7% trichloroacetic acid containing 1% sodium pyrophosphate was added to the reaction mixture at the end of incubation. After standing in an ice bath for 10 minutes, the mixture was centrifuged at 3000 rpm for 10 minutes, and the pellet was suspended in 2 ml of 5% trichloroacetic acid containing 1% sodium pyrophosphate using a tightly fitting glass pestle. The acid insoluble material was collected on Whatman glass-fiber disk (GF/C) and washed successively with 25 ml of 5% trichloroacetic acid–1% sodium pyrophosphate and 10 ml of ethanol.

For analysis of *in vitro* synthesized RNA, the amount of incubation mixture was increased proportionately. After incubation, 1/10 volume each of 10% sodium dodecyl sulfate and 1 *M* sodium acetate (pH 5.5) were added to the mixture, which was extracted with phenol as described previously (*26,27*).

### 2. Requirement for a High Concentration of Substrate

When various concentrations of pyrimidine nucleotides were tested for the successful synthesis of 45 S RNA, it was found that a certain concentration of these substrates was required to make a maximum amount of 45 S RNA (Fig. 1). For both CTP and UTP, 0.006 m*M* was not enough. At concentrations 0.012 m*M* or higher, maximum amounts of 45 S RNA were synthesized judging from the sedimentation profile. This is reminiscent of a situation found in the case of reverse transcriptase (*29*) and probably indicates a premature termination of transcription under low substrate concentrations. In any event, we include at least 0.012 m*M* of either CTP or UTP as a labeled substrate.

### 3. Effect of RNase Inhibitors

Nucleoli possess both endo- and exoribonuclease which are capable of degrading endogenous as well as exogenous RNA (*14,17*). In order to suppress the RNase activity in this system, RNase inhibitor of rat liver

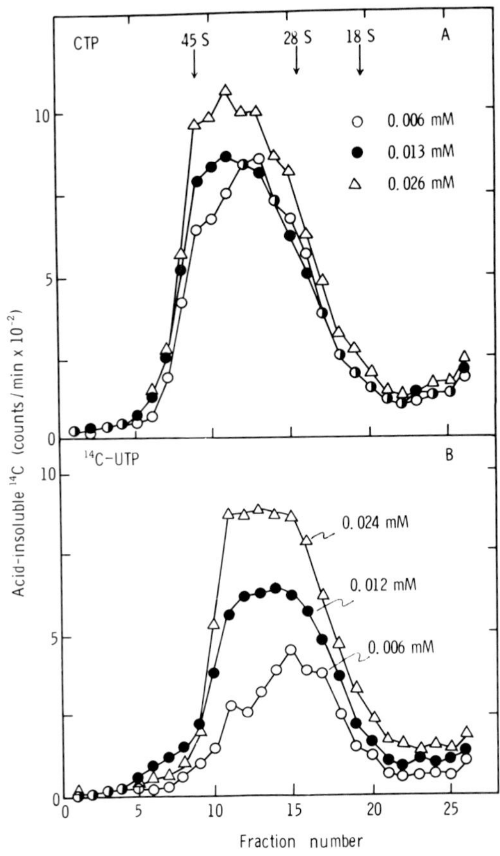


FIG. 1. Effect of concentrations of CTP and [$^{14}$C]UTP on the synthesis of 45 S RNA with isolated nucleoli. Reaction mixtures scaled up 3-fold from the standard mixture contained: 900 units of rat liver RNase inhibitor, 15 $\mu$g of DNA equivalent rat liver nucleoli, 0.024 m*M* [$^{14}$C]UTP, and various concentrations of CTP (A) or different amounts of [$^{14}$C]UTP (B). [One unit of inhibitor is defined as that amount which gives 50% inhibition on 0.5 ng of pancreatic RNase.] After incubation at 25°C for 15 minutes, 1/10 volume each of 10% SDS and 1 *M* sodium acetate (pH 5.5) were added to the reaction mixture, and RNA was extracted by the hot phenol procedure as described by Muramatsu *et al.* (*26*). The deproteinized RNA was dissolved in water and layered onto 5 ml of a linear 10 to 30% sucrose gradient containing 10 m*M* Tris-HCl (pH 7.4)–10 m*M* NaCl–10 m*M* EDTA–0.2% SDS–0.1% sodium deoxycholate. The gradient was centrifuged at 40,000 rpm for 5 hours at 5°C, and fractions of 0.2 ml were collected from the bottom of the tube. Acid-insoluble radioactivity was determined as described previously (*28*).

cytoplasm was purified essentially according to Gribnau *et al.* (*30*). As a modification, we have used DEAE-Sephadex rather than DEAE-cellulose, and this fraction was concentrated with 65% ammonium sulfate precipitation followed by dialysis against 50 m*M* Tris-HCl (pH 7.6)–5 m*M* $\beta$-mercaptoethanol. The final preparation could be stored at −80°C for months without any decrease in activity.

Addition of this RNase inhibitor to the incubation mixture enhanced incorporation of labeled UMP into rat liver as well as MH 134 nucleolar RNA (Fig. 2). At a time when incorporation of nucleoside triphosphates had reached a plateau, or rather a decrease in acid-insoluble radioactivity had nearly begun, a supply of RNase inhibitor again stimulated incorporation of the labeled precursor as shown in Fig. 3. These results indicate that both synthesis and degradation of RNA are simultaneously occuring in the nucleolus in the absence of RNase inhibitor. As shown in Fig. 4A, RNA synthesized in the absence of RNase inhibitor was of small size with a modal sedimentation value of only 9 S, even if incubation was carried out under optimal conditions, i.e., with $Mg^{2+}$ at 25°C (see Section II,B,4), whereas in the presence of the RNase inhibitor, the sedimentation value increased remarkably, especially in the presence of $Mg^{2+}$. Under optimal conditions, the main peak of radioactivity came to almost 45 S, and increasing the incubation period resulted in an accumulation of high-molecular-weight RNA. Figure 3 also shows that this RNase inhibitor is rather heat labile under incubation conditions. When the inhibitor was supplemented again at 15 minutes after incubation, the incorporation continued almost linearly for additional 7–8 minutes, whereas nonsupplemented nucleoli no longer incorporated labeled precursors. As will be discussed later, this inhibitor seems more stable at 25°C than at 37°C.

Other RNase inhibitors, such as dextran sulfate, polyvinyl sulfate, and bentonite, all improve the recovery of labeled 45 S pre-rRNA and also somewhat stimulate apparent RNA polymerase activity at certain low concentrations. However, they are reported to be inhibitory to methylase activity (*13*). Potassium chloride, at a concentration as high as 0.5 *M*, causes accumulation of 45 S RNA but inhibits RNA polymerase activity by about 85% (*13*). Heparin also inhibits RNase (*31,32*), but this polyanionic inhibitor, even at a concentration as high as 2 mg/ml, cannot completely inhibit RNase activity associated with nucleoli from both rat liver and MH 134 cells (our unpublished data). This reagent at higher concentrations appears to stimulate RNA synthesis by removing a part of nuclear proteins, especially histones, from the nuclear as well as nucleolar chromatin [unpublished results and Cox (*33*)]. It has also been established that heparin inhibits RNA synthesis at the level of initiation of eukaryotic as well as prokaryotic RNA polymerases (*21a,33–35*).

### 4. Effect of Temperature and Divalent Cations

The effect of temperature on the *in vitro* nucleolar RNA synthesis is shown in Fig. 2 together with effect of rat liver cytosol RNase inhibitor. As

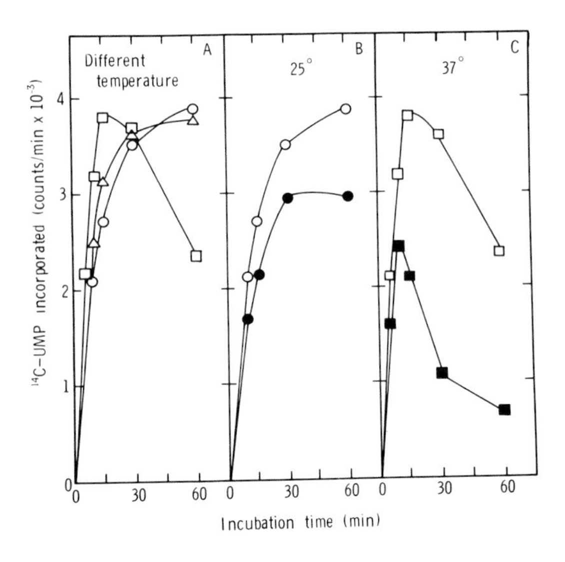


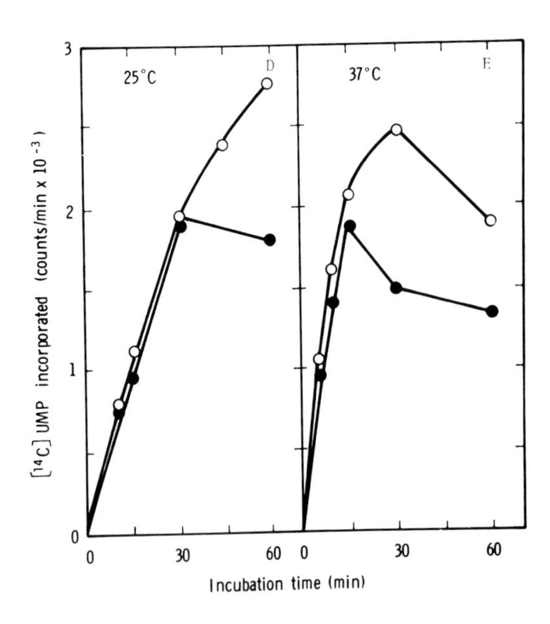


FIG. 2. Effect of temperature and the RNase inhibitor on *in vitro* nucleolar RNA synthesis. The standard reaction mixture was incubated in the presence and in the absence of 20 units of RNase inhibitor isolated from rat liver, at 25° C (○, ●), 30° C (△), or 37° C (□, ■) for indicated periods. (A, B, C) Experiments with rat liver nucleoli; (C, D) with MH 134 nucleoli. ○, △, □, Experiments in the presence of the RNase inhibitor; ●, ■, those in the absence of the RNase inhibitor.

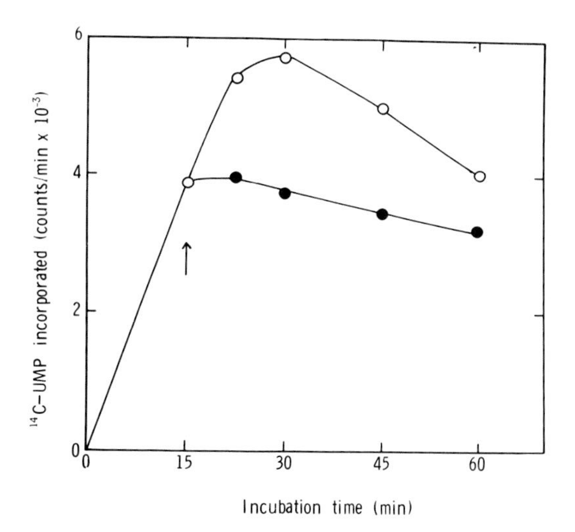


FIG. 3. Effect of a supply of RNase inhibitor on incorporation of labeled nucleotides into acid-insoluble material. After incubation for 15 minutes at 37° C of standard reaction mixture containing 20 units of RNase inhibitor, an additional 20 units of the inhibitor were supplied and incubation was continued for indicated time periods at 37° C. ○—○, RNase inhibitor supplied; ●—●, not supplied.

seen clearly, the higher the reaction temperature, the earlier was the time to reach the plateau. Thus, at 37°C, incorporation reached a maximum at 15 minutes and then began to decrease fairly rapidly, most likely owing to RNase activity. At 25°C, it reached a plateau after 30–60 minutes, and never again decreased so rapidly. The behavior at 30° C was similar to that at 25° C, but a little faster. The effect of RNase inhibitor is also apparent in each case, especially at 25°C, where the incorporation was more efficient and continued much longer. This is probably due to the heat lability of the inhibitor, as shown in Fig. 3. Similar effects were demonstrated on MH 134 nucleoli (Fig. 2D and E).

Divalent cation is essential for *in vitro* RNA synthesis with nucleoli. $Mg^{2+}$ and $Mn^{2+}$ were equally effective for overall incorporation of [$^{14}C$]UMP into acid-insoluble material (Fig. 5). However, sedimentation profiles of RNA synthesized with different divalent cations were very different, as shown in Fig. 4. In the presence of RNase inhibitor and $Mg^{2+}$, labeled RNA accumulated in high-molecular-weight regions with a maximum sedimentation coefficient of 45 S as incubation proceeded at 25°C (Fig. 4B). On the other hand, any high-molecular-weight RNA was not detected when incubation was performed in the presence of $Mn^{2+}$ (Fig. 4A and B). This is most likely due to the protection of newly synthesized RNA by $Mg^{2+}$ ions as demonstrated by the following experiment. After nucleoli were allowed to

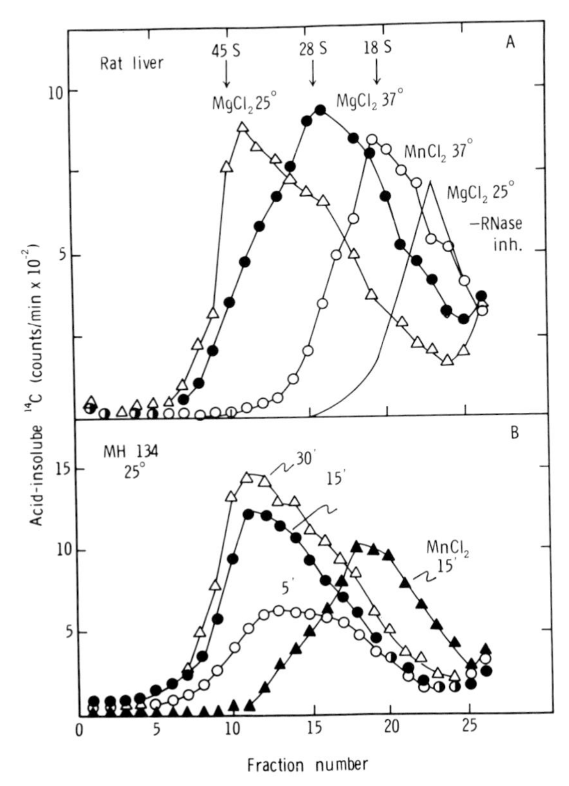


FIG. 4. Sedimentation profiles of *in vitro* synthesized RNA. Reaction mixtures scaled up 3-fold from the standard mixture contained: 900 units of RNase inhibitor, 15 μg nucleoli (as DNA) isolated from rat liver (A) or MH 134 (B), and 5 m*M* $MgCl_2$ or 1 m*M* $MnCl_2$. In A, the reaction mixture was incubated for 15 minutes under specified conditions. In B, it was incubated at 25°C for specified periods in the presence of $Mg^{2+}$, except for one where $Mn^{2+}$ was used (▲—▲). Thereafter RNA was deproteinized and analyzed on a sucrose gradient as described in the legend to Fig. 1.

incorporate nucleoside triphosphates for 10 minutes at 25°C, they were separated by centrifugation and resuspended in RNA synthesis medium, omitting nucleoside triphosphates. They were incubated for an additional 20 minutes at 25°C in the presence of RNase inhibitor with either $Mg^{2+}$ or $Mn^{2+}$. Sucrose gradient analysis of the resulting RNA showed that the RNA in the presence of $Mn^{2+}$ was considerably smaller (28 S vs 43 S) than that incubated in the medium containing $Mg^{2+}$ (data not shown). Observation by Winicov and Perry (*20*) that endoribonuclease purified from nucleoli was inhibited by $Mg^{2+}$ may be compatible with our observation, although no data are available as to the multiplicity of endoribonuclease in the nucleolus.

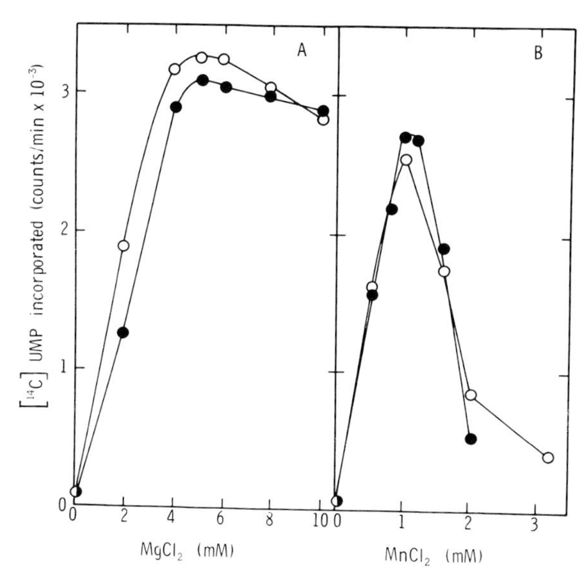


FIG. 5. Effect of $Mg^{2+}$ and $Mn^{2+}$ on RNA polymerase activity in isolated nucleoli. The standard reaction mixture contained various concentrations of $MgCl_2$ (A) or $MnCl_2$ (B). Incubation was carried out at 25°C for 15 minutes. ○—○, Rat liver nucleoli; ●—●, MH 134 nucleoli.

Incubation temperature of nucleoli also affects the accumulation of high-molecular-weight RNAs. In the presence of $Mg^{2+}$, RNA synthesized at 25°C contains twice as much 45 S RNA as that synthesized at 37°C (Fig. 4A). This may be both due to the higher stability of the RNase inhibitor and to the lesser activity of RNase at 25°C than at 37°C.

From these observations, we have chosen $Mg^{2+}$ as the divalent cation and 25°C as the incubation temperature for the efficient synthesis of 45 S RNA.

## III. Characterization of *in Vitro* System by Product Analysis

### A. RNA Size

We have shown elsewhere (*21a, 25*) that the size of the DNA in isolated nucleoli is of the order of $1.7 \times 10^4$ nucleotide length, which is sufficient for the coding of at least one 45 S molecule. This provides a theoretical basis for the attempt to synthesize 45 S RNA with isolated nucleoli. When nucleoli were incubated under optimal conditions described above, there was a broad distribution of radioisotope-labeled RNA after 5 minutes' incubation (Fig. 4B). At this time, a small amount of label was already detectable in the region of 45 S. More substantial amounts of radioactivity appeared near the 45 S region with longer labeling periods. The position of 45 S RNA was determined with a cold 45 S RNA purified from rat liver nucleoli and included as an internal standard. Although the highest peak was usually a little shifted toward lower sedimentation values from the 45 S peak, RNA species with the size of 45 S was apparently accumulated as the incubation proceeded, judging from the shape of the radioactivity in the high-molecular-weight region. A similar pattern was obtained when the incubation time was extended to 60 minutes. These kinetics of sedimentation patterns appear to indicate that most, if not all, of the incorporation represents elongation of preinitiated RNA chains, and that little, if any, reinitiation occurs in isolated nucleoli under the present incubation conditions.

Grummt and her co-workers observed radioactive peaks larger than 45 S RNA in their *in vitro* system of rat liver nucleoli (*9,21*). They considered that these large molecules were possibly synthesized by a readthrough of a few 45 S transcription units (rRNA cistrons). Such long rRNA molecules had also been reported to exist in nucleoli by Scharman *et al.* (*36*) and Quagliarotti *et al.* (*37*). In our system, however, no distinct peaks sedimenting faster than 45 S RNA species are detected under the present conditions. Whether or not any processing of 45 S RNA into more mature rRNA precursors occurs in this system could not be answered with this sedimentation analysis alone. However, structural analysis with a fingerprinting technique suggests that the processing may indeed be taking place.

## B. RNA Chain Elongation Rate

More than 80% of RNA-synthesizing activity in isolated nucleoli seems to represent elongation of *in vivo* preinitiated RNA chains (see above) (*21a*). Since RNase activities could be suppressed almost completely by a combination of RNase inhibitor of rat liver, suitable temperature, and $Mg^{2+}$, the rate of chain elongation in this system was calculated from the peaks of pulse-labeled RNA. At a 3-minute pulse, there was distribution of growing chains with a modal sedimentation coefficient around 23 S. When the incubation time was extended to 6 minutes, the growing chains were elongated resulting in a shift of the 23 S peak to 28 S. If the molecular weights of the 23 S and 28 S peak are assumed to be $1.1 \times 10^6$ and $1.6 \times 10^6$, respectively, then about 1400 nucleotides had been polymerized during the incubation period of 3 minutes, which yields a chain elongation rate of approximately 8 nucleotides per second. This velocity is considerably slower than the value 43 nucleotides per second at 37°C estimated by Grummt (*38*) and that calculated from RNA chain growth *in vivo* by Greenberg and Penman (*39*), but is in good agreement with that observed in isolated HeLa cell nuclei in the presence of $\alpha$-amanitin at 25°C (*40*).

With this step time, 45 S RNA with 14,000 nucleotides could be synthesized in full length in about 30 minutes. Experimental data that indicate a saturation of counts in 45 S RNA at about 30 minutes are compatible with this estimation of the step time.

## C. Hybridization Analysis of the Product

The fidelity of transcription in isolated nucleoli *in vitro* could be tested by a molecular hybridization-competition experiment. However, there are in fact some difficulties in performing this type of experiment, when one uses isolated nucleoli for the synthesis of labeled RNA, because they contain a large amount of *in vivo* synthesized pre-rRNA in comparison with *in vitro* synthesized RNA. This leads to a rather big dilution of labeled RNA with unlabeled RNA, resulting in a low specific activity of *in vitro* synthesized RNA sequences. The best way to avoid this difficulty may be to use, for instance, mercury-substituted nucleotides as a substrate for *in vitro* synthesis, followed by separation of this RNA by affinity chromatography (*41,42*). However, by carefully purifying 45 S RNA on a sucrose density gradient, Grummt (*38*) obtained *in vitro*-synthesized RNA with a specific activity sufficiently high for performing successful hybridization-competition experiments. The data have shown that *in vitro*-synthesized 45 S RNA in isolated nucleoli could be competed with *in vivo* 45 S RNA to about 95%, whereas it was competed with 24 S + 18 S rRNA only by 60–70% (*38*). This is in accord with the expectation that the rRNA genes are transcribed rather faithfully in isolated nucleoli.

## D. Fingerprinting Analysis of the Product

The most powerful tool for identification of an RNA is the fingerprint analysis of large oligonucleotides derived from RNase digestion. In order to see the sequences of RNA synthesized in isolated nucleoli, a homochromatography pattern of 45 S RNA synthesized *in vitro* was compared with that labeled *in vivo* (*21b*). Figure 6B (*27,43*) shows an RNase T1 fingerprint of 45 S RNA synthesized with isolated nucleoli of MH 134 cells. For comparison, a fingerprint of *in vivo* labeled 45 S RNA is also presented (Fig. 6B). It is apparent from the pattern that discrete segments of the DNA in the nucleolus are being transcribed in this system and that the segments include a considerable portion of 45 S RNA. The fact that no spots derived from 18 S RNA could be found in *in vitro* synthesized RNA (e.g., spot s *5, 11, 16, 21* and *22*) may be explained by the short labeling period, during which only about $\frac{1}{6}$ of

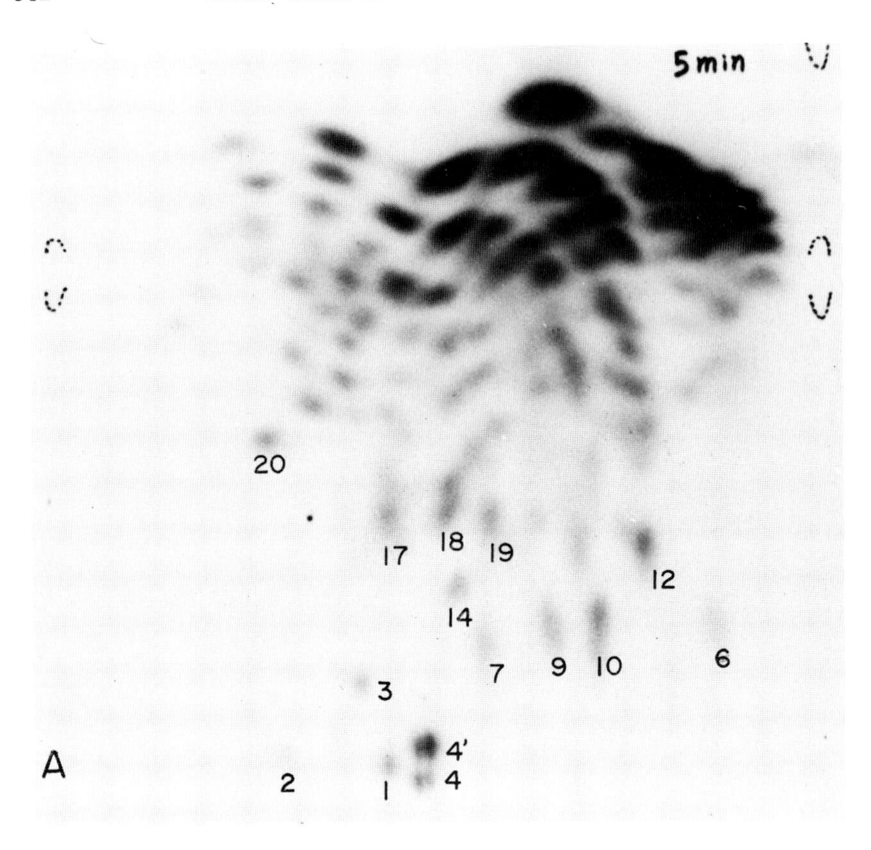


FIG. 6. Fingerprints of 45 S RNA synthesized *in vivo* and *in vitro* on DEAE-homochromatography.

(A). *In vitro* RNA was labeled in a reaction mixture scaled up 20-fold from the standard condition in the presence of 0.66 m*M* nonradioactive UTP, 0.013 m*M* [$\alpha$-$^{32}$P]CTP, 6000 units of RNase inhibitor, and 100 $\mu$g of DNA-equivalent MH 134/C nucleoli. After incubation at 25° C for 5 minutes, RNA was extracted and fractionated in a sucrose gradient as described in the legend to Fig. 1, except that the gradient was centrifuged in a Hitachi RPS-40T rotor at 26,000 rpm for 15 hours.

(B) *In vivo* RNA was labeled by incubating MH 134/C cells collected from ascites fluid in a phosphate-depleted MEM medium containing 0.2 mCi/ml of $^{32}$P-labeled orthophosphate for 5 hours at 37°C. RNA was extracted from isolated nucleoli and fractionated in a sucrose gradient (*27*). 45 S RNA was completely digested with RNase $T_1$, and the digest was fingerprinted using a homochromatography procedure essentially as described by Brownlee and Sanger (*43*).

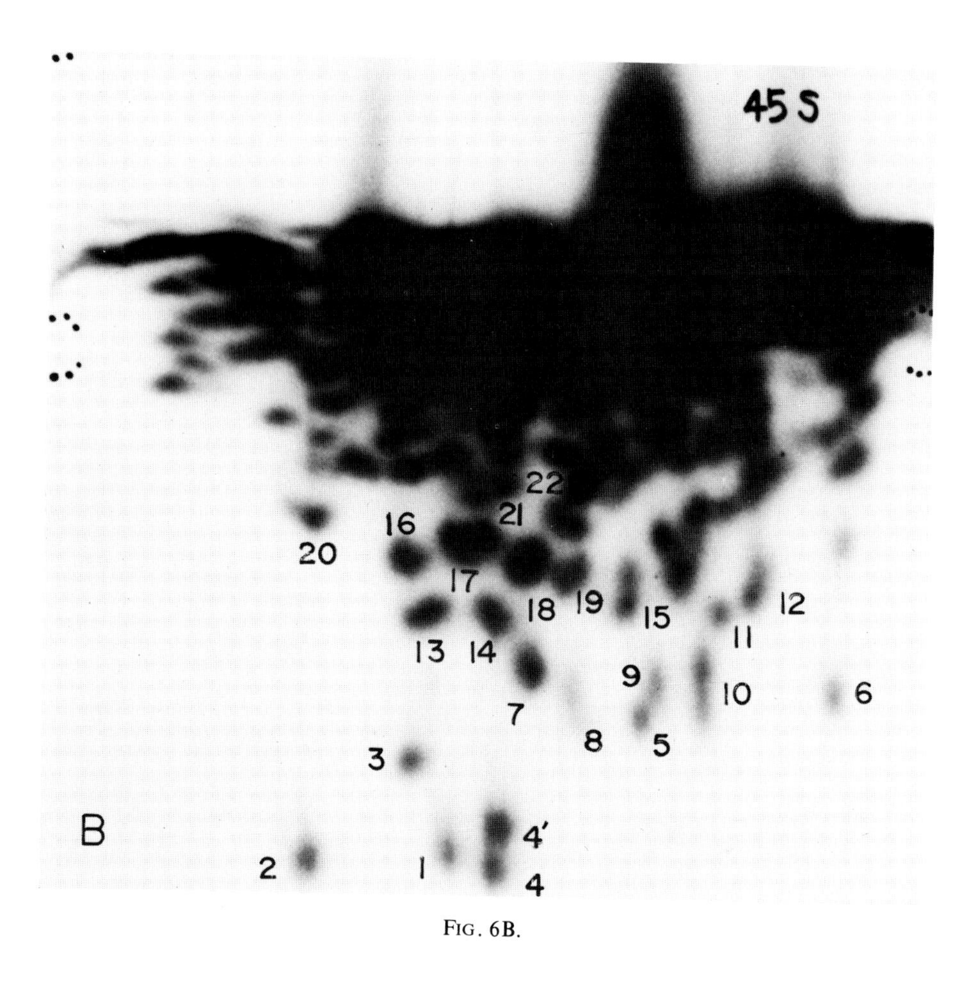


FIG. 6B.

the 45 S RNA molecule was labeled from the 3′ end. The intense labeling of spot 4′ may indicate the location of this sequence near the 3′ end of the 28 S RNA. Spots 1 and 2 are clearly derived from the transcribed spacer, since they are found only in 45 S RNA, but not in 28 S or in 18 S RNA (data not shown). These spots are found in *in vitro* synthesized 45 S RNA, which is labeled presumably at 3′ one sixth of the molecule. Spots 13 and 16, which are derived from 28 S RNA, are missing in *in vitro*-synthesized RNA. This implies that the labeling began at somewhere in the middle of 28 S RNA molecule. If this is the case, the spacer containing spots 1 and 2 must be attached to the 3′ end of the 28 S RNA. Although these topographical reasonings require more detailed analysis of the fingerprints at different sedimentation values and labeling times, this *in vitro* system will provide an unique opportunity for the structural analysis of 45 S RNA and its processing.

## IV. Conclusions

We have described in this chapter optimal conditions for the *in vitro* synthesis of 45 S preribosomal RNA with isolated nucleoli. It was possible with the use of RNase inhibitor from rat liver cytoplasm to synthesize a maximum amount of 45 S RNA in the nucleoli of both rat liver and mouse ascites hepatoma, MH 134. The presence of $Mg^{2+}$ ions was found to be necessary, and the recovery of 45 S RNA was higher at 25°C than 37°C in this system. From the movement of radioactive peak in the sucrose gradient (Fig. 4B), it may be inferred that elongation of preinitiated chains is predominant under these conditions. Recent experiments with heparin, however, suggest that 10–20% of the incorporation may be due to newly initiated chains (*21a*), although more detailed analysis is required to establish this point. In any event, the control mechanisms of the initiation of transcription in isolated nucleoli will be one of the most exciting subjects in future studies.

In view of the chain elongation rate, it will take 30 minutes to label the entire length of 45 S RNA in this system. By using different labeling times and by taking peaks with different S values, it may be possible to map the 45 S RNA with respect to the 18 S, 28 S, and spacer sequences. If the initiation can also take place, then one could determine whether *in vivo* 45 S RNA is really the primary transcript or not, and also find an approach to the structure of the promoter of rRNA genes in mammalian cells.

In this way, isolated nucleolar system may be utilized for a number of studies which aim at elucidation of molecular mechanisms of transcription and processing of rRNA in eukaryotic organisms.

### Acknowledgments

The authors wish to thank Dr. I. Grummt for kindly showing us her techniques of 45 S RNA synthesis in isolated nucleoli. We also thank all our colleagues who collaborated with us in the studies presented here, especially Drs. K. Niyi and R. Kominami for discussion and for permitting us the use of unpublished data. The work presented here was supported in part by grants from Ministry of Education, Science and Culture, and from the Naito Foundation.

### References

*1.* Busch, H., and Smetana, K., "The Nucleolus". Academic Press, New York (1970).
*2.* Matsui, T., Onishi, T., and Muramatsu, M., *Eur. J. Biochem.* **71**, 361 (1976).
*3.* Ro, T. S., Muramatsu, M., and Busch, H., *Biochem. Biophys. Res. Commun.* **14**, 149 (1964).
*4.* Weiss, S. B., *Proc. Natl. Acad. Sci. U.S.A.* **46**, 1020 (1960).
*5.* Ro, T. S., and Busch, H., *Cancer Res.* **24**, 1630 (1964).
*6.* Liau, M. C., Hnilica, L. S., and Hurlbert, R. B. *Proc. Natl. Acad. Sci. U.S.A.* **53**, 626 (1965).

7. Jacob, S. T., Sajdel, E. M., and Munro, H. N., *Eur. J. Biochem.* **7**, 449 (1969).
8. Culp, L. A., and Brown, G. M., *Arch. Biochem. Biophys.* **137**, 222 (1970).
9. Grummt, I., Loening, U. E., and Slcak, J. M. W., *Eur. J. Biochem.* **59**, 313 (1975).
10. Liau, M. C., Flatt, N. C., and Hurlbert, R. B., *Biochim. Biophys. Acta* **224**, 282 (1970).
11. Liau, M. C., Hunt, J. B., Smith, D. W., and Hurlbert, R. B., *Cancer Res.* **33**, 323 (1973).
12. Liau, M. C., and Hurlbert, R. B., *J. Mol. Biol.* **98**, 321 (1975a).
13. Liau, M. C., and Hurlbert, R. B., *Biochemistry* **14**, 127 (1975b).
14. Liau, M. C., Craig, N. C., and Perry, R. P., *Biochim. Biophys. Acta* **169**, 196 (1968).
15. Kelley, D. E., and Perry, R. P., *Biochim. Biophys. Acta* **238**, 357 (1971).
16. Mirault, M. E., and Scherrer, K., *FEBS Lett.* **20**, 233 (1972).
17. Villalobos, J., Jr., Steele, W. J., and Busch, H., *Biochim. Biophys. Acta* **103**, 195 (1965).
18. Prestayko, A. W., Lewis, B. C., and Busch, H., *Biochim. Biophys. Acta* **269**, 90 (1972).
19. Prestayko, A. W., Lewis, B. C., and Busch, H., *Biochim. Biophys. Acta* **319**, 323 (1973).
20. Winicov, I., and Perry, R. P., *Biochemistry* **13**, 2908 (1974).
21. Grummt, I., and Lindigkeit, R., *Eur. J. Biochem.* **36**, 244 (1973).
21a. Onishi, T., Matsui, T. and Muramatsu, M., *J. Biochem.* **82**, 1109 (1977).
21b. Onishi, T., Niyi, K., Kominami, R., Urano, Y. and Muramatsu, M. (1977).
22. Grummt, I., Smith, V. A., and Grummt, F., *Cell* **7**, 439 (1976).
23. Higashinakagawa, T., Muramatsu, M., and Sugano, H., *Exp. Cell Res.* **71**, 65 (1972).
24. Muramatsu, M., Hayashi, Y., Onishi, T., Sakai, M., Takai, K., and Kashiyama, T., *Exp. Cell Res.* **88**, 345 (1974).
25. Muramatsu, M., and Onishi, T., *Methods Cell Biol.* **15**, 221 (1977).
26. Muramatsu, M., Shimada, N., and Higashinakagawa, T., *J. Mol. Biol.* **53**, 91 (1970).
27. Muramatsu, M., *Methods Cell Biol.* **7**, 23 (1973).
28. Onishi, T., Shimada, K., and Takagi, Y., *Biochim. Biophys. Acta* **312**, 248 (1973).
29. Efstratiadis, A., Maniatis, T., Kafatos, F. C., Jeffrey, A., and Vournakis, J. N., *Cell* **4**, 367 (1975).
30. Gribnau, A. A. M., Schenmakers, J. G. G., and Bloemendal, H., *Arch. Biochem. Biophys.* **130**, 48 (1969).
31. Rhoads, R. E., McKnight, G. S., and Schimke, R. T., *J. Biol. Chem.* **248**, 2031 (1973).
32. Palmiter, R. D., *J. Biol. Chem.* **248**, 2095 (1973).
33. Cox, R. F., *Eur. J. Biochem.* **39**, 49 (1973).
34. Ferencz, A., and Seifart, K. H., *Eur. J. Biochem.* **53**, 605 (1975).
35. Walter, G., Zilling, W., Palm, P., and Fuchs, E., *Eur. J. Biochem.* **3**, 194 (1967).
36. Scharman, O. K., Hidvégi, E. J., Marks, F., Prestayko, A. W., Smetana, K., and Busch, H., *Physiol. Chem. Phys.* **1**, 185 (1969).
37. Quagliarotti, G., Hidvégi, E., Wickman, J., and Busch, H., *J. Biol. Chem.* **245**, 1962 (1970).
38. Grummt, I., *Eur. J. Biochem.* **57**, 159 (1975).
39. Greenberg, H., and Penman, S., *J. Mol. Biol.* **21**, 527 (1966).
40. Udvardy, A., and Seifart, K. H., *Eur. J. Biochem.* **62**, 353 (1976).
41. Smith, M. M., and Huang, R. C. C. *Proc. Natl. Acad. Sci. U.S.A.* **73**, 775 (1976).
42. Crouse, G. F., Fodor, E. J. B., and Doty, P., *Proc. Natl. Acad. Sci. U.S.A.* **73**, 1564 (1976).
43. Brownlee, G. G., and Sanger, F., *Eur. J. Biochem.* **11**, 395 (1969).

# Chapter 27

# *Transcription of RNA in Isolated Nuclei*

WILLIAM F. MARZLUFF, JR.

*Department of Chemistry,*
*Florida State University,*
*Tallahassee, Florida*

## I. Introduction

The goal of research in control of gene activity is to reproduce in a cell-free system the controls that operate *in vivo*. Here I describe a system that fulfills many of the requirements for study of gene control *in vitro* (*1–3*). In addition the system makes it possible to study the primary transcription event separate from later maturation and transport events in RNA synthesis. While the results here are described for myeloma cells, lines derived from the MPC-11 tumor, our preliminary results with chick embryos and those of others with HeLa cells (*4,5*) and other myeloma cells (*6*) suggest that these methods may be generally applicable. Recently, we have shown (D. Price and W. F. Marzluff, unpublished results) that identical results are obtained with rat mammary tumor cells grown *in vitro*.

## II. Growth of Cells

Under our conditions (Dulbecco's Minimal Eagle's Medium +10% horse serum) cells grow logarithmically up to $1.5 \times 10^6$/ml with a doubling time of 16–18 hours. Optimal activity is obtained from cells growing exponentially at less than $6 \times 10^5$ ml. Nuclei isolated from cells growing at $10^6$/ml have reduced activity, approximately one half of the activity per microgram of DNA. Normally cells are diluted to 5 to $7 \times 10^4$/ml and grown for 48 hours prior to harvesting. The nuclei reflect the activity of the cells from which they are isolated. Thus one can isolate nuclei from cells in a particular growth condition (i.e., amino acid starvation, synchronization) with a characteristic activity.

## III. Preparation of Nuclei

All glassware and solutions are routinely sterilized either by autoclaving or rinsing with a dilute solution of diethylpyrocarbonate. The key to successful preparation of nuclei (and particularly ability to store active nuclei) is speed in preparation. All steps are carried out at 0°–4°C. The cells are centrifuged for 2 minutes at 800 *g* in the cold, resuspended in cold lysis buffer [0.32 *M* sucrose–3 m*M* $CaCl_2$–2 m*M* $Mg(CH_3COO)_2$–0.01 *M* Tris-HCl (pH 8)–1 m*M* dithiothreitol–0.1% Triton X-100] at a concentration of 1 to $5 \times 10^7$/ml and homogenized with 10–15 strokes with the B (tight) pestle in a Dounce homogenizer. The homogenate is mixed with 1/2 volume of suspension buffer [2 *M* sucrose–5 m*M* $Mg(CH_3COO)_2$–0.01 *M* Tris (pH 8)–1 m*M* dithiothreitol]. Then 3.5 ml is layered over a 2-ml pad of the same suspension buffer and centrifuged for 45 minutes at 20,000 rpm in the SW 50 rotor. Larger buckets have been used, but unless large amounts of nuclei ($> 5 \times 10^8$ cells) at high concentrations are used, the yield is much lower. The nuclear pellet (yield 30–70% of total DNA) is drained and suspended by gentle homogenization (0.5 to 1 ml per $10^8$ cells) in storage buffer [25% glycerol–5 m*M* Mg $(CH_3COO)_2$–0.05 *M* Tris-HCl (pH 8)–5 m*M* dithiothreitol]. The nuclei may be quick-frozen in Dry Ice and stored in liquid $N_2$ for at least a month without loss of activity.

For maximal activity, nuclei are incubated in a 0.2 ml reaction mixture containing 0.1 ml of nuclei, 0.04 ml of salt solution [0.6 *M* KCl (or $NH_4Cl$)–12.5 m*M* Mg $(CH_3COO)_2$], 0.05 ml of triphosphates [1.6 m*M* ATP–0.8 m*M* CTP–0.8 m*M* UTP–0.16 m*M* $^3H$ or $\alpha$-$^{32}PO_4$-GTP], and 0.01 ml of $H_2O$ (or inhibitors of RNA synthesis). The labeled GTP is dried under $N_2$ and dissolved in the triphosphate solution directly before use. The individual triphosphates are stored at −20°C at a concentration of 20–40 m*M* in sterile water. The radioactive triphosphate solution loses activity on freezing within a week.

The nuclei are incubated at 25°C for up to 60 minutes. Incorporation is assayed by spotting 5–10 $\mu$l directly onto Whatman No. 1 filter paper at desired times (or dissolving the aliquot in 5–10 volumes of 1% SDS–10 m*M* EDTA and then spotting the viscous solution directly on filter paper.) The filters are washed in 5% trichloroacetic acid (TCA–0.01 *M* sodium pyrophosphate at 4°C for 10 minutes (15 ml/filter) in a beaker with gentle stirring, washed twice more for 5 minutes in the same solution, rinsed with acetone, dried thoroughly, and counted in toluene-based scintillation fluor at 20% efficiency. Lower backgrounds (zero time) are found if the aliquot is dissolved in sodium dodecyl sulfate (SDS). The background observed is strongly dependent on DNA concentration. If high specific activities of [$^3$H]GTP (10 Ci/mmole) are used, 5 $\mu$l of reaction mix (300–500 $\mu$g of DNA/per milliliter) incorporates 50 to 100 $\times$ $10^3$ cpm (10–20 pmoles of RNA) in 30 minutes over a background of 1 to 5 $\times$ $10^3$ cpm.

Incorporation is routinely expressed in terms of picomoles of GMP incorporated per microgram of DNA. Up to 10% of the labeled GTP is incorporated into RNA. To ensure that no whole cells were present $\alpha$-$^{32}PO_4$-GTP was used as a precursor, and the RNA product was treated with venom phosphodiesterase. Only $^{32}PO_4$-GMP was found, indicating that only nuclei were active. Whole cells will incorporate [$^3$H]GTP into RNA (unpublished results), presumably by cleaving the substrate to guanine or guanosine.

## IV. Characterization of the Reaction

Incubation at 25°C gives a larger net incorporation than at 37°C with approximately the same initial rate. The reason for this is not known, but the phenomenon has been observed by others (*6, 7*) and also is common in *in vitro* protein synthesis systems (*8*). It is not due to a defect in initiation of RNA synthesis, as the same extent of initiation is seen at both temperatures after 10 minutes (*1*). The kinetics of RNA synthesis are somewhat variable. Linear kinetics are observed about half the time, with incorporation continuing up to 60 minutes. Often the rate for the first 10–15 minutes is much higher (up to 3–5 pmoles of GMP per microgram of DNA in 10 minutes) with a more gradual increase (1–2 pmoles per microgram of DNA in 10 minutes after that.

In original experiments 1 m*M* $Mn^{2+}$ was included in the assay mix (*1–3*). However, if the $Mn^{2+}$ is omitted, the rate is increased about 50% and no difference in the product is detectable by any criteria (size, distribution, relative activity of the three polymerases, etc.) (Cooper and Marzluff, unpublished observations).

The rates of incorporation are highly reproducible ($\pm$10–20%) in parallel independent nuclear preparations with aliquots from the same culture of cells. However, greater variability is seen with preparations from different cultures, and the activity probably parallels the growth state of the particular culture.

## V. Integrity of the Nucleus during Incubation

There is little evident change in any nuclear constituents during the incubation. Over 80% of the RNA synthesized remains associated with the nuclei after brief centrifugation (800 *g* for 5 minutes) at the end of the reaction. Analysis by gel electrophoresis of the nuclear RNA, nonhistone chromosomal proteins, and histones shows no detectable degradation. In particular, normal amounts of histone I are found after incubation (the histone most sensitive to proteolysis), and the 6 distinguishable low-molecular-weight nuclear RNAs (*9*) are unchanged in mobility and relative amounts.

In addition, there is no apparent change in the DNA molecular weight. The isolated DNA has the same molecular weight on sedimentation (both alkali and neutral) as the DNA prepared from zero time nuclei (Dr. Henry Berger, unpublished results). I feel that this stability of nuclear constituents is the prime reason for the success of the system.

## VI. Identification of RNA Polymerase Activities

The RNA polymerase activities are separated by the differential effects of specific inhibitors $\alpha$-Amanitin inhibits RNA polymerase II at low concentrations ($< 5$ $\mu$g/ml), and polymerase III at high concentrations ($> 100$ $\mu$g/ml) (*10*). At low concentrations it inhibits 40–50% of the activity in isolated nuclei, comparable to the relative amounts of this polymerase present in the nuclei ( unpublished observations).

The ribosomal RNA precursor (45 S rRNA) may be the only product of RNA polymerase (*11, 12*). Its synthesis may be specifically inhibited by incubating cells in low concentrations (0.04 $\mu$g/ml) of actinomycin D for 1 hour before preparing nuclei (*13*). Nuclei prepared from cultures treated this way have 40–50% of the activity of control cells (Fig. 1). Of this activity, 90% is inhibited by $\alpha$-amanitin (1 $\mu$g/ml), and the residual activity is RNA polymerase III synthesizing primarily the 4.5 S rRNA precursor and 5 S rRNA (*2*).

Interestingly, $Mn^{2+}$ is required for maximal activity of RNA polymerase II with DNA (*11*), but in our system partially inhibits the activity of polymerase II in isolated nuclei.

Using this approach, different enzyme activities have been assigned to different RNA products. It should be emphasized that any RNA polymerase I that might be involved in transcribing genes other than 45 S rRNA would not be inhibited by the treatment with low concentrations of actinomycin D.

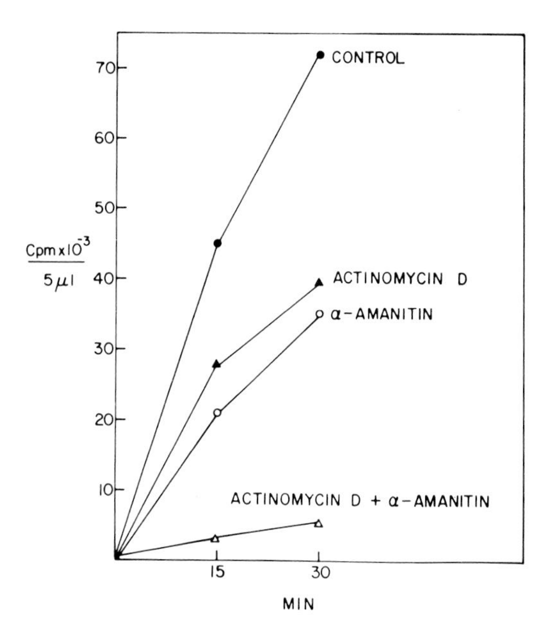


FIG. 1. Activity of different polymerases in isolated nuclei. Nuclei were prepared as described and incubated at a concentration of 300 μg of DNA per milliliter with [$^3$H]GTP (12 Ci/mMole, 1 pmole = 5000 cpm). Half of the culture was preincubated with 0.04 μg of actinomycin D per milliliter for 30 minutes before preparation of the nuclei. Each reaction mix contained 0.3 ml, and 5-μl aliquots were removed at 15 and 30 minutes for determination of TCA-precipitable radioactivity. α-amanitin was present at 1 μg/ml where indicated. Background of 2 to 3 × $10^3$ cpm (0 time) has been subtracted.

## VII. Isolation of RNA

RNA has been isolated from the nuclei be several different methods. Routinely, the hot phenol procedure of Penman is used (*14*). The nuclear reaction is diluted with 5–10 volumes of 1% SDS–10 m*M* EDTA, and the viscous solution is mixed thoroughly and adjusted to 0.1 *M* NaAc (pH 5). Then 0.7 volume of phenol (80% in water) is added, the solution is mixed well, 0.3 volume of chloroform is added, and the mixture is headed to 55°C for 3 minutes. The samples are cooled for 10 minutes and centrifuged; the aqueous phase is adjusted to 0.2 *M* NaCl, and the RNA is precipitated with 2–3 volumes of ethanol.

RNA has been prepared with identical results by digesting the nuclei briefly with DNase I (100 μg/ml for 3 minutes at 25°C) after the reaction (*15*), then extracting as above without heating. Routinely, RNA recovery is 80–95%. We have seen no evidence of loss of specific RNA classes in this procedure.

The isolated RNA is dissolved in sterile water, treated with DNase I [10 μg/ml from a 1 mg/ml stock in 0.1 *M* Tris (pH 7.5)–0.1 *M* $MgCl_2$], adjusted to 0.5% SDS, 0.01 *M* EDTA, and analyzed by sucrose gradient centrifugation.

## VIII. Analysis of RNA

RNA is analyzed by standard techniques of sucrose-gradient centrifugation and gel electrophoresis. Gradients are prepared in 0.1 *M* NaCl–0.01 *M* Tris-HCl (pH 7.5)–.001 *M* EDTA–0.1% SDS. We routinely use two types of gradients: (a) Using the SW-27 rotor we analyze RNA on a 10 to 50% sucrose gradient for 16–18 hours at 25 K at 20°C (a convenient overnight run); (b) using a SW-41 rotor, we use a gradient of 5 to 50% sucrose for 6 hours at 39 K at 20°C. (Gradients may be prepared the day before, allowing the centrifugation and fractionation to be conveniently completed the next day.) Because of the wide range of RNA sizes in the sample (4 S to > 45 S) analysis of specific regions by gel electrophoresis is desirable (*16*).

The following general comments can be made about the RNA product:

1. The bulk (> 50%) of the material is larger than 28 S rRNA (Fig. 2). This is true even after treating the RNA with either formaldehyde or formamide. There are discrete bands in the same position as RNA species found *in vivo*, which are the precursors to 28 S and 18 S rRNA, and this RNA hybridizes specifically to ribosomal DNA (Table I). Large RNA is made both by polymerase I (Fig. 2B) and polymerase II (Fig. 2C), but only small RNA is synthesized by polymerase III (Fig. 2D). Polymerase II synthesizes much of the RNA made in the 7–18 S size range (Fig. 2A–C).
2. There is 5–10% of the total incorporation into small RNAs (< 6 S).
3. There is no obvious incorporation into 18 S and 28 S rRNA.
4. Some of the product is retained by oligo(dT) cellulose (particularly in the 7–18 S size class RNAs), suggesting the presence of oligo-or poly(A) (*1*). This synthesis is greatly inhibited by α-amanitin.
5. There is no great change in the various RNA species accumulated during a chase period. Chase experiments are very conveniently done by adding unlabeled GTP to a final concentration of 0.5–1 m*M* at a desired time without affecting RNA synthesis.

Taken together, these results suggest that there is minimal posttranscriptional maturation taking place in this system. In agreement with this conclusion is the fact that normally there is no 4 S tRNA formed, but only

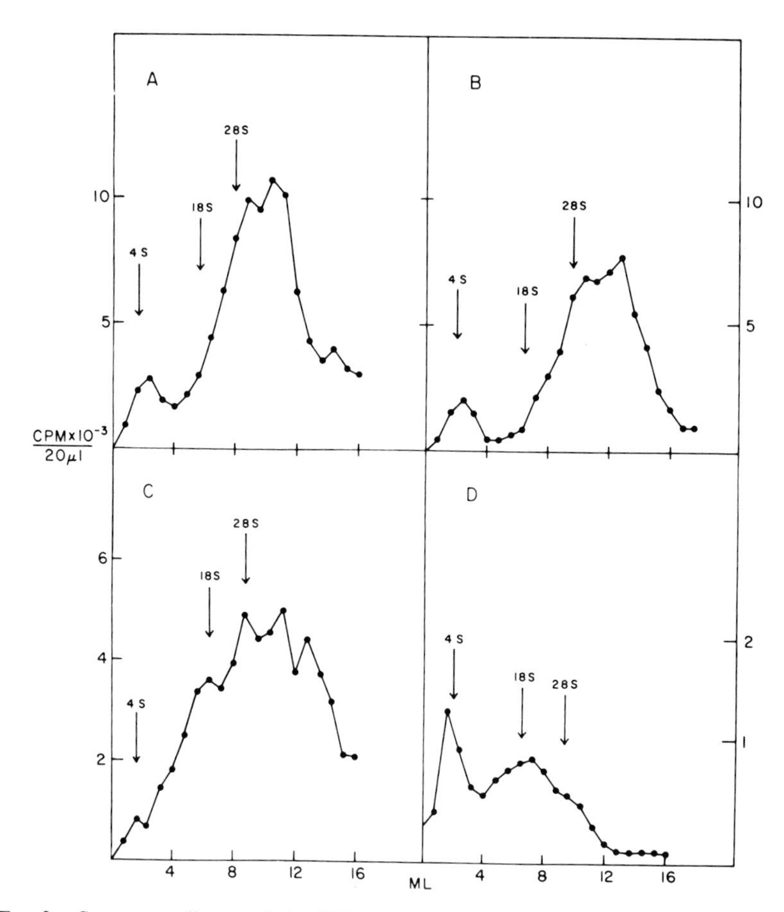


FIG. 2. Sucrose gradient analysis of RNA. RNA was prepared by the hot phenol procedure from each of the four reactions shown in Fig. 1 after incubation for 30 minutes. The RNA was treated with DNase (10 μg/ml), heated to 60°C, and analyzed on a 10 to 50% sucrose gradient [0.1 *M* NaCl–1 m*M* EDTA–10 m*M* Tris (pH 7.5)] for 16 hours at 25 K in the AH 827 rotor (17-ml buckets) in a Sorvall OTD-2 ultracentrifuge. Fractions of 0.8 ml were collected, and 0.02-ml aliquots were taken for determination of radioactivity. Recovery of radioactivity from both the reaction and the gradient was greater than 90%. The position of marker RNA included in the gradient is shown. (A) Control; (B) + 1 μg of α-amanitin per milliliter; (C) preincubated with 0.04 μg of actinomycin D per milliliter; (D) preincubated with actinomycin D + 1 μg of α-amanitin per milliliter.

the 4.5 S precursor to tRNA, which is stable in the *in vitro* system (Fig. 3). No detectable RNA methylation or "capping" (*17*) takes place in these purified nuclei, in contrast to cruder systems (*18*). These maturation steps normally take place in the nuclei, but the enzymes are washed out during preparation of nuclei.

TABLE I

HYBRIDIZATION OF *in Vitro* SYNTHESIZED RNA TO PURIFIED DNAs[a]

| Input RNA | DNA | $^{3}H$ input | $^{32}PO_4$ standard input | Counts of bound $^{3}H$ | Counts of bound $^{32}PO_4$ | Counts of 10 μg of bound and competitor $^{3}H$ | Counts of 10 μg of bound and competitor $^{32}PO_4$ | Efficiency of hybridization (%) | Percent of sample as indicated by hybridization | Percent by gel electrophoresis |
|---|---|---|---|---|---|---|---|---|---|---|
| 1. 4–8 S[b] | 5 S | 35,000 | 2,000 | 97 | 138 | 7 | 8 | 7 | 4 | 5.5 |
| 2. 5 S[c] | 5 S | 3,000 | 4,000 | 86 | 173 | 0 | 0 | 4 | 70 | 90 |
| 3. 4–8 S[d] | 5 S+ | 33,000 | 3,000 | 80 | 50 | 5 | 0 | 1.6 | 14 | 10 |
| 4. 4–8 S[d] | 5 S− | 33,000 | 3,000 | 3 | 0 | 0 | 0 | 0 | — | — |
| 5. 4.5 S[e] | tRNA | 8,000 | 19,000 | 176 | 164 | 36 | 16 | 1 | 100 | 100 |
| 6. 4.5 S[f] | tRNA | 4,000 | 5,500 | 44 | 30 | 7 | 5 | 0.5 | 100[i] | 100 |
| 7. 30–50 S[g] | Ribosomal | 86,000 | 5,100 | 924 | 199 | 68 | 6 | 4 | 27.6[j] | — |
| 8. 30–50 S[h] | Ribosomal | 48,000 | 4,400 | 514 | 159 | 0 | 0 | 3.6 | 34[j] | — |

[a] RNA was hybridized to the indicated DNA immobilized on nitrocellulose filters. In lines 3 and 4, the separated strands of 5 S DNA were used, the + strand being the one coding for 5 S RNA. The 5 S DNA and tRNA–DNA were isolated from *Xenopus* erythrocytes, and the ribosomal DNA from urchin sperm (*Arbacia punctulata*). Hybridization was in 4 × SSC–50% formamide–0.1% sodium dodecyl sulfate (pH 8) for 18 hours (SSC = 0.15 *M* NaCl–0.015 *M* sodium citrate).

[b] RNA made in the presence of α-amanitin (Fig. 2B).

[c] Same as in footnote *b* except that the 5 S RNA was isolated by gel electrophoresis.

[d] RNA made in the presence of α-amanitin + actinomycin D (Fig. 2D).

[e] RNA labeled *in vivo* for 10 minutes, isolated by gel electrophoresis.

[f] Same as in footnote *b* except the 4.5 S RNA was isolated by gel electrophoresis.

[g] RNA made *in vitro* and fractionated on a sucrose gradient (Fig. 2A).

[h] RNA made *in vitro* in the presence of α-amanitin (Fig. 2B).

[i] Assumed on the relative hybridization of $^{3}H/^{32}PO_4$ for the *in vivo* sample. The higher efficiency for $^{3}H$ is probably due to the lack of modified bases in the precursor, which results in a more stable hybrid (confirmed by melting analysis).

[j] The amount shown in a minimum one, as the standard $^{32}PO_4$ was 3000 cpm (28 S) and 1500 cpm (18 S). There is a higher efficiency of hybridization of 18 S due to the higher concentration of endogenous sequences containing 28 S RNA in the nuclei. On gel electrophoresis, the samples contain discrete bands presumably corresponding to the stages of processing of the rRNA precursor (not shown).

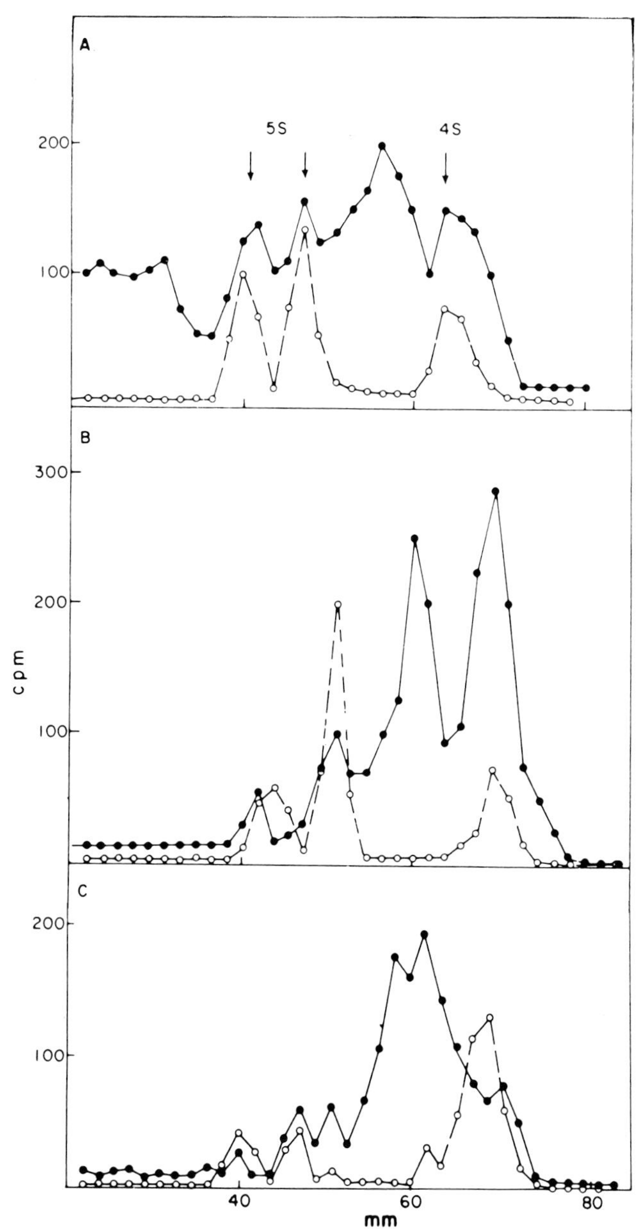


FIG. 3. Low-molecular-weight RNA (equivalent to fractions 2–6 from the gradients in Fig. 2) was analyzed by gel electrophoresis on 10% polyacrylamide gels. $^{32}PO_4$-labeled 5 S and 4 S RNA was included in the sample. The gels were sliced into 1.5 mm fractions, and radioactivity was determined after eluting the RNA in 1 ml of 0.3 *N* NaOH at 37°C for 18 hours; 0.5 ml of 1 *N* acetic acid and 10 ml of Triton–toluene fluor were added. (A) Control; (B) + 5 $\mu$g of $\alpha$-amanitin per milliliter; (C) preincubated with actinomycin D (0.04 $\mu$g/ml) and incubated with $\alpha$-amanitin (5 $\mu$g/ml). From Marzluff *et al.* (*2*).

Analysis of the RNA is limited by the necessity of having a specific assay for a particular RNA species. These assay procedures are developing rapidly.

## IX. Analysis of RNA Products by Hybridization with Specific DNA Sequences

Hybridization is carried out by the general procedure of Reeder and Roeder (*12*) using an internal standard of pure RNA labeled with a different isotope. Purified specific DNAs are used. DNA complementary to ribosomal RNA, 5 S rRNA, or tRNA prepared from *Xenopus laevis* by the methods of Brown and co-workers (*19*) cross-reacts well with the same mouse species and shows very low binding for other sequences. This allows data to be interpreted quantitatively (Table I) with great accuracy. The unknown sample and a small sample of pure RNA (1000–2000 cpm) are mixed and hybridized in 4 × SSC (0.6 NaCl–0.06 *M* sodium citrate adjusted to pH 8)–50% formamide–0.1% SDS for 18 hours at 45°C. Under these conditions no DNA is lost from the filter (*9*). The filters are washed 3 times in 4 × SSC, dried, and counted. Usually 100–400 counts of the standard are bound (zero background). Because of the high efficiency of hybridization using purified DNA, only a small amount of input is necessary [an input of 50,000 cpm with only 2% of the sample the desired RNA (1000 cpm) will result in specific hybridization of 50–100 cpm over a background of generally less than 10 cpm]. Both melting analysis of the hybrids (*2*) and appropriate competition experiments have established the fidelity of the RNA product. This assay technique should be readily extendable to other gene products using an immobilized reverse transcript of mRNA (*20,21*). In my experience it is worth the extensive effort necessary to purify the DNA species necessary for the hybridization. The assay works well with heterologous DNAs for many RNA species (ribosomal, 5 S rRNA, tRNA, histone mRNA) allowing choice of a species for DNA abundance and also reducing nonspecific hybridization with contaminating DNA sequences. Use of competitor RNA is essential to assure specificity of the assay.

## X. Initiation of RNA Synthesis

If the system is to be used to study changes in the pattern of transcription, it must be able to initiate RNA chains. The extended kinetics of synthesis suggest that this is the case, although since the rate is substantially lower

(2–5% as fast) than the *in vivo* rate, this does not necessarily imply that the chains are initiated *de novo*. Consideration of specific gene products, particularly the small RNAs, allows one to conclude that some 5 S and 4.5 S pre-tRNA are totally synthesized *de novo*.

First, these genes are small enough that they probably accommodate only one polymerase molecule at a time. Thus knowing the number of genes coding for 5 S and tRNA species (*9*) and the number of RNA molecules synthesized, one can conclude that there is net synthesis and hence necessarily chain initiation. The same argument cannot be used for the longer RNAs due to the packing of many polymerase molecules onto each gene which leads to net synthesis without chain initiation (*12*).

In addition the 5′ nucleotide of 5 S RNA, the initiator ppp Gp has been identified after incubation with [$\gamma$-$^{32}PO_4$]GTP, indicating initiation of this species. Some $\gamma$-$^{32}PO_4$ was also incorporated into 4.5 S tRNA. Proof of initiation of larger species will be difficult owing to the unusual 5′terminal. Incorporation of GTP into the capped, 7-methylguanosine pppXmp has been demonstrated by Groner and Hurwitz (*18*). However, this obviously is not dependent on chain initiation. Incorporation of label into the X nucleotide is also not necessarily an indication of initiation, as this base is not necessarily the initiator. In fact, since X can be any base (*21*), it is unlikely that it is the initiator, as most if not all chains are initiated with pppAp or pppGp.

Thus initiation has been shown for 5 S and 4.5 S tRNA, both of which are transcribed by polymerase III. This polymerase is likely to be most efficient in initiation of the three polymerases, as it makes only small RNAs. There is no reason to conclude from this that polymerases I and II do initiate RNA synthesis in the isolated nuclei. It will be extremely difficult to demonstrate initiation of larger RNA species directly.

## XI. Discussion

The method of preparation may be varied somewhat without altering activity. In particular, detergent may be omitted if cells are first washed and swollen 5 minutes in a hypotonic buffer: 10 m*M* Tris HCl–3 m*M* $CaCl_2$–2 m*M* $Mg(CH_3COO)_2$–25 m*M* KCl. However, the cells are still difficult to break, as they do not swell to the same extent as other cells (e.g., HeLa cells). However, nuclei prepared this way have essentially the same activity and the RNA product is identical.

These nuclei may be lysed gently by removal of divalent cations (washing with 10% glycerol–0.001 *M* DTT) and centrifugation (*3*). The resultant viscous chromatin retains much of the activity of isolated nuclei even though the "soluble" constituents of nuclei have been removed. This method may not be generally applicable as fibroblast nuclei (3T3 cells) are not lysed by this procedure.

The high activity of the system approaches that seen *in vivo*. First, the proportion of different types of RNA is similar to that found *in vivo*. (Table II) (*9,22–25*). Second, there is initiation of most of the low molecular weight (5 S and pre-tRNA) synthesized. The actual activity is calculated to be 2–5% of the *in vivo* rate (based on the estimate of absolute rate of RNA synthesis measured by Brandhorst and McConkey (*24*) or on a calculation of the necessary rate of rRNA synthesis to double the cellular rRNA content per generation) rising to 20% for the initial rate in some experiments. Note that at the accepted maximal rate of rRNA synthesis, 30 nucleotides per second, it takes 5 minutes normally to make a 45 S precursor (*26*). Thus no complete 45 S molecules may be made *in vitro* before 50 minutes of incubation has elapsed. However, it is possible that complete synthesis does eventually take place.

Only a few of the *in vitro* RNA species have been definitively identified, all using various RNA–DNA hybridization methods. rRNA, 5 S RNA, and tRNA have all been identified by hybridization to purified DNAs complementary to those RNA species (Table I). These experiments have demonstrated correct strand selection as well in the case of rRNA (*12*) and 5 S RNA (*2*).

Recently, Smith and Huang (*25*) have demonstrated transcription of the κ-chain messenger RNA *in vitro* using a reverse transcript of the mRNA as a probe. These experiments depended on a quantitative method for separation of newly made RNA from the large amounts of RNA already in the nuclei. This was accomplished by using 5 Hg-UTP as a substrate (*27*) and absorbing the newly made mercurated RNA on a sulphydryl–Sephadex column followed by elution with mercaptoethanol (*28*). This removes all RNA which was not actually synthesized in the nuclei and promises to be an excellent general technique to follow the fate of the newly synthesized RNA (in terms of methylation, polyadenylation, etc.) separate from the RNA already present in the nuclei.

These experiments have pointed out the fact that restriction of transcription follows the *in vivo* pattern quantitatively. The genes for rRNA, 5 S RNA, tRNA (*1,2*) and myeloma κ-chain mRNA (*25*) are transcribed at a rate 2–3 orders of magnitude higher than their representation in the DNA of the genome (Table II). This quantitative restriction has not generally been observed in transcription experiments with isolated chromatin and RNA polymerases (*29–31*).

TABLE II

PROPORTION OF DIFFERENT RNA SPECIES SYNTHESIZED *in Vitro*

| Species | Percent of genome corresponding to RNA species | Amount made *in vivo* in 10-min pulse (%) | Fold amplification | Amount made in isolated nuclei (%) | Fold amplification |
|---|---|---|---|---|---|
| | | | | 0.25–0.5 | |
| | | | | 0.5–1.0[e] | |
| 5 S rRNA | 0.002[a] | 0.8–1 | 400 | 5[f] | 100 |
| | | | | 2–5 | |
| | | | | 5–10[e] | |
| 4 S tRNA (4.5 S precursor) | 0.02[a] | 4–8 | 400 | 50[f] | 200 |
| | | | | 40 | |
| rRNA | 0.1[a] | 40[c] | 400 | 80–90[e] | 400 |
| $\kappa$ chain mRNA | 0.0001 | 0.045[b] | 500 | 0.03[d] | 300 |

[a]Taken from Marzluff *et al.* (*9*).

[b]Calculated assuming a molecular weight of 300,000 and two copies per genome (*22*), and a maximal rate of transcription of 30 mRNAs/minute (*23*) using the rate of RNA synthesis given by Brandhorst and McConkey (*24*).

[c]Calculated from Brandhorst and McConkey (*24*).

[d]Smith and Huang (*25*).

[e]RNA made in the presence of $\alpha$-amanitin.

[f]RNA made from nuclei prepared from cells preincubated in actinomycin D and incubated in the presence of $\alpha$-amanitin.

The major products are not completely matured. Thus the system provides an opportunity to study maturation and also to isolate normally unstable precursors. This is particularly true of the tRNA precursor, which is readily labeled *in vitro* with [$\alpha$-$^{32}PO_4$]GTP but not *in vivo* with $^{32}PO_4$. This RNA could then provide the substrate to use in maturation studies. Methylation and polyadenylation of newly transcribed RNA can also be studied conveniently, especially as newly made RNA can now be separated from endogenous nuclear RNA.

Finally, the system provides a starting point for demonstrating control *in vitro*. It should be possible to predictably alter the pattern of RNA synthesis by adding specific extracts (e.g., extracts from cell variants or cells in different growth stages) and thus identify control elements using the nuclei as an assay system. The nuclei are permeable to proteins; RNA polymerase from *E. coli* stimulated net RNA synthesis severalfold, although partially purified myeloma polymerases do not (unpublished results), suggesting that all the normal polymerase sites are occupied.

## XII. Concluding Remarks

This type of system comes closest to reproducing accurately transcription in the *in vivo* situation. Further dissection of the system will help to clarify the normal transcription process. In addition, it provides a tool for studying many other nuclear events, particularly RNA maturation and transport, and presumably the many intranuclear protein modification systems could be studied. This system is superior to an isolated chromatin-exogenous RNA polymerase system for studying the fine points of transcription and control at the many possible posttranscriptional levels.

The general application of this type of system is unclear at the present time. It is obvious that the key is the removal of all degradative activities that are not normally active *in vivo*. Thus great difficulties will be experienced with systems rich in these enzymes, as may be expected from nuclei that autodigest their DNA (*32,33*). There are many examples of systems of isolated nuclei where the RNA product is small (*34–36*). Grummt (*37*) and Grummt and Lindigkeit (*38*) have overcome these problems with rat liver by fractionating the nuclei. The isolated nuclei synthesize a very small product, but the nuclear subfraction they isolate (nucleoli) synthesizes very large RNA. However, we have recently shown (D. Price and W. F. Marzluff, unpublished results) that nuclei isolated from rat mammary tumors grown in culture by the same methods behave similarly to the myeloma nuclei, both with regard to kinetics of synthesis and the RNA product.

Similarly we have found that crude homogenates and crude nuclei have minimal RNA synthetic activity (unpublished results), but activity becomes evident on further purification. Thus in each system careful evaluation of conditions for optimal activity will be required and they may well vary extensively.

## Acknowledgments

I am grateful to Dr. Ru Chih Huang, in whose laboratory this project was originally developed, for guidance and for many helpful suggestions and to Dr. Edwin Murphy, who collaborated in developing this system. I also thank Dr. Mitchell Smith for communicating results prior to publication.

This work was supported by a USPHS grant AG-00413-02 and Career Development Award 1 KO4 CA00178-01 from the National Institute of Health.

## References

1. Marzluff, W. F., Murphy, E. C., and Huang, R. C., *Biochemistry* **12**, 3440 (1973).
2. Marzluff, W. F., Murphy, E. C., and Huang, R. C., *Biochemistry* **13**, 3689 (1974).
3. Marzluff, W. F., and Huang, R. C., *Proc. Natl. Acad. Sci. U.S.A.* **72**, 1082 (1975).
4. Price, R., and Penman, S., *J. Virol.* **9**, 621 (1972).
5. Zylber, E., and Penman, S., *Proc. Natl. Acad. Sci. U.S.A.* **68**, 2861 (1971).
6. Weinmann, R., and Roeder, R. G., *Proc. Natl. Acad. Sci. U.S.A.* **71**, 1790 (1974).
7. Sarma, M. H., Feman, E. M., and Baglioni, C., *Biochim. Biophys. Acta* **418**, 29 (1976).
8. Gross, M., and Rabinovitz, M., *Biochim. Biophys. Acta* **299**, 472 (1973).
9. Marzluff, W. F., White, E. L., Benjamin, R. B., and Huang, R. C., *Biochemistry* **14**, 3715 (1975).
10. Schwartz, T. B., Sklar, V. E. F., Jaehning, J. A., Weinmann, R., and Roeder, R. G., *J. Biol. Chem.* **249**, 5889 (1974).
11. Roeder, R. G., and Rutter, W. J., *Proc. Natl. Acad. Sci. U.S.A.* **65**, 675 (1970).
12. Reeder, R. H., and Roeder, R. G., *J. Mol. Biol.* **67**, 433 (1972).
13. Perry, R. P., *Proc. Natl. Acad. Sci. U.S.A.* **48**, 2179 (1962).
14. Wagner, E. K., Katz, L., and Penman, S., *Biochem. Biophys. Res. Commun.* **28**, 152 (1967).
15. Penman, S., *J. Mol. Biol.* **17**, 117 (1966).
16. Loenig, U. E., *J. Mol. Biol.* **38**, 355 (1968).
17. Adams, J. M., and Cory, S., *J. Mol. Biol.* **99**, 519 (1975).
18. Groner, Y., and Hurwitz, J., *Proc. Natl. Acad. Sci. U.S.A.* **72**, 2930 (1975).
19. Brown, D. D., Jordan, E., and Wensink, P., *Proc. Natl. Acad. Sci. U.S.A.* **68**, 3380 (1971).
20. Pemberton, R. E., and Baglioni, C., *J. Mol. Biol.* **81**, 255 (1974).
21. Saxinger, W. C., Ponnanperuma, C., and Gillespie, D., *Proc. Natl. Acad. Sci. U.S.A.* **69**, 2975 (1972).
22. Stavnezer, J., Huang, R. C., Stavnezer, E., and Bishop, J. M., *J. Mol. Biol.* **88**, 43 (1974).
23. Palmiter, R. D., *Cell* **4**, 189 (1975).
24. Brandhorst, B. P., and McConkey, E. H., *J. Mol. Biol.* **85**, 451 (1974).
25. Smith, M. M., and Huang, R. C., *Proc. Natl. Acad. Sci. U.S.A.* **73**, 775 (1976).
26. Emerson, C. P., *Nature (London), New Biol.* **232**, 101 (1971).
27. Dale, R. M. K., Livingston, D. C., and Ward, D. C., *Proc. Natl. Acad. Sci. U.S.A.* **70**, 2238 (1973).
28. Dale, R. M. K. and Ward, D. C., *Biochemistry* **14**, 2458 (1975).
29. Reeder, R. H., *J. Mol. Biol.* **80**, 229 (1974).

30. Honjo, T., and Reeder, R. H., *Biochemistry* **13**, 1896 (1974).
31. Wilson, G. N., Steggles, A. W., and Nienhuis, A. W., *Proc. Natl. Acad. Sci. U.S.A.* **72**, 4835 (1975).
32. Hewish, D. R., and Burgoyne, L. A., *Biochem. Biophys. Res. Commun.* **52**, 504 (1972).
33. Paul, S., and Duerksen, J. D., *Biochem. Biophys. Res. Commun.* **68**, 97 (1975).
34. Wu, G.-J., and Zubay, G., *Proc. Natl. Acad. Sci. U.S.A.* **71**, 1803 (1974).
35. Beebee, T. J. C., and Butterworth, P. H. W., *Eur. J. Biochem.* **51**, 537 (1975).
36. Schaefer, K., *Biochem. Biophys. Res. Commun.* **68**, 219 (1976).
37. Grummt, I., *Eur. J. Biochem.* **57**, 159 (1975).
38. Grummt, I., and Lindigkeit, R., *Eur. J. Biochem.* **36**, 244 (1973).

# Chapter 28

# *Transcription of Chromatin with Heterologous and Homologous RNA Polymerases*

RONALD H. REEDER

*Department of Embryology, Carnegie Institution of Washington, Baltimore, Maryland*

This article lists some recipes that have been used for synthesizing RNA transcripts from chromatin templates *in vitro*.

## I. Heterologous RNA Polymerase

### A. Reagents

The reaction mixture (in a total volume of 0.4 ml) contains:
Tris-HCl (pH 8), 50 m*M*
$MgCl_2$, 5 m*M*
Unlabeled GTP, UTP, and ATP, 0.6 m*M* each
[$^3H$]CTP, 50 $\mu$Ci (20 Ci/mmole)
DNA, 50 $\mu$g, as chromatin
*Escherichia coli* RNA polymerase, 200 units
Reagents used to assay the reaction are:
Filter paper disks, 2.2 cm (Whatman No. 1)
Trichloracetic acid, 5%, ice-cold
Ethanol, 95%
Ether
NCS Tissue solubilizer (Amersham/Searle)
Toluene liquid scintillation fluid (16 gm of PPO and 0.2 gm of POPOP per gallon of toluene)
Reagents used to purify the RNA product are:
Ethylenediaminetetraacetate (EDTA) (pH 8), 0.5 *M*
Sodium dodecyl sulfate (SDS), 10%

*E. coli* tRNA in $H_2O$, 10 mg/ml
Phenol, redistilled and water saturated
Sephadex G-25
Ethanol, 100% (at $-20°C$)

## B. Procedure

[$^3H$]CTP is generally supplied (and should be stored) in 50% ethanol to minimize radiation damage to itself. The ethanol severely inhibits polymerase and can be removed by taking an appropriate amount of isotope nearly to dryness in a vacuum desiccator (preferably in the tube to be used for the subsequent synthesis reaction). Do not leave the isotope as a dry film, for it degrades rapidly at such high concentration. Ingredients are added to the reaction tube on ice, and the reaction is started by addition of RNA polymerase and transferring the tube to a water bath set at 37°C. Under these ionic conditions, the reaction stops after about 20–30 minutes, presumably owing to failure of the polymerase to release and reinitiate (*1*). These ionic conditions also cause much of the chromatin to precipitate in the reaction. Therefore, it is probably beneficial to occasionally shake the tube throughout the reaction to resuspend the template.

RNA polymerase may be prepared according to any of several published procedures (*2, 3*). For transcription of eukaryotic chromatin, the presence or the absence of the sigma subunit makes little difference. The polymerase should be purified free of RNase and polynucleotide phosphorylase activities. One unit of polymerase activity is defined as 1 nmole of CTP incorporated per minute at 37°C using deproteinized DNA as template. Two hundred units is nearly saturating for 50 μg of chromatin DNA.

Chromatin preparation is outside the scope of this article. However, there is some evidence that the more rapid and gentle chromatin preparation methods yield a more biologically relevant template than do the more lengthy procedures.

## C. Assay of Acid-Insoluble Radioactivity

Reactions are terminated by addition of SDS to 0.5%, and a 10-μl aliquot is pipetted onto a filter paper disk for determination of acid insoluble radioactivity. A number of such disks can then be processed together (*4*). The disks are washed four times with 200-ml portions of cold 5% TCA (5 minutes in each wash), then washed once with 95% ethanol, once with ether, airdried, and put in a scintillation vial. They are counted in a scintillation mix of 0.5 ml of NCS solubilizer per 10 ml of toluene fluor.

In some cases addition of SDS causes the reaction to become so viscous that pipetting is difficult. This can be prevented by adding 0.1 volume of 1 mg/ml RNase-free DNase before the SDS, continuing the incubation at 37°C for 1 minute, then adding SDS and proceeding as usual.

In order to convert acid-insoluble counts per minute incorporated into amount of RNA synthesized, it is useful to determine the total radioactivity in the reaction. For this purpose, 1 $\mu$l of the reaction is spotted on a filter disk, dried under a heat lamp, and counted in toluene–NCS fluor without TCA washing. The RNA amount can then be calculated from the equation:

$$\frac{\text{TCA insoluble cpm in reaction}}{\text{total cpm in reaction}} \times \frac{\text{Curies added to reaction}}{\text{specific activity of CTP (Ci/mmole)}} = \text{mmoles RNA synthesized}$$

Tritium dried on paper in the form of a nucleoside triphosphate is counted with a lower efficiency than when it is precipitated on the filter as part of an RNA chain. In order to count both at the same efficiency, they must be brought into solution with a solubilizing agent such as NCS.

## D. Purification of the RNA Product

After the reaction is stopped by adding SDS to 0.5%, EDTA is added to 10 m*M* plus 100 $\mu$g of *Escherichia coli* tRNA as carrier. The mixture is then shaken with an equal volume of water-saturated phenol, and the aqueous layer is put on a 2 × 30 cm column of Sephadex G-25 equilibrated with 0.1 *M* NaCl–0.05 *M* Tris-HCl (pH 8)–1 m*M* EDTA–0.5% SDS. RNA and DNA is eluted in the void volume and can be located either by counting an aliquot of each fraction or by following the $A_{260}$ of the carrier tRNA. Phenol and unreacted triphosphates are retarded by the column. Once the column is washed until the $A_{260}$ due to phenol returns to base line, it is ready for another sample. The void volume is pooled, mixed with 2 volumes of cold ethanol, and left at −20°C for 30 minutes. The precipitate may then be collected by centrifuging at 20,000 *g* for 10 minutes.

## E. Comments

The reaction conditions described here are designed to yield a high specific activity RNA product under conditions of low ionic strength that should minimize exchange of chromatin proteins. The amount of RNA made is usually less than the amount of chromatin DNA added. For some purposes it may be desirable to obtain net synthesis. This may be accomplished by adding KCl to 0.15 *M* and raising the level of each triphosphate to 4 m*M* (*5*).

In the presence of chromatin proteins, *E. coli* RNA polymerase transcribes an RNA product that more closely resembles the *in vivo* spectrum of transcripts than does the transcript of deproteinized DNA (*6,7*). Since *E. coli* RNA polymerase reads the missense as well as the sense strand of the DNA and appears not to recognize eukaryotic promotors (*8*), it is likely that the selective transcription of chromatin is due to a masking of some regions of the DNA by chromatin proteins. How this selective masking is affected by changes in the various constitutents of the *in vitro* transcription reaction is poorly understood. At the ionic strength normally used for RNA synthesis, chromatin forms a precipitate. By lowering the ionic strength to the bare minimum needed for polymerase activity it is possible to keep most of the chromatin in solution, but it has been shown (*9*) that the number of binding sites for polymerase is similar whether the chromatin is soluble or not. Therefore, most workers use the higher ionic strengths, which allow more RNA synthesis.

## II. Homologous RNA Polymerase

### A. Reagents

1. The reaction mixture (in a total volume of 0.4 ml) contains:
   Tris·HCl (pH 8), 50 m*M*
   $MgCl_2$, 5 m*M*
   $MnCl_2$ 2 m*M*
   KCl, 150 m*M*
   ATP, UTP, and GTP, 0.6 m*M* each
   [$^3H$]CTP 50 $\mu$Ci (20 Ci/mmole)
   DNA, 50 $\mu$g, as chromatin
   RNA polymerase, 30 units
2. Reagents used to assay the reaction and to purify the RNA products are the same as for heterologous polymerase.

### B. Procedure

Each of the three major classes of eukaryotic RNA polymerases has different optima for ionic strength and metal ion concentration. For example, polymerases I, II, and III from *Xenopus laevis* have KCl optima of 100 m*M*, 200 m*M* and 120 m*M* and 120 m*M*, respectively (*10*). The 150 m*M* KCl suggested in the protocol is a reasonable compromise, which allows assay of all three. All three enzymes are active with $Mg^{2+}$ alone, and it is arguable whether the use of $Mn^{2+}$ is physiological. However, since polymerases II and III show several fold higher maximal activity in the presence of $Mn^{2+}$, many workers include it in the reaction.

The ratio of polymerase to chromatin DNA suggested here is one in which the chromatin is nearly saturating the polymerase, and thus is making the most efficient use of the enzyme.

So far attempts to detect correct initiation by eukaryotic RNA polymerase *in vitro* have not been very successful, and it is not known whether the polymerase, the template, or both are at fault. This necessarily means that it is still unknown whether the reaction conditions described here are optimal for correct promotor recognition.

## References

*1.* Fuchs, E., Millette, R. L., Zillig, W., and Walter, G. *Eur. J. Biochem.* **3**, 183 (1967).
*2.* Burgess, R., *J. Biol. Chem.* **244**, 6160 (1969).
*3.* Chamberlin, M., and Berg, P. *Proc. Natl. Acad. Sci. U.S.A.* **48**, 81 (1962).
*4.* Bollum, F. J. *Methods Enzymol.* **12B**, 169 (1968).
*5.* Astrin, S. M., *Biochemistry* **14**, 2700 (1975).
*6.* Axel, R., Cedar, H., and Felsenfeld, G., *Proc. Natl. Acad. Sci. U.S.A.* **70**, 2029 (1973).
*7.* Gilmour, R. S., and Paul, J. *Proc. Natl. Acad. Sci. U.S.A.* **70**, 3440 (1973).
*8.* Reeder, R. H., *J. Mol. Biol.* **80**, 229 (1973).
*9.* Cedar, H., and Felsenfeld, G., *J. Mol. Biol.* **77**, 237 (1973).
*10.* Roeder, R. G., *J. Biol. Chem.* **249**, 241 (1974).

# Chapter 29

# *Transcription of Chromatin with Prokaryotic and Eukaryotic Polymerases*

GOLDER N. WILSON, ALAN W. STEGGLES, W. FRENCH ANDERSON, AND ARTHUR W. NIENHUIS

*Molecular Hematology Branch, National Heart and Lung Institute, National Institutes of Health, Bethesda, Maryland*

## I. Introduction

The process of transcription in eukaryotic cells remains poorly defined. One of the reasons for this is that considerable metabolism of newly synthesized RNA occurs in the nucleus before RNA reaches the cytoplasm. Eukaryotic DNA exists in close association with chromosomal proteins. This nucleoprotein complex is a restrictive and tissue-specific template for purified RNA polymerases, and it may allow separation between those factors that influence the transcription of RNA and those that influence RNA metabolism and transport. Assessing the relevance of chromatin transcription to cellular transcription is difficult, however, since the properties of nascent nuclear RNA are speculative and because no precise criteria exist for defining the integrity of chromatin during its isolation and transcription. Several laboratories (*1–12*) are approaching this dilemma by studying the chromatin primed transcription of specific RNA sequences in the hopes that precise correlation of cellular and cell-free transcription will be possible.

In this chapter we describe and compare our methods for the cell-free transcription of globin mRNA sequences from rabbit marrow chromatin by bacterial and mammalian RNA polymerases and point out certain problems which have complicated these studies. Our approach is to incubate RNA polymerase and chromatin under well defined conditions, to purify the transcript RNA free of DNA, protein, and other contaminants, and to measure the proportion of globin mRNA sequences in the transcript by hybridization to cDNA.[1] RNA transcribed from the DNA strand complementary to the

[1] cDNA, DNA complementary to globin mRNA; cRNA, RNA complementary to cDNA.

globin gene (antistrand transcript) is detected by hybridization to cRNA.[1] We also describe some preliminary results that examine the symmetry and selectivity of globin mRNA transcription by various preparations of RNA polymerase.

## II. The Transcription Reaction

### A. Preparation of Chromatin

#### 1. Buffer Solutions

Buffer A: 0.3 *M* sucrose (Schwarz-Mann ultrapure)–10 m*M* Tris-HCl (pH 7.9)–2 m*M* $MgCl_2$–3 m*M* $CaCl_2$–0.1% Triton X-100–0.5 m*M* dithiothreitol.

Buffer B: 2 *M* sucrose–10 m*M* Tris-HCl (pH 7.9)–5 m*M* $MgCl_2$–0.5 m*M* dithiothreitol.

Buffer C: 1.6 *M* sucrose–10 m*M* Tris-HCl (pH 7.9)–5 m*M* $MgCl_2$–0.5 m*M* dithiothreitol.

Buffer D: 80 m*M* NaCl–20 m*M* EDTA (pH 6.3).

Buffer E: 2 m*M* Tris-HCl (pH 7.9)–0.1 m*M* EDTA.

All solutions are filtered through Millipore HA filters and stored at 4°C.

#### 2. Isolation of Nuclei

The method of Marzluff *et al.* (*12a*) has been modified to allow the rapid preparation of rabbit marrow and liver nuclei with minimal cytoplasmic contamination. Young rabbits (1–2 kg) are treated with phenylhydrazine as described previously (*13, 14*) and exsanguinated by cardiac puncture. Marrow is scraped from the leg bones and suspended in NCTC-109 medium (Grand Island Biological Co.) at 25°C with a plastic syringe. Fat is removed by filtration through one layer of chessecloth (Chickopee Mills), and the marrow cells are collected as a loose pellet by centrifugation at room temperature for 2 minutes at 2000 rpm in an International clinical centrifuge. Liver is perfused with 50 ml of cold buffer A, and a preliminary separation of crude liver nuclei is performed by mincing 40 gm of perfused tissue in 50 ml of buffer A, homogenizing at high speed with a Teflon pestle (10 strokes), filtering through four layers of cheese-cloth, and centrifuging for 10 minutes at 10,000 rpm (12,000 *g*) in the Sorvall SS-34 rotor.

Marrow cells (1 gm, or approximately 1 ml of pellet volume) or crude liver nuclei (2 gm) are suspended in 80 ml of buffer A, homogenized at high speed (20 strokes), and diluted with 1 volume of buffer B (for marrow) or buffer C (liver). Aliquots (28 ml) of each suspension are layered over 10-ml cushions of buffer B (for marrow) or buffer C (for liver) and centrifuged for 45 minutes at 19,000 rpm (60,000 *g*) in an SW-27 rotor. The pelleted nuclei, examined microscopically after staining with methylene blue, are free of visible cytoplasmic adhesions and include less than 1% intact cells. Both marrow and liver nuclei actively incorporate UTP into RNA when incubated in the *in vitro* system described by Marzluff and Huang (*11*).

#### 3. Isolation of Chromatin

The isolation of chromatin is similar to that described by Spelsberg and Hnilica (*15*). The pellets obtained in the isolation step, containing approximately $10^9$ rabbit marrow or rabbit liver nuclei, are suspended in 40 ml of buffer D with a Teflon-glass homogenizer and homogenized by hand (10 strokes) before centrifugation at 10,000 rpm (12,000 *g*) for 10 minutes in the Sorvall SS-34 rotor. The pellets are washed twice more with buffer D, once with 0.35 *M* NaCl, and twice with buffer E, each time using homogenization for 10 strokes at full speed. The final pellets are suspended in 5–10 ml of buffer E at 10–15 $A_{260}$ units/ml (0.7–1 mg per milliliter of DNA) by gentle homogenization. The chromatin is sonicated before use with a Branson W 185 sonufier for 30 seconds at 50 W. We have found this to be a necessary preliminary step for transcriptional studies with mammalian RNA polymerase. Sonication does not appear to alter the transcription of globin mRNA sequences from chromatin by *Escherichia coli* RNA polymerase (*16, 17*) even though shearing of chromatin (*18, 19*) is known to alter the digestion of chromatin by certain nucleases. Chromatin is always prepared rapidly at 4°C from fresh tissue and transcribed within 2 hours of isolation.

## B. Preparation of RNA Polymerases

#### 1. Buffer Solutions

Buffer F: 10 m*M* Tris-HCl (pH 7.9)–5% glycerol (v/v)–10 m*M* $MgCl_2$–0.1 m*M* EDTA–0.1 m*M* dithiothreitol;

Buffer G: 10 m*M* Tris-HCl (pH 7.9)–50% glycerol–10 m·*M* $MgCl_2$–0.1 *M* KCL–0.1 m*M* EDTA–0.1 m*M* mercaptoethanol–0.1 m*M* EDTA–0.1 m*M* dithiothreitol;

Buffer H: 50 m*M* Tris-HCl (pH 7.9)–25% glycerol–0.1 m*M* EDTA–0.5 m*M* dithiothreitol.

Buffer $I_{50}$: buffer I containing 50 m*M* $(NH_4)_2SO_4$;

Buffer $I_{75}$: buffer I containing 75 m*M* $(NH_4)_2SO_4$; similarly buffer $I_{100}$, $I_{300}$, etc.

## 2. Assays for RNA Polymerase and RNase Activity

Both *E. coli* and sheep liver RNA polymerases are assayed under the following conditions. Each 250-μl reaction mixture contains: 30 m*M* Tris–HCl (pH 7.9)–8 m*M* KCl–3 m*M* $MnCl_2$–1.2 m*M* 2-mercaptoethanol–1–5% glycerol–2.0 m*M* ATP, CTP, and GTP (Schwarz-Mann)–0.4 m*M* [$^3$H]UTP (Schwarz-Mann) at 0.1 Ci/mmole, 10 μg of calf thymus DNA (Worthington), and 50 μl of the enzyme fraction to be tested. After incubation at 30°C for 10 minutes, the trichloroacetic acid-precipitable radioactivity is collected on nitrocellulose filters (Millipore), washed with 10% trichloroacetic acid, and counted by liquid scintillation spectrometry. One unit of RNA polymerase activity is defined as that amount of enzyme catalyzing the incorporation of 1 nmole of UMP into RNA in 10 minutes at 30°C. Identical conditions are used for the assay of RNase activity except that 50 ng (25,000 cpm) of [$^3$H]poly(U) (Miles) is substituted for the DNA and the four nucleoside triphosphates. The hydrolysis of greater than 500 cpm of [$^3$H]poly(U) in 10 minutes at 30°C is considered to be significant RNase contamination.

## 3. Preparation of *E. coli* RNA Polymerase

*E. coli* RNA polymerase is prepared by the procedure of Burgess (*20*) from 100 gm of frozen late log phase cells (Miles). The procedure is followed exactly through batch DEAE-cellulose chromatography (step 4). Those DEAE-cellulose fractions containing greater than 50 units of RNA polymerase activity per milliliter are combined, mixed with 370 gm of ammonium sulfate per liter of solution, and centrifuged for 30 minutes at 15,000 rpm (27,000 *g*) in the Sorvall SS-34 rotor. The pellets are suspended in 5 ml of buffer F and applied to a 35 × 2.6 cm column of Sepharose 6B (Pharmacia) which has been equilibrated with buffer F. Fractions (5 ml) are collected at a flow rate of 30 ml per hour, and those containing greater than 20 units of RNA polymerase activity per milliliter are pooled and precipitated with ammonium sulfate as described above. The pellets are suspended in 2 ml of buffer G and stored at 4°C. Preparations of RNA polymerase which contain significant RNase activity or RNA polymerase activity with template omitted (greater than 5% of the activity in the presence of DNA) are again subjected to Sepharose 6B chromatography. We have employed *E. coli* polymerase purchased from Grand Island Biological Company for certain experiments (*9,21,22*) although each batch should be checked for total activity, presence of sigma factor by SDS polyacrylamide gel electrophoreses, and contamination by DNase or RNase.

## 4. Preparation of Sheep Liver RNA Polymerase II

The isolation procedure for RNA polymerase II is similar to that described by Kedinger *et al.* (*24*) for the calf thymus enzyme. Since sheep liver

RNA polymerase II activity has a half-life of about 36 hours at 4°C even in buffers containing 25–30% glycerol, all procedures must be performed rapidly and dialysis must be avoided. Sheep liver polymerase II has the characteristic sensitivity to α-amanitin as shown in Fig. 1 (*25*): liver polymerases III and I are less sensitive.

*a. Step 1: Whole-Cell Homogenization.* Sheep liver (300 gm of fresh tissue) is minced and homogenized in 600 ml of buffer H. The mixture is homogenized first for 1 minute at 40% of full power in a Waring Blender, then 2 to 3 strokes with a tight-fitting Teflon-glass homogenizer in a 50-ml vessel. After filtering through one layer of cheesecloth, 80 ml per liter of saturated ammonium sulfate (4.3 *M* ammonium sulfate in water adjusted to pH 7 with ammonium hydroxide) are added with vigorous stirring. The viscous solution is sonicated in 60-ml batches using three 30-second bursts at 100 W. The sonicate is clarified by centrifugation at 19,000 rpm (54,000 *g*) for 90 minutes in a type 19 rotor.

*b. Step 2: Ammonium Sulfate Precipitation.* The supernatant fraction

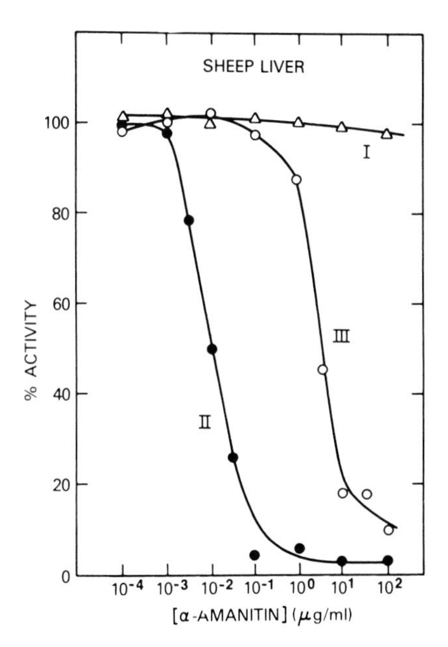


FIG. 1. Inhibition of sheep liver RNA polymerases by α-amanitin. RNA polymerases were extracted from 10 gm of sheep liver nuclei as described in the text and then were applied to a 2 × 20 cm column of DEAE-Sephadex and eluted with a 200-ml linear gradient of 50 to 400 m*M* ammonium sulfate in buffer I. The three peaks of RNA polymerase activity were assayed in the presence of $10^{-4}$ to $10^{2}$ μg of α-amanitin per milliliter in the RNA polymerase assay. Activity corresponding to 100% was 0.085 unit for fraction I and 0.09 unit for fractions II and III. Results are plotted in the fashion of Weinmann and Roeder (*25*).

from step 1 is brought to 50% saturation with solid ammonium sulfate (313 gm per liter) and stirred for 1 hour. After centrifugation for 2 hours at 18,000 rpm (48,000 *g*) in the type 19 rotor, the precipitate is dissolved in a minimum volume of buffer H (500–600 ml).

*c. Step 3: Protamine Sulfate Precipitation.* Buffer G is added to the step 2 material to a final volume of 1 liter. Thirty milliliters of protamine sulfate (Calbiochem) is slowly added, and the mixture is stirred for 1 hour. After centrifugation for 90 minutes at 33,000 rpm in the type 35 rotor (130,000 *g*), the supernatant fraction was collected and stored overnight in liquid nitrogen.

*d. Step 4: First DEAE-Cellulose Chromatography.* Material from step 3 is thawed and diluted with buffer H until the ammonium sulfate concentration is 75 m*M* (4–5 liters) as determined by conductivity. After stirring with 25 gm (dry weight) of fibrous DEAE-cellulose (Whatman DE-23) which has been equilibrated with buffer $I_{75}$, the mixture is filtered through another 25 gm of DEAE-cellulose, which has been collected as a moist cake in a 4 × 5 inch Buchler funnel. After washing with 2 liters of buffer $I_{75}$ to remove RNA polymerases I and III, the DEAE cellulose is transferred to a 100 × 5 cm column. RNA polymerase II is eluted as a single peak with buffer $I_{300}$, precipitated with ammonium sulfate as described in Step 2 (10–15 ml final volume), and stored in liquid nitrogen.

*e. Step 5: Second DEAE-Cellulose Chromatography.* Material from step 4 is diluted with buffer I to an ammonium sulfate concentration of 75 m*M* and applied to a 1.6 × 60 cm column of DEAE-cellulose (Whatman DE-52) which has been equilibrated with 4 volumes of buffer $I_{50}$. The column is washed with 150 ml of buffer $I_{50}$ and then eluted with a 500-ml linear gradient of 75 to 600 m*M* ammonium sulfate in buffer I at a flow rate of 40 ml per hour. Most of the RNase activity, as assayed by poly (U) hydrolysis, elutes at the beginning of the gradient and is well separated from RNA polymerase II activity (*26*). Fractions (8 ml) with greater than 2 units of RNA polymerase activity per milliliter are pooled, stirred with an equal volume of saturated ammonium sulfate, centrifuged for 90 minutes in the type 35 rotor at 33,000 rpm (48,000 *g*) suspended in buffer I, and stored in liquid nitrogen.

*f. Step 6: Phosphocellulose Chromatography.* Material from step 5 is again diluted to an ammonium sulfate concentration of 75 m*M* and applied to a 10 × 1.6 cm column of phosphocellulose which has been equilibrated with buffer $I_{50}$. The column is washed with 30 ml of buffer $I_{50}$, then with 30 ml of buffer $I_{120}$, and finally with 30 ml of buffer $I_{330}$. Those fractions (5 ml) containing greater than 4 units of RNA polymerase activity per milliliter are pooled and precipitated with ammonium sulfate as described in step 5.

*g. Step 7: DNA Sepharose Chromatography.* DNA-Sepharose was prepared as described by Poonian *et al.* (*23*) using sonicated, heat-denatured

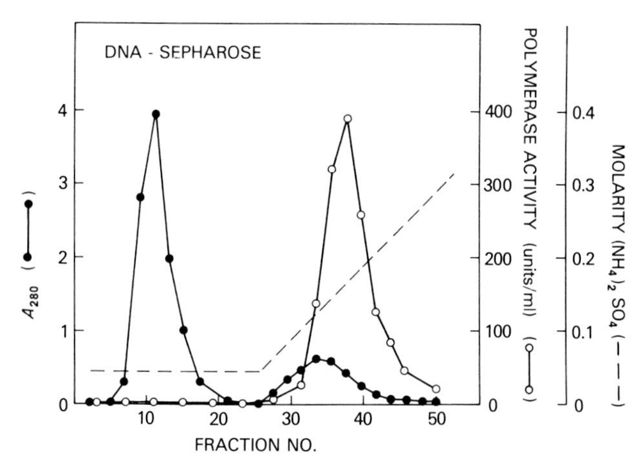


FIG. 2. Purification of sheep liver RNA polymerase II by DNA–Sepharose chromatography (step 7). Details of the chromatography appear in the text. For this experiment, no bovine serum albumin was added to the elution buffer. Polymerase activity (○—○) refers to the RNA polymerase assay but is multiplied by 100 in this figure. Protein (●—●) was determined by $A_{280}$ readings, and the molarity of ammonium sulfate (----) was determined by conductivity measurements.

calf thymus DNA and cyanogen bromide-activated Sepharose 4B (Pharmacia). Material from step 6 is diluted with buffer I to an ammonium sulfate concentration of 50 m*M* and applied to a 2.6 × 12 cm column of DNA–Sepharose equilibrated with buffer $I_{50}$. The column is washed with 100 ml of Buffer $I_{50}$ and eluted with a 200-ml linear gradient of 50 to 500 m*M* ammonium sulfate in buffer I. The recovery of enzyme is improved if the elution buffer contains 0.5 mg of bovine serum albumin per milliliter. However, accurate protein readings are difficult to obtain against this background. Those fractions (2 ml) containing greater than 1 unit of RNA polymerase activity per milliliter are pooled and precipitated with ammonium sulfate as described previously. The elution of RNA polymerase II from DNA–Sepharose is shown in Fig. 2. Greater than 90% of this activity was sensitive to α-amanitin.

Table I summarizes the 1800-fold purification of sheep liver RNA polymerase II with 22% yield of activity. A simple batch chromatography step is used to separate RNA polymerase II from I and III and to reduce the amount of protein so that contaminating RNase activity may be removed by DEAE-cellulose gradient chromatography (*26*). The DNA–Sepharose fraction is only 20–30% pure as judged by sodium dodecyl sulfate–polyacrylamide electrophoresis. Further purification by glycerol gradient sedimentation has not yielded enough RNA polymerase activity to test in the chromatin transcription reaction.

TABLE I

PURIFICATION OF SHEEP LIVER RNA POLYMERASE II

| Step | Polymerase activity (units)[a] | Protein (mg) | Polymerase specific activity (units/mg) | Fold | Yield (%) |
|---|---|---|---|---|---|
| 1. Crude extract | 500 | 25,000 | 0.020 | 1 | 100 |
| 2. 7.5–50% $(NH_4)_2SO_4$ | 620 | 23,000 | 0.027 | 1.4 | 124 |
| 3. Protamine sulfate | 700 | 21,200 | 0.033 | 1.7 | 140 |
| 4. DEAE-cellulose-1 | 270 | 242 | 1.1 | 55 | 54 |
| 5. DEAE-cellulose-2 | 210 | 75 | 2.8 | 140 | 42 |
| 6. Phosphocellulose | 160 | 25 | 6.4 | 320 | 32 |
| 7. DNA–Sepharose | 110 | 3 | 37 | 1800 | 22 |

[a] 1 unit = 1 nmol of UMP incorporated into RNA in 10 minutes at 30° C.

## C. Transcription of Chromatin

Conditions for the transcription of rabbit marrow and liver chromatin by RNA polymerase were first studied by using a defined RNA polymerase assay and substituting chromatin equivalent to 10–50 μg of DNA for calf thymus DNA. Preliminary experiments established that the inclusion of both $MgCl_2$ and $MnCl_2$ was necessary for maximal sheep liver RNA polymerase activity with chromatin as template. The reaction also contained 100 m*M* KCl. This concentration was chosen to enhance polymerase activity while minimizing the removal of chromosomal proteins by salt dissociation (*27*). Chromatin was titrated with increasing amounts of sheep liver or *E. coli* RNA polymerase to ensure the addition of excess RNA polymerase to the reaction. Under these conditions, RNA synthesis by either polymerase is linear for only 20–30 minutes at 37°C and reaches a plateau after 60 minutes.

The optimal transcription reaction (usually 2 ml) as defined by these studies, contained 30 m*M* Tris-HCl (pH 7.9)–100 m*M* KCl–3 m*M* $MgCl_2$–3 m*M* $MnCl_2$–1.2 m*M* 2-mercaptoethanol–10% glycerol–2 m*M* ATP, GTP, and CTP–0.4 m*M* [$^{32}$P]UTP (0.1 mCi/mmole) chromatin containing 500 μg of DNA and 500 units of *E. coli* RNA polymerase or 30 units of sheep liver RNA polymerase II. Incubation was for 1 hour at 37°C even though the original activities of the polymerases were determined at 30°C.

Because globin mRNA sequences are present in RNA associated with chromatin (endogenous RNA), the accurate comparison of transcription reactions with or without added RNA polymerase (necessary for the estimation of net globin mRNA transcription) requires that a similar recovery of RNA be obtained from each reaction. For this purpose, an amount of

[$^{32}$P]RNA carrier (5–100 μg) is added to control reactions equal to the amount of RNA synthesis in the corresponding reaction with RNA polymerase. The [$^{32}$P]RNA carrier (200 cpm/μg) is isolated from a transcription reaction containing 50 μg of heat-denatured sonicated *E. coli* DNA (Worthington) and 200 units of *E. coli* RNA polymerase. Each reaction thus contains radioactive RNA so that the recovery of RNA during the purification procedure can be calculated.

## D. Purification of RNA Transcripts

The extraction and purification of RNA from transcription reactions containing 5–15 μg of RNA synthesized by sheep liver RNA polymerase II, 80–100 μg of RNA synthesized by *E. coli* RNA polymerase, or 5–100 μg of [$^{32}$P]RNA carrier, is modified from the procedures of (*28*) and Gilmour *et al.* (*29*). After the incubation, 80 μl of 2 *M* KCl are added and the reactions are centrifuged for 10 minutes at 4°C at 10,000 rpm (12,000 g) in the Sorrall SS-34 rotor. This removes 80–90% of the DNA but regains greater than 80% of the acid-insoluble radioactivity. The supernatant is mixed with 40 μl of 1 *M* $MgCl_2$, 100 μl of 1 mg/ml solution of DNase I (Worthington) and incubated for 30 minutes at 37°C. After a second incubation for 30 minutes at 37°C with 30 μl of 10 mg of Pronase per milliliter (predigested at 37°C for 2 hours) (Calbiochem), 80 μl of 12.5% SDS, 0.4 ml of 0.2 *M* EDTA, and 2 ml of phenol saturated with 0.5 *M* KCl are added. The solution is mixed vigorously at 60°C for 5 minutes, combined with 3 ml of 24:1 chloroform:isoamyl alcohol (v/v), and again mixed at 60°C for 5 minutes. The phases are separated by centrifugation, the aqueous phase is removed taking care to exclude the interface, and the extraction with chloroform–isoamyl alcohol is repeated until no further interface remains. The aqueous phase is then applied to a 1.6 × 100 cm column of Sephadex G-50 (Pharmacia) at a flow rate of 200 ml/hour. Fractions (6 ml) are collected and analyzed for absorbance at 260 nm ($A_{260}$) or acid-insoluble radioactivity. The transcript RNA elutes in the void volume and is separated by 5 or 6 fractions from the low-molecular-weight nucleotides and DNA fragments. The high-molecular-weight material is lyophilized and stored at −20°C for hybridization studies.

The purified RNA contains 50–70% of the initial acid-insoluble radioactivity and less than 0.2 μg of DNA. RNA transcribed by *E. coli* RNA polymerase from rabbit marrow or liver chromatin migrates as 5–6 S species when examined by SDS–polyacrylamide gel electrophoresis (*22*). RNA synthesized by sheep liver RNA polymerase II is slightly larger (7–10 S). Removal of chromatin DNA by centrifugation simplifies subsequent purification of RNA by eliminating the need for prolonged DNase and protease digestion with the attendant risk of degradation of the transcript (*30*). An important test of the extraction procedure was devised by adding increasing amounts of globin mRNA to RNA transcribed from liver chromatin (*22*). The number of globin mRNA sequences detected in the purified RNA transcripts was proportional to the amount of globin mRNA added to the incubation mixture.

## E. Preparation of Hybridization Probes

Complementary DNA (cDNA) was synthesized and prepared for use as a hybridization probe by methods which have been described in detail elsewhere (*21,22*). We routinely isolated a probe ranging from 400–600 nucleotides in length by preparative alkaline sucrose gradient centrifugation. Ribonucleic acid complementary to rabbit cDNA (cRNA) was synthesized by an adaption of the procedures of Poon *et al.* (*31*) and Marotta (*32*) and has been described previously (*22*). We have used this probe to study the transcription of the DNA strand complementary to the globin gene (antistrand transcript) from rabbit marrow chromatin (*22*).

## F. Hybridization of RNA Transcripts

The use of cDNA or related hybridization probes and the analysis of hybridization reactions with single-strand specific nucleases in finding widespread application. It is important to mention several experimental problems which may arise when analyzing RNA transcripts by this method. First, cDNA synthesis begins at the 3′ end of globin mRNA. However, the 5′ end of the globin mRNA molecule may be overrepresented in globin mRNA sequences transcribed from chromatin; consequently, a complete cDNA copy is highly desirable to analyze chromatin-directed transcripts. We have demonstrated that estimates of globin mRNA sequences in chromatin-directed transcripts are up to 50% greater when complete cDNA is used (*33*). Second, considerable degradation of cDNA and transcribed RNA may occur when the annealing reaction is carried out for long periods of time in the presence of SDS at high temperatures. We have confirmed the results of Young *et al.* (*34*) who found that the use of 50% formamide and low temperatures (43 °C) for hybridization reactions minimizes nucleic acid degradation (Fig. 3). Also helpful is the use of saturation analysis as described by Young *et al.* (*34*) which avoids the need for an excess of globin mRNA sequences over cDNA required for kinetic hybridization analysis. Finally, single-strand specific nucleases may be inhibited by phenol, SDS, or even large amounts of single-stranded DNA or RNA (Fig. 4). Inadequate dilution of the hybridization mix into the buffer used for nuclease digestion

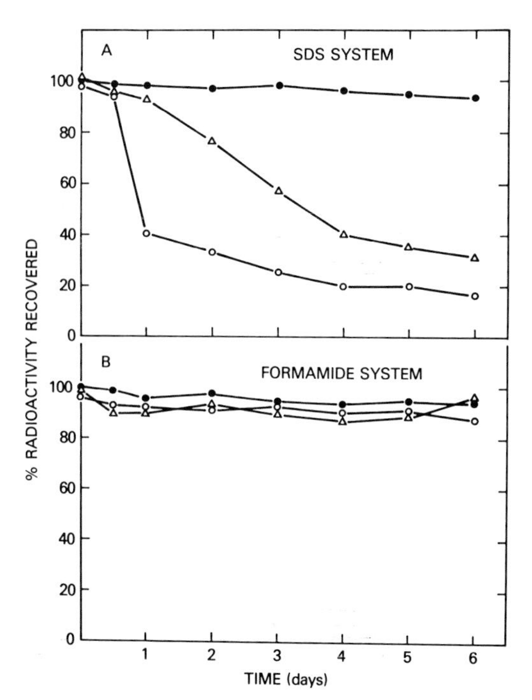


FIG. 3. Stability of cDNA, $cRNA_d$, and rabbit liver transcript in the sodium dodecyl sulfate (SDS) (A) or formamide hybridization systems (B). Hybridization reactions contained 1 ng (3000 cpm) of [$^3$H]cDNA (●), 1 ng (2000 cpm) of [$^3$H]$cRNA_d$ (△), or 1 $\mu$g (5000 cmp) of $^{32}$P-labeled rabbit liver transcript (○), and were carried out at 68° C or 43° C, respectively.

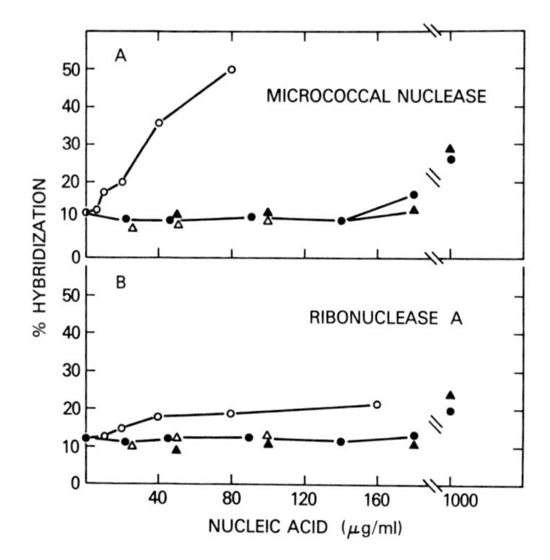


FIG. 4. Inhibition of micrococcal nuclease and RNase A by various nucleic acids. Hybridization reactions (formamide system) containing 1 ng (3000 cpm) of [$^3$H] cDNA (A) or 1 ng (20,000 cpm) of [$^3$H]$cRNA_d$ (B) were incubated 24 hours at 43°C. The reactions were analyzed with 40 $\mu$g of micrococcal nuclease per milliliter (A), or 10 $\mu$g of RNase A per milliliter (B) in the presence of the specified concentrations of single-stranded calf thymus DNA (○), native calf thymus DNA (●), *E. coli* 5 S RNA (▲), or rabbit liver transcript (△).

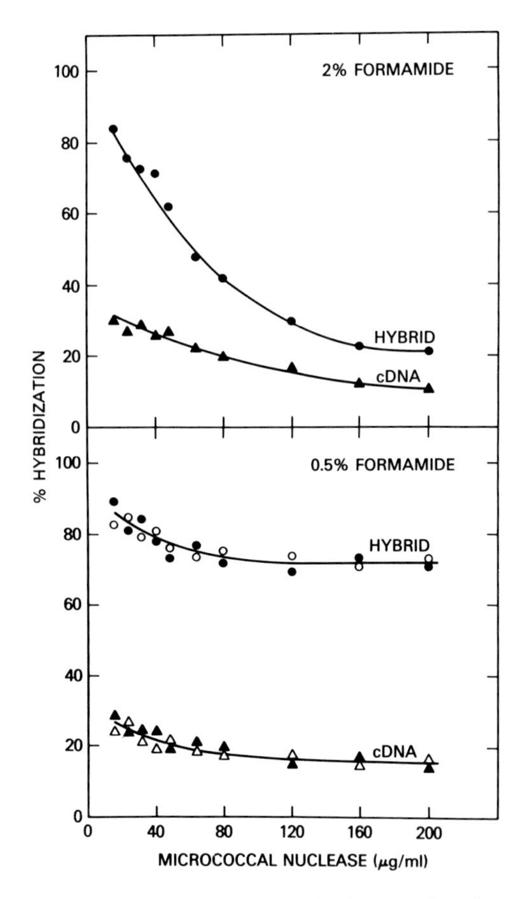


FIG. 5. Effect of formamide on the specificity of micrococcal nuclease for single-stranded cDNA. Upper panel: Hybridization reactions containing 1 ng (3000 cpm) of [$^3$H]cDNA (▲—▲) or cDNA and 6 ng of globin mRNA (●—●) were incubated 24 hours at 43°C in the formamide system. After incubation, the reactions were analyzed with the specified concentrations of micrococcal nuclease in the presence of 2% formamide. Lower panel: Hybridization reactions identical to those described above (●—●, ▲—▲) or containing cDNA and 10 μg of rabbit liver transcript (△—△), or cDNA, globin mRNA, and rabbit liver transcript (○—○) were incubated 24 hours at 43°C in the formamide system. The reactions were analyzed with the specified concentrations of micrococcal nuclease in the presence of 0.5% formamide.

may result in a concentration of formamide that renders the probe sensitive to nuclease digestion (Fig. 5). Under appropriate conditions (Fig. 6) micrococcal nuclease has identical specificity for cDNA and RNase for cRNA in the presence or in the absence of excess RNA transcribed from rabbit liver chromatin.

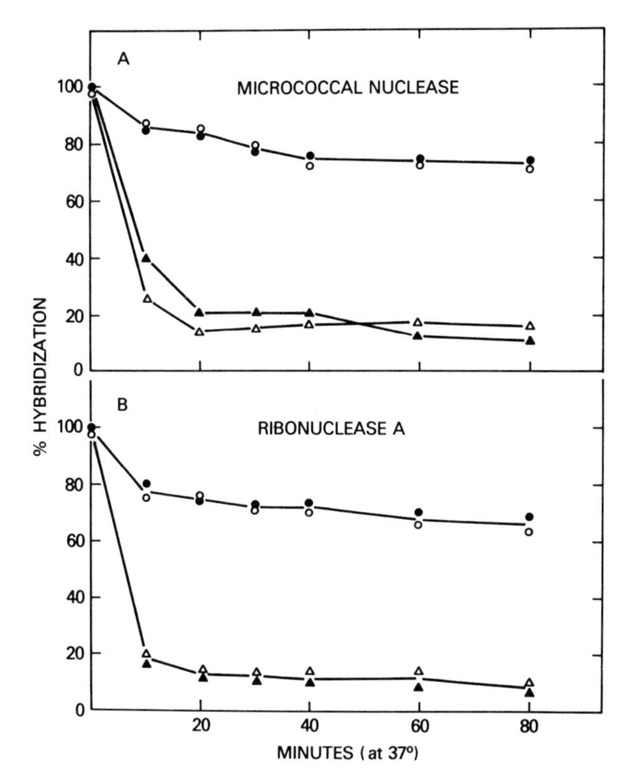


FIG. 6. Time course of cDNA and $cRNA_d$ digestion in the presence and in the absence of rabbit liver transcript. (A) Hybridization reactions (formamide system) containing 1 ng (3000 cpm) of [$^3$H]cDNA (○—○), cDNA, and 10 $\mu$g of rabbit liver transcript (●—●), cDNA and 6 ng of globin mRNA (△—△), or cDNA, globin mRNA, and 10 $\mu$g of rabbit liver transcript (▲—▲) were incubated for 24 hours at 43°C and digested with 40 $\mu$g of micrococcal nuclease per milliliter for the specified times. (B) Hybridization reactions (formamide system) containing 1 ng (2000 cpm) of [$^3$H]$cRNA_d$ (○—○), $cRNA_d$ and 10 $\mu$g of rabbit liver transcript (●—●), $cRNA_d$ and 2 ng of cDNA (△—△), or $cRNA_d$, cDNA, and 10 $\mu$g of rabbit liver transcript (▲—▲) for 48 hours at 43°C and digested with 10 $\mu$g of RNase A per milliliter for the specified times.

# III. Interpretation of Results

## A. Selectivity of Globin Gene Transcription

The hybridization of cDNA with RNA transcribed from rabbit marrow or liver chromatin by *E. coli* RNA polymerase is shown in Fig. 7 (also in Fig. 8, upper right panel). The proportion of globin mRNA sequences in the transcribed RNA may be estimated by comparing the initial slope of the

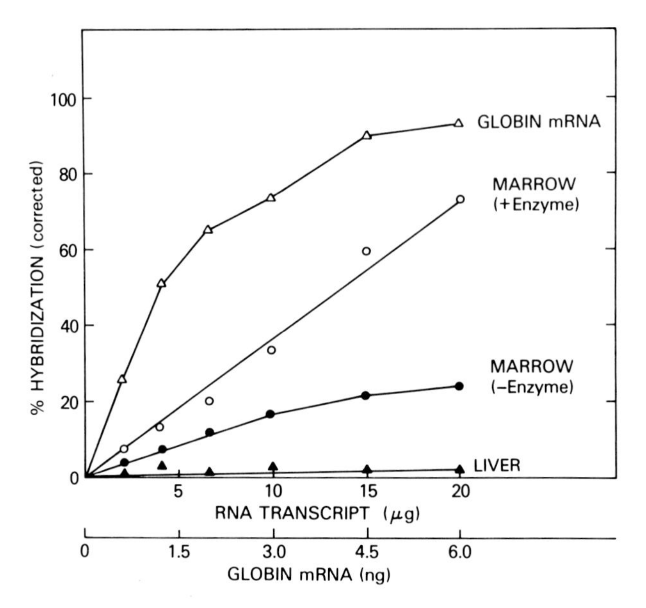


FIG. 7. Hybridization of cDNA to RNA transcribed from rabbit marrow or liver chromatin. Standard transcription reactions with or without *Escherichia coli* RNA polymerase were extracted as described in the text. Hybridization reactions contained 1 ng of [$^3$H]cDNA with 0.6–6.0 ng of globin mRNA (△), 2–20 μg of RNA extracted from a transcription reaction containing rabbit marrow chromatin without *E. coli* RNA polymerase (●), or 2–20 μg of RNA transcribed from rabbit liver chromatin (▲). The reactions were analyzed with micrococcal nuclease and the fraction of cDNA hybridized was corrected for self-annealing of the cDNA as described previously (*22*). From Wilson *et al.* (*21*).

hybridization curve to that of the globin mRNA standard (*21*). Although less than 0.1 ng of globin mRNA sequences could be detected for each 100 μg of RNA isolated from the reaction of *E. coli* RNA polymerase and rabbit liver chromatin, the values for rabbit marrow chromatin were 8.7 ng of globin mRNA sequences in the presence of *E. coli* RNA polymerase and 4.3 ng in the control reaction without enzyme. Thus, whereas globin mRNA transcription by *E. coli* RNA polymerase is tissue-specific, only 0.005% of the RNA is transcribed from globin genes.

Figure 7 also demonstrates that approximately one-half the globin mRNA sequences detected in the transcript of rabbit marrow chromatin are present in chromatin-associated RNA. Attempts to remove this background of globin mRNA sequences by using α-amanitin to inhibit chromatin-associated RNA polymerase II and by removing chromatin-associated globin mRNA sequences which contain poly (A) have been described in detail (*21*). The differentiation of globin mRNA sequences which are newly synthesized from those present at the start of the reaction remains a serious problem, and

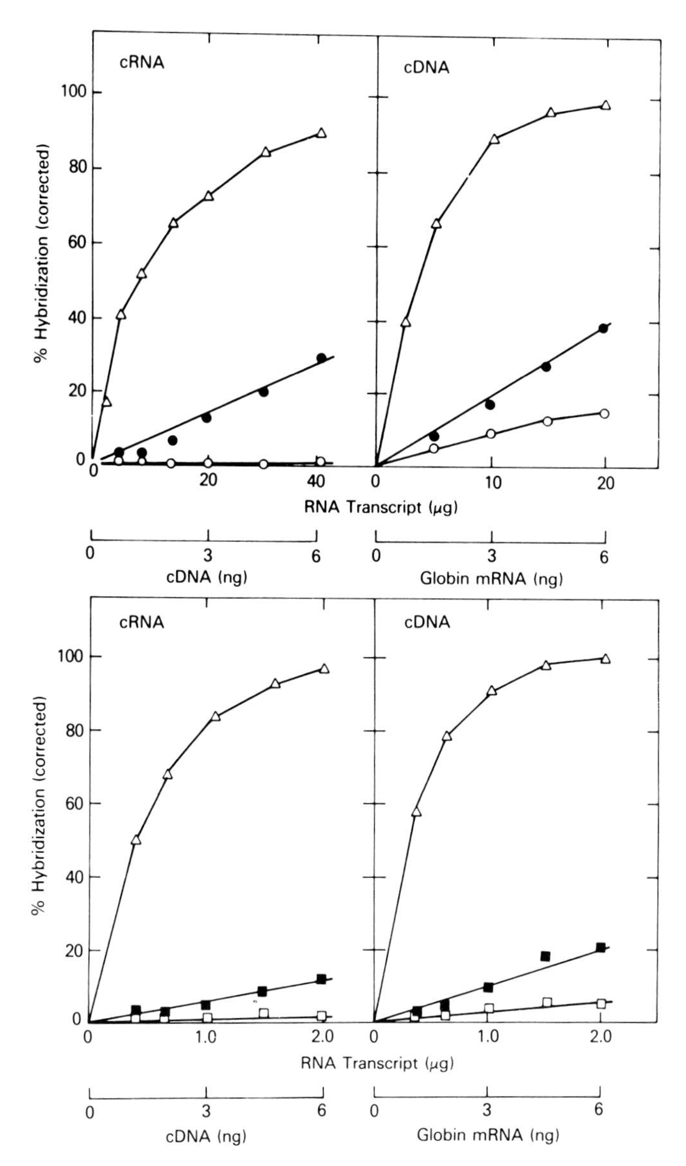


FIG. 8. Hybridization of chromatin-primed RNA to strand-specific probes. The upper right and left panels give the results with RNA transcribed by *Escherichia coli* RNA polymerase, and the lower right and left panels give results with mammalian polymerase. Hybridization reactions contained 1 ng of [$^{3}$H]cRNA or [$^{3}$H]cDNA with 1.2–6 ng of rabbit globin mRNA or [$^{32}$P]cDNA (△), 8–40 μg or 4–20 μg of RNA transcribed from rabbit marrow chromatin by *E. coli* RNA polymerase (●), 0.4–2 μg of RNA transcribed from rabbit marrow chromatin by sheep liver RNA polymerase II (■), or the specified amount of RNA extracted from the transcription reaction without RNA polymerase (○, □). The hybridization reactions were analyzed with micrococcal nuclease or RNase as described previously (*21*,*22*).

it is reassuring to demonstrate that some of the globin mRNA sequences detected by hybridization contain radioactivity. We have isolated globin mRNA–cDNA hybrids from reactions of RNA transcript with cDNA by cesium sulfate density equilibrium centrifugation and have shown that the appearance of [$^{32}$P]RNA in the hybrid requires the presence of *E. coli* RNA polymerase in the initial transcription reaction (*21*).

The proportion of globin mRNA sequences in RNA transcribed from rabbit marrow chromatin by sheep liver RNA polymerase II (fraction 5) can be estimated from the data in Fig. 8 (lower right panel) to be 0.03%. More selective globin mRNA transcription by mammalian RNA polymerase may relate to its saturation of chromatin at low levels of activity (50 units per 200 $\mu$g of chromatin DNA) or less likely to an increased affinity for globin gene promotor sites. Similar results have been described by Tsai *et al.* (*35*).

## B. Symmetry of Globin Gene Transcription

The cRNA probe was used to search for evidence of transcription of the DNA strand complementary to the globin gene (antistrand). Figure 8, upper and lower left panels, demonstrates such transcription by *E. coli* and sheep liver RNA polymerases. Comparison of these results to standard curves indicates that transcription of globin mRNA sequences from rabbit marrow chromatin is strand-selective by an approximate 2:1 ratio. Since antistrand transcript is not present in rabbit marrow cells or nuclei (*22*), these results may indicate that certain factors required for asymmetrical globin gene transcription may not be present in the cell-free system. There appear to be no significant differences between the prokaryotic and eukaryotic polymerases with respect to symmetry of transcription *in vitro*. The presence of antistrand transcript may cause underestimation of the proportion of globin mRNA sequences in RNA transcribed from chromatin.

## C. Comparison of Globin mRNA Transcription by Partially Purified Preparations of Sheep Liver Polymerase

Comparison of globin mRNA transcription from rabbit marrow chromatin by sheep liver RNA polymerase II purified through steps 5, 6, or 7 is shown in Fig. 9. Despite the addition of equivalent amounts of RNA polymerase to each reaction (50 units), purification of sheep liver RNA polymerase markedly decreases the selectivity and asymmetry of globin mRNA transcription from rabbit marrow chromatin. Although these results might suggest that factors required for globin gene transcription are removed by phosphocellulose and DNA–Sepharose chromatography, they must be regarded as highly preliminary. We include them here to demonstrate that, after contaminating RNase activity has been removed, further purification

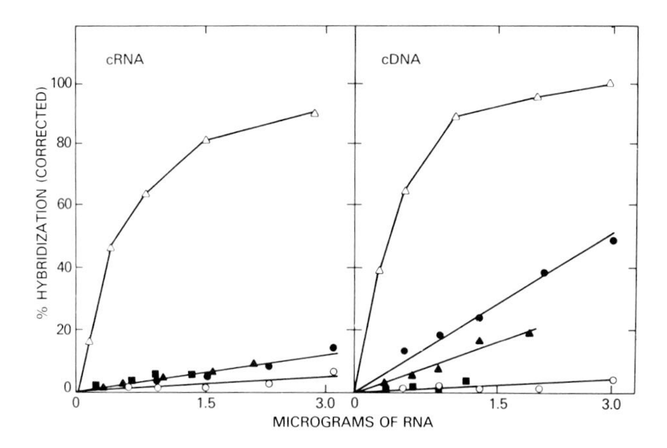


FIG. 9. Hybridization of chromatin-primed RNA synthesized by partially purified preparations of sheep liver RNA polymerase to strand-specific probes. Hybridization reactions contained 1 ng of [$^3$H]cRNA or [$^3$H]cDNA with 1.2–6 ng of rabbit globin mRNA or [$^{32}$P]cDNA (△), 0.6–3 $\mu$g of RNA transcribed by step 5 (●), 0.4–2.0 $\mu$g of RNA transcribed by step 6 (▲), or 0.28–1.4 $\mu$g of RNA transcribed by step 7 (■) sheep liver RNA polymerase from rabbit marrow chromatin. The control reaction (○) contained RNA extracted from the standard transcription reaction containing 50 units of step 5 sheep liver RNA polymerase II and 10 $\mu$g of $\alpha$-amanitin per milliliter to which had been added carrier RNA. Hybridization reactions were analyzed with micrococcal nuclease or RNase as described previously (*21*,*22*).

of eukaryotic RNA polymerases may not be necessary for the cell-free transcription of chromatin.

## IV. Concluding Remarks

We describe here an approach to the cell-free transcription of chromatin by bacterial and mammalian RNA polymerases. Batch chromatography procedures are required to obtain enough mammalian RNA polymerase for transcription experiments. Mammalian RNA polymerase is more selective for globin gene transcription than bacterial RNA polymerase, and this selectivity appears to be lost with purification of the enzyme. These results are consistent with other data showing differences in chromatin template activity for bacterial and mammalian RNA polymerases (*25*,*35–41*). Many advances in theory and technique will be needed to define the roles of chromatin and polymerase structure in gene transcription and to establish the relevance of cell-free transcription to the study of gene regulation. The reward of such efforts will be an invaluable assay for the fractionation and characterization of native eukaryotic genes.

## REFERENCES


*1.* Astrin, S., *Proc. Natl. Acad. Sci. U.S.A.* **70**, 2304 (1973).
*2.* Axel, R., Cedar, H., and Felsenfeld, G., *Proc. Natl. Acad. Sci. U.S.A.* **70**, 2029 (1973).
*3.* Gilmour, R. S., and Paul, J., *Proc. Natl. Acad. Sci. U.S.A.* **70**, 3440 (1973).
*4.* Shih, T. Y., Khoury, G., and Martin, M. A., *Proc. Natl. Acad. Sci. U.S.A.* **70**, 3506 (1973).
*5.* Barrett, T., Maryanka, D., Hamlyn, P. H., and Gould, H. J., *Proc. Natl. Acad. Sci. U.S.A.* **71**, 5057 (1974).
*6.* Honjo, T., and Reeder, R. H., *Biochemistry* **13**, 1896 (1974).
*7.* Jacquet, M., Groner, Y., Monroy, G., and Hurwitz, J., *Proc. Natl. Acad. Sci. U.S.A.* **71**, 3045 (1974).
*8.* Janowski, M., Baugnet-Mahieu, L., and Sassen, A., *Nature* (*London*) **251**, 347 (1974).
*9.* Steggles, A. W., Wilson, G. N., Kantor, J. A., Picciano, D. J., Falvey, A. K., and Anderson, W. F., *Proc. Natl. Acad. Sci. U.S.A.* **71**, 1219 (1974).
*10.* Chiu, J. F., Tsai, Y. H., Sakuma, K., and Hnilica, L. S., *J. Biol. Chem.* **250**, 9431 (1975).
*11.* Marzluff, W. F., and Huang, R. C. C., *Proc. Natl. Acad. Sci. U.S.A.* **72**, 1082 (1975).
*12.* Stein, G. S., Park, W. D., Thrall, C. L., Mans, R. J., and Stein, J. L., *Biochem. Biophys. Res. Commun.* **63**, 945 (1975).
*12a.* Marzluff, W. F. Murphy, E. C., and Huang, R. C. C., *Biochemistry* **12**, 3440 (1973).
*13.* Falvey, A. K., Kantor, J. A., Robert-Guroff, M. G., Picciano, D. J., Weiss, G. B., Vavich, J. M., and Anderson, W. F., *J. Biol. Chem.* **249**, 7049 (1974).
*14.* Nienhuis, A. W., Falvey, A. K., and Anderson, W. F., *Methods Enzymol.* **30**, Part F, 621 (1974).
*15.* Spelsberg, T. C., and Hnilica, L. S., *Biochim. Biophys. Acta* **228**, 202 (1971).
*16.* Axel, R., *Biochemistry* **14**, 2921 (1975).
*17.* Steggles, A. W. (1975). Unpublished results.
*18.* Noll, M., Thomas, J. O., and Kornberg, R. D., *Science* **187**, 1203 (1975).
*19.* Sollner-Webb, B., and Felsenfeld, G., *Biochemistry* **14**, 2915 (1975).
*20.* Burgess, R. R., *J. Biol. Chem.* **244**, 6160 (1969).
*21.* Wilson, G. N., Steggles, A. W., and Nienhuis, A. W., *Proc. Natl. Acad. Sci. U.S.A.* **72**, 4835 (1975).
*22.* Wilson, G. N., Steggles, A. W., Kantor, J. A., Nienhuis, A. W., and Anderson, W. F., *J. Biol. Chem.* **250**, 8604 (1975).
*23.* Poonian, M. S., Schlabach, A. J., and Weissbach, A., *Biochemistry* **10**, 424 (1971).
*24.* Kedinger, C., Gissinger, F., Gniazdowski, M., Mandel, J., and Chambon, P., *Eur. J. Biochem.* **28**, 269 (1972).
*25.* Weinmann, R., and Roeder, R. G.,*Proc. Natl. Acad. Sci. U.S.A.* **71**, 1790 (1974).
*26.* Anderson, W. F., Steggles, A., Wilson, G., Kantor, J., Velez, R., Picciano, D., Falvey, A., and Nienhuis, A., *Ann. N. Y. Acad. Sci.* **241**, 262 (1974).
*27.* Kostraba, N. C., Montagna, R. A., and Wang, T. Y., *J. Biol. Chem.* **250**, 1548 (1975).
*28.* Penman, S., *J. Mol. Biol.* **17**, 117 (1966).
*29.* Gilmour, R. S., Harrison, P. R., Windass, J. D., Affara, N. A., and Paul, J., *Cell Differ.* **3**, 9 (1974).
*30.* Zimmerman, S. B., and Sandeen, G., *Anal Biochem.* **14**, 269 (1966).
*31.* Poon, R., Paddock, C. V., Heindell, H., Whitcome, P., Salser, W., Kacian, D., Bank, A., Gambins, R., and Ramirez, F., *Proc. Natl. Acad. Sci. U.S.A.* **71**, 3502 (1974).
*32.* Marotta, C. A., Forget, B. A., Weissman, S. M., Verma, I. M., McCaffrey, R. P., and Baltimore, D., *Proc. Natl. Acad. Sci. U.S.A.* **71**, 2300 (1974).
*33.* Weiss, G. B., Wilson, G. N., Steggles, A. W., and Anderson, W. F., *J. Biol. Chem.* (1978).

*34.* Young, B. D., Harrison, P. R., Gilmour, R. S., Birnie, G. D., Hell, A., Humphries, S., and Paul, J., *J. Mol. Biol.* **84**, 555 (1974).

*35.* Tsai, M. J., Towle, H. C., Harris, S. E., and O'Malley, B. W., *J. Biol. Chem.* **251**, 1960 (1976).

*36.* Cox, R. F., *Eur. J. Biochem.* **39**, 49 (1973).

*37.* Maryanka, D., and Gould, H., *Proc. Natl. Acad. Sci. U.S.A.* **70**, 1161 (1973).

*38.* Meilhac, M., and Chambon, P., *Eur. J. Biochem.* **35**, 454 (1973).

*39.* Cedar, H., *J. Mol. Biol.* **95**, 257 (1975).

*40.* Magee, B. B., Paoletti, J., and Magee, P. T., *Proc. Natl. Acad. Sci. U.S.A.* **72**, 4830 (1975).

*41.* Tsai, M. J., Schwartz, R. J., Tsai, S. Y., and O'Malley, B. W., *J. Biol. Chem.* **250**, 5165 (1975).

# Chapter 30

# *Transcription of Chromatin from Cells Transformed by SV40 Virus*

SUSAN M. ASTRIN

*The Institute for Cancer Research,*
*The Fox Chase Cancer Center,*
*Philadelphia, Pennsylvania*

## I. Introduction

Evidence is accumulating in support of the concept that the structure of chromatin plays an important role in regulation of transcription in eukaryotic cells. Much of the support for this concept comes from *in vitro* studies in which the transcription of individual genes in chromatin by *Escherichia coli* RNA polymerase has been monitored (*1–9*). The general finding in these experiments is that the *E. coli* polymerase preferentially transcribes those genes that are being transcribed in the cell, presumably by recognizing some aspect of chromatin structure that is unique to active genes.

One system that may be used to advantage to study transcriptional regulation is the integrated SV40 genes in SV-3T3 cells, a line of mouse fibroblasts transformed by the small DNA virus SV40. The SV40 genome is covalently integrated into cellular DNA (*10*) at a frequency of about one copy per cell (*11*). The entire viral genome (molecular weight $3 \times 10^6$) is present as demonstrated by the fact that virus can be rescued by fusion with a sensitive cell (*12*). The major transcript, which apparently codes for T antigen (*13*), an antigen specific for SV40-transformed cells, is transcribed from about 50% of the "minus" strand of the integrated genome (*14*). This pattern of transcription has been very accurately defined through the use of probes which consist of specific fragments of SV40 DNA produced by cleavage of the DNA with site-specific restriction endonuclease (*15*). The relative positions on the SV40 genome of the fragments produced by a large number of restriction endonucleases have been mapped, and the "maps" have been useful not only in defining the pattern of transcription observed during lytic infection and in transformed cells, but also in characterizing the transcripts produced in *in vitro* systems.

This article describes the methods used for the analysis of the SV40-specific sequences transcribed with *E. coli* RNA polymerase from SV3T3 chromatin. The isolation and transcription of the chromatin is described as well as the analysis of the *in vitro* transcription pattern, which resembles to a high degree the pattern observed in the transformed cells. Special emphasis is placed on methods for preparing the probes used to analyze the transcripts, as it is these probes and the method of analysis that allow great precision in the determination of the transcription pattern.

## II. Preparation of Chromatin from SV40-Transformed Cells

SV-3T3 cells (*10*) are grown in Dulbecco's modification of Eagle's medium (*16*) containing 500 units of penicillin (Charles Pfizer and Co.) and 100 $\mu$g of streptomycin (Charles Pfizer and Co.) per milliliter and 10% calf serum (Microbiological Associates). The cells are grown to confluency in 100-mm plastic petri dishes. Generally $2 \times 10^9$ cells (about 100 dishes) are used for the preparation of chromatin.

Chromatin is prepared from isolated nuclei by the method of Seligy and Miyagi (*17*). The cell layers are washed, and the cells are removed by trypsinization, then washed three times with ice-cold TD buffer [TD = 0.8% NaCl–0.038% KCl–0.01% $Na_2HPO_4$–0.3% Tris (pH 7.4)]. The cells are resuspended in ice-cold 0.4 × RSB (RSB = 0.01 *M* Tris (pH 7.4)–0.01 *M* NaCl–0.0015 *M* $MgCl_2$), allowed to swell for 3 minutes, and then homogenized by hand using a loose-fitting Dounce homogenizer. The homogenate is centrifuged at 1000 *g* for 5 minutes, and the pellet is resuspended in ice-cold 0.4 × RSB and homogenized again until at least 95% of the cells have broken to yield nuclei. The nuclei are pelleted by centrifuging at 1000 *g* for 5 minutes, resuspended in 10 ml of ice-cold 0.01 *M* Tris (pH 7.4) and homogenized using a tight-fitting Dounce homogenizer. The resulting suspension is centrifuged at 10,000 *g* for 15 minutes (4°C). The pellet is resuspended in ice-cold 0.01 *M* Tris (pH 7.4) and homogenized again until all nuclei have broken. This suspension is centrifuged at 10,000 *g* for 15 minutes, and the pellet is resuspended in 10 ml of ice-cold 0.01 *M* Tris (pH 7.4) and layered onto 28 ml of 1.7 *M* sucrose–0.01 *M* Tris (pH 7.4). The upper two-thirds of the tube is gently mixed, and the sample is centrifuged for 3 hours at 25,000 rpm in a Spinco SW-27 rotor at 4°C. The pellet is washed by resuspending in 10 ml of ice-cold 0.01 *M* Tris (pH 7.4) and collected by centrifuging at 10,000 *g* for 15 minutes. The washing procedure is repeated once, and the resulting pellet, considered purified chromatin, is resuspended in ice-cold 0.01 *M* Tris (pH 7.9) at a DNA concentration of 1 mg/ml.

The chromatin, cooled in ice, is sheared for 30 seconds at a setting of 6 in a Sorvall Omni-mix fitted with a micro cup. The size of the DNA in the sheared chromatin is distributed in a rather broad molecular weight range with a mean MW of $10^7$ as estimated by sedimentation in neutral sucrose gradients. Less than 5% of the DNA has a molecular weight lower than that of intact SV40 DNA ($3 \times 10^6$). The DNA content of the chromatin is determined by the method of Burton (*18*), and protein is determined by the method of Lowry *et al.* (*19*). RNA content is determined by incubating the chromatin in 1 *M* NaOH to hydrolyze the RNA, precipitating the DNA and protein with acid, and measuring the UV absorption spectra of nucleotides in the supernatant. The DNA: protein: RNA ratio of purified SV3T3 chromatin is about 1:2:0.05.

## III. Transcription of Chromatin and DNA by RNA Polymerase from *Escherichia coli*

RNA is synthesized from a chromatin template in incubation mixtures containing equal weights of DNA as chromatin and *E. coli* RNA polymerase containing sigma factor (purified according to Burgess, *20*). A 10-ml standard reaction mixture contains 150 m*M* KCl–40 m*M* Tris (pH 7.9)–0.1 m*M* dithiothreitol–5 m*M* $MgCl_2$–1 m*M* $MnCl_2$–4 m*M* each ATP, GTP, CTP, and UTP (P-L Biochemicals) and 4 mg each of DNA as chromatin and RNA polymerase. The reaction mixture is incubated at 37°C for 4 hours. The time course of the transcription reaction is followed by adding 5 μCi of [$^3$H]UTP (17 Ci/mmole) to 100 μl of the standard reaction mixture. At appropriate time intervals, duplicate aliquots are removed for the determination of total and of acid-precipitable counts per minute. At the end of a 4-hour reaction, about 8 mg of RNA will be produced from 4 mg of chromatin DNA template. Thus, on the average, each region of the chromatin DNA is transcribed twice. However, since there are fewer binding sites for polymerase on chromatin than on naked DNA (*21*), and all sequences of the DNA in chromatin are not available for transcription by *E. coli* polymerase (*22*), each available region of the chromatin must have been transcribed many times.

To isolate RNA from the reaction, chromatin is first removed by centrifugation at 1000 *g* for 15 minutes. EDTA is then added to the supernatant to a final concentration of 0.1 *M*, and sodium dodecyl sulfate (SDS) is added to a concentration of 1%. The mixture is deproteinized by extracting twice with phenol–chloroform (1:1) and once with chloroform. The RNA is precipitated with cold 5% trichloroacetic acid to remove triphosphates, redissolved in 1 *M* Tris (pH 7.9)–0.1 *M* NaCl, and precipitated with ethanol overnight at −20°C. The precipitate is dissolved in 0.1 *M* sodium acetate (pH 5.3)–0.01 *M* $MgCl_2$ and treated with RNase-free DNase (Worthington Biochemicals) at a concentration of 20 μg/ml for 0.5 hour. EDTA is added to a final concentration of 0.1 *M*, and SDS is added to a concentration of 1%; the reaction is phenol–chloroform extracted and ethanol precipitated as described above. The RNA is dissolved in distilled water, and its concentration determined by absorbance at 260 nm. Recovery of *in vitro* synthesized RNA is generally greater than 90%.

SV3T3 DNA, to be used as a template for transcription, is prepared from sheared chromatin by addition of sodium perchlorate to a final concentration of 1 *M* and extraction four times with phenol–chloroform and once with chloroform. The DNA is precipitated with ethanol and dissolved in 0.01 *M* Tris (pH 7.9). DNA concentration is determined by absorbance at 260 nm. RNA transcribed from SV-3T3 DNA is synthesized and purified as described above for SV-3T3 chromatin except that the DNase treatment is carried out in the reaction mixture at the end of the incubation period.

## IV. Assay of SV40-Specific Sequences

The SV40-specific transcripts contained in *in vitro* synthesized RNA are detected and analyzed by annealing the RNA to the minus strands of purified fragments of SV40 DNA produced by cleavage of the DNA with a restriction endonuclease isolated from *Hemophilus aegyptius*. Fragments produced by other site-specific restriction endonucleases can be used in the same way. DNA–RNA annealing is carried out under conditions of probe excess to allow quantitation of the frequency of transcription from the region represented by each fragment. The method for synthesis of the probes and for the hybridization are described below. In an analysis of this sort, it is important to distinguish between SV40-specific RNA that is synthesized *in vitro* by *E. coli* polymerase and any SV40-specific RNA that may be contained in the chromatin as isolated or that is synthesized by the endogenous RNA polymerase present in the isolated chromatin. To determine the amount of SV40-specific RNA that is *not* synthesized by *E. coli* RNA polymerase, a control reaction is used. This reaction is identical to the standard reaction described above except that the *E. coli* RNA polymerase is omitted. The reaction is incubated as above, and yeast carrier RNA is added in an amount equal to the amount of RNA synthesized in the reaction which includes polymerase.

RNA is isolated from the control reaction as described above. In the SV3T3 system, analysis reveals that the amount of SV40-specific RNA contaminating the isolated chromatin or synthesized by the endogenous polymerase accounts for less than 5% of the SV40-specific RNA that is isolated from a standard reaction containing *E. coli* polymerase. Thus greater than 95% of the SV40-specific RNA is synthesized by the added *E. coli* polymerase. A method is available for physically separating RNA synthesized *in vitro* from RNA present in the isolated chromatin. This method involves the use of Hg-labeled triphosphates in the reaction mixture and isolation of Hg containing RNA by ion exchange chromatography on sulfhydryl–Sepharose columns. This method is described in detail elsewhere (*8, 23–25*).

## A. *In Vitro* Synthesis of Radiolabeled SV40 DNA

High specific activity $^{3}H$-labeled DNA probes ($10^{7}$ cpm/$\mu$g) are obtained by the method of Summers *et al.* (*26*) using DNA synthesized *in vitro* by *E. coli* DNA polymerase on a template of single-stranded circular SV40 DNA primed by oligonucleotides.

For preparation of SV40 DNA for use as template, confluent cultures of BSC-1 cells in 100-mm plastic petri dishes are infected with 0.4 ml of an SV40 stock preparation ($2.3 \times 10^{5}$ plaque-forming units/ml; input multiplicity of infection = 0.5 plaque-forming units/cell). After 8 days, when cells show cytopathic effect, SV40 DNA is selectively extracted by the procedure of Hirt (*27*). Cell layers are washed, and 1 ml of 0.01 *M* Tris (pH 8.0)–0.01 *M* EDTA–0.5% SDS is added to each petri dish. After 10 minutes at room temperature, 0.2 ml of 5 *M* NaCl is added and mixed with the lysate by gently rotating the dish. The lysate is gently scraped into a centrifuge tube with a rubber policeman and stored for 8 hours at 4°C. A white precipitate, containing cellular DNA, forms and is removed by centrifugation at 10,000 *g* for 30 minutes. The supernatant, containing the SV40 DNA, is extracted three times with phenol saturated with 1 *M* Tris (pH 7.4) once with chloroform–isoamyl alcohol (24:1, v/v), and precipitated with 2 volumes of 95% ethanol at −20°C. The DNA is dissolved in 0.1 *M* Tris (pH–7.4)–1 m*M* EDTA, and centrifuged in an ethidium bromide–CsCl gradient (400 $\mu$g of ethidium bromide per milliliter, density = 1.58 gm/cm$^{3}$) at 43,000 rpm for 48 hours in a Spinco type 50 rotor. The lower band, containing the closed circular form I SV40 DNA as well as mitochondrial form I DNA, is extracted three times with an equal volume of isopropanol: water (9:1, v/v), and dialyzed extensively against 0.01 *M* Tris (pH 7.4)–1 m*M* EDTA. Form I DNA is then sedimented in a 5 to 20% (w/v) neutral sucrose gradient containing 0.15 *M* NaCl–0.015 *M* sodium citrate–0.1 *M* Tris (pH 7.4)–1 m*M* EDTA for 17.5 hours at 25,000 rpm in a Spinco SW-27 rotor at 4°C. The 21 S fraction, containing the SV40 DNA, is precipitated with two volumes of ethanol at −20°C. DNA concentration is determined by the diphenylamine procedure of Burton (*18*) or the absorbance at 260 nm.

Single-stranded circular SV40 DNA is produced after limited endonucleolytic digestion of component I SV40 DNA with pancreatic DNase. Component II DNA thus produced is sedimented for 8 hours through a 5 to 20% alkaline sucrose gradient in a Beckman SW-40 rotor at 40,000 rpm, 20°C. The faster sedimenting single-stranded circular molecules are collected, dialyzed to neutral pH, and concentrated. Single-stranded DNA is denatured with 0.2 *M* NaOH immediately prior to its use.

Purified circular SV40 DNA is used as a template for *E. coli* DNA polymerase I [purified according to Jovin *et al.* (*28*)] with the addition of random oligonucleotide primers (6 per circle). These primers are produced by the method of Dumas *et al.* (*29*) by exhaustive digestion of calf thymus DNA with pancreatic DNase. Degraded calf thymus primer does not stimulate incorporation by *E. coli* DNA polymerase I of deoxyribonucleoside triphosphates into DNA in the absence of a high molecular weight template. The DNA polymerase reaction mixture contains 0.1 *M* Tris (pH 7.5)–10 m*M* $MgCl_2$–100 μ*M* each of dATP, dGTP, and dCTP–20 μ*M* [$^3H$]dTTP (50 Ci/mmole) and 1 μg of *E. coli* DNA polymerase I, 6 μg of oligonucleotide primers, and 1 μg of single-stranded circular SV40 DNA in a final volume of 0.1 ml. The reaction is incubated for 4 hours at 15°C. The product consists of 2 μg of double-stranded DNA containing one nonradioactive strand (the template strand) and one newly synthesized radioactive strand. The product sediments as SV40 component II DNA (double-stranded open circular form) in neutral sucrose gradients and has a specific activity of $1.0 \times 10^7$ cpm of $^3H$ per microgram.

## B. Restriction Enzyme Cleavage of SV40 DNA

$^3H$-labeled SV40 DNA prepared as described above is digested with endonuclease R·HaeIII [purified from *H. aegyptius* ATCC 11116 according to Middleton *et al.* (*30*) except that a streptomycin sulfate precipitation is substituted for the Bio-Gel chromatography step]. Digestions are carried out in a reaction volume of 0.02 ml containing 10 m*M* $MgCl_2$–100 m*M* Tris (pH 7.5)–5 m*M* dithiothreitol, 2 μg of DNA, and 5 μg of enzyme. DNA is digested for 12 hours at 37°C. The resulting fragments are separated by electrophoresis through 17 × 0.6 cm cylindrical gels of 3.5% polyacrylamide. The electrode and gel buffer is the following: 0.04 *M* Tris–0.02 *M* sodium acetate–0.001 *M* disodium EDTA adjusted to pH 7.2 with about 2 ml of acetic acid per liter. The sample, containing 2% sucrose and 0.04% bromophenol blue, is electrophoresed at 3 mA per gel for such a time that the tracking dye migrates to within 1 cm of the bottom of the gel. Digests of 2 μg of SV40 DNA are visualized by staining the DNA with 5 μg/ml of ethidium bromide in distilled water for 15 minutes, then viewing under a long wavelength UV light. The digest contains 10 large fragments, in easily visible bands, as well as 6 very small fragments. Individual DNA fragments are well resolved allowing the portions of the gel containing the DNA to be cut out with a razor blade and crushed by passing them through an orifice prepared by cutting off a 23 gauge needle. The DNA is eluted from the crushed gel slices with 0.05 *M* EDTA (pH 7.0) and stored at 0°C. The endonuclease R·HaeIII cleavage map of SV40 is shown in Fig. 1 (*31*). The positions and relative sizes of the 10 large fragments (lettered A through H) are shown. In addition, the region transcribed in the transformed cell is indicated and is comprised of regions A and D (*14*).

## C. Separation of Strands of SV40 DNA

In order to prepare probes to assay for the strand specificity of the *in vitro* transcripts (the minus strand is transcribed *in vivo*), it is necessary to separate the "plus" and "minus" strands of the restriction fragments of SV40 DNA. The strand separation procedure is based on the observation (*32*) that when purified form I SV40 DNA is used as a template for *E. coli* RNA polymerase, the viral RNA produced (termed complementary RNA) is asymmetric (transcribed from only the minus DNA strand) and can be used to separate the plus and minus strands of SV40 DNA. The SV40 DNA is denatured, annealed to an excess (50-fold or greater) of the asymmetric complementary RNA, and the DNA–RNA hybrids are separated from single-stranded DNA on hydroxyapatite. The DNA–RNA hybrid contains the SV40 minus DNA

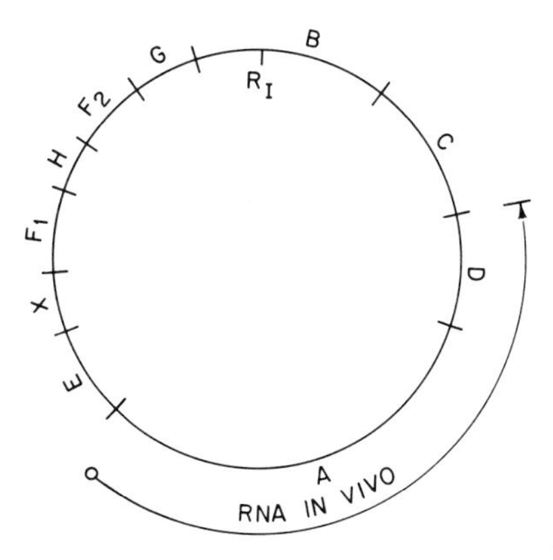


FIG. 1. Map of the HaeIII fragments of SV40 DNA as determined by Lebowitz *et al.* (*31*).

strand, and the plus DNA strand remains single-stranded. The technical details of the procedure are described below.

Asymmetric complementary RNA is synthesized in a 2-ml reaction volume containing 1 m*M* each of ATP, GTP, UTP, and CTP–0.04 *M* Tris (pH 7.9)–0.1 m*M* dithiothreitol–10 m*M* $MgCl_2$–0.15 *M* KCl; 100 μg of SV40 component I DNA; and 50 μg of *E. coli* RNA polymerase [purified according to Burgess (*20*), on Bio-Gel columns]. The reaction mixture is incubated for 2 hours at 37°C, treated with RNase-free DNase (Worthington Biochemical) at 20 μg/ml for 0.5 hour, and then extracted twice with an equal volume of phenol–chloroform (1:1) and once with chloroform. The RNA is applied to a Sephadex G-50 column equilibrated with 0.1 *M* NaCl, collected in the void volume, precipitated with 2 volumes of cold ethanol, and finally dissolved in distilled water. RNA concentration is determined by absorbance at 260 nm. The above reaction produces about 1 mg of complementary RNA.

The separation of the plus and minus strands of the SV40 DNA fragments is performed as described by Khoury *et al.* (*33*). SV40 complementary RNA, at a concentration of 2 μg/ml (prepared as described above), is incubated with denatured SV40 [$^3$H]DNA fragments, at a molar ratio of 100:1, in a reaction mixture containing 0.5 *M* NaCl–0.01 *M* Tris (pH 7.5) for 1 hour at 68°C. The reaction mixture is applied to a hydroxyapatite column equilibrated with 0.01 *M* sodium phosphate buffer (pH 6.8) at 60°C. The DNA is eluted with a gradient of 0.1 *M* to 0.3 *M* sodium phosphate buffer (pH 6.8) at 60°C. Two peaks of DNA are obtained, the first being single-stranded DNA plus strands, the second being DNA minus strands hybridized to complementary RNA. SV40 complementary RNA present in the eluates is removed by treatment with 0.5 *N* NaOH for 6 hours at 37°C after passage of the 2 fractions through Chelex 100 (Bio-Rad Laboratories) to remove calcium ions. Both DNA fractions from the hydroxyapatite column are allowed to "self-associate" at a sodium ion concentration of 1.0 *M* for 24–48 hours at 60°C. Any residual DNA reassociation products are removed by passage through hydroxyapatite columns. EDTA is added to samples containing DNA plus and minus strands to achieve a final concentration of 0.01 *M*, and samples are dialyzed against 1.0 *M* NaCl–0.01 *M* Tris (pH 7.4) followed by extensive dialysis against 0.01 *M* Tris (pH 7.4)–0.001 *M* EDTA.

## D. Annealing of Probes to *in Vitro* Transcripts

Minus strands of purified restriction fragments of $^3$H-labeled SV40 DNA are used as probes for the detection of SV40 specific transcripts in the RNA transcribed *in vitro*. As previously mentioned, the minus strands are used as probes because *E. coli* RNA polymerase transcribes only the minus strand of SV40 DNA and because it is the minus strand of the integrated SV40 DNA that is transcribed *in vivo* in transformed cells. In order to measure the frequency of transcription from each region of the integrated SV40 genome, a small amount of nonradiolabeled RNA transcribed *in vitro* from a template of SV-3T3 chromatin or SV-3T3 DNA is annealed to an excess of the $^3$H-labeled minus DNA strand of each of the following HaeIII fragments: A, B, C, D, and an equimolar mixture of the small fragments E, $F_1$, and $F_2$. (These fragments in sum constitute greater than 85% of the SV40 genome). The technical details of the annealings are given below. Under the conditions of probe excess employed in these experiments all the RNA transcribed from a given region of the DNA should hybridize to the probe representing that region and may be quantitated by determining the amount of DNA probe resistant to the single-strand specific nuclease S-1.

Annealing reactions are performed in sealed capillary tubes in 1 *M* NaCl–0.001 *M* EDTA–0.01 *M* Tris (pH 7.4) at 65°C for 72 hours. Annealings of RNA transcribed from chromatin are performed in a total volume of 25 μl; annealings of RNA transcribed from DNA are performed in a total volume of 10 μl. The time course of annealing in a typical experiment is such that the time period chosen (72 hours) is sufficient to allow the annealing reaction to go to at least 95% of completion in all cases. More than 70% of the [$^3$H]DNA probe remains hybridizable after 72 hours at 65°C. The quantity of RNA added to each annealing reaction is chosen so that the DNA probe will be in excess over the quantity of SV40-specific RNA added. This determination is made by annealing various quantities of RNA to two different concentrations of probe and choosing conditions such that addition of more RNA gives additional annealing, but addition of probe does not. In a typical experiment these conditions are met by the use of 500 and 1000 cpm of probe ($5 \times 10^{-5}$ and $10^{-4}$ μg) and 120 μg of input RNA. In the case of the annealing of RNA transcribed from chromatin to the fragment A probe, however, conditions of probe excess are achieved only at a much lower input of RNA (6 μg). After hybridization, samples are diluted 10-fold into buffer containing 0.05 *M* sodium acetate (pH 4.0)–0.005 *M* $ZnSo_4$–0.2 *M* NaCl. Half of the sample is spotted onto glass-fiber filters and precipitated with 0.5 *N* HCl to determine total counts present. To the remainder is added 900 units of S1 nuclease (*34*) and 250 cpm of denatured $^{14}$C-labeled mouse kidney DNA (added as a control for the S1 digestion). After digestion at 37°C for 1 hour the samples are spotted onto glass-fiber filters and acid precipitated. In a typical experiment, the extent of digestion of the control DNA is always greater than 95% whereas less than 1% of double-stranded DNA is digested.

The experimental design and data for a typical experiment in which RNA transcribed from SV3T3 chromatin is annealed to the probes is shown in Table I. The results, tabulated in the right-hand column, indicate that

TABLE I

ANNEALING OF RNA TRANSCRIBED FROM SV3T3 CHROMATIN TO ENDO R·HaeIII FRAGMENTS OF SV40 DNA[a]

| Fragment | Probe: Cpm | Probe: μmoles of fragment × $10^{12}$ [b] | Percent S1 resistance[c] | μmoles of transcript × $10^{12}$/120 μg of RNA |
|---|---|---|---|---|
| A | 500 | 104 | 10 | 210 |
| A | 1000 | 208 | 5 | 210 |
| B | 500 | 208 | 13 | 27 |
| B | 1000 | 416 | 6 | 25 |
| C | 500 | 303 | 20 | 61 |
| C | 1000 | 606 | 11 | 67 |
| D | 500 | 476 | 25 | 120 |
| D | 1000 | 952 | 15 | 142 |
| E, $F_1$, $F_2$ | 500 | 175 | 16 | 28 |
| E, $F_1$, $F_2$ | 1000 | 350 | 9 | 31 |
| Control A | 500 | 104 | 6 | 7 |

[a] The table shows the results of annealing the minus strand of labeled genome fragments of SV40 with RNA transcribed from chromatin. Annealings were performed under conditions of probe excess as described in the text. In each case, 120 μg of input RNA was used except in the case of fragment A, in which 6 μg of RNA was used. The annealing of fragment A to 120 μg of RNA gave 80% S1 resistance at both probe concentrations, indicating that the RNA was in excess over the probe. Only at a much lower input of RNA (6 μg) were conditions of probe excess achieved allowing a calculation of the absolute amount of RNA complementary to fragment A. The right-hand column gives the absolute amount of each fragment that annealed to 120 μg of input RNA. In the case of fragment A, this number is taken as 20 times the μmoles of transcript in 6 μg of RNA. The data designated control are an annealing to 120 μg of RNA isolated from a reaction to which no polymerase had been added. Yeast carrier RNA was added to this reaction in an amount equal to the amount of RNA synthesized in a reaction which included polymerase.

[b] These numbers were calculated from the specific activity of the *in vitro* labeled DNA serving as probe and from the nucleotide length of each fragment as determined by Lebowitz *et al.* (*31*). The radiolabeled nucleotide may not be equally distributed in the cleavage products, causing slight errors in these figures. From the data of Danna and Nathans (*15*), it can be calculated that values for the A + T content for the HaeIII fragments range from 58 to 64%. The error is therefore probably small and does not significantly affect the conclusions.

[c] Initial S1 resistance of each probe (less than 1%) has been subtracted.

regions A and D are transcribed preferentially; regions B, E, $F_1$, and $F_2$ are transcribed eight times less frequently than the A region. As the data show, the absolute amount of RNA hybridized to each probe does not change even when the amount of probe is doubled. Thus, the hybridization reactions

TABLE II

ANNEALING OF RNA TRANSCRIBED FROM SV-3T3 DNA TO ENDO R·HaeIII FRAGMENTS OF SV40 DNA[a]

| | Probe | | | |
|---|---|---|---|---|
| Fragment | Cpm | μmoles of fragment × $10^{12}$ [b] | Percent S1 resistance[c] | μmoles of transcript × $10^{12}$/120 μg of RNA |
| A | 500 | 104 | 20 | 21 |
| A | 1000 | 208 | 11 | 23 |
| B | 500 | 208 | 9 | 19 |
| B | 1000 | 416 | 5 | 21 |
| C | 500 | 303 | 12 | 36 |
| C | 1000 | 606 | 7 | 42 |
| D | 500 | 476 | 10 | 48 |
| D | 1000 | 952 | 6 | 57 |
| E, $F_1$, $F_2$ | 500 | 175 | 7 | 12 |
| E, $F_1$, $F_2$ | 1000 | 350 | 4 | 14 |

[a]The table shows the results of annealing the minus strand of labeled genome fragments of SV40 with RNA transcribed from SV-3T3 DNA. In each case, 120 μg of input RNA was used. The right-hand column gives the absolute amount of each fragment that annealed to the RNA.

[b]These numbers were calculated as described in Table I, footnote *b*.

[c]Initial S1 resistance of each probe (less than 1%) has been subtracted.

have gone to completion and all the complementary RNA sequences have been detected. Preferential transcription *in vitro* of the A and D regions is in good agreement with the transcription pattern observed *in vivo* in this line of transformed cells (see Fig. 1).

Also shown in Table I are the results of a typical control experiment in which RNA isolated from a reaction to which no polymerase had been added was annealed to fragment A. A low level of annealing of the control RNA to fragment A is observed. As mentioned previously, this annealing represents the level of *in vivo* synthesized RNA contaminating the isolated chromatin as well as any RNA synthesized by the endogenous RNA polymerase *in vitro*. The value for the quantity of A region specific transcripts in the control experiment is about 3% of the value for the reaction with added polymerase indicating that the added polymerase is responsible for 97% of the observed synthesis.

In order to determine the extent to which chromatin structure determines the transcriptional specificity observed in the *in vitro* system, the transcription pattern from SV-3T3 chromatin can be compared to the pattern obtained when SV3T3 DNA, extracted from chromatin, is used as a template for transcription. Purified SV3T3 DNA is transcribed with *E. coli* polymerase and the SV40-specific transcripts quantitated as described for the transcripts from chromatin. The experimental design and data for such an experiment are shown in Table II. As can be seen from the figures in the right-hand column, there is no longer preferential transcription of region A. In fact, all regions are transcribed with about equal frequency. Thus, the major transcripts from SV-3T3 chromatin closely resemble the *in vivo* transcripts, but the transcripts of DNA come from all regions of the SV40 genome.

## V. Concluding Remarks

A number of systems have been developed in which the *in vitro* transcription of a specific gene can be monitored. Such experiments have already shown that there is a correlation between the structure of chromatin and transcriptional specificity. In general, the transcripts from chromatin resemble the *in vivo* transcripts much more closely than do the transcripts from DNA. The type of analysis described above allows a very precise comparison of the sequences transcribed *in vitro* and those made *in vivo*. Such an analysis is possible in any system for which purified genes can be obtained, a goal made possible by the development of the technology for inserting eucaryotic genes into bacterial plasmids (*35–37*).

Mapping techniques such as the one described here can be used to obtain information about both the *in vivo* and *in vitro* transcription products of a given gene. Initiation and termination sites can be identified and one will be able to determine definitively whether the same initiation sites are used in *in vitro* systems as are used *in vivo*. Such information will be useful for determining the mechanism by which regulatory molecules act and for future purification of such molecules.

### Acknowledgments

This work was supported by U.S. Public Health Service Grants RR-05539 and CA-06927, a grant from the National Science Foundation, PCM-76-09721, and an appropriation from the Commonwealth of Pennsylvania.

### References

*1.* Astrin, S. M., *Proc. Natl. Acad. Sci. U.S.A.* **70**, 2304 (1973).
*2.* Shih, T. H., Khoury, G., and Martin, M. A., *Proc. Natl. Acad. Sci. U.S.A.* **70**, 3506 (1973).
*3.* Axel, R., Cedar, H., and Felsenfeld, G., *Proc. Natl. Acad. Sci. U.S.A.* **70**, 2029 (1973).

4. Gilmour, R. S., and Paul, J., *Proc. Natl. Acad. Sci. U.S.A.* **70** 3440 (1973).
5. Steggles, A. W., Wilson, G. N., Kanter, J. A., Picciano, D. J., Falvey, A. K., and Anderson, W. F., *Proc. Natl. Acad. Sci. U.S.A.* **71**, 1219 (1974).
6. Barrett, T., Maryanka, D., Hamlyn, P. H., and Gould, H. J., *Proc. Natl. Acad. Sci. U.S.A.* **71**, 5057 (1974).
7. Astrin, S. M., *Biochemistry* **14**, 2700 (1975).
8. Smith, M. M., and Huang, R. C. C., *Proc. Natl. Acad. Sci. U.S.A.* **73**, 775 (1976).
9. Kleinsmith, L. J., Stein, J., and Stein, G., *Proc. Natl. Acad. Sci. U.S.A.* **73**, 1174 (1976).
10. Sambrook, M., Westphal, H., Srinivasan, P. R., and Dulbecco, R., *Proc. Natl. Acad. Sci. U.S.A.* **60**, 1288 (1968).
11. Gelb, L. D., Kohne, D. E., and Martin, M. A., *J. Mol. Biol.* **57**, 129 (1971).
12. Watkins, J. R., and Dulbecco, R., *Proc. Natl. Acad. Sci. U.S.A.* **58**, 1396 (1967).
13. Roberts, B. E., Gorecki, M., Mulligan, R. C., Danna, K. J., Rozenblatt, S., and Rich, A., *Proc. Natl. Acad. Sci. U.S.A.* **72**, 1922 (1975).
14. Khoury, G., Martin, M. A., Lee, T. N., and Nathans, D., *Virology* **63**, 263 (1975).
15. Danna, K. J., and Nathans, D., *Proc. Natl. Acad. Sci. U.S.A.* **69**, 3097 (1972).
16. Vogt, M., and Dulbecco, R., *Proc. Natl. Acad. Sci. U.S.A.* **49**, 171 (1963).
17. Seligy, V., and Miyagi, M., *Exp. Cell Res.* **58**, 27 (1969).
18. Burton, K., *Biochem. J.* **62**, 315 (1956).
19. Lowry, O. H., Rosebrough, M. J., Farr, A. L., and Randall, R. J., *J. Biol. Chem.* **193**, 265 (1951).
20. Burgess, R. R., *J. Biol. Chem.* **244**, 6160 (1969).
21. Cedar, H., and Felsenfeld, G., *J. Mol. Biol.* **77**, 237 (1973).
22. Biessmann, H., Gjerset, R. A., Levy, W. B., and McCarthy, B. J., *Biochemistry* **15**, 4356 (1976).
23. Dale, R. M. K., Livingston, D. C., and Ward, D. C., *Proc. Natl. Acad. Sci. U.S.A.* **70**, 2238 (1973).
24. Dale, R. M. K., Martin, E., Livingston, D. C., and Ward, D. C., *Biochemistry* **14**, 2447 (1975).
25. Dale, R. M. K., and Ward, D. C., *Biochemistry* **14**, 2453 (1975).
26. Summers, J., O'Connell, A., and Millman, I., *Proc. Natl. Acad. Sci. U.S.A.* **72**, 4597 (1975).
27. Hirt, B., *J. Mol. Biol.* **26**, 365 (1967).
28. Jovin, T. M., Englund, P. T., and Bertsch, L. R., *J. Biol. Chem.* **244**, 2996 (1969).
29. Dumas, L. B., Darby, G., and Sinsheimer, R. L., *Biochim. Biophys. Acta* **228**, 407 (1972).
30. Middleton, J. H., Edgell, M. H., and Hutchinson, C. A., III., *J. Virol.* **10**, 42 (1972).
31. Lebowitz, P., Siegel, W., and Sklar, J., *J. Mol. Biol.* **88**, 105 (1974).
32. Westphal, H., *J. Mol. Biol.* **50**, 407 (1970).
33. Khoury, G. Byrne, J. C., and Martin, M. A., *Proc. Natl. Acad. Sci. U.S.A.* **69**, 1925 (1972).
34. Ando, T., *Biochim. Biophys. Acta* **114**, 158 (1966).
35. Jackson, D. A., Symons, R. H., and Berg, P., *Proc. Natl. Acad. Sci. U.S.A.* **69**, 2904 (1972).
36. Mertz, J. E., and Davis, R. W., *Proc. Natl. Acad. Sci. U.S.A.* **69**, 3370 (1972).
37. Lobban, P. E., and Kaiser, A. D., *J. Mol. Biol.* **78**, 453 (1973).

# Part E. Chromatin Reconstitution

# Chapter 31

## *Chromatin Reconstitution*

R. S. GILMOUR

*The Beatson Institute for Cancer Research, Bearsden, Glasgow, Scotland*

### I. Background

The study of the *in vitro* transcription of reconstituted chromatin offers the only meaningful approach at present for determining which aspects of chromatin composition and organization are essential for the maintenance of tissue-specific gene expression.

Reconstitution was first described by Huang *et al.* (*1*) for the formation of DNA–histone complexes. Later, Marushige *et al.* (*2*) used a similar method to reconstitute DNA and nonhistone proteins (NHP) by mixing the components in 1 *M* NaCl (where both are dissociated) and reassociating by gradient dialysis to low ionic strength. This procedure was used to dissociate and reconstitute whole chromatin, and the effect of transcriptional specificity was measured. Evidence was obtained for organ-specific reconstitution of chromatin by combining acidic proteins of different tissues with DNA and histones (*3*). Transcription of reconstituted chromatin with bacterial polymerase yielded RNA essentially identical to *in vivo* nuclear RNA as judged by competition experiments. Earlier reconstitution experiments carried out on whole chromatin by Bekhor *et al.* (*4*) and Huang and Huang (*5*) also demonstrated tissue-specific reconstitution; however, these workers presented data to show that a species of chromatin-bound RNA, chromosomal RNA (cRNA), was responsible for directing specificity. The inclusion of 5 *M* urea in the reconstitution procedure was considered mandatory for the participation of cRNA.

Subsequent investigations have placed the existence and properties of cRNA in serious doubt. Artman and Roth (*6*) suggested that cRNA was a degradation product of nuclear RNA. The same conclusion was reached by Heyden and Zachau (*7*), (degradation of tRNA), Szeszac and Pihl (*8*), Arnold and Young (*9*) and Getz and Saunders (*10*) (degradation of HnRNA), Tolstoshev and Wells (*11*) (degradation of rRNA), and Scharpe and van Parijs (*12*) (degradation of tRNA and rRNA). At about the same time a number of laboratories published data to suggest that the NHP fraction of chromatin was responsible for directing the specificity of reconstitution (*13–17*). In many cases isolated DNA histones and crude NHP preparations were reconstituted to show that tissue specificity of RNA transcripts correlated with the source of NHP. However, because of the type of hybridization analysis employed, all these results are open to the criticism that only transcripts from repetitive DNA sequences are measured. The relationship of these sequences to mRNA is still unknown; however, it is interesting to note that a tissue-specific pattern of repetitive DNA transcription is found in chromatin templates and appears to be under the control of NHP. Significantly, Crain and Saunders (*18*) have found that *E. coli* RNA polymerase preferentially initiates RNA synthesis on repetitive DNA sequences in chromatin.

## II. Present Status

Recently, a number of authors have reexamined the specificity of reconstitution using gene-specific cDNA probes. Barrett *et al.* (*19*) reconstituted chicken reticulocyte chromatin from separately purified DNA, histones, and NHP and showed that *E. coli* polymerase was capable of transcribing globin-specific RNA sequences from the template. This effect was not observed when liver NHP were substituted.

Gilmour *et al.* (*20*) performed similar experiments with mouse fetal liver and mouse brain NHP purified on hydroxyapatite. The separate NHP preparations were reconstituted with identical DNA and histone preparations. Only in the case of chromatin reconstituted with fetal liver NHP were *E. coli* transcripts found to contain globin mRNA sequences as judged by globin cDNA hybridization. Chiu *et al.* (*21*) prepared extractable NHP from chicken reticulocyte chromatin by treatment with 5 *M* urea/50 m*M* sodium phosphate and showed that this fraction when reconstituted with chicken brain chromatin components activated the repressed globin genes. Tsai *et al.* (*22*) fractionated the chromatin from hormone-stimulated chick oviduct (ovalbumin, synthesizing) and withdrawn oviduct by the same method. When NHP from stimulated chromatin was reconstituted with

DNA and histones from unstimulated chromatin, the resulting template supported the *in vitro* synthesis of ovalbumin mRNA sequences as judged by hybridization to ovalbumin cDNA. Stein *et al.* (23) and Park *et al.* (*24*) prepared cDNA to histone mRNA and probed for histone mRNA transcripts in various HeLa cell chromatins. A comparison of RNA transcribed from chromatins isolated from S phase and $G_1$ phase of the HeLa cell cycle show marked synthesis of histone mRNA sequences in the case of S-phase transcripts, but not for $G_1$-phase transcripts. Isolation of NHP from S-phase chromatin and reconstitution with $G_1$-phase chromatin results in an enhancement of the previously inactive histone genes. A similar conclusion is derived from experiments where NHP from $G_1$- and S-phase chromatins are reconstituted separately with identical DNA and histone preparations. S-phase chromatin contains an NHP which has the ability to render histone genes available for transcription.

Many of these experiments show background levels of endogenous RNA. In some cases controls are performed to eliminate the possibility that this contamination is responsible for the data. Gilmour and Paul (*25*) purified fetal liver NHP using CsCl gradients, a procedure that selectively removes endogenous RNA. In a series of experiments mouse fetal liver NHP were first fractionated on HAP by differential phosphate elution; after removal of endogenous RNA on CsCl, the individual nonhistone fractions were reconstituted to purified DNA and histones (Fig. 1) (*25,26*). Transcription of the reconstituted chromatins with *E. coli* RNA polymerase reveals that most of the globin gene activation is associated with the HAP2 fraction. This fraction of NHP, however, still contains about 60% of the original protein and a comparatively large number of the individual NHP species, as seen on two-dimensional gels.

## III. Problems and Perspectives

It is clear from the examples discussed that reconstitution can be used effectively to determine whether the *in vitro* activity of a specific gene is under the control of the NHP component in chromatin. However, it has been noted from personal communication with workers in this field that the current methods of reconstitution do not always give fully active chromatin preparations. The reason for this inconsistency is not known, mainly because no attempt has been made to define clearly the optimal conditions of dissociation and reassociation with respect to biological activity.

The experiments of the type described by Kleiman and Huang (*27*) and

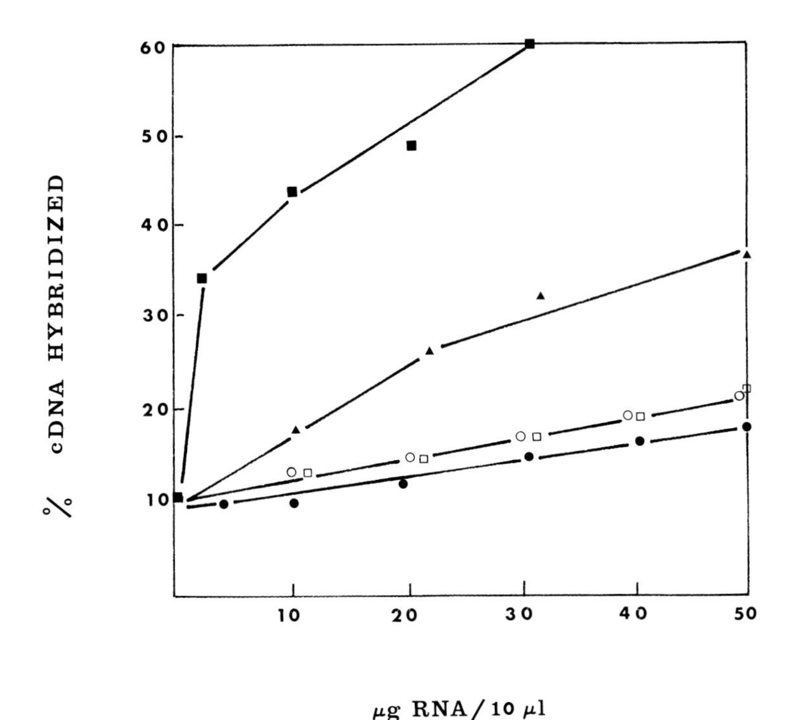


FIG. 1. Hybridization of globin cDNA to RNAs transcribed from mouse embryonic liver chromatins reconstituted from DNA, histones, and nonhistone proteins (NHP). NHP fractionation was carried out according to Rickwood and MacGillivray (*26*) by hydroxyapatite (HAP) chromatography. Histones (HAP 1) are not adsorbed by HAP. Three fractions of NHP (HAP 2, 3, and 4) were then eluted from the column in buffers containing in addition to 5 *M* urea, 50 m*M* phosphate, 200 m*M* phosphate and 200 m*M* phosphate; 2 *M* guanidinium chloride, respectively. The proteins were purified further on CsCl (*25*) to remove endogenous RNA. Equivalent amounts of each NHP fraction were reconstituted to purified mouse DNA and histones (HAP 1) in stoichiometric amounts by mixing in 2 *M* NaCl–5 *M* urea–0.01 *M* Tris·HCl (pH8) and dialyzing successively against the same buffer containing 0.6, 0.4, 0.2, and 0.1 *M* NaCl for 4 hours each. Urea was omitted during 0.2 and 0.1 *M* NaCl dialyses; $10^{-4}$ *M* phenylmethylsulfonyl chloride was present throughout.

The resulting chromatin was suspended in distilled water to form a gel and incubated *in vitro* as described by Gilmour and Paul (*25*). Purified RNA was titrated against 0.5 ng of cDNA, and hybrids were assayed by Sl nuclease digestion. In controls chromatin was incubated without polymerase, and an appropriate amount of *E. coli* tRNA was added. ●——●, control □——□, DNA + HAP1; ■——■, DNA + HAP1 + HAP2; ▲——▲, DNA + HAP1 + HAP3; ○——○, DNA + HAP1 + HAP4.

Gadski and Chae (*28*) will help to define the conditions which allow chromosomal proteins to reconstitute efficiently with DNA. However, these studies have to be correlated with transcriptional activity in order to be completely relevant. The criticism that only a small part of the transcription product is analyzed when single-gene cDNA probes are used is particularly relevant here. The use of more complete sequence analyses of the types already described (*29,30*) will provide a more general diagnostic test for the fidelity of reconstitution in relation to the experimental conditions employed.

However, in addition to the problems concerning the present usage of reconstitution, the future applications of the technique are worth considering. The author feels that it is not a practical proposition to use chromatin reconstitution in its present form for discriminating either the individual NHP protein molecules or DNA sequences regulating the activity of a single gene. The reasons are 2-fold. First, an extension of current NHP fractionation methods based on chemical or physical separation techniques will inevitably generate a large number of individual groups of proteins, which would then have to be assayed individually by reconstitution. This is both a tedious and unreliable approach. There is no guarantee that a single protein will be responsible for regulation. To attempt to bring together a small number of proteins or an individual protein from a separation spectrum of this complexity is unrealisitc. It also has to be admitted that even the most sensitive analytical technique may not detect regulatory protein elements, but rather the more abundant protein species, many of which are known to be enzymes or structural components. Second, the reconstitution of chromatin with total genomic DNA must involve a multitude of nucleic acid–protein interactions. It is not certain to what extent the currently employed reconstitution conditions favor specific binding in all instances, although with certain single-gene systems a fair measure of success has been seen. The fact that reconstitution is not always reproducible may be a reflection on the limitations imposed by the complexity of the interactions and the experimental conditions. Finally, it is also clear that even although a gene regulatory event can be recognized in this system, the possibility of isolating the relevant DNA sequence is negligible.

The most promising approach, which circumvents many of these criticisms, is to take advantage of the recent developments in genetic engineering. If one makes the initial assumption that the proposed DNA regulatory sequences which bind NHP are contiguous with the structural gene, then it should be possible to obtain a cloned fragment of the genomic DNA which contains both a specific gene and its regulatory sequences. Given a more detailed knowledge of the parameters which promote faithful reconstitution, it is then feasible to construct a "minichromosome" in which only a few NHP interact specifically with a DNA of much reduced sequence complexity.

## REFERENCES


1. Huang, R. C., Bonner, J., and Murray, K. *J.,Mol. Biol.* **8**, 54 (1964).
2. Marushige, K., Brutlag, D., and Bonner, J., *Biochemistry* **7**, 3149 (1968).
3. Gilmour, R. S., and Paul, J., *FEBS Lett.* **9**, 242 (1970).
4. Bekhor, I., Kung, G. M., and Bonner, J., *J. Mol. Biol.* **39**, 351 (1969).
5. Huang, R. C., and Huang, P. C., *J. Mol. Biol.* **39**, 365 (1969).
6. Artman, M., and Roth, J. S., *J. Mol. Biol.* **60**, 291 (1971).
7. Heyden, H. W., and Zachau, H. G., *Biochim. Biophys. Acta* **232**, 651 (1969).
8. Szeszac, F., and Pihl, A., *FEBS Lett.* **20**, 177 (1972).
9. Arnold, E. A., and Young, K. E., *Biochim. Biophys. Acta* **269**, 252 (1972).
10. Getz, H. J., and Saunders, G. F., *Biochim. Biophys. Acta* **312**, 555 (1973).
11. Tolstoshev, P., and Wells, J., *Biochemistry* **13**, 103 (1974).
12. Scharpe, A., and van Parijs, R., *Biochim. Biophys. Acta* **353**, 45 (1974).
13. Paul, J., and Gilmour, R. S., *J. Mol. Biol.* **34**, 305 (1968).
14. Ananieva, L. V., Kozlov, Y. V., Ryskov, A. P., and Georgiev, G. P., *Mol. Biol.* **2**, 588 (1968).
15. Spelsberg, T. C., and Hnilica, L. S., *Biochem. J.* **120**, 435 (1970).
16. Spelsberg, T. C., Hnilica, L. S., and Ansevin, A. T., *Biochim. Biophys. Acta* **228**, 550 (1971).
17. Kostraba, N. C., and Wang, T. Y., *Exp. Cell Res.* **80**, 291 (1973).
18. Crain, W. R., and Saunders, G. F., *Cell Differ.* **3**, 209, (1974).
19. Barrett, T., Maryanka, D., Hamlyn, P. H., and Gould, H. J.,*Proc. Natl. Acad. Sci. U.S.A.* **71**, 5057 (1974).
20. Gilmour, R. S., Windass, J. D., Affara, N., and Paul, J., *J. Cell. Physiol.* **85**, 449 (1975).
21. Chiu, J. F., Tsai, Y. H., Sakuma, K., and Hnilica, L. S.,*J. Biol. Chem.* **250**, 9431 (1975).
22. Tsai, S. Y., Harris, S. E., Tsai, M. J., and O'Malley, B. W., *J. Biol. Chem.* **15**, 4713 (1976).
23. Stein, J. L., Reed, K., and Stein, G. S., *Biochemistry* **15**, 3291 (1976).
24. Park, W. D., Stein, J. L., and Stein, G. S., *Biochemistry* **15**, 3296 (1976).
25. Gilmour, R. S., and Paul, J., *in* "Chromosomal Proteins and Their Role in the Regulation of Gene Expression" (G. S. Stein and L. J. Kleinsmith, eds.), p. 19. Academic Press, New York, 1975.
26. Rickwood, D., and MacGillivray, A. J., *Eur. J. Biochem.* **51**, 593 (1975).
27. Kleiman, L., and Huang, R. C., *J. Mol. Biol.* **64**, 1 (1972).
28. Gadski, R. A., and Chae, C. B., *Biochemistry* **15**, 3812 (1976).
29. Birnie, G. D., MacPhail, E., Young, B. D., Getz, M. J., and Paul, J., *Cell Differ.* **3**, 221 (1974).
30. Bacheler, L. T., and Smith, K. D., *Biochemistry* **15**, 3281 (1976).

# Chapter 32

# *Methods for Dissociation, Fractionation, and Selective Reconstitution of Chromatin*

G. S. STEIN AND J. L. STEIN

*Department of Biochemistry and Molecular Biology and Department of Immunology and Medical Microbiology, University of Florida, Gainesville, Florida*

## I. Introduction

Chromatin reconstitution provides a direct approach for assaying the contribution of various genome-associated macromolecules toward defining the availability of specific genetic sequences for transcription. This procedure, by which chromatin is dissociated, fractionated into its principal components, and then reconstituted in a selective manner, was developed by Bonner and co-workers in the early 1960s (*1,2*). While there are several modifications of the technique, generally chromatin is dissociated in the presence of high salt–urea and then reconstituted by gradually removing the salt, followed by removal of the urea. In the present article we will describe the method used in our laboratory for chromatin dissociation and reconstitution and discuss evidence for the level of fidelity attained by the procedure.

## II. Preparation of Chromatin

The procedures employed for isolation of chromatin have a very direct bearing on the final product. In fact, chromatin could be appropriately defined by the isolation protocol. It therefore follows that the validity of chromatin as representative of the genome found in the nucleus of intact cells, as well as the feasibility of dissociating, fractionating, and reconstituting chromatin, is dependent on the method used for chromatin isolation.

Although a broad spectrum of approaches has been utilized for isolation of chromatin, there is no single procedure that appears to be optimal for all organisms, tissues, and cell types. Rather, modifications in protocols are necessary to accommodate specific biological situations. The following are general considerations which should determine the acceptability of a particular technique: (a) It is best to utilize nuclei free of cytoplasmic material and stripped of the outer aspect of the nuclear envelope. This can be best achieved by treatment with a citric acid-buffer or with nonionic detergents, such as Triton X-100 or NP-40. The concentrations of detergent which can be tolerated vary depending on the cell type. (b) An effort should be made to eliminate material present in the nuclear sap. (c) Caution must be exercised to avoid extraction of proteins bound to chromatin. (d) An attempt should be made to shear the DNA as little as possible. (e) Nuclease and protease activity should be minimized. Described below are the procedures which we use for preparation of chromatin from two lines of tissue culture cells, HeLa $S_3$ cells (human cervical carcinoma cells grown in suspension culture) (*3*) and WI-38 human diploid fibroblasts (embryonic lung cells grown in monolayer culture) (*4*), as well as the procedure we utilize for the preparation of chromatin from mouse and rat liver (*5*).

## A. Tissue Culture Cells

All procedures are carried out at 4°C. Cells grown in suspension culture are harvested by centrifugation at 1000 *g* for 5 minutes, and cells grown in monolayers are scraped from the culture vessel with a rubber policeman and then collected by centrifugation. Treatment with trypsin should be avoided since the enzyme can utilize chromosomal proteins as substrates. The harvested cells are washed three times with Earle's balanced salt solution to remove serum proteins—each wash step followed by centrifugation at 1000 *g* for 5 minutes. Cells are lysed by resuspension in 80 volumes of 80 m*M* NaCl–20 m*M* EDTA–1% Triton X-100 (pH 7.2) and agitated with a Vortex mixer for 20 seconds at maximal speed. Nuclei are pelleted by centrifugation at 1000 *g* for 5 minutes in a swinging-bucket rotor and then washed twice in 80 m*M* NaCl–20 m*M* EDTA–1% Triton X-100. The nuclei should now be free of visible cytoplasmic material when examined by phase contrast microscopy—this should be a routine procedure. Electron microscopic examination of the nuclei should reveal the absence of the outer aspect of the nuclear envelope, and often the inner component of the nuclear envelope is also removed by detergent treatment. Nuclei are washed twice with 80 volumes of 0.15 *M* NaCl–10 m*M* Tris (pH 8.0) and centrifuged at 1000 *g* for 5 minutes to remove the detergent. It is important to remove the supernatant completely following the last wash step since the salt will interfere with nuclear lysis. Nuclei are lysed by resuspension in double-distilled water (at a concentration of 250–500 μg of DNA per milliliter) by gentle agitation with a Vortex mixer. Nuclear lysis results in a marked increase in viscosity, and after swelling in an ice bath for 20 minutes, the material is clear and gelatinous. Incomplete removal of cytoplasm during nuclear isolation, incomplete nuclear lysis, and protein denaturation will be reflected by a "cloudy" or "milky" appearance of the gel. Chromatin is pelleted by centrifugation at 12,000 *g* for 20 minutes in a fixed-angle rotor, and then the chromatin is again washed in distilled water and pelleted at 12,000 *g*. The chromatin pellets should be clear to slightly opalescent and extremely gelatinous. Chromatin prepared by this method has a protein:DNA ratio of 1.8–2.0. The histone:DNA ratio is 1.1, and the nonhistone chromosomal protein:DNA ratio is 0.89.

All procedures can be carried out in the tube in which cells were initially harvested, resulting in 85–95% recovery of chromatin. Avoiding sonication or homogenization steps results in minimal shearing of the DNA. We have observed that shearing can result in a significant increase in chromatin template activity assayed with exogenous RNA polymerase.

## B. Mouse and Rat Liver

Animals are sacrificed by cervical dislocation. All lobes of the liver are excised, placed in a plastic boat on ice, and weighed. Nuclei are prepared by a modification of the method of Cheveau *et al.* (*6*). The liver is immediately minced with surgical scissors and suspended in 20 volumes of 2.2 *M* sucrose–4 m*M* $MgCl_2$. All procedures are carried out at 4°C. The liver is homogenized to homogeneity in a motor-driven Potter–Elvehjem homogenizer with a wide-clearance Teflon pestle. The homogenate is filtered through one layer of Miracloth (Chicopee Mills, New York) and centrifuged in a Beckman SW-27 rotor for 60 minutes at 25,000 rpm. The supernatant is discarded and, after removal of material adhering to the walls of the centrifuge tube, the nuclear pellet is resuspended in 80 volumes of 0.15 M NaCl–10 m*M* Tris (pH 8.0). After centrifugation at 1500 *g* in a swinging-bucket rotor for 3 minutes, the nuclei are again washed in 0.15 *M* NaCl–10 m*M* Tris (pH 8.0). The latter two washing steps deplete the nuclei of sucrose. Nuclear lysis and recovery of chromatin are carried out as described above for tissue culture cells.

## III. Dissociation, Fractionation, and Reconstitution

### A. Dissociation

All procedures are carried out at 4°C. Solid NaCl and urea are added to the chromatin preparation to final concentrations of 3 *M* and 5 *M*, respectively. The salt–urea gel mixture is briefly agitated on a Vortex mixer, and then the appropriate volume is achieved by addition of 10 m*M* Tris (pH 8.0). The presence of NaCl should promote dissociation of chromatin components which are held together by electrostatic bonds, and urea should promote dissociation of components held together by hydrophobic and hydrogen bonding. Dissociation is achieved by intermittent agitation with a Vortex mixer for 2–3 hours. We have found that the optimal concentration of chromatin for dissociation under these conditions is 500 μg/ml DNA; higher concentrations of "minimally sheared chromatin" result in a solution which is unmanageably viscous. The dissociated chromatin is centrifuged in a fixed-angle rotor at 180,000 *g* for 36 hours, after which time 95% of the chromosomal proteins are recovered in the supernatant. The supernatant should contain less than 0.1% nucleic acid. If the chromatin is excessively sheared, the supernatant may contain a high level of nucleic acid, which can in some instances be removed by additional centrifugation. Oligonucleotides which may result from extreme shearing or nuclease digestion will not pellet, and, if the level of nucleic acid in the supernatant remains high, the preparation may best be discarded.

Sodium chloride is removed from the chromosomal proteins prior to ion exchange chromatography by dialysis against 100 volumes of 5 *M* urea–10 m*M* Tris (pH 8.3); three changes of the dialysis buffer is adequate. The urea (before addition of Tris buffer) should be run over a mixed-bed ion exchange resin (for example, Bio-Rad resin AG501-X8), and the pH should be checked just prior to use. This is necessary to ensure the absence of cyanates and ammonia, which can accumulate during storage, as well as to remove other impurities from the urea.

In reconstitution experiments in which purified DNA is used, the DNA is isolated from intact cells by the method of Marmur (*7*). The DNA preparations are then treated with RNase (50 μg/ml, heated for 10 minutes at 80°C to eliminate DNase activity) and Pronase (50 μg/ml, self-digested at 37°C for 60 minutes), followed by a series of phenol and chloroform–isoamyl alcohol (24:1) extractions to remove the enzymes.

We have not observed proteolytic activity during isolation, dissociation, fractionation, and reconstitution of HeLa $S_3$ cell and WI-38 chromatin. However, in systems where protease activity occurs, fidelity of reconstitution can be achieved by carrying out these procedures in the presence of protease inhibitors such as phenylmethylsulfonylfluoride and diisopropylfluorophosphonate. Appropriate protease inhibitors have been summarized by Carter and Chae (*8*).

## B. Fractionation of Chromosomal Proteins

While a number of approaches have been pursued for the fractionation of chromosomal proteins into histones and nonhistone chromosomal proteins, the procedure that we most often utilize is ion-exchange chromatography on QAE Sephadex (*9, 10*) a technique reported by Gilmour and Paul (*11*).

### 1. Batch Separation

QAE Sephadex A-25 or A-50 is equilibrated against 5 *M* urea–10 m*M* Tris (pH 8.3). The Sephadex is hydrated in 50 volumes of buffer and at least three changes of buffer are made over a period of 24 hours. A-25 and A-50 will yield similar separation; however, we have found the A-25 to be preferable since A-50 undergoes significant osmotic shrinkage during salt fractionation. This is particularly a problem in the column separation procedure described below. The amount of Sephadex required is 1 gm/25 mg of chromosomal proteins.

Chromosomal proteins in 5 *M* urea–10 m*M* Tris (pH 8.3) are added to the QAE Sephadex and mixed intermittently for approximately 45 minutes at 4°C. The slurry is then poured into a porcelain Büchner funnel lined with Whatman No. 1 filter paper or into a sintered-glass filter (medium porosity). The filtrate, consisting of greater than 98% of the histones and approximately 5% of the nonhistone chromosomal proteins, is collected by vacuum filtration. The slurry is then washed with 50 volumes of 5 *M* urea–10 m*M* Tris (pH 8.3) to elute any residual histones. Nonhistone chromosomal proteins are eluted as a single fraction by washing the QAE Sephadex with 3 *M* NaCl–5 *M* urea–10 m*M* Tris (pH 8.3). Alternatively, several nonhistone chromosomal protein fractions can be collected by eluting the resin with increasing concentrations of NaCl in 5 *M* urea–10 m*M* Tris (pH 8.3).

If it is necessary to concentrate the chromosomal protein fractions, this can be readily achieved by any one of several methods: (a) transferring the protein solution to a dialysis bag and then covering the dialysis tubing with dry Sephadex; a rapid decrease in the fluid volume can be achieved by this method if the Sephadex directly in contact with the dialysis tubing is changed every 30 minutes; (b) dialyzing against a concentrated sucrose solution, followed by subsequent dialysis against 5 *M* urea–10 m*M* Tris to remove the sucrose; (c) concentrating in an Amicon pressure filtration apparatus using a UM-10 filter.

### 2. Column Chromatography

A column of QAE Sephadex can be prepared, and the chromosomal proteins equilibrated against 5 *M* urea–10 m*M* Tris (pH 8.3) can be fractionated as in the batch procedure described above. It is strongly recommended that only QAE Sephadex A-25 be used with column fractionation because the osmotic shrinkage of QAE Sephadex A-50 will break the continuity of the column. The sample is applied to the column in 5 *M* urea–10 m*M* Tris (pH 8.3), and histones are eluted in the void volume. The nonhistone chromosomal proteins are eluted with a NaCl gradient in a buffer of 5 *M* urea–10 m*M* Tris (pH 8.3) or by a stepwise batch procedure.

## C. Reconstitution

Chromosomal proteins and DNA are combined in 3 *M* NaCl–5 *M* urea–10 m*M* Tris (pH 8.3). We have found that the optimal range of DNA concentration is between 200 and 800 $\mu$g/ml, and the protein : DNA ratio should not exceed 4:1. The reconstitution mixture is dialyzed for 3 hours against 20 volumes of 3 *M* NaCl–5 *M* urea–10 m*M* Tris (pH 8.3). Then every 3 hours the NaCl concentration is progressively decreased. The sequential NaCl steps which we utilize in our gradient dialysis procedure are 3 *M*, 2.5 *M*, 2 *M*, 1.5 *M*, 1 *M*, 0.8 *M*, 0.7 *M*, 0.6 *M*, 0.5 *M*, 0.4 *M*, 0.2 *M*, 0.1 *M*. The NaCl is then completely removed by dialysis against several changes of 5 *M* urea–10 m*M* Tris (pH 8.3), and the reconstituted chromatin is pelleted by centrifugation at 12,000 *g* for 30 minutes. The reconstituted chromatin preparation is washed twice in distilled water or 10 m*M* Tris (pH 8.0)–each wash followed by centrifugation at 12,000 *g*. Recovery of DNA as chromatin is 75%.

# IV. Fidelity of Reconstitution

Several lines of evidence have suggested the equivalence of native and reconstituted chromatin. We have compared native and reconstituted chromatin preparations from HeLa $S_3$ cells and found them to be indistinguishable with respect to the following parameters (*12*): (a) The banding patterns of histones and nonhistone chromosomal proteins when fractionated electrophoretically according to charge and molecular weight on a high-resolution polyacrylamide gel; (b) The binding of histones and nonhistone chromosomal proteins in chromatin when assayed by extractability with dilute mineral acids and ionic detergents; (c) The extent of $\gamma$-ray-induced thymine damage in the DNA; (d) The availability of sites for binding of "reporter molecules" which exhibit specificity for intercalculation in the major or minor grooves of the DNA helix; (e) the circular dichroism spectra; (f) the *in vitro* transcription under conditions which prohibit reinitiation. Additional evidence for fidelity of chromatin reconstitution can be gleaned from studies which have demonstrated that transcription of globin (*13–15*), histone (*10, 16–19*), and ovalbumin (*20*) genes from chromatin remains unaltered following dissociation, fractionation, and reconstitution. Other studies in which fidelity of transcription was assayed by hybridization to DNA complementary to total poly(A)-containing polysomal RNA suggest the similarity of native and reconstituted chromatin (*21*; C. B. Chae, private communication). However, to view the question of fidelity of chromatin reconstitution from a realistic perspective, one must continue to assay additional parameters of the native and reconstituted material.

## ACKNOWLEDGMENT

These studies were supported by grants from the National Institutes of Health (GM 20535, CA 18875, DA 01188) and the National Science Foundation (BMS 75-18583).

## REFERENCES

*1.* Bekhor, I., Kung, G. M., and Bonner, J. *J. Mol. Biol.* **39**, 351 (1969).
*2.* Huang, R. C., and Bonner, J., *Proc. Natl. Acad. Sci. U.S.A.* **54**, 960 (1965).
*3.* Stein, G. S., and Borun, T. W., *J. Cell Biol.* **52**, 292 (1972).
*4.* Stein, G. S., Chaudhuri, S. C., and Baserga, R., *J. Biol. Chem.* **247**, 3918 (1972).
*5.* Thomson, J. A., Stein, J. L., Kleinsmith, L. J., and Stein, G. S., *Science* **194**, 428 (1976).
*6.* Cheveau, J., Moule, Y., and Rouiller, C., *Exp. Cell Res.* **11**, 317 (1956).
*7.* Marmur, J., *J. Mol. Biol.* **3**, 208 (1961).
*8.* Carter, D. B., and Chae, C.-B., *Biochemistry* **15**, 180 (1976).
*9.* Stein, G. S., and Farber, J. L., *Proc. Natl. Acad. Sci. U.S.A.* **69**, 2918 (1972).
*10.* Stein, G. S., Park, W. D., Thrall, C. L., Mans, R. J., and Stein, J. L., *Nature (London)* **257**, 764 (1975).
*11.* Gilmour, R. S., and Paul, J., *FEBS Lett.* **9**, 242 (1970).
*12.* Stein, G. S., Mans, R. J., Gabbay, E. J., Stein, J. L., Davis, J., and Adawadkar, P. D., *Biochemistry* **14**, 1859 (1975).
*13.* Paul, J., Gilmour, R. S., Affara, N., Birnie, G., Harrison, P., Hell, A., *Cold Spring Harbor Symp. Quant. Biol.* **38**, 885 (1974).
*14.* Barrett, T., Maryanka, D., Hamlyn, P., and Gould, H., *Proc. Natl. Acad. Sci. U.S.A.* **71**, 5057 (1974).
*15.* Chiu, J.-F., Tsai, Y.-H., Sakuma, K., and Hnilica, L. S., *J. Biol. Chem.* **250**, 9431 (1975).
*16.* Park, W. D., Stein, J. L., and Stein, G. S., *Biochemistry* **15**, 3296 (1976).
*17.* Stein, J. L., Reed, K., and Stein, G. S., *Biochemistry* **15**, 3291 (1976).
*18.* Jansing, R. L., Stein, J. L., and Stein, G. S. *Proc. Natl. Acad. Sci. U.S.A.* **74**, 173 (1977).
*19.* Stein, G. S., Stein, J. L., Kleinsmith, L. J., Thomson, J. L., Park, W. D., and Jansing, R. L., *Cancer Res.* **36**, 4307 (1976).
*20.* Tsai, S., Tsai, M.-J., Harris, S., and O'Malley, B. W., *J. Biol. Chem.* **251**, 6475 (1976).
*21.* Biessmann, H., Gjerset, R. A., Levy, W. B., and McCarthy, B. J., *Biochemistry* **15**, 4356 (1976).

# Chapter 33

# *The Assay of Globin Gene Transcription in Reconstituted Chromatin*

H. J. GOULD, D. MARYANKA, S. J. FEY, G. J. COWLING, AND J. ALLAN

*Department of Biophysics,*
*King's College,*
*London University, London, England*

## I. Introduction

The reconstitution of chromatin from selected components offers a rational approach to the problem of the mechanism of transcriptional control in eukaryotic systems. The complete functional reconstitution of chromatin from all the required components, e.g., DNA, histones, and chromatin nonhistone proteins, must be judged by an appropriate assay. This assay should measure the differential synthesis of at least one tissue-specific RNA species when the chromatin, or the reconstituted chromatin, is transcribed *in vitro*.

Before 1972, studies on the "fidelity" of transcription using chromatin templates were limited to the characterization of populations of RNA, defined operationally in terms of hybridization specificity (*1–4*). The introduction of complementary DNA (cDNA) probes (*5–7*) at this time offered the promise of the necessary specific assay. Hence it was not long before cDNAs were used to measure the levels of specific RNAs in transcripts of "native" (*8–11*) and reconstituted (*12–14*) chromatin, first in the globin system and subsequently with several others. A number of technical difficulties in the application of cDNA probes to the assay of chromatin transcripts have since come to light. In retrospect, the history of chromatin reconstitution and transcription since 1972 looks like a minefield littered with dead conclusions. Six years on, one finds that practically every aspect of the assay system has had to be reexamined in order to obtain meaningful results with cDNAs. Despite this effort, it is still not possible to prove, or disprove, the claims of successful chromatin reconstitution. Our primary objective here is to set out rigorous criteria by which the results of reconstitution attempts should be judged, and to describe in detail the experimental procedures and their pitfalls as well as the technical uncertainties that still remain. We refer throughout to the chicken globin system.

The experimental system must satisfy a number of exacting requirements: The preparative procedures, reconstitution, transcription, extraction, fractionation, and analysis of RNA must be workable on a routine basis. The assay itself makes stringent demands at once of specificity, sensitivity, and precision, which in the past it was not possible to meet. The form of the results must be such as to allow a direct comparison of the activity of reconstituted chromatin with that of an appropriate standard of reference, if not native chromatin isolated from the tissue of interest, then chromatin *in vivo*. The procedures developed and described here by and large fulfill these conditions.

On all the above methodological points, the previous experimental systems fell short. Experiments required the assembly of numerous components after long and arduous preparative procedures, the consistent success of which, in terms of purity and high yields, could not be guaranteed. The hybridization assays require probes of adequate purity, preferably complementary to the full length of the messenger RNA (mRNA), but these until recently were unavailable; moreover the sensitivity of the assays suffered from limitations of specific activity of the probes. At the same time the major problem besetting chromatin transcription studies is a still unresolved ambiguity arising out of the newly developed methods. The difficulty is that endogenous RNA sequences introduced together with the chromatin or any of the fractions used in reconstitution compete with the newly synthesized RNA in the hybridization assay. Thus direct methods involve the measurement of a small increment of newly synthesized RNA against a very high background of endogenous RNA.

We first consider the characteristics of the chicken globin system for the measurement of specific transcription. The choice of this system was dictated by a number of advantageous features, primarily that synthesis of globin predominates overwhelmingly over that of any other single protein, being more than 40% of the total. It is known that globin RNA is the most abundant messenger species in the cytoplasm of reticulocytes so that control mechanisms must operate at the transcriptional and/or the posttranscriptional levels. The fraction of the total RNA synthesis represented by this messenger *in vivo* is variously estimated as lying between 0.01 (*15*) and 0.24% (*16*). There is good evidence from the mouse erythroid system that this level is indeed significantly greater than that in the undifferentiated tissue (*17*), indicating that transcriptional control *in vivo* is no mere illusion.

The likelihood that globin mRNA may constitute at most 0.01–0.24% of the total transcript of reticulocyte chromatin *in vitro* at the same time, however, underlines the need for a very sensitive assay. Hybridization with DNA complementary to the globin messenger is in fact entirely satisfactory for this purpose. The isolation of chicken globin mRNA of high purity is straightforward. It is now possible to obtain full-length cDNA transcripts of this RNA of high specific activity ($5 \times 10^7$ cpm/$\mu$g), and the use of cDNA excess hybridization with this probe allows the detection of one part per million of globin RNA in a transcript of 1 mg of chromatin DNA, as will be seen later.

The interference of endogenous RNA in the hybridization assay remains the most intractable problem. Because of this, earlier evidence for selective globin gene expression, using either isolated native chromatin (*8–11*) or reconstituted chromatin (*12–14*), must be regarded as inadequate. The strategy of measuring endogenous RNA and subtracting this from the total globin RNA following transcription has proved generally unsatisfactory for this system. The levels of endogenous globin RNA, present as a contaminant of the nonhistone "protein" fraction, are generally in the same range as the expected synthesis, based on *in vivo* estimates (*15,16*). To take a specific example, 0.6 mg of nonhistone proteins may contain about 10 ng of globin RNA. Transcription of reconstituted chromatin, containing 1 mg DNA, 1 mg of histones, and 0.6 mg of nonhistone protein, yields perhaps 50 $\mu$g RNA, depending among other things on the amount of polymerase added. If this template resulted in differential synthesis of globin RNA at the *in vivo* target level, then a minimum of 0.01% of 50 $\mu$g, or 5 ng, would have to be detected in a total of 15 ng. The measurement of an increment of 50% is not beyond the precision and accuracy of the hybridization assay as such, but the determination of the baseline values of endogenous RNA here is the source of difficulty, as will be discussed.

The ideal solution to this problem would be to separate the endogenous RNA and nonhistone proteins, but we have not yet found a reliable method. Isopycnic gradient centrifugation in CsCl has been used for this purpose (*18*), but this will not be very effective if the RNA is small or tightly bound to protein. Destruction of the RNA is an alternative to separation. This has been achieved by preincubating chromatin prior to addition of polymerase, taking advantage of endogenous nucleases in the system (*18,19*), but this approach has even less appeal.

Separation of the endogenous and newly synthesized RNA after transcription is an alternative solution to the problem. Three available methods are based on this principle. The RNA may be synthesized using high-specific-activity $^{32}$P-labeled nucleoside triphosphates, hybridized to cDNA in cDNA excess, and the hybrids quantitatively recovered in a CsCl gradient and assayed for radioactivity (*9,18,19*). Or, it is possible to incorporate high-density nucleoside triphosphate analogs, such as bromodeoxyuridine triphosphate, or mercurated nucleoside triphosphate derivatives, into the newly synthesized RNA, and use this as a basis for separation in the CsCl gradient (*19,20*).

A more attractive method for fractionating newly synthesized and endogenous RNA has recently been adopted in a number of laboratories. This involves the transcription of chromatin with mercurated nucleoside triphosphates, followed by the isolation of the mercurated RNA on an affinity column of thiolated Sepharose (*20*). Huang and co-workers (*21*) first introduced this technique to assay immunoglobulin mRNA in transcripts of nuclei and chromatin isolated from mouse myeloma cells, and Crouse *et al.* (*23*) and Towle *et al.* (*24*) similarly studied globin gene transcription in chicken reticulocyte chromatin. The use of mercurated nucleotides was then naturally extended to the study of reconstituted chromatin using chicken and mouse globin systems.

Using mercurated nucleotide precursors, Gilmour (*25*) and ourselves (unpublished results) obtained evidence of globin RNA synthesis from reconstituted chromatin templates at levels approaching those found *in vivo*. However, these results were again thrown into doubt by the discovery of a surprising artifact: Zasloff and Felsenfeld (*26*) observed that *Escherichia coli* RNA polymerase transcribed the endogenous RNA in experimental conditions commonly employed. Our results suggest that it is possible to obtain stoichiometric amounts of "antistrands" synthesized in the RNA-dependent reaction. The mercurated product remains paired with its single-stranded template and, consequently, up to 100% of the endogenous globin RNA can be bound in the form of a duplex to the Thiol-Sepharose column.

Zasloff and Felsenfeld (*26*) found that they could obtain a satisfactory separation of the mercurated and unmercurated RNA species after heating the sample to dissociate the complementary strands. However, it appears that heating is not a universal remedy, for several groups have found that the mercurated RNA fraction was still contaminated with endogenous RNA when the customary column fractionation procedure was used. Konkel and Ingram (*27*) overcame this problem by performing the fractionation in denaturing solvents at elevated temperature. Orkin and Swerdlow (*17*) screened a number of batches of commercially available Thiol-Sepharose until they found one which showed an acceptably low level of non-specific adsorption of unmercurated RNA.

We have attempted to eliminate the residual contamination of endogenous RNA is a simpler and more reliable manner. This involves, first, improving the procedure of extracting the RNA to obtain a product free of DNA and protein contaminants, and, second, using Thiopropyl-

Sepharose in place of Thiol-Sepharose to fractionate the mercurated and unmercurated RNA. The rationale and merits of this modified procedure are discussed below.

The combination of procedures fully described in Section II has allowed us to evaluate unambiguously, for the first time to our satisfaction, the fidelity of globin gene transcription in reconstituted chromatin.

## II. Methods

Figure 1 summarizes the steps involved in the assay for globin gene transcription in reconstituted chromatin: preparation of globin mRNA and cDNA; preparation of HgUTP; preparation of chromatin constituents, reconstitution of chromatin, transcription of reconstituted chromatin, extraction and fractionation of the RNA, and hybridization analysis.

### A. Preparation of Chicken Globin Messenger RNA

Pure intact mRNA is required for the subsequent synthesis of the pure full-length cDNA probe. The biggest hazard in the preparation of the mRNA is the ubiquitous nuclease. From the first moment of bleeding the chickens to the final stage of messenger purification, every precaution is observed to keep the preparation cold and free of nuclease contamination. It is important to work quickly and without interruption to the stage of phenol extraction. The preparation is usually carried out on a large scale, sacrificing 10 or more chickens, to provide ample RNA for characterization and synthesis of cDNA. Preparation and characterization require about a week's work, but once this is done the RNA is stable for at least a year at −20°C.

Twelve 2–3 kg White Leghorn chickens are made anemic by 4 consecutive daily intraperitoneal injections with 0.6 ml 2.5% neutralized phenylhydrazine (recrystallized from ethanol), which is stored frozen and thawed once. Some chickens do not survive the course of injections and others do not become anemic, as indicated by the failure of the combs to turn pale, and are not used. A different injection schedule that produces "the most severe anemia with the least mortality" has been described by Longacre and Rutter (*28*).

The cooperation of two workers facilitates handling the chickens. Before bleeding, the chickens are anesthetized with ether and the neck feathers are plucked. The skin at the base of the neck is disinfected with alcohol, cut in a

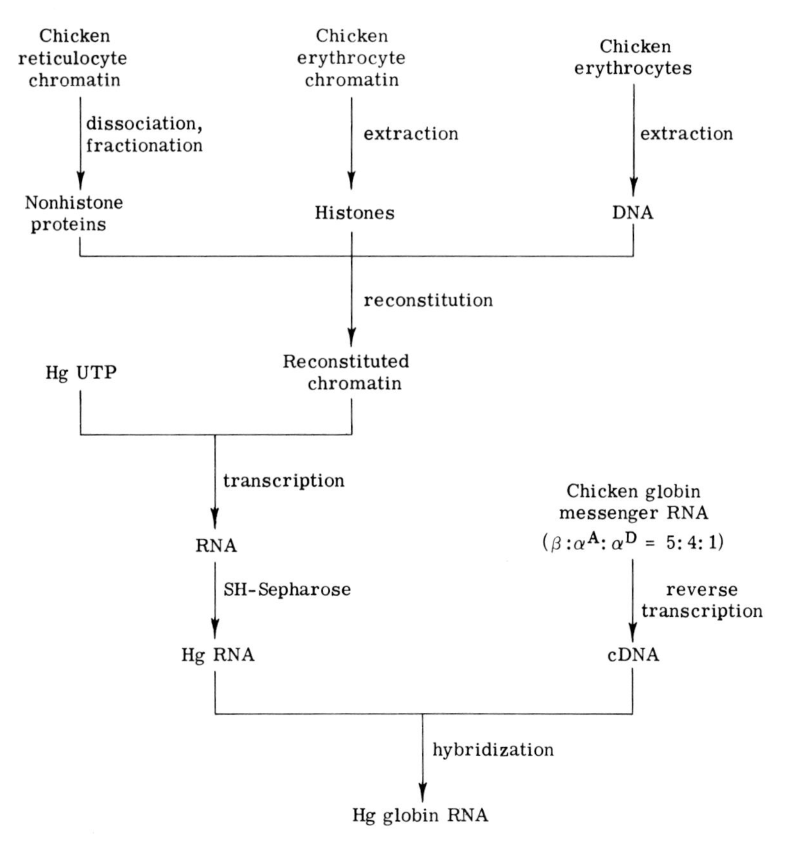


FIG. 1. Stages in the assay for globin gene transcription in reconstituted chromatin. The symbols: $\beta$, $\alpha^A$, and $\alpha^D$ = 5:4:1 serve as a reminder (*55*) that adult chicken globin consists of $\alpha$ and $\beta$ polypeptide chains in equal amounts, and that there is a major and a minor hemoglobin population in a 4:1 ratio, differing in the identity of the $\alpha$ chain ($\alpha^A$ in the major and $\alpha^D$ in the minor population). This implies that globin mRNA is a heterogeneous mixture of 3 species. Preliminary studies (*28,31*; our unpublished data) indicate that there are approximately equal amounts of the $\alpha$ and $\beta$ types, as in other species.

circle and pulled back over the head like a sleeve, exposing the jugular veins. While one worker holds the body of the chicken, the other holds up the head, teases the veins free and cuts them with scissors. The blood is collected through a filter funnel covered with 4 layers of surgical gauze into 250 ml centrifuge bottles standing in ice. The gauze is first wetted by pouring 50 ml phosphate-buffered saline [PBS: 0.15 *M* NaCl–0.015 *M* sodium phosphate buffer (pH 7.6)] containing 25 units per milliliter of heparin plus 0.2 m*M* cycloheximide into the bottle, and the blood is mixed with this solution to prevent clotting. This procedure requires less than an hour and yields up to 500 ml blood.

Unless otherwise specified, all centrifugations are performed in the Sorvall RC2-B refrigerated centrifuge. The blood cells are sedimented in the GSA rotor at 2000 rpm (650 *g*) for 5 minutes. The thin white cell layer at the top is removed by aspiration, and the red cells are resuspended in PBS containing 0.2 m*M* cycloheximide, resedimented and washed twice more with this solution, using the original blood volume each time. The final sediment of packed cells is lysed by resuspension in 4 volumes 5 m*M* $MgCl_2$–20 m*M* 2-mercaptoethanol, then isotonicity is restored by the addition of 1 volume 50% sucrose–0.15 *M* KCl. Nuclei and membranes are removed by centrifugation in the HB-4 rotor at 10,000 rpm (16,300 *g*) for 15 minutes. Polysomes are then recovered from the supernatant by centrifugation in the Spinco 30 rotor at 30,000 rpm (105,000 *g*) for 2 hours.

The polysome pellets are resuspended in approximately 20 ml 10 m*M* Tris-HCl (pH 7.2)–3% sodium dodecyl sulfate (SDS), by gentle hand homogenization. The volume is made up to 50 ml and the solution is then subjected to the following phenol extraction scheme:

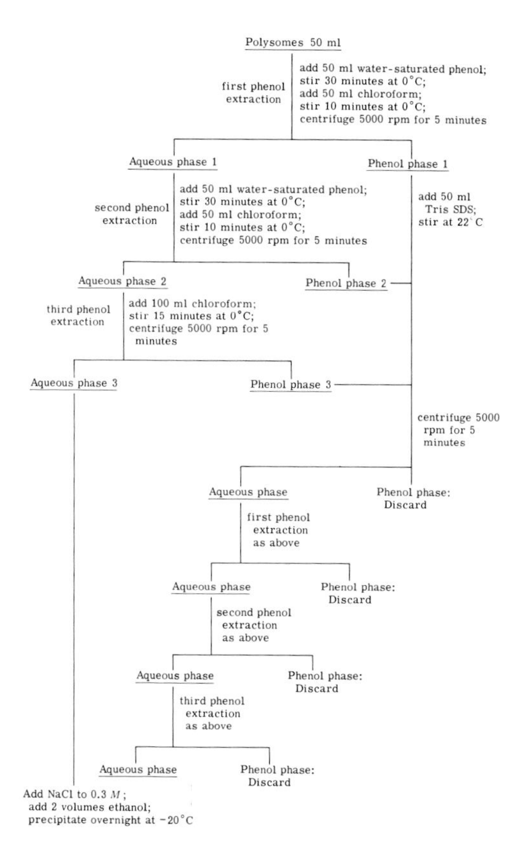

The RNA is pelleted at 5000 rpm for 20 minutes. The pellet is washed twice with 70% ethanol, dried under vacuum, and dissolved in 20 ml 10 m*M* Tris-HCl (pH 7.2)–0.5% SDS. The yield estimated from the spectrum ($A_{260}$ for 1 mg/ml = 20) is about 75 mg total RNA.

The RNA is fractionated according to size by two successive sucrose gradient centrifugation steps. A crude fractionation is first obtained by zonal centrifugation using a Beckman Ti XIV zonal rotor. The RNA is loaded on a sucrose gradient formed by mixing 30% sucrose–10 m*M* Tris-HCl (pH 7.2)–0.5% SDS into a 325 ml fixed volume reservoir of 5% sucrose–10 m*M* Tris-HCl (pH 7.2)–0.5% SDS, and centrifuged at 30,000 rpm for 17.5 hours at 25°C. Fractions of 25 ml are collected and the absorbance profile at 260 nm is determined. A pool of RNA smaller than 18 S RNA is precipitated as before.

This RNA is dissolved in 1.0 ml 10 m*M* Tris-HCl (pH 7.2)–1 m*M* ethylenediamine tetraacetic acid (EDTA)–0.1% SDS, and loaded on two concave 5–30% sucrose gradients containing the same buffer. The gradients are centrifuged in the Beckman SW 41 rotor at 40,000 rpm for 17 hours at 15°C. The 9 S peak is located by comparison with 28 S, 18 S, and 4 S RNA markers run on a separate gradient, pooled, and precipitated as before, with a recovery of about 3 mg.

Polyadenylated 9 S RNA is obtained by oligo (dT)-cellulose chromatography. An oligo (dT)-cellulose (Collaborative Research, T-3) column of 1 gm is previously washed with 0.1 *N* NaOH and then equilibrated with 0.5 *M* lithium acetate–10 m*M* Tris-HCl (pH 7.5)–0.5% lithium dodecyl sulfate (LDS). The RNA is dissolved in 0.5 ml of this high salt buffer, loaded onto the column, and washed with the same buffer until a peak of nonadenylated RNA has been eluted. This RNA is then reloaded and the procedure repeated. The adenylated RNA is then eluted with 10 m*M* Tris-HCl (pH 7.5)–0.5% LDS, with a yield of about 0.5 mg.

Preparative formamide gel electrophoresis is used to purify the 9 S RNA further (*29,30*). The sample is prepared by precipitating 150 μg of the polyadenylated 9 S fraction as before, centrifuging at 4000 *g* for 20 minutes, washing the pellet in 70% ethanol, and drying *in vacuo*. The RNA is allowed to dissolve in 50 μl freshly deionized formamide (BDH), buffered with 0.02 *M* barbituric acid, nominal pH 9, for about 16 hours. A set of 0.6 × 8 cm 4% polyacrylamide gels is prepared with the buffered formamide the day before use, and preelectrophoresed at 2 mA/gel for 30 minutes just prior to loading the samples. The reservoir buffer is changed after preelectrophoresis. A dye marker, 20 μl 0.2% bromophenol blue–50% sucrose, is added to the RNA sample, and 10 μl is loaded onto each of 7 gels. Electrophoresis at 2 mA/gel constant current is carried out until the bromophenol blue marker reaches the bottom of the gel, about 2.5–3 hours, with continuous buffer circulation between the two reservoir compartments. The gels are "rimmed" with a needle and, using a water-filled rubber teat, gently expelled into test tubes containing 0.02% methylene blue–0.4 *M* sodium acetate–0.4 *M* acetic acid, and stained for 45 minutes. The gels are destained by soaking overnight in water at 4°C. The gels are placed on a clean glass plate on a light box, and the two well-separated zones corresponding to the $\alpha$- and $\beta$-chain mRNAs (*28,30*) are cut out with a razor blade. The slices of gel, about 0.5 cm thick, are pooled and placed in a 2 ml syringe fitted with a 21 gauge needle, together with 1.5 ml 0.5 *M* sodium acetate buffer (pH 5.4)–0.5% SDS–1 m*M* EDTA, and passed twice through the needle. The resulting suspension is incubated for 2 hours at 37°C in a 30 ml Sorvall centrifuge tube, and then centrifuged at 10,000 rpm (16,300 *g*) for 15 minutes. The supernatant is retained and the gel pellet is reextracted as above and centrifuged. The combined supernatants are extracted with 1 ml of phenol:chloroform 1:1, and the phases separated by centrifugation at 5000 rpm for 5 minutes. The lower phase is removed and the aqueous phase and interface are reextracted with chloroform 4 times. The final aqueous phase is removed to a clean tube and the RNA precipitated with 2 volumes of ethanol. A cloudy precipitate forms immediately. This is kept at −20°C overnight and then centrifuged at 5000 rpm for 20 minutes. The pellet is washed with 70% ethanol and dried *in vacuo*. The RNA is then dissolved in 0.5 *M* NaCl–10 m*M* Tris-HCl (pH 7.5), and passed through the oligo (dT)-cellulose column, as described above. Polyacrylate from the gel elutes first and the RNA is then eluted at the lower salt concentration. We have not consistently obtained the good (30–50%) recoveries of RNA from formamide gels reported by Forget *et al.* (*30*).

The purified RNA is characterized by its electrophoretic profile in a formamide gel and by its hybridization kinetics with cDNA in RNA and cDNA excess. The electrophoretic pattern exhibits only the two 9 S zones, as might be expected. Back-hybridization of the RNA to its cDNA (see below) reveals a single kinetic component hybridizing 90% of the cDNA in two decades of the $R_0t$ curve, and exhibiting a $R_0t_{1/2}$ value of $5.8 \times 10^{-4}$, as shown in Fig. 2. This $R_0t_{1/2}$ value is appropriate for a mixture of equal amounts of two RNA species, each about 600 nucleotides in length, and is similar to the values obtained by other workers (*8*).

## B. Preparation of Complementary DNA

The DNA probe is obtained by copying the purified chicken globin mRNA with reverse transcriptase isolated from avian myeloblastosis virus. This enzyme, having a specific activity of 29,677 units per milligram, was

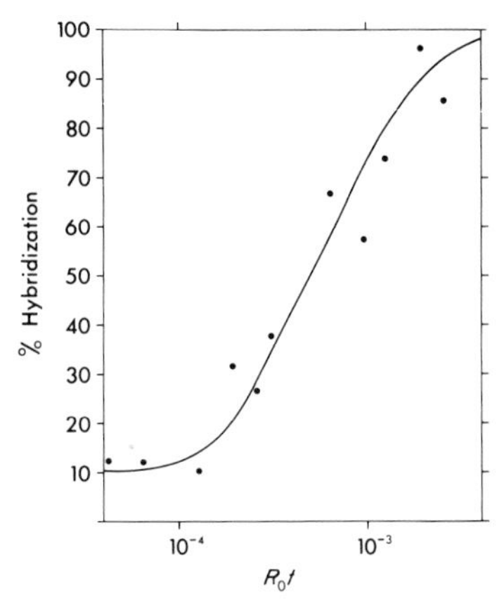


FIG. 2. Hybridization of chicken globin mRNA to cDNA in RNA excess. To present a $R_0t$ curve extending over several decades, two separate mixes were prepared. Each contained, in 20 μl, 10 μl formamide, 2 μl 5 *M* NaCl, 2 μl Hepes buffer, 0.25 *M* Hepes–5 m*M* EDTA (pH 6.8), 1 μl cDNA (0.26 ng), either 2.3 ng or 0.46 ng RNA, and water. Capillaries were filled with 2 μl of the hybridization mix, sealed, heated in a boiling water bath for 5 minutes, cooled rapidly on ice, and incubated at 43°C for the appropriate time intervals. Capillaries containing the lesser amount of RNA were used for the early time points, while those containing the greater amount of RNA were used for the later time points. Capillaries were expelled into 0.45 ml micrococcal nuclease buffer, 0.4 *M* NaCl–10 m*M* $MgCl_2$–0.1 m*M* $CaCl_2$–0.1 *M* Tris-HCl (pH 7.9). Two 200 μl aliquots were pipetted into separate tubes. One aliquot, to which 150 μg of *E. coli* tRNA was added as carrier, was precipitated with 1 ml 20% TCA–5% sodium pyrophosphate. The other aliquot was digested with 10 μl of micrococcal nuclease (11,000 units per milliliter from Worthington) at 37°C for 1 hour, and then precipitated as above. After precipitation for at least 15 minutes on ice, the samples were filtered through Whatman GF/C filters, washed with 5% TCA–3% sodium pyrophosphate, dried at 50°C, incubated with 0.5 ml Soluene 350 for 1 hour at room temperature, and counted with 6 ml Permablend scintillant for 10 minutes. The percentage hybridization was calculated as the ratio of counts (minus background) in the nuclease digest sample to the counts (minus background) in the undigested sample, multiplied by 100.

generously supplied by W. J. Beard. The higher substrate concentrations recommended by Efstratiadis *et al.* (*32*) lead to the synthesis of full-length copies of the RNA. The addition of rat liver ribonuclease inhibitor (recommended by Spence Emtage, Searle Laboratories, High Wycombe, England) prevents the degradation of RNA during incubation and improves the yields of cDNA.

The three labeled precursors used to obtain high specific activity, $[^3H]$-dGTP, $[^3H]$-dCTP, and $[^3H]$-dATP (Amersham), are dried down under nitrogen and redissolved in water to give a final concentration of 1 m*M*.

The incubation mixture, in a total volume of 109 μl, contains the following ingredients:

| | | |
|---|---|---|
| Salt mixture | | 20 μl |
| 1.0 *M* NaCl | 2 μl | |
| 0.6 *M* Mg acetate | 2.5 μl | |
| 1.0 *M* Tris-HCl (pH 7.8) | 10 μl | |
| 1.0 *M* KCl | 4 μl | |
| $H_2O$ | 3 μl | |
| Dithiothreitol, 6 mg/ml | | 5 μl |
| Oligo $(dT)_{10-12}$ (Miles), 1 mg/ml | | 5 μl |
| Actinomycin D (Merck), 1 mg/ml | | 5 μl |
| dTTP (Sigma), 20 m*M* | | 1 μl |
| [$^3$H]-dCTP (Amersham), 1 m*M* at 15.5 Ci/mmole | | 20 μl |
| [$^3$H]-dATP (Amersham), 1 m*M* at 25 Ci/mmole | | 20 μl |
| [$^3$H]-dGTP (Amersham), 1 m*M* at 11.6 Ci/mmole | | 20 μl |
| mRNA, 1 mg/ml | | 2 μl |
| Reverse transcriptase, 4600 units/ml | | 10 μl |
| Rat liver RNase inhibitor (Searle), $10^3$ units/ml | | 1 μl |

The reaction mixture is incubated at 37°C for 1 hour and then stopped by the addition of 0.1 ml of 1.0 *M* NaCl, 10 μl 10% SDS, and 0.1 ml phenol saturated with buffer [0.1 *M* NaCl–0.1 m*M* EDTA–10 m*M* Tris-HCl (pH 7.2)]. Chloroform: isoamyl alcohol, 500:1 (0.15 ml), is added to the mixture with Vortex mixing, and the emulsion is centrifuged at 2500 rpm for 5 minutes. The lower phenol phase is reextracted with 0.1 ml of the buffer and 5 μl 10% SDS and, after centrifuging at 2500 rpm for 5 minutes, the aqueous phase is combined with the aqueous phase from the first extraction. The bulked aqueous phase is reextracted with 0.15 ml chloroform: isoamyl alcohol, 500:1, and centrifuged as before. The phenol lower phase is discarded and the aqueous phase is reextracted 3 times with chloroform: isoamyl alcohol. The final aqueous phase is made 0.3 *N* NaOH using a solution of 10 *N* NaOH, and incubated at 37°C for 3 hours. The cDNA is transferred to narrow dialysis tubing and dialysed overnight at 4°C using 4 changes of 250 ml 0.1 *M* NaCl–1 m*M* EDTA–10 m*M* Tris-HCl (pH 7.2).

The cDNA is centrifuged on a 5 ml linear alkaline sucrose gradient, 5–20% sucrose–0.2 *N* NaOH–0.1 *M* NaCl–10 m*M* EDTA, in the Beckman SW65 rotor at 60,000 rpm for 5 hours at 20°C. The gradient is fractionated by upward displacement with 50% sucrose into 200 μl fractions. Aliquots (2 μl) from each fraction are spotted on Whatman filter paper squares which are dried in air and counted with 6 ml Permablend/toluene scintillant.

To estimate the length of the cDNA in the fractions, 1 μl of the fraction is mixed with 2 μl (about 3 μg) marker RNAs (4 S and 9 S) in buffered

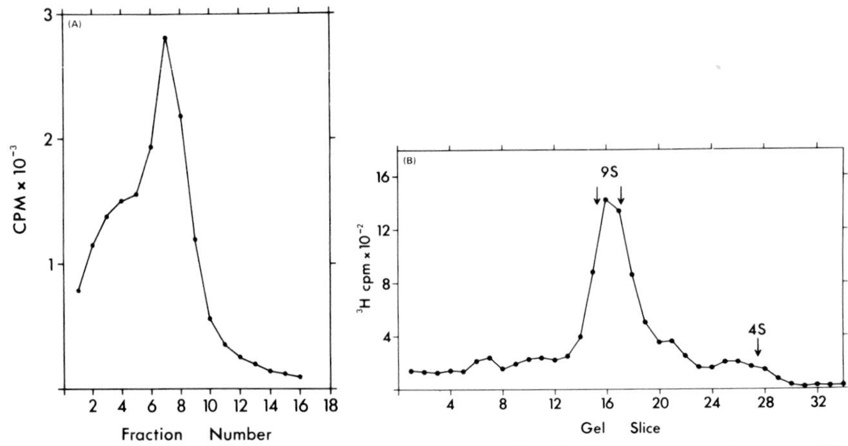


FIG. 3. Fractionation of globin cDNA: (A) Preparative alkaline sucrose gradient centrifugation of globin cDNA prepared as described in the text; fractions 4–8 were pooled, having established the DNA size to be greater than 7 S by formamide gel electrophoresis. (B) Analytical formamide gel electrophoresis of the pooled cDNA fractions, showing the 9 S size and homogeneity. Arrows indicate the positions of 9 S rabbit $\beta$- and $\alpha$-globin mRNA and 4 S transfer RNA markers.

formamide and 7 $\mu$l buffered formamide containing bromophenol blue and 10% sucrose, and subjected to electrophoresis in 4% polyacrylamide–99% formamide gels, 0.5 $\times$ 6.5 cm. The gels are stained with methylene blue for 1 hour, destained with water, and the positions of the markers noted. The gels are frozen at $-70$°C for 30 minutes and then sliced. The slices are treated with Soluene at 50°C for 3 hours and counted with 3 ml Permablend/toluene scintillant.

Alkaline sucrose gradient fractions containing cDNA larger than 7 S, as indicated by formamide gel electrophoresis, are pooled, dialysed against 4 changes of 250 ml 5 m*M* NaCl, and concentrated by freeze-drying to give a 0.25 ng/$\mu$l solution. The final cDNA preparation has the same electrophoretic mobility as the messenger template, as shown in Fig. 3. For the cDNA used in the experiments described in Section III, the yield was 321 ng (16% of the input mRNA) and the specific activity was 4.8 $\times$ $10^7$ cpm/$\mu$g.

## C. Preparation of Chicken Erythrocyte DNA

The method of Gross-Bellard *et al.* (*33*) for extracting high molecular weight rat liver DNA has been adapted for chicken erythrocyte DNA.

All glassware, buffer solutions, and dialysis tubing are sterilized. The dialysis tubing is first boiled in 5% $NaHCO_3$ and washed with 5 m*M* EDTA and water. "Treated" dialysis tubing is obtained by soaking the tubing first in 64% $ZnCl_2$ and then in 0.01 *N* HCl, each for several hours, and then washing with 5 m*M* EDTA and water. This treatment increases the effective pore size.

DNA is extracted from 10 ml of blood from normal (untreated) chickens. The cells are washed twice with PBS as described for mRNA preparation. The cells are lysed by adding 20 ml PBS–0.5% Triton X-100. The nuclei are pelleted by centrifuging at 5000 rpm for 5 minutes. The nuclei are lysed by the addition of 30 ml 0.01 *M* NaCl–0.01 *M* EDTA–0.01 *M* Tris-HCl (pH 8.0)–0.5% SDS, containing 50 μg/ml proteinase K (Merck), and the viscous lysate is incubated for 12 hours at 37°C with gentle shaking. The digest is extracted with an equal volume of phenol, saturated with 0.01 *M* NaCl–0.01 *M* EDTA–0.5 *M* Tris-HCl (pH 8.0)–0.5% SDS, with gentle shaking for 10 minutes at room temperature. The mixture is centrifuged in 30 ml Sorvall centrifuge tubes at 500 *g* for 10 minutes and the viscous aqueous layer is transferred to treated dialysis tubing and dialyzed against several changes of 0.01 *M* NaCl–0.01 *M* EDTA–0.5 *M* Tris-HCl (pH 8.0) at room temperature until all the phenol is removed ($A_{270nm}$ of the dialysate less than 0.05). The viscous solution is then transferred to untreated dialysis tubing and digested with $T_1$ ribonuclease (Sigma) at 2 μg/ml and pancreatic ribonuclease (Sigma) at 50 μg/ml, while continuing dialysis for 12 hours at 37°C. The solution is then dialyzed against 0.05 *M* Tris (pH 8.0)–0.01 *M* EDTA–0.01 *M* NaCl for 4 hours. Further digestion with proteinase K at 50 μg/ml in the presence of 0.5% SDS is carried out in the dialysis tube for the next 12 hours at 37°C, and then two phenol extractions are carried out as before. The viscous DNA solution is again transferred to treated dialysis tubing and dialyzed against 0.01 *M* NaCl–0.5 m*M* EDTA–0.01 *M* Tris-HCl (pH 8.0) until the phenol is removed. This requires about 4 days with frequent buffer changes.

In our earlier work the DNA was finally dialyzed against reconstitution buffer, the concentration measured from the spectrum of small aliquots, and 1 mg samples dispensed, using an Eppendorf pipette with the tip cut off, into the dialysis tubing for reconstitution. We found that aliquoting the viscous DNA solution, both for spectrophotometric analysis and for reconstitution, too inaccurate for the purpose of reconstitution experiments. Instead we now use the following procedure: At the end of dialysis, the DNA is precipitated with two volumes of ethanol, washed 5 times with 70% ethanol, and dried *in vacuo*. Aliquots of 1.0 mg ± 5% (to give 0.8 mg) are weighed out and dissolved in 0.3 ml distilled water for 5 days or more at 0°C and made up to the reconstitution buffer immediately before use. The concentration of DNA in weighed aliquots is determined spectrophotometrically after sonication to reduce the viscosity of the solution. The accuracy of the aliquoting procedure is determined by the accuracy of weighing.

The total yield of erythrocyte DNA from 10 ml blood is about 50 mg. The maximum levels of contaminating RNA and protein, revealed by chemical analysis, are shown in Table I. The DNA molecules have been spread on electron microscope grids to obtain an estimate of the molecular weight distribution, using the Kleinschmidt technique (*34*). The mean molecular weight was found to be too large (greater than $10^8$) to be measured by this technique. We are indebted to M. Walmsley for carrying out this analysis in our laboratory.

## D. Preparation of Chicken Erythrocyte Histones

The method is modified from Panyim *et al.* (*35*). Nuclei are obtained from erythrocytes, chromatin from nuclei, and histones from chromatin, as follows:

Erythrocytes from 3 normal (untreated) chickens are washed 3 times in PBS, as described for the preparation of globin mRNA above. The packed cells are resuspended in an equal volume of 0.25 *M* sucrose buffer: 6 m*M* $MgCl_2$–0.05 *M* Tris-HCl (pH 7.5)–0.5 m*M* PMSF, and the cell suspension is added dropwise to 10 volumes of the same buffer containing 1% Triton X-100 in a beaker, stirring for 10 minutes at 0°C, and then centrifuged in the Sorvall at 5000 rpm for 10 minutes. The pellets are resuspended in two-thirds of the previous volume of 1.7 *M* sucrose buffer. The suspension is transferred to centrifuge tubes, underlayered with an equal volume of 2 *M* sucrose buffer, and centrifuged in the Sorvall at 10,000 rpm for 70 minutes. The pellets are resuspended in 5 blood volumes of 0.25 *M* sucrose buffer and gently homogenized using a VirTis homogenizer (using two 1-minute steps at 20 mV, with cooling in between).

The nuclei are lysed by gentle homogenization in the VirTis homogenizer in 2 blood volumes of 25 m*M* EDTA–10 m*M* Tris-HCl (pH 7.5)–0.5 m*M* PMSF. The chromatin is centrifuged in the Sorvall at 7000 rpm for 10 minutes. The pellet is resuspended and homogenized as in the previous step. After centrifugation at 7000 rpm for 10 minutes, the pellet is resuspended in

FIG. 4. Electrophoretic patterns of proteins used in chromatin reconstitution. (A) "Panyim and Chalkley" gels showing the separation of 5 histone fractions of chicken erythrocytes. (B) "Laemmli" gels showing lack of cross-contamination between histone and nonhistone proteins and analytical separation of the nonhistone proteins. Details of the electrophoretic conditions are given in the references quoted in the text. The gels were stained with 0.025% Coomassie Blue G-250 in methanol:acetic acid:water 45:10:45 and destained in an 8:5:88 solution.

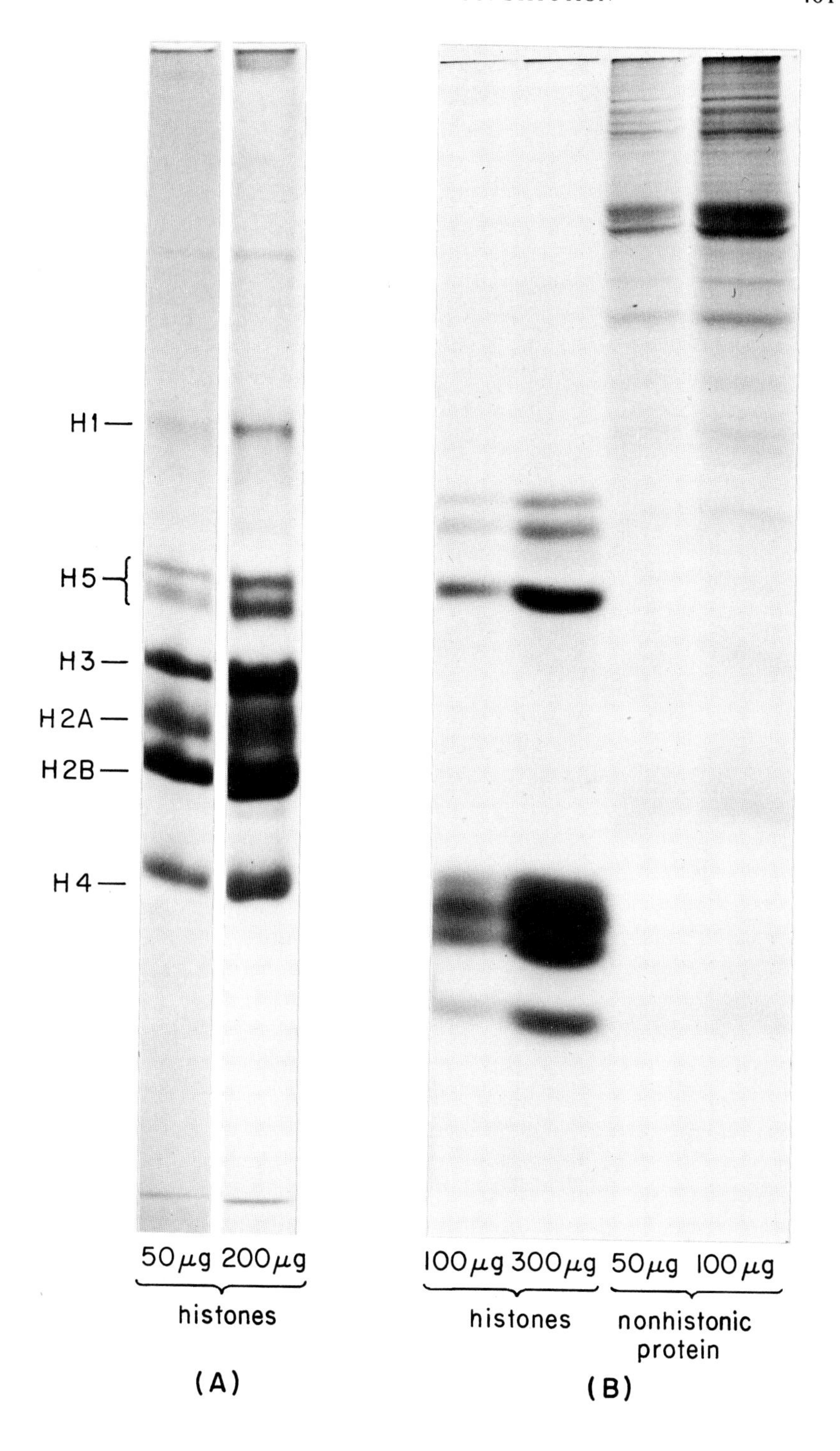
H1
H5
H3
H2A
H2B
H4
50μg 200μg
histones
(A)
100μg 300μg 50μg 100μg
histones
nonhistonic protein
(B)

100–150 ml distilled water, homogenized in the VirTis, and centrifuged in the Sorvall at 10,000 rpm for 20 minutes. The water layer is removed and replaced with an equal volume (50–100 ml) of cold distilled water. The swollen chromatin is homogenized and centrifuged as before.

To extract histones, an equal volume of 0.8 *M* sulfuric acid is added to the chromatin gel, and the suspension is vigorously homogenized in the VirTis homogenizer (twice for 1 minute at 60–70 mV). The suspension is centrifuged at 10,000 rpm for 10 minutes and the supernatant is stored on ice while the pellet is reextracted with 0.4 *M* sulfuric acid and centrifuged as before. The two supernatants are combined and the histones are precipitated with 3 volumes of 95% ethanol and left to stand overnight at −20°C. The histones are collected by centrifugation at 10,000 rpm for 10 minutes. The pellet is dried *in vacuo*, and stored in a desiccator in the dark at room temperature. For reconstitution, histone is dissolved in 2 *M* NaCl–0.01 *M* Tris-HCl (pH 8.3)–5 *M* urea–0.5 m*M* PMSF at 5 mg/ml and stored at −70°C.

The chemical purity of the protein is demonstrated by the data shown in Table I. The histones are also analyzed by polyacrylamide gel electrophoresis in two different solvent systems. Electrophoresis according to Panyim and Chalkley (*36*) in acid–urea solution was carried out to check on the recoveries of the 5 principal histone types, as shown on the left in Fig. 4. Electrophoresis, according to Laemmli (*37*), in Tris-glycine gels at pH 8.8 demonstrated the lack of cross-contamination with nonhistone proteins, as shown on the right in Fig. 4.

## E. Preparation of Reticulocyte Nonhistone Proteins

Nonhistone proteins are prepared from nuclei that have been isolated as described by Levy *et al.* (*38*), except that fractionation of the histones from nonhistone proteins is carried out using hydroxyapatite columns (cf. Ref. *39*) instead of Bio-Rex 70 as originally recommended (*38*).

The nuclei from 6 anemic chickens are suspended in 50 ml 0.35 *M* NaCl–10 m*M* Tris-HCl (pH 7.5)–0.5 m*M* PMSF, homogenized in the Potter homogenizer, poured into a beaker containing 40 ml of the salt–buffer, and stirred gently for 30 minutes at 4°C. The suspension is centrifuged at 5000 rpm for 5 minutes, and the salt treatment is repeated, stirring for 20 minutes, and centrifuging at 5000 rpm for 5 minutes. The pellet is Potter homogenized in 80 ml cold distilled water and centrifuged for 15 minutes at 10,000 rpm. If the pellet is not swollen and gelatinous the homogenization and centrifugation steps are repeated. Proteins are dissociated from the DNA by adding 120 ml of 10 *M* urea–0.58 *M* guanidinium chloride–0.17 *M* sodium phosphate buffer (pH 7.0)–20 m*M* 2-mercaptoethanol–0.5 m*M*

TABLE I
CHEMICAL ANALYSIS OF PURIFIED CHROMATIN COMPONENTS[a]

| Assays: | DNA[b] | RNA[c] | Protein[d] |
|---|---|---|---|
| DNA | 1.0 | >0.03 | >0.01 |
| Histones | >0.008 | >0.002 | 1.0 |
| Nonhistone proteins | >0.013 | 0.067 | 1.0 |

[a] All values expressed on a w/w ratio relative to the major constituent.
[b] Burton assay (Ref. *45*), using salmon sperm DNA as a standard.
[c] Ceriotti method (Ref. *53*), using *E. coli* tRNA as a standard.
[d] Lowry method (Ref. *54*), using bovine serum albumin (Sigma, 5 × recrystallized) as a standard.

PMSF to the pellets and making the volume up to 200 ml with water. The suspension is homogenized with a Potter homogenizer, with a tight-fitting pestle, using 4 up and down strokes. Undissolved chromatin is removed by centrifugation at 5000 rpm for 5 minutes and the supernatant is then centrifuged in the Spinco Ti 60 rotor at 50,000 rpm for 19 hours at 4°C to pellet the DNA, leaving the protein in the supernatant. The supernatant is dialyzed against two changes of column buffer: 2.0 *M* NaCl–5 *M* urea–1 m*M* sodium phosphate buffer (pH 7.0)–0.2 m*M* EDTA–0.2 m*M* phenylmethyl sulphonylfluoride (PMSF). This reduces the sodium phosphate concentration to below 10 m*M* required for binding the nonhistone proteins to hydroxyapatite.

The hydroxyapatite was prepared by M. Spencer of this laboratory. The sample is applied to the column (25 ml packed volume) previously equilibrated with the column buffer. Histones are eluted in the void volume, and any residual histones are eluted by washing the column with 100 ml of loading buffer. The nonhistone proteins are then eluted with 2.0 *M* guanidinium chloride –5 *M* urea–200 m*M* sodium phosphate buffer (pH 7.0)–0.2 m*M* EDTA–0.2 m*M* PMSF. Protein for reconstitution is dialyzed against reconstitution buffer: 2 *M* NaCl–5 *M* urea–0.01 *M* Tris-HCl (pH 8.3)–0.5 m*M* PMSF, diluted to 1–2 mg/ml, and stored at −70°C.

The nucleic acid (RNA and DNA) contamination of the nonhistone protein fraction is revealed by chemical analysis, as shown in Table I. RNA is also extracted with phenol and the globin mRNA content is determined by hybridization with cDNA, as shown in Table II. The electrophoretic pattern of nonhistone proteins fractionated on a Bio-Rex column is shown on the right in Fig. 4. There is negligible cross-contamination with histones.

Nonhistone proteins fractionated by hydroxyapatite chromatography show qualitatively similar patterns.

The hydroxyapatite column is preferred because the nonhistone proteins emerge in sufficiently high concentration from the column to use directly in reconstitution. We found that lengthy concentration steps, required if Bio-Rex chromatography is used, i.e., filtration on Amicon membranes or vacuum dialysis, reduce the yields and appear to cause some proteolytic degradation, even if the protease inhibitor PMSF is present throughout. The modified procedure speeds up the preparation of the nonhistone proteins, and gives dependable yields, about 35 mg from 6 chickens.

## F. Reconstitution of Chromatin

DNA (0.8 mg) dissolved in 0.3 ml water, as described above, is mixed with 0.7 ml 3 *M* NaCl and 8 *M* urea–0.01 *M* Tris-HCl (pH 8.3)–5 m*M* PMSF. This solution, together with the histone (0.8 mg) and nonhistone proteins (0.6 mg), both dissolved in reconstitution buffer, i.e., 2.0 *M* NaCl, 5 *M* urea (freshly deionized)–10 m*M* Tris-HCl (pH 8.3)–0.5 m*M* PMSF, plus buffer to make up the volume to 2.0 ml, are carefully put into dialysis tubing and gently mixed.

The reconstitution mixture is dialyzed on a rocking dialyzer at 4°C against 1 liter of 2.0 *M* NaCl–5 *M* urea–10 m*M* Tris-HCl (pH 8.3)–0.1 m*M* PMSF for 1 hour; then overnight against 1 liter of 0.6 *M* NaCl–10 m*M* Tris-HCl (pH 8.3)–0.1 m*M* PMSF; and then 3 hours against 1 liter of 0.4 *M* NaCl–10 m*M* Tris-HCl (pH 8.3)–0.1 m*M* PMSF; and finally for 3 hours against 1 liter of 0.2 *M* NaCl–10 m*M* Tris-HCl (pH 8.3)–0.1 m*M* PMSF.

This regime differs from the ones principally used by other workers (*3,18,19,24*), based on the recommendation of Bekhor *et al.* (*4*), in that the urea is removed in the first dialysis step (from 2.0 *M* NaCl to 0.6 m*M* NaCl). The other workers maintain the urea concentration at 5 *M* until the final salt concentration is reached and remove the urea from the chromatin generally by centrifugation. In our procedure, the appearance of a stringy precipitate in 0.4 *M* NaCl is one indication that reconstitution has proceeded normally.

## G. Preparation of HgUTP

The method used for the mercuration of uridine triphosphate is similar to that described by Dale *et al.* (*21*).

The UTP (0.5 mmole, Sigma) is dissolved in 25 ml of 0.1 *M* sodium acetate buffer (pH 6.0) and is added to 25 ml of the same buffer containing mercuric acetate (2.5 mmole). After heating the reaction mixture at 50°C for 3 hours, it is applied to a column of Chelex-100 (1.1 × 23 cm) that has been previously equilibrated with 0.1 *M* sodium acetate buffer (pH 6.0). This step removes unreacted mercuric ions and the column bed volume used is calculated to bind 10 times the amount of $Hg^{2+}$ present, based on the capacity (0.33 mmole/ml) specified by the manufacturers (Bio-Rad). The sample is eluted quickly to avoid demercuration (*21*) and the fractions containing nucleotide, monitored by absorbance, are pooled and precipitated with 4 volumes of acetone at −20°C. The precipitate is collected by centrifugation at 5000 rpm for 5 minutes.

Sephadex G-10 gel filtration provides a further guarantee against unreacted mercuric ions in the product of this reaction. Batches of 40 mg of HgUTP, dissolved in 0.1 *M* sodium acetate buffer (pH 6.0), are applied to a column (2.6 × 15 cm) and eluted with the same buffer. The HgUTP is precipitated and collected as before and dried *in vacuo*.

To ensure that mercuration of UTP is complete (100% HgUTP), a sample of the product (5 $A_{268nm}$ units) is applied to a column of reduced Thiol-Sepharose (Pharmacia) in 0.1 *M* sodium acetate buffer (pH 6.0) and allowed to equilibrate for 30 minutes. Elution of the column with the same buffer gives fractions that contain any unreacted UTP. Bound HgUTP is then eluted with the same buffer containing 0.1 *M* 2-mercaptoethanol, and the absorbance at 268 nm of the fractions is measured. Between 97 and 99% of the reaction product binds to the Thiol-Sepharose column. The HgUTP is not further purified by chromatography on sulfhydryl Sepharose because the elution with 2-mercaptoethanol may cause demercuration.

## H. Preparation of Mercurated Globin Messenger RNA

Globin mRNA is chemically mercurated using a method similar to that described by Dale *et al.* (*21*) for the mercuration of yeast transfer RNA.

Purified globin mRNA (3 μg) in 100 μl 5 m*M* mercuric acetate, 5 m*M* lithium acetate buffer (pH 6.0) is heated at 50°C for 16 hours, and the reaction stopped by the addition of an equal volume of 2.0 *M* NaCl–0.2 *M* EDTA–10 m*M* Tris-HCl (pH 7.5). The reaction product is separated from inorganic ions by centrifugation and gel filtration on Sephadex G-50. Preswollen gel, first equilibrated against 10 m*M* lithium acetate buffer (pH 6.0), is packed into a 1 ml disposable syringe barrel. After gentle centrifugation in a conical centrifugation tube (1000 *g* for 0.5 minutes) to remove any excess buffer, the gel is washed 4–5 times with 100 μl of buffer. The washing cycle is continued until the excluded volume equals that of the applied wash. Aliquots of the reaction mixture (100 μl) are then applied to separate gels and centrifuged. One or two further washes with buffer (50 μl) may be necessary to recover all the Hg mRNA, and washing should be continued until the excluded volume equals the applied volume.

To check for mercuration of globin mRNA, a sample of the excluded material (10 μl) is diluted 5 times with 10 m*M* lithium acetate buffer (pH 6.0)–0.1% LDS and applied to a column of Thiopropyl-Sepharose 6B (0.3 ml, Pharmacia) in the same buffer. The column is first washed with 6 column volumes of 10 m*M* lithium acetate buffer (pH 6.0)–0.1% LDS (2 ml) to remove any unbound material, and then washed with a similar volume of buffer containing 25 m*M* 2-mercaptoethanol to recover the bound material. Aliquots of the total RNA and the bound and nonbound fractions are hybridized to globin cDNA to measure the concentration of globin mRNA, as described below. The hybridization analysis (Table II) shows that 85% of the globin mRNA was sufficiently mercurated to bind to the Thiopropyl-Sepharose column.

## I. Transcription of Reconstituted Chromatin

The total contents of the dialysis tubing are transferred to 30 ml Sorvall centrifuge tubes containing the other components of the transcription mixture. The final concentrations are: 0.4 m*M* each of ATP, CTP, GTP, and UTP/HgUTP; a labeled substrate, [$\alpha$-$^{32}$P]CTP of specific activity 1–3 Ci/mmole; 0.15 *M* NaCl, 6 m*M* $MgCl_2$, 1.6 m*M* $MnCl_2$, 100 m*M* Tris-HCl (pH 7.9), 15 m*M* 2-mercaptoethanol (*22*); 250 units of *E. coli* RNA polymerase of specific activity 690 units per milligram (Sigma); water to a final volume of 5 ml. The following protocol is convenient for assembling the transcription mixture:

| Component of reaction mixture | Constant | Variable |
|---|---|---|
| 1.0 *M* Tris-HCl (pH 7.9) | 0.5 ml | |
| 60 m*M* $MgCl_2$, 16 m*M* $MnCl_2$ | 0.5 ml | |
| 5.0 *M* NaCl | 0.1 ml | |
| 10 m*M* each ATP, CTP, and GTP (Sigma) | 0.2 ml | |
| 10 m*M* HgUTP in 50 m*M* sodium acetate buffer (pH 6.0) | | 0.2 ml or 0.05 ml |
| 10 m*M* UTP (Sigma) in 50 m*M* sodium acetate buffer (pH 6.0) | | 0.15 ml or none |
| 1 mCi/ml [$\alpha$-$^{32}$P]CTP (1 Ci/mmole, Amersham) | | 0.005–0.05 ml |
| 2-Mercaptoethanol | 0.005 ml | |
| 1000 units/ml *E. coli* RNA polymerase (690 units/mg, Sigma) | 0.25 ml | |
| Water | | To 5 ml final |
| Reconstituted chromatin | 2.0 ml | |

The mixture is incubated in a shaking water bath at 37°C for 1 hour and the reaction is terminated by one of the extraction procedures described in the next section.

The use of the monovalent cation (0.15 *M* NaCl) in the transcription mixture may be noted. Our conditions are based on an earlier study of chain initiation on transcription of reticulocyte chromatin with *E. coli* RNA polymerase (*40*). Many workers carry out the transcription effectively at zero ionic strength. The *E. coli* RNA polymerase used in the present study was subjected to electrophoresis in 4% polyacrylamide gels in SDS, and sigma factor, molecular weight ca. 90,000, identified in roughly stoichiometric amounts with the holoenzyme subunits (*41*). This contrasts to Crouse *et al.* (*23*) who deliberately chose to use the bacterial polymerase without initiation factors to obtain random transcription.

## J. The Extraction of RNA

Two procedures have been used, our "improved" procedure, Method A, and our previous procedure, Method B.

### 1. Method A

This procedure involves the following sequence: Pronase digestion to hydrolyse protein, the first phenol extraction to remove pronase, DNase I digestion to hydrolyze DNA, and the second phenol extraction to remove DNase I.

After transcription, 0.6 ml 0.5 *M* EDTA (pH 7.4), 0.3 ml 10% SDS, and 0.5 ml proteinase K (1 mg/ml, BDH) are added to the transcription mixture, and this is incubated for a further 20 minutes at 37°C. Five milliliters of water-saturated phenol, 0.3 ml 10% SDS, and 0.3 ml 5.0 *M* NaCl are added, and the mixture is shaken for 15 minutes at room temperature; then a volume of chloroform : isoamyl alcohol (500 : 1) equal to that of the phenol is added. After shaking, the phases are separated by centrifugation at 1000 *g* for 2–3 minutes and the bottom phenol-chloroform layer is discarded. The aqueous phase is reextracted twice with the same volume of chloroform : isoamyl alcohol. The final aqueous layer is removed, taking care not to disturb any of the chloroform : isoamyl alcohol layer, and the nucleic acid is precipitated with two volumes of ice-cold 95% ethanol at −20°C. The nucleic acid precipitate is collected by centrifugation at 10,000 rpm for 10 minutes and is resuspended in 5–10 ml of 70% ethanol at −20°C to remove NaCl and SDS and again collected by centrifugation. This is repeated two or three times and finally the precipitate is dried *in vacuo*.

The precipitate is resuspended in 0.5 ml 5 m*M* Tris-HCl (pH 7.5)–50 m*M* NaCl–4 m*M* $MgCl_2$, and 50 μg DNase I (RNase-free, Worthington) is added and the mixture incubated for 30 minutes at 37°C. The reaction is stopped by the addition of 0.05 ml 5.0 *M* NaCl, 0.12 ml 10% SDS, 12 μl 0.5 *M* EDTA, and 0.5 ml water-saturated phenol. The RNA is extracted as described above.

The final aqueous phase is directly applied to a column of Sephadex G-50 (1.1 × 30 cm) previously equilibrated in 10 m*M* lithium acetate (pH 6.0)–0.1% LDS. The sample is eluted with a flow rate of 120 ml/hour and 0.5 ml fractions are collected and Cerenkov counted. RNA is eluted in the excluded volume (7–10 ml), whereas the unreacted nucleotides elute after 15–20 ml. The RNA is stored at −20°C until it is fractionated by Thiol-Sepharose chromatography. The amount of RNA synthesized is calculated from the radioactivity, correcting for [$^{32}$P] decay, assuming that CTP and the other three nucleotides are incorporated in equal amounts.

### 2. Method B

This procedure is similar to that described by Penman (*42*) and Zylber and Penman (*43*), and involves, first, DNase I digestion and, second, hot phenol extraction of protein.

Transcription is stopped by the addition of 0.3 ml 5.0 *M* NaCl. Then 100 μg DNase I (RNase-free, Worthington) is added and the mixture is further incubated for 30 minutes at 37°C. A total of 0.6 ml 10% SDS, 0.15 ml 0.2 *M* EDTA, and 5 ml of melted phenol are added and the mixture is heated to 55°C for 15 minutes; then 5 ml of chloroform : isoamyl alcohol (500 : 1) is added and the mixture heated for a further 5 minutes at 55°C. The phases are separated by centrifugation, as in Method A, and the aqueous phase and interface are twice reextracted with chloroform : isoamyl alcohol at 55°C. The final aqueous phase is removed and the nucleic acid is precipitated by the addition to two volumes of ethanol. The nucleic acid is collected by centrifugation at 10,000 rpm for 30 minutes and redissolved in 0.05 ml 10 m*M* sodium acetate (pH 6.0). It is separated from nucleotides by gel filtration on Sephadex G-50 column, as described for Method A, except that the column is eluted with 10 m*M* sodium acetate buffer (pH 6.0)–0.1% SDS.

### 3. Comparison of Methods A and B

The two extraction procedures A and B give similar recoveries of RNA transcript. Recoveries were determined by adding [$^{125}$I]-labeled globin mRNA ($1.6 \times 10^5$ cpm, $10^7$ cpm/μg), labeled by the Commerford procedure (*44*), to the transcription mixture before extracting the RNA, and by measuring the recovery after Sephadex G-50 gel filtration. [$^{125}$I] radioactivity was measured using a gamma counter. For both Methods A and B the recovery was between 60 and 75%.

Contamination of the RNA with DNA was estimated by the Burton colorometric assay (*45*). The RNA extract from Method A had less than 3% of the chromatin DNA input, whereas that from Method B had 95% of the input DNA.

Method A has the experimental advantage that there is no interface containing nucleoprotein complexes in the first phenol extraction, owing to the preceding protease step. This eliminates the need to carry out phenol extractions at elevated temperatures (*42,43*).

## K. Purification of Mercurated RNA by Affinity Chromatography

We use Thiopropyl-Sepharose to isolate the mercurated RNA extracted by Method A, while retaining the use of Thiol-Sepharose for that of the old Method B. Before describing the two methods of separation, we comment on the reasons for changing from Thiol-Sepharose to Thiopropyl-Sepharose.

### 1. Rationale of the Method

From the foregoing history, it can be seen that it is important to denature the RNA before affinity chromatography, and to minimize aggregation both before and during chromatography. For protein-free RNA (especially in Method A), this requires maintaining ionic strength as low as possible. In preliminary experiments we found, using Thiol-Sepharose, that we obtained far less efficient binding of HgRNA at 10 m*M* than at 100 m*M* sodium acetate, even when 100% HgUTP was used in transcription. Elevation of the salt concentration from 10 m*M* (used in the heating step) to 100 m*M* for chromatography is to be avoided if possible. Yet most workers have carried out Thiol-column chromatography using 50–100 m*M* sodium acetate (*26,27,47,56*), perhaps unnecessarily. Thiol-Sepharose is prepared by coupling glutathione to CNBr-activated Sepharose 4B, and the need for the higher salt concentration may be due to the nature of the spacer group, which is negatively charged at pH 6.0, and may therefore repel nucleic acids at low ionic strength. The majority of workers have employed Thiol columns made by the method of Cuatracasas (*46*), using a resin that has a similar spacer arm as activated Thiol-Sepharose, but does not have a charged group.

Our reasons for choosing Thiopropyl-Sepharose were the following: (1) We indeed found, as expected, that the efficiency of binding HgRNA was not ionic-strength-dependent in the range of 10–100 m*M* sodium acetate. (2)

Thiopropyl-Sepharose is not prepared from CNBr-activated Sepharose. The ligand is connected to the gel directly by an ether link through a noncharged spacer arm and is therefore not susceptible to the slow leakage of active groups, as found in CNBr-activated gels. (3) The Thiol-binding capacity (35 $\mu$moles SH/ml swollen gel) is more than 10 times greater than either Thiol-Sepharose (1 $\mu$mole SH/ml) or the resins used by other authors (less than 3 $\mu$moles SH/ml). This should enhance the ratio of specific binding to non-specific adsorption. This new "breed" of Thiol gels should also allow the isolation of RNA transcripts containing much lower input levels of HgUTP and thus help answer the question arising out of our results concerning the validity of using HgUTP in chromatin transcription. Thiopropyl-Sepharose has the disadvantage that the spacer arm is shorter than those in Thiol-Sepharose or the Cuatracasas type, and this may increase the steric hindrance between the support and HgRNA molecules. At the HgUTP levels used in this study, this does not appear to be a problem. For future work, the preparation of gels with longer spacers than as 2-hydroxypropyl-Sepharose, may be necessary to improve the efficiency of separation.

## 2. Method A

The Thiopropyl-Sepharose 6B is prepared for chromatography using the manufacturer's recommended washing procedure. The lyophilized gel is suspended in 10 m*M* Tris-HCl (pH 7.0)–1 m*M* EDTA (200 ml/gm) and allowed to swell for 1 hour at room temperature. The swollen gel is collected by Millipore filtration (to avoid damage to the gel particles) and converted to the free Thiol form by suspending in 0.5 *M* 2-mercaptoethanol–0.3 *M* sodium bicarbonate–1 m*M* EDTA, adjusted to pH 8.4 (4 ml/gm of the original dry gel) for 1 hour at room temperature. The gel is collected by Millipore filtration and washed continuously with 0.1 *M* acetic acid–0.5 *M* sodium chloride–1 m*M* EDTA (400 ml/gm dry gel). The gel can be stored at 4°C in this buffer without loss of free Thiol groups.

Columns of Thiopropyl-Sepharose are poured in 0.1 *M* acetic acid–0.5 *M* sodium chloride–1 m*M* EDTA, and washed with at least 20 column volumes of 10 m*M* lithium acetate buffer (pH 6.0)–0.1% LDS (degassed under reduced pressure). Samples of RNA in the same buffer (1–2 ml) are heated at 100°C for 5 minutes where specified, rapidly cooled in ice, and allowed to warm to room temperature before they are loaded onto the column (0.8 × 10 cm) and allowed to permeate the gel. After washing the sample into the gel with a further 0.5 ml of buffer, the RNA is left on the column to bind for 1 hour. The column is eluted with 10 column volumes of the loading buffer, and about ten 2 ml fractions are collected before the bound RNA is eluted with buffer containing 50 m*M* 2-mercaptoethanol, and five 2 ml fractions are collected. Cerenkov radiation in the bound and unbound fractions is monitored, and the peak fractions containing RNA are pooled and precipitated by adding 15 μg *E. coli* transfer RNA (tRNA)/ml, NaCl to a final concentration of 0.3 *M* and 2 volumes of ethanol at −20°C. RNA is collected by centrifugation at 10,000 rpm for 15 minutes and the precipitate is washed with 5 ml 70% ethanol and dried *in vacuo*. The RNA is dissolved in 100 μl of water and 10 μl aliquots are spotted onto Whatman GF/C filters, dried, and counted in the presence of Permablend/toluene scintillant.

#### 3. Method B

Mercurated transcripts extracted by method B are separated on activated Thiol-Sepharose 4B (Pharmacia). The procedure is identical to that described above, except that chromatography is carried out in 0.1 *M* sodium acetate buffer (pH 6.0)–0.1% SDS. The RNA, dissolved in 10 m*M* sodium acetate buffer (pH 6.0)–0.1% SDS is heated to 100°C for 5 minutes where specified, and cooled rapidly on ice. Before it is applied to the column, 0.1 volume 1.0 *M* sodium acetate buffer (pH 6.0) is added and the unbound material is eluted with 0.1 *M* sodium acetate buffer (pH 6.0)–0.1% SDS. The bound fraction is eluted with the same buffer containing 0.1 *M* 2-mercaptoethanol. The RNA is precipitated as described above.

### L. Hybridization Assay for Globin RNA

The amounts of globin RNA in chromatin transcripts are determined from the hybridization kinetics of the extracted RNA in cDNA excess (*48*).

A calibration curve is first constructed using the purified globin mRNA. A standard reaction mixture, "Mix A," is added to varying amounts of RNA:

Mix A

| Component | Volume (parts) |
|---|---|
| 1.0 *M* sodium phosphate buffer (pH 6.8) | 2 |
| 5% SDS | 1 |
| 1 m*M* EDTA (pH 7.0) | 1 |
| 7.5 mg/ml *E. coli* tRNA | 1 |
| 0.25 ng/μl cDNA | 1 |

Fifteen microliters of Mix A are added to 10 μl RNA stock solutions ranging from 0–0.025 ng/μl. Two 10 μl aliquots are taken from each of the 25 μl hybridization mixtures and sealed in siliconized capillaries ("microcaps").

Each capillary contains 0.25 ng of cDNA and 0–0.1 ng RNA. The capillaries are placed in a boiling NaCl-saturated water bath for 15 minutes and then cooled rapidly on ice. The denatured RNA and cDNA are hybridized by incubating the capillaries at 68°C for 20 hours. Then the capillaries are opened and the contents are expelled into 0.5 ml $S_1$ nuclease buffer: 50 m*M* NaCl–30 m*M* sodium acetate buffer (pH 4.5)–1 m*M* $ZnSO_4$–10 $\mu$g/ml denatured salmon sperm DNA. Three microliters of $S_1$ nuclease (10 IU/ml, Calbiochem) are added, and the digestion is carried out at 45°C for 1 hour. Digestion is terminated by placing the samples on ice and adding 50 $\mu$l *E. coli* tRNA at 1 mg/ml and 0.3 ml ice cold 30% trichloroacetic acid (TCA). The samples are left on ice for at least 10 minutes and are then collected by filtration onto Whatman GF/C filters. The filters are washed with absolute ethanol, dried, and counted in 6 ml Permablend/toluene scintillant.

Since the amount of RNA in the chromatin transcripts is unknown, an inspired guess is required to perform the hybridization in the appropriate concentration range. A portion of the transcript, in distilled water, is dried down *in vacuo* and dissolved in 10 $\mu$l of water if unmercurated, or, if mercurated, in 250 m*M* dithiothreitol (DTT)–0.1% SDS. To this is added 15 $\mu$l Mix A, and capillaries are prepared as for the standard curve. The heating step at 107°C for 15 minutes is sufficient to demercurate the mercurated RNA in 100 m*M* DTT (*26*). This is important in case mercurated uridine residues should affect the hybridization kinetics (*26,49*).

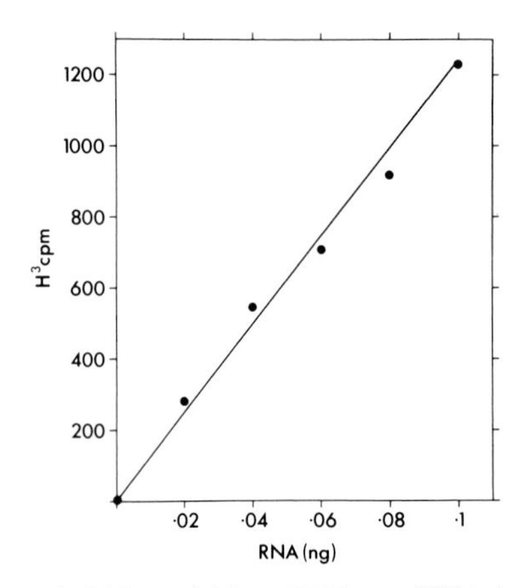


FIG. 5. Hybridization of chicken globin mRNA to cDNA in DNA excess. This curve served as a calibration for the amount of mRNA in transcripts presented in Table II. The abscissa refers to the amount of purified globin mRNA in each capillary. The cDNA by itself hybridized to a maximum of 15% in the longest time period.

After hybridization, $S_1$ nuclease digestion and counting, the concentration of RNA is determined from the standard curve, an example of which is shown in Fig. 5. A sample without any RNA is always hybridized in parallel with the samples containing unknown amounts of globin RNA, and the background counts obtained from this "blank" are subtracted from the sample counts.

## III. Results

We describe three experiments of heuristic interest. The first illustrates the dangers of taking the results of transcription experiments at face value; these arise from the intrusion of RNA-dependent RNA synthesis, first described by Zasloff and Felsenfeld with the aid of mercurated nucleotides (*26*). The second experiment demonstrates unambiguously that the globin gene is not selectively transcribed from our particular chromatin template in the presence of 100% HgUTP. The third experiment draws attention to the possibility of a further problem that has not been previously considered. This is brought to light by the comparison of results obtained in transcription experiments with 100% and 25% HgUTP.

In the first experiment, we merely substituted HgUTP for UTP in the transcription mixture. The "control" sample, 1 RCT in Table II, contained no other additives; while the "experimental" sample, 2 RCT, had 30 ng of purified globin mRNA added just before transcription. After extracting the RNA (by Method B) and separating the RNA on a Thiol-Sepharose column, 2.1 ng globin RNA was detected in the bound fraction in the first case and 40.3 in the second. The comparison reveals that virtually all the added globin RNA is converted to a form that binds to the affinity column in the conditions of the experiment. The results are reminiscent of those of a similar experiment described by Zasloff and Felsenfeld (*26*). These authors adduced evidence for the existence of an RNA-dependent synthesis of globin antistrands by the *E. coli* RNA polymerase during chromatin transcription. They concluded that mercurated antistrands remain associated with the endogenous RNA template, which results in binding of the latter to the affinity column. We suggest that the present results are most likely a reflection of the same phenomenon.

In the second experiment, we compare the results of transcribing free chicken DNA (carried through the reconstitution procedure) and reconstituted chromatin with 100% HgUTP. The DNA transcript, 3 RDT in Table II, contains two parts per million (0.0002%) of globin RNA. This cor-

TABLE II

ASSAY OF GLOBIN RNA IN THIOL-SEPHAROSE COLUMN FRACTIONS[a]

| Sample | Additions | HgUTP (%) | Extraction procedure | Heat | Thiol-Sepharose fraction ($\mu$g) | | Globin RNA ($\mu$g) | Globin RNA (%) |
|---|---|---|---|---|---|---|---|---|
| 1 RCT | — | 100 | B | – | Bound | 64.3 | 2.1 | 0.003 |
| | | | | | Nonbound | 18.6 | 1.5 | |
| | | | | | Total | 82.9 | 3.6 | |
| 2 RCT | 30 ng mRNA | 100 | B | – | Bound | 105 | 40.3 | 0.058 |
| | | | | | Nonbound | 11.6 | 2.0 | |
| | | | | | Total | 117 | 42.3 | |
| 3 RDT | — | 100 | B | – | Bound | 141 | 0.26 | 0.00019 |
| | | | | | Nonbound | 8.9 | | |
| | | | | | Total | 150 | | |
| 4 RCT | — | 100 | A | + | Bound | 50.4 | 0.11 ± 0.01 | 0.00020 |
| | | | | | Nonbound | 8.4 | 6.1 ± 1.1 | |
| | | | | | Total | 58.8 | 6.2 | |
| 5 RCT(1/4) | — | 25 | B | – | Bound | 9.6 | 0.7 ± 0.01 | 0.0072 |
| | | | | | Nonbound | 10.7 | 3.5 | |
| | | | | | Total | 20.3 | 4.2 | |

| | | | | | | | | |
|---|---|---|---|---|---|---|---|---|
| 6 RCT(1/4) | — | 25 | B | + | Bound | 6.2 | 0.22 ± 0.01 | 0.0035 |
| | | | | | Nonbound | 11.3 | 3.3 + 0.4 | |
| | | | | | Total | 17.5 | 3.5 | |
| 7 RCT(1/4) | — | 25 | A | – | Bound | 10.2 | 0.55 ± 0.01 | 0.0054 |
| | | | | | Nonbound | 6.1 | 5.0 ± 0.7 | |
| | | | | | Total | 16.3 | 5.6 | |
| 8 RCT(1/4) | — | 25 | A | + | Bound | 7.4 | 0.21 | 0.0028 |
| | | | | | Nonbound | 17.4 | 3.8 ± 0.9 | |
| | | | | | Total | 24.8 | 4.0 | |
| 5–8 | | | | | Sum total | 78.9 | | |
| 9 RCT | 90 ng Hg mRNA | 25 | A | + | Bound | 29.5 | 51.0 | — |
| | | | | | Nonbound | 59.1 | 13.8 | |
| | | | | | Total | 88.6 | 64.8 | |
| 10 RC | — | — | A | — | — | | 14.1 ± 0.2 | — |

[a] Abbreviations: RCT = Reconstituted chromatin transcript; RDT = reconstituted DNA transcript; RCT(1/4) = quarter-sized transcripts: transcribed, reconstituted chromatin split into two parts for processing by Methods A and B, and the two RNA extracts split into two further parts after the Sephadex G50 column, one part applied directly to the Thiol-Sepharose column, and the other heated as described in Section II before applying to the column; RC = reconstituted chromatin (not transcribed); Hg mRNA = mercurated globin mRNA.

responds precisely to the amount expected in a random DNA transcript, considering the number of nucleotides in the "structural" gene, i.e., the part that codes for the mature globin mRNA, about $10^3$ nucleotides, and the size of the entire chicken genome, about $10^9$ nucleotides. The reconstituted chromatin transcript, 4 RCT in Table II, contained the same percentage, and we conclude therefore that the globin gene is not selectively transcribed from either DNA or chromatin templates.

The third experiment was designed to test the effects of different RNA extraction procedures and of heating on the amount of globin RNA detected in the RNA fraction bound to the affinity column. The heating procedure adopted was shown to melt out the duplexes between the endogenous RNA and the mercurated antistrands and allow an effective separation on a Thiol column (*26*). The new extraction procedure (Method A) should eliminate residual aggregation artifacts. Two further controls were performed: First, mercurated globin RNA was added to one chromatin sample after transcription, 9 RCT in Table II, and its recovery after RNA extraction by Method A, heating and Thiol-Sepharose chromatography, was determined by hybridization. Second, RNA was extracted from a sample of reconstituted chromatin that was not transcribed, 10 RC in Table II, to determine the amount of endogenous RNA. Comparing the results of 9 RCT and 10 RC in Table II, and taking into account the loss of 25–40% of the RNA in extraction (see Section II), we see that the recovery of mercurated globin RNA sequences in column-bound fraction is close to 100%. Thus significant amounts of newly synthesized globin RNA should be detected.

To compare the two RNA extraction methods and determine the effect of heating the extracted RNA samples from the *same* reconstituted chromatin transcript, the following experiment was done: Transcription of reconstituted chromatin was carried out on the usual scale with 0.8 mg chromatin DNA. After transcription, the chromatin sample was divided into two (roughly equal) parts for processing by Methods A and B. After purifying the extracts on the Sephadex G-50 column, each of the half portions was further split into two equal parts. One part was heated as described in Section II and the other was not heated prior to Thiol-Sepharose chromatography. The results show that in fact the extraction procedure made little difference to the amount of globin RNA detected on the columns (cf. 5 RCT and 7 RCT, and 6 RCT and 8 RCT). Furthermore, heating only halved the amount of globin RNA detected in the bound fractions (Methods A or B), the remainder then amounting to approximately 0.003% of the newly synthesized RNA. This may represent a greater proportion of DNA-dependent synthesis if some (i.e., half) of the labeled RNA is transcribed from endogenous RNA.

The third experiment also differs from the second in that 25% rather than 100% HgUTP was used in transcription. Slightly more (20%) RNA was synthesized at the lower concentration of HgUTP, and slightly less (50% as against 80%) of the RNA product bound to the affinity column. The most provoking observation, however, was the increase in the amount of globin RNA detected in the bound fraction, from 0.0002% in the case of 100% HgUTP to 0.003% in the case of 25% HgUTP. Further work will be needed to determine whether this difference represents globin RNA synthesized using 25% HgUTP in transcription, or residual endogenous RNA that the preventative measures for some reason failed to eliminate.

## IV. Discussion

In our earlier work (*13*), we obtained evidence for selective globin gene transcription in reconstituted chromatin, using procedures not dissimilar in principle to those described here. The apparent level of globin RNA synthesis was close to that found with native chromatin about 0.01% of the total RNA synthesis. The earlier experimental system was unsuitable as the basis of a routine functional assay for a number of reasons. On the practical side, we found the available preparative procedures unreliable and the analytical techniques barely sensitive enough to detect the amount of globin RNA in native chromatin transcripts.

More seriously, we are now compelled to question the earlier inference because of a possible ambiguity in the determination of the "background" of endogenous RNA from the minus polymerase and minus histone controls. Recent work with HgUTP has shown the existence of an RNA-dependent synthesis in the transcription of chromatin with *E. coli* RNA polymerase (*26*). Our present work is consistent with these findings. The formation of a duplex between endogenous globin RNA and its antistrand copy may afford protection of RNA from nucleases present in the system. We cannot exclude the possibility that the observed difference between the plus and minus enzyme samples represents the difference in recovery of endogenous RNA, rather than globin RNA synthesized *in vitro*. Similarly, the omission of histones, which are believed to inhibit DNA transcription (*50,51*), may alter the balance between RNA- and DNA-dependent synthesis, so that the endogenous RNA is less protected in the incomplete than in the complete system. In view of unresolved difficulties in measuring endogenous RNA levels in chromatin transcribed with normal as opposed to mercurated substrates, we are no longer satisfied that the earlier results in the field provide sufficient proof for the synthesis of globin RNA *in vitro*.

We have overcome the practical limitations of experiments with reconstituted chromatin by the combination of improved preparative and analytical methods described here. In this regard we consider the following developments to be critical: (1) improvement in the preparative procedure for nonhistone proteins to give high yields of protein, undegraded by proteases and uncontaminated with histones; (2) purification of globin mRNA by formamide gel electrophoresis; (3) synthesis of a full-length cDNA of high specific activity in good yields; (4) the use of weight aliquots of high-molecular-weight DNA in reconstitution; and (5) the use of cDNA excess hybridization for the assay of globin RNA. The new procedures permit the reliable detection of one part per million of globin RNA in a transcript from 1 mg chromatin DNA.

Mercurated nucleotides were introduced to overcome the need to obtain an accurate estimate of endogenous RNA with its attendant difficulties. The unexpected binding of endogenous RNA to the affinity column following transcription required the further development of special methods of sample preparation and chromatography and independent criteria for *de novo* synthesis of RNA. We have succeeded finally in obtaining an unambiguous result for the transcription of reconstituted chromatin in the presence of 100% HgUTP. In these conditions it seems certain that the globin gene is not selectively transcribed (Table II). We know from the control experiment with Hg globin mRNA that newly synthesized globin RNA should bind to the column.

The results obtained in the transcription of reconstituted chromatin with 25% HgUTP are less clear-cut. The amount of globin RNA detected on the affinity column, 0.003% of the labeled RNA, is intermediate between the background due to random transcription, 0.0002%, and the (minimum) target level for native chromatin, 0.01% (*15*). We hesitate to assert that bound globin RNA was synthesized *in vitro*, since the various independent tests have not yet been completed. These include: (1) the demonstration that added globin mRNA is not bound to the column; (2) demonstration that synthesis is inhibited by actinomycin D; and (3) formation of a cDNA hybrid and rebinding to the affinity column (*57*). We regard the difference between the results with 25% and 100% HgUTP as significant, but we do not yet know the reason for it. Crouse *et al.* (*23*) using 50% HgUTP, reported a level of *de novo* globin RNA synthesis (0.0025% of total synthesis) from native chicken reticulocyte chromatin, closely similar to the apparent value observed in this study (0.003%) for reconstituted chromatin. This level, however, is two orders of magnitude less than Fodor and Doty (*16*) observed on incubating chicken reticulocyte nuclei with HgUTP *in vitro*.

The validity of mercurated nucleotide substrates as analogs for natural nucleotides in chromatin transcription studies has yet to be justified. We feel we have still not eliminated the possibility that the cause of the failure of globin gene transcription in some of the above experiments may be due to the use of the mercurated nucleotide, the very innovation designed to resolve the previous uncertainty in distinguishing between newly synthesized and endogenous RNA. In order to settle this question once and for all, further information is required about the effects of the artificial substrate on the ability of *E. coli* RNA polymerase to transcribe the globin gene *in vitro*. The negative results obtained with native reticulocyte chromatin (*26*), following earlier positive results in experiments with the natural substrate (*8*), can be interpreted in one of two ways: (1) selective globin gene transcription does not occur with either substrate; or (2) HgUTP inhibits selective globin gene transcription in the chromatin systems. In their seminal paper on the transcription of myeloma chromatin with mercurated nucleotides, Smith and Huang (*22*) already noted that transcripts made with HgUTP were smaller than normal. Schäfer has presented evidence that the mercurated nucleotides cause premature chain termination (*47*). To the extent that we do not yet have answers to these residual crucial questions, this review must be acknowledged as all too far from definitive.

## V. General Remarks

The purpose of explorations such as those described here is to use chromatin reconstitution to study transcriptional control in higher organisms. The initial requirement, indispensable for success, is an assay for specific gene transcription. This must be able to distinguish unequivocally between newly synthesized and endogenous RNA, which indeed is the main problem to which we have addressed our attention. Conceivably, at some future stage of the work, it may be appropriate to employ more than one specific genetic probe, or to apply more stringent criteria than cDNA hybridization to analyze the fidelity of transcription. However, the next task will be to establish conditions for chromatin transcription in which the appropriate transcriptional control signals that operate *in vivo* can be recognized and genes selectively transcribed *in vitro*. It is not known, for example, to what extent *E. coli* RNA polymerase, which is often used for reasons of convenience (but see, however, Refs. *10,11,24*), reproduces the properties of the endogenous enzyme. If the enzyme were found to transcribe eukaryotic genes selectively, even without recognizing the "correct" start and stop signals in the DNA, this would in itself be interesting, for it would point to gross structural differences between the active and inactive genes, as a number of authorities have in fact suggested (e.g., Ref. *52*). Having established conditions for faithful transcription, one may then begin to make informed guesses about reconstitution regimes most likely to restore full activity. In this endeavor, one may be guided by the large corpus of work on chromatin structure, and not least by the optimum conditions for reconstituting nucleosomes, which have been reasonably well defined. It is also obvious from Fig. 1 that the experimental baseline, as it were, of the present study, i.e., the choice of three fractions from which to reconstitute functional chromatin, is in a sense arbitrary. It stems from the period before 1972 (*1–4*) when highly specific criteria for the structure and function of chromatin were not yet available. Only when a "total" reconstitution is achieved, judged by a rigorous functional assay, will systematic variations in reconstitution conditions be capable of yielding interpretable information about the role of the several constituent molecular species present, and the nature of the control mechanism.

## Acknowledgments

We thank especially R. C. C. Huang and G. Felsenfeld for the communication of relevant unpublished results. We also thank M. Walmsley, M. Spencer, and W. J. Beard for their contributions to this work as mentioned in the text. This work was supported by a Medical Research Council Programme Grant (G.969/509) and a London University postgraduate fellowship to S. J. Fey.

## References

*1.* Paul, J., and Gilmour, R. S., *J. Mol. Biol.* **34,** 305 (1968).
*2.* Gilmour, R. S., and Paul, J., *J. Mol. Biol.* **40,** 137 (1969).
*3.* Gilmour, R. S., and Paul, J., *FEBS Lett.* **9,** 242 (1970).
*4.* Bekhor, I., Kung, G. M., and Bonner, J., *J. Mol. Biol.* **39,** 351–364 (1969).
*5.* Ross, J., Aviv, H., Scolnick, E., and Leder, P., *Proc. Natl. Acad. Sci. U.S.A.* **69,** 264 (1972).
*6.* Verma, I. M., Temple, G. F., Fan, H., and Baltimore, D., *Nature* (*London*) *New Biol.* **235,** 163 (1972).
*7.* Kacian, D. L., Spiegelman, S., Bank, A., Terada, M., Metafora, S., Dow, L., and Marks, P. A., *Nature* (*London*) *New Biol.* **235,** 167 (1972).
*8.* Axel, R., Cedar, H., and Felsenfeld, G., *Proc. Natl. Acad. Sci. U.S.A.* **70,** 2029 (1973).
*9.* Gilmour, R. S., and Paul, J., *Proc. Natl. Acad. Sci. U.S.A.* **70,** 3440 (1973).
*10.* Steggles, A. W., Wilson, G. N., Kantor, J., Picciano, D. J., Falvey, A. K., and Anderson, W. F., *Proc. Natl. Acad. Sci. U.S.A.* **7,** 1219 (1974).
*11.* Wilson, G. N., Steggles, A. W., Kantor, J. A., Nienhuis, A. W., and Anderson, W. F., *J. Biol. Chem.* **250,** 8604 (1975).
*12.* Paul, J., Gilmour, R. S., Affara, N., Birnie, G., Harrison, P., Hell, A., Humphries, S., Windass, J., and Young, B., *Cold Spring Harbor Symp. Quant. Biol.* **38,** 885 (1973).

13. Barrett, T., Maryanka, D., Hamlyn, P. H., and Gould, H. J., *Proc. Natl. Acad. Sci. U.S.A.* **71,** 5057 (1974).
14. Chiu, J. F., Tsai, Y. H., Sakuma, K., and Hnilica, L. S., *J. Biol. Chem.* **250,** 9431 (1975).
15. Imaizumi, R., Diggelman, H., and Scherrer, K., *Proc. Natl. Acad. Sci. U.S.A.* **70,** 1122 (1973).
16. Fodor, E. J. B., and Doty, P., *Biochem. Biophys. Res. Commun.* **77,** 1478 (1977).
17. Orkin, S. H., and Swerdlow, P. S., *Proc. Natl. Acad. Sci. U.S.A.* **74,** 2475 (1977).
18. Gilmour, R. S., and Paul, J., *in* "Chromosomal Proteins and Their Role in the Regulation of Gene Expression" (G. S. Stein and L. J. Kleinsmith, eds.), p. 19. Academic Press, New York (1975).
19. Gilmour, R. S., Windass, J. D., Affara, N., and Paul, J., *J. Cell. Physiol.* **85,** 449 (1975).
20. Dale, R. M. K., and Ward, D. C., *Biochemistry* **14,** 2458 (1976).
21. Dale, R. M. K., Martin, E., Livingston, D. C., and Ward, D. C., *Biochemistry* **14,** 2447 (1976).
22. Smith, M. M., and Huang, R. C. C., *Proc. Natl. Acad. Sci. U.S.A.* **73,** 775 (1976).
23. Crouse, G. F., Fodor, E. J. B., and Doty, P., *Proc. Natl. Acad. Sci. U.S.A.* **73,** 1564 (1976).
24. Towle, H. C., Tsai, M. J., Tsai, S. Y., and O'Malley, B. W., *J. Biol. Chem.* **252,** 2396 (1977).
25. Gilmour, R. S., *Proc. Roy. Soc.*, in press, 1978.
26. Zasloff, M., and Felsenfeld, G., *Biochem. Biophys. Res. Commun.* **75,** 598 (1977).
27. Konkel, D. A., and Ingram, V. M., *Nucleic Acids Res.* **4,** 1979 (1977).
28. Longacre, S. S., and Rutter, W. J., *J. Biol. Chem.* **252,** 2742 (1977).
29. Bishop, J., and Freeman, K. B., *Cold Spring Harbor Symp. Quant. Biol.* **38,** 707 (1973).
30. Forget, B. G., Housman, D., Benz, E. J., Jr., and McCaffrey, R. P., *Proc. Natl. Acad. Sci. U.S.A.* **72,** 984 (1975).
31. Knochel, W., Lange, D., and Hendrick, D., *Mol. Biol. Reports* **3,** 143 (1976).
32. Efstratiadis, A., Kafatos, F. C., Maxam, A. M., and Maniatis, T., *Cell* **7,** 279 (1976).
33. Gross-Bellard, M., Oudet, P., and Chambon, P., *Eur. J. Biochem.* **36,** 32 (1976).
34. Davis, R. W., Simon, M., and Davidson, N., *Methods in Enzymol.* **21,** 413 (1971).
35. Panyim, S., Bilek, D., and Chalkley, R., *J. Biol. Chem.* **246,** 4206 (1971).
36. Panyim, S., and Chalkley, R., *Arch. Biochem. Biophys.* **130,** 337 (1969).
37. Laemmli, U. K., *Nature (London)* **227,** 680 (1970).
38. Levy, S., Simpson, R. T., and Sober, H. A., *Biochemistry* **11,** 1547 (1972).
39. Rickwood, D., and MacGillivray, A. J., *Eur. J. Biochem.* **51,** 593 (1975).
40. Morris, M., and Gould, H. J., *Proc. Natl. Acad. Sci. U.S.A.* **68,** 481 (1971).
41. Burgess, R. R., and Travers, A. A., *Fed. Am. Soc. Esp. Biol. Fed. Proc.* **29,** 1164 (1970).
42. Penman, S., *J. Mol. Biol.* **17,** 117 (1966).
43. Zylber, E. A., and Penman, S., *Proc. Natl. Acad. Sci. U.S.A.* **68,** 2861 (1971).
44. Commerford, S. L., *Biochemistry* **10,** 1993 (1971).
45. Burton, K., *Methods in Enzymol.* **12B,** 163 (1968).
46. Cuatracasas, P., *J. Biol. Chem.* **245,** 3059 (1970).
47. Schäfer, K. P., *Nucleic Acids Res.* **4,** 3109 (1977).
48. Young, B. D., Harrison, P. R., Gilmour, R. S., Birnie, G. D., Hell, A., Humphries, S., and Paul, J., *J. Mol. Biol.* **34,** 555 (1974).
49. Beebe, T. J. C., and Butterworth, P. H. W., *Eur. J. Biochem.* **66,** 543 (1976).
50. Cedar, H., and Felsenfeld, G., *J. Mol. Biol.* **77,** 236 (1973).
51. Cedar, H., Solage, A., and Zurucki, F., *Nucleic Acids Res.* **3,** 1659 (1976).
52. Yamammoto, K. R., and Alberts, B. M., *Ann. Rev. Biochem.* **45,** 722 (1976).
53. Ceriotti, G., *J. Biol. Chem.* **214,** 59 (1955).

*54.* Lowry, O. H., Rosenbrough, N. J., Farr, A. L., and Randall, R. J., *J. Biol. Chem.* **193,** 265 (1951).
*55.* Moss, B. A., and Thompson, E. O. P., *Aust. J. Biol. Sci.* **22,** 1455 (1969).
*56.* Giesecke, K., Sippel, A. E., Nguyen-Huu, M. C., Groner, B., Hynes, N. E., Wurtz, T., and Schütz, G., *Nucleic Acids Res.* **4,** 3943 (1977).
*57.* Zasloff, M., and Felsenfeld, G., *Biochemistry* **16,** 5135 (1977).

# SUBJECT INDEX

# CONTENTS OF PREVIOUS VOLUMES

## *Volume I*

# Volume II



# Volume III

# *Volume IV*

## Volume V



## Volume VI

# *Volume VII*

# Volume VIII

# *Volume IX*

# *Volume X*

# Volume XI



# Volume XII

# *Volume XIII*

# *Volume XIV*

## *Volume XV*

# *Volume XVI*

# *Volume XVII*

# Volume XVIII

A B C 8 D 9 E 0 F 1 G 2 H 3 I 4 J 5